HYPOTENSIVE PEPTIDES

SYMPOSIUM CHAIRMEN: F. Sicuteri and E.G. Erdös

ORGANIZING COMMITTEE: N. Back, E.G. Erdös and F. Sicuteri

Assistants to the Organizing Committee: G. Franchi and M. Fanciullacci

HONORARY COMMITTEE

G. G. ARCHI	- Rector of University, Florence
D. BOVET	- Institute of Pharmacology, University of Sassari
L. CALIFANO	- Institute of General Pathology, University of Naples
J. H. GADDUM *	- Depatrment of Pharmacology, University of Edinburgh
E. GREPPI	- Department of Medicine, University of Florence
E. KATCHALSKI	- Weizmann Institute, Israel
L. LAGORIO	- Mayor of Florence
A. LUNEDEI	- Dean of the School of Medicine of the University of Florence
D. H. MURRAY	- Dean of the School of Pharmacy, Buffalo, State U. of N.Y.
M. ROCHA e SILVA	- Chairman, Dept. of Pharmacol. U. of S. Paulo, Ribeirao Preto, S. Paulo, Brazil

* Deceased

SPONSORED BY The State University of New York at Buffalo

and

The University of Florence

HYPOTENSIVE PEPTIDES

Proceedings of the International Symposium

October 25–29, 1965, *Florence, Italy*

Ervin G. Erdös
Professor of Pharmacology
University of Oklahoma School of Medicine
Editor-in-Chief

Nathan Back
Professor of Biochemical Pharmacology
State University of New York at Buffalo

Federigo Sicuteri
Professor of Medicine
University of Florence

and

Anne F. Wilde
New Haven, Connecticut
Technical Editor

SPRINGER-VERLAG • BERLIN • HEIDELBERG • NEW YORK – 1966

ISBN-13: 978-3-642-94967-8 e-ISBN-13: 978-3-642-94965-4
DOI: 10.1007/978-3-642-94965-4

Library of Congress Catalog Card Number 66-21083

Title No. 1379

Softcover reprint of the hardcover 1st edition 1966

List of Contributors

AKIMOTO, SETSUKO, Laboratory of Physiological Chemistry, Tokyo College of Science, Tokyo, Japan

AMARO, J., Department of Pharmacology, Tulane University School of Medicine, New Orleans, Louisiana

AMUNDSEN, EGIL, Institute of Physiology, University of Oslo, Norway

ANASTASI, A., Farmitalia Research Laboratories, Milan, Italy

ARMSTRONG, DESIRÉE, Middlesex Hospital Medical School, University of London, England

ATSUMI, T., Institute for Cardiovascular Diseases and Department of Medicine, Tokyo Medical and Dental University, Tokyo, Japan

AUNGST, C. W., Division of Medicine, Roswell Park Memorial Institute, Buffalo, New York

AUSTEN, K. FRANK, Massachusetts General Hospital, Boston, Massachusetts

BACK, N., Department of Biochemical Pharmacology, The State University of New York at Buffalo, School of Pharmacy, Buffalo, New York

BARDOS, T., The State University of New York at Buffalo, School of Pharmacy, Buffalo, New York

BARLOW, B., The State University of New York at Buffalo, School of Pharmacy, Buffalo, New York

BARROSO, JADER, Departamento de Bioquimica, Faculdade de Medicina, Universidade de Minas Gerais, Belo Horizonte, Brazil

BAUER, G., Pharmakologisches Institut der Universität Innsbruck, Austria

BEACHAM, J., The Robert Robinson and Physiological Laboratories, Liverpool University, England

BENTLEY, P. H., The Robert Robinson and Physiological Laboratories, Liverpool University, England

BERETTA ANGUISSOLA, A., Institute of Medical Pathology and Clinical Methodology, University of Turin, Italy

BERNARDI, L., Istituto Ricerche Farmitalia, Milan, Italy

BERT, G., Institute of Medical Pathology and Clinical Methodology, University of Turin, Italy

BERTACCINI, GIULIO, Institute of Pharmacology, University of Parma, Italy

BOBBIN, R., Department of Pharmacology, Tulane University School of Medicine, New Orleans, Louisiana

BOREHAM, P. F. L., Nuffield Institute of Comparative Medicine, The Zoological Society of London, England

BORGHETTI, A., Instituto de Patologia Speciale Medica, Universita di Parma, Italy

CAMPUS, S., Institute of Medical Pathology and Clinical Methodology, University of Turin, Italy

CANO, G., Department of Pharmacology, Tulane University School of Medicine, New Orleans, Louisiana

CEI, J. M., Institute of Biology, National University of Cuyo, Mendoza, Argentina

CHIANDUSSI, L., Institute of Medical Pathology and Clinical Methodology, University of Turin, Italy

COLLIER, H. O. J., Department of Pharmacological Research, Parke-Davis & Co., Hounslow, Middlesex, England

CRAIG, LYMAN C., The Laboratories of The Rockefeller University, New York, New York

DE CARO, G., Istituto di Patologia Speciale Medica, University of Parma, Italy

DEL BIANCO, P. L., Center for the Study of Headaches, General Medical Clinic of the University of Florence, Italy

DI MATTEI, PIETRO, Institute of Pharmacology, University of Rome, Italy

DINIZ, CARLOS R., Departamento de Bioquimica, Faculdade de Medicina, Universidade de Minas Gerais, Belo Horizonte, Brazil

DYER, DONALD C., Department of Pharmacology, Kansas University Medical Center, Kansas City, Kansas

EDERY, H., Israel Institute for Biological Research, Ness Ziona, Israel

EISEN, V., Department of Pharmacology, The Middlesex Hospital Medical School, London, England

ERDÖS, ERVIN G., Department of Pharmacology, University of Oklahoma School of Medicine, Oklahoma City, Oklahoma

ERSPAMER, V., Institute of Pharmacology of the University of Parma, Italy

FANCIULLACCI, M., Center for the Study of Headaches, General Medical Clinic of the University of Florence, Italy

FERREIRA, S. H., Department of Pharmacology, Faculty of Medicine, University of São Paulo, Ribeirao Preto, São Paulo, Brazil

FERUGLIO, F. S., Institute of Medical Pathology and Clinical Methodology, University of Turin, Italy

FRANCHI, G., Center for the Study of Headaches, General Medical Clinic of the University of Florence, Italy

FREGNAN, G. B., Farmitalia Research Laboratories, Milan, Italy

FRITZ, H., Klinisch-Chemisches Institut an der Chirurgischen Klinik der Universität München, Germany

FUJITA, T., Institute for Cardiovascular Diseases and Department of Medicine, Tokyo Medical and Dental University, Tokyo, Japan

FUKUSHIMA, HIROKO, Laboratory of Physiological Chemistry, Tokyo College of Science, Tokyo, Japan

FURANO, E., Laboratory of Clinical Biochemistry, National Heart Institute, Bethesda, Maryland

GASPARRI, F., Istituto di Clinica Ostetrica della Universita degli Studi di Firenze, Florence, Italy

GLADNER, J. A., Laboratory of Biophysical Chemistry, National Institute of Arthritis and Metabolic Diseases, Bethesda, Maryland

GLASSER, A. H., Farmitalia Research Laboratories, Milan, Italy

GOODWIN, L. G., Nuffield Institute of Comparative Medicine, The Zoological Society of London, England

GREENBAUM, LOWELL M., Department of Pharmacology, College of Physicians and Surgeons, Columbia University, New York, New York

GREGORY, R. A., The Robert Robinson and Physiological Laboratories, Liverpool University, England

GREPPI, E., Clinica Medica dell' Universita, Florence, Italy

GUTH, P. S., Department of Pharmacology, Tulane University School of Medicine, New Orleans, Louisiana

HABER, EDGAR, Department of Medicine, Harvard Medical School, Boston, Massachusetts

HABERMANN, E., Institut für Pharmakologie der Universität, Würzburg, Germany

HAEFELY, W., Department of Experimental Medicine, F. Hoffmann-La Roche & Co. Ltd., Basle, Switzerland

HAUGE, A., Institute of Physiology, University of Oslo, Norway

HEIZMANN, A., Department of Pharmacology, University of Tübingen, Germany

HILTON, S. M., Department of Physiology, University of Birmingham, England

HOCHSTRASSER, K., Klinisch-Chemisches Institut an der Chirurgischen Klinik der Universität München, Munich, Germany

HÜRLIMANN, A., Department of Experimental Medicine, F. Hoffmann-La Roche & Co. Ltd., Basle, Switzerland

ISHIOKA, T., Institute for Cardiovascular Diseases and Department of Medicine, Tokyo Medical and Dental University, Tokyo, Japan

IWANAGA, SADAAKI, The Institute for Protein Research, Osaka University, Osaka, Japan

KATCHALSKI, EPHRAIM, Department of Biophysics, Weizmann Institute of Science, Rehovoth, Israel

KEELE, C. A., Department of Pharmacology, The Middlesex Hospital Medical School, London, England

KENNER, G. W., The Robert Robinson and Physiological Laboratories, Liverpool University, England

KONZETT, H., Pharmakologisches Institut der Universität Innsbruck, Austria

KRIVOY, W., Department of Pharmacology, Baylor University College of Medicine, Texas Medical Center, Houston, Texas

KROEGER, D. C., Department of Pharmacology, University of Texas Dental Branch, Houston, Texas

LA GRECA, G., Istituto di Farmacologia dell' Universita di Parma, Italy

LANZ, P., F. Hoffmann-La Roche & Co. Ltd., Basle, Switzerland

LEMBECK, F., Department of Pharmacology, University of Tübingen, Germany

LÜBKE, KLAUS, Schering AG, Hauptlaboratorium, Berlin, West Germany

LUNDE, P. K. M., Institute of Physiology, University of Oslo, Norway

MAC LEOD, J. K., The Robert Robinson and Physiological Laboratories, Liverpool University, England

MARES-GUIA, MARCOS, Departamento de Bioquimica, Faculdade de Medicina, Universidade de Minas Gerais, Belo Horizonte, Brazil

MARGOLIS, J., Children's Medical Research Foundation, Royal Alexandra Hospital for Children, Camperdown, New South Wales, Australia

MEINARDI, HARRY, The Laboratories of The Rockefeller University, New York, New York
MELMON, KENNETH L., Departments of Medicine and Pharmacology, The Section of Clinical Pharmacology and Cardiovascular Research Institute, University of California, San Francisco, California
MERRIFIELD, R. B., The Rockefeller University, New York, New York
MILLS, G. L., The Middlesex Hospital Medical School, University of London, England
MORIWAKI, CHIAKI, Laboratory of Physiological Chemistry, Tokyo College of Science, Tokyo, Japan
MORIYA, HIROSHI, Laboratory of Physiological Chemistry, Tokyo College of Science, Tokyo, Japan
NAGASAWA, SHIGEHARU, The Institute for Protein Research, Osaka University, Osaka, Japan
NUMANO, F., Institute for Cardiovascular Diseases and Department of Medicine, Tokyo Medical and Dental University, Tokyo, Japan
OATES, JOHN A., Departments of Medicine and Pharmacology, The Division of Clinical Pharmacology, Vanderbilt University, Nashville, Tennessee
PANDOLFO, G., Institute of Medical Pathology and Clinical Methodology, University of Turin, Italy
PEREIRA, ANIBAL A., Departamento de Bioquimica, Faculdade de Medicina, Universidade de Minas Gerais, Belo Horizonte, Brazil
PERITI, P., Istituto di Farmacologia della Universita degli Studi di Firenze, Florence, Italy
PIERCE, JACK V., Laboratory of Metabolism, National Heart Institute, Bethesda, Maryland
PISANO, J. J., Laboratory of Clinical Biochemistry, National Heart Institute, Bethesda, Maryland
PRADO, J. L., Escola Paulista de Medicina, São Paulo, Brazil
SARDESAI, V. M., The Robert S. Marx Laboratories of the Department of Surgery, Wayne State University School of Medicine, Detroit, Michigan
SATO, TADASHI, The Institute for Protein Research, Osaka University, Osaka, Japan
SCHACHTER, M., Department of Physiology, University of Alberta, Edmonton, Alberta, Canada
SCHRÖDER, E., Schering AG, Hauptlaboratorium, Berlin, West Germany

SEIDEL, G., Department of Pharmacology, University of Tübingen, Germany

SHEPPARD, R. C., The Robert Robinson and Physiological Laboratories, Liverpool University, England

SHIMAMOTO, T., Institute for Cardiovascular Diseases and Department of Medicine, Tokyo Medical and Dental University, Tokyo, Japan

SICUTERI, F., Center for the Study of Headaches, General Medical Clinic of the University of Florence, Italy

SPRAGG, JOCELYN, Department of Medicine, Harvard Medical School, Boston, Massachusetts

STEGER, R., The Roswell Park Memorial Institute, Buffalo, New York

STERN, P., Department of Pharmacology, Medical Faculty, Sarajevo, Yugoslavia

STEWART, JOHN MORROW, The Rockefeller University, New York, New York

STÜRMER, E., Biological and Medical Research Division, Sandoz Ltd., Basle, Switzerland

SUZUKI, TOMOJI, Institute for Protein Research, Osaka University, Osaka, Japan

TAMURA, Z., Laboratory of Clinical Biochemistry, National Heart Institute, Bethesda, Maryland

THAL, A. P., The Robert S. Marx Laboratories of the Department of Surgery, Wayne State University School of Medicine, Detroit, Michigan

THOENEN, H., Department of Experimental Medicine, F. Hoffmann-La Roche & Co. Ltd., Basle, Switzerland

TRAUTSCHOLD, I., Klinisch-Chemisches Institut an der Chirurgischen Klinik der Universität München, Munich, Germany

TSUKADA, G. A., The State University of New York at Buffalo, School of Pharmacy, Buffalo, New York

UDENFRIED, S., Laboratory of Clinical Biochemistry, National Heart Institute, Bethesda, Maryland

VICK, ROBERT L., Department of Physiology, Baylor University College of Medicine, Texas Medical Center, Houston, Texas

VOGLER, K., F. Hoffmann-La Roche & Co. Ltd., Basle, Switzerland

VOGT, W., Department of Pharmacology, Max-Planck Institut fur Experimentelle Medizin, Göttingen, Germany

VON EULER, U. S., Fysiologiska Institutionen 1, Karolinska Institutet, Stockholm, Sweden

WAALER, B. A., Institute of Physiology, University of Oslo, Norway

WALASZEK, EDWARD J., Department of Pharmacology, Kansas University Medical Center, Kansas City, Kansas

WEBSTER, MARION E., Laboratory of Cardiovascular Physiology, National Heart Institute, Bethesda, Maryland

WERLE, E., Klinisch-Chemisches Institut an der Chirurgischen Klinik der Universität München, Munich, Germany

WILKENS, H., Department of Biochemical Pharmacology, School of Pharmacy, The State University of New York at Buffalo, New York

WOOLLEY, D. W., The Rockefeller University, New York, New York

YAMAFUJI, KEIKO, Department of Pharmacology, College of Physicians and Surgeons, Columbia University, New York, New York

YAMAZAKI, KEIKO, Laboratory of Physiological Chemistry, Tokyo College of Science, Tokyo, Japan

YANG, HSIU Y. T., Department of Pharmacology, University of Oklahoma School of Medicine, Oklahoma City, Oklahoma

ZETLER, G., Pharmakologisches Institut, Medizinische Akademie Lübeck, Lübeck, Germany

ZÖLLNER, GEORGE, Schering AG, Hauptlaboratorium, Berlin, West Germany

ZUBER, H., Laboratory for Molecular Biology, Federal Institute of Technology, Zurich, Switzerland

ZWEIFACH, BENJAMIN W., Department of Pathology, New York University Medical Center, New York, New York

Preface

The papers collected in this volume were presented at the International Symposium on Hypotensive Peptides held in Florence, Italy, on October 25-29, 1965. The purpose of the meeting was to bring together scientists of fourteen countries ranging from the chemist to the surgeon, who usually do not speak a common language and for whom it would be quite difficult to follow the latest developments in fields allied to, but still somewhat removed from their basic interest. Their reports summarized their recent progress and current achievements, and gave indication of the trend that future research on hypotensive peptides may take.

Various aspects of the structure and function of bradykinin, kallidin, kallikrein, gastrin, eledoisin, physalaemin, substance P, and related substances were discussed during the nine sessions. The first session was devoted to the chemistry of peptides. This was followed in loose order by reports dealing with the structure of some newer peptides, with kininogen, and with enzymes that release or inactivate peptides. Subsequently, reports on the physiological, pharmacological, pathological, and clinical importance of kinins, kallikrein, and eledoisin were presented. One entire session was devoted to substance P. At the closing session two motion pictures demonstrated the effect of the peptides on the circulation. The appendix of this volume contains the report of the international Committee on Nomenclature.

The meeting could not have been held without the generous help of the following pharmaceutical houses: C. H. Boehringer Sohn, Belfanti, CIBA, Crinos, Farmitalia, Merck, Sharp & Dohme, Menarini, Pfizer, Sandoz, Schering, and Sclavo. The Presidenza del Consiglio and the Ministero della Sanita also contributed toward the cost of the conference.

The participants of the symposium are grateful for the wonderful hospitality extended to them in Florence and for the remarkable work of the local organizing committee headed by Professor F. Sicuteri, Dr. E. Buongiorno, Dr. G. Franchi, and Dr. M. Fanciullacci.

The auditorium and the Villa Lemmi were made available through the courtesy of Istituto Nazionale Assicurazioni Infortuni sul Lavoro.

It is a pleasant duty to report here the great assistance of Mrs. S. G. Wolf in preparing this publication and of Mrs. Irene Meyerhof Erdös.

Oklahoma City, Oklahoma
February, 1966

Ervin G. Erdös

Contents

In Memoriam

The International Symposium on Hypotensive Peptides wishes to pay respectful homage to the memory of a colleague who was to have taken an important part in this meeting.

SIR JOHN GADDUM

The announcement of the death of Sir John Gaddum on June 30, 1965, caused deep distress among his many friends and colleagues and meant a severe loss to contemporary pharmacology and physiology.

Born at Hale in Cheshire in 1900, he matriculated at Trinity College in Cambridge with a mathematical scholarship in 1919 after having spent six months in the army at the end of World War I. His mathematical interest and skill brought him First Class honors in Part I of the Mathematical Tripos.

After preclinical medical studies 1919-1922, followed by clinical studies at University College Hospital in London, he received his medical qualification in 1924. In the same year he began his scientific work under J. W. Trevan at Burroughs & Wellcome's Laboratories in Beckenham and moved two years later to the National Institute for Medical Research at Hampstead under H. H. (now Sir Henry) Dale. During the following six years he took an active and successful part in the important work led by Dale on the physiological role of acetylcholine and the new aspects of the autonomic nervous system. With W. Feldberg, Gaddum made the important discovery (1934) that acetylcholine acts as transmittor at preganglionic synapses.

During the later half of 1930 I had the privilege of working in Dale's laboratory with Gaddum as my immediate supervisor and teacher. The finding of a potent gut-stimulating factor in extracts of rabbit intestine was systematically followed up with Gaddum, whose experience and skill were invaluable in the further work on purification and characterization of the new substance, which also was found in brain and was later referred to as substance P.

In 1932 Gaddum was appointed Professor of Pharmacology in Cairo and remained there for two years; he returned to London as Professor of Pharmacology at University College until 1937, when he took over the chair in pharmacology at the School of Pharmacy. During World War II he was assigned to work for the government, and from 1942 he divided his efforts between this work and as Professor of Materia Medica at Edinburgh University, where he stayed until 1958. At that time he accepted an offer to succeed I. de Burgh Daly as Director of the Institute of Animal Physiology at Babraham, Cambridge. He held this position until the end of 1964 when he resigned, owing to ill health.

All his life Gaddum maintained a vivid interest in naturally occurring active substances, including substance P, which was shown to belong to the polypeptides. At the suggestion of M. Rocha e Silva and the writer, a symposium was arranged at the Physiological Congress in Montreal in 1953, dealing with biologically active polypeptides. Gaddum took a very active part in this symposium, which was the first of its kind, and shortly afterward he edited a volume containing the chief contributions from this meeting. In the succeeding ten years he made a number of important contributions toward the understanding of the properties and functions of this and other active substances. An occasion of special significance was the First International Symposium on Substance P in Sarajevo in 1961, organized by P. Stern, of which Gaddum was one of the principal chairmen. It was also planned that he should take a leading part in the Florence meeting.

Gaddum's wide interests and solid basic knowledge in many fields led him to consider the problem of various kinds. His important contributions to the quantitative aspects of drug action are a good example of his clear mind and mathematical abilities. He also enjoyed improving techniques, and several apparatuses carry his name, including the ingenious microbath for studies of biologically active substances.

Gaddum was an admirable personality with high ethical and moral standards; he also had much intellectual and personal charm. His unconventionality and his modesty were genuine and always seemed to serve good purposes. He generously shared knowledge and ideas with his co-workers. Those who had the privilege of enjoying his friendship will always remember Jack Gaddum as a noble friend and a distinguished scientific personality.

Gaddum received many honors and tokens of appreciation. He became a Fellow of The Royal Society in 1945 and received an honorary Ll.D. from Edinburgh in 1964. He was knighted in the same year.

In 1929 he married a former fellow student, Dr. Iris Mary Harmer, who shared his interests during many happy years and accompanied him on many of his journeys and now mourns him, together with their three married daughters.

Professor U. S. von Euler
Fysiologiska Institutionen 1
Karolinska Institutet
Stockholm, Sweden

Introductory Remarks

E. Greppi

Clinica Medica dell'Universita di Firenze

Since 1953, Montreal, London, and New York have set the stage for successive steps in the presentation and development of new aspects concerning polypeptides that are active as vasodilators, pain-inducers, and hypotensives. The late Gaddum, and our friends Schachter and Erdös were the editors for the three meetings in which the participation of British and American experts was especially noted.

Today Florence extends her welcome to the scientists participating in this symposium especially devoted to hypotensive polypeptides. I see quite a few of the same experts on polypeptides who, however, did not entrench themselves within the rigid biochemical and pharmacodynamic boundaries of the subject. I see disciples and continuers of the same schools, more scholars from different places, and last, as an old clinician, I take special pleasure in noting the presence of numerous physicians motivated by a keen interest in the physiopathology of vascular tonus and motility and in the chemical and nervous regulation of the systemic blood pressure and peripheral small vessels.

Right after this introductory address, we have before us five days of concentrated scientific activities, with only one afternoon devoted to culture and art, as it were, to balance the abundance of biochemical and pharmacological knowledge.

Florence seems to have been built specially for blending art and the sciences, from Brunelleschi's Cupolone (a pure theorem in the Pythagorean sense) to the famous motto, "provando e riprovando" of Galileo's followers, the Accademici del Cimento, whose emblem graces the front page of our program. We, the hosts of this symposium, are grateful to our foreign colleagues for recommending Florence for this meeting. We are also confident that this atmosphere of culture and art and warm hospitality will be both comfortable and fruitful to all our guests.

Our crowded program is full of titles and documents clearly indicative of the leit motiv peptides, including vasodilators, pain-producing factors, bradykinin, substance P, and extracts from polyps and amphibia such as eledoisin and physalaemin, investigated by Erspamer.

Roche e Silva's bradykinin, historically derived from Frey's kallikrein, is well known to Florentine scholars through the initiative of Sicuteri and his associates at the Center for the Study of Headaches, who in 1962 organized a symposium on plasmakinins—an unusual subject for cultivators of clinical physiopathology. Thus it is fitting that Sicuteri should be the editor of the volume Bradykinin and Its Precursors, which may be regarded as a lesser symposium in this series.

Our present meeting, again in Florence, has a broader scope. Our subject matter was born within very specialized confines but is rapidly acquiring new aspects and far-reaching implications. Among our subjects the "kinins" are again preeminent in their chemical aspects all the way to chemical synthesis; these include not only bradykinin but also other recently discovered peptides, such as those of animal origin mentioned above, and research on enzymology concerning the "kinin-forming system" and the "kallikrein-kallidin system."

The pharmacodynamic section deals with glandular secretions, the bronchial musculature, pulmonary arterial pressure, ganglionic transmission, pregnancy, experimental and human arteriosclerosis, pancreatitis, carcinoid, and many other aspects of physiology and pathology. One session will be devoted to substance P, currently of the greatest interest, with particular reference to its methods of purification and to its relationships with the nervous system. Bradykinin, eledoisin, and physalaemin will conclude the symposium with some new concepts of hemodynamics. Those capable of surviving so massive an imbibition of science, and interested in the vasoactive aspects of inflammation, will meet again on October 30th in Milan at a symposium representing the logical "seal" to our current meetings.

As a senior man and as a clinician I am happy to take this opportunity to thank all guests for gathering here and to wish everybody a pleasant stay in Florence, plentiful scientific satisfaction, and spiritual uplift.

AUTOMATED PEPTIDE SYNTHESIS

R. B. MERRIFIELD

The Rockefeller University, New York, N. Y., U.S.A.

The value of synthetic organic chemistry to our understanding of the action of hypotensive peptides, and other biologically active peptides, has been well demonstrated. For example, the structure of bradykinin was originally established not only by degradative procedures (Elliott et al., 1960) but also by synthesis (Boissonnas et al., 1960). Through further synthetic work on bradykinin (Nicolaides and De Wald, 1961; Guttmann et al., 1962), kallidin (Nicolaides et al., 1961; Pless et al., 1962), methionyl-lysyl-bradykinin (Schröder, 1964; Merrifield, 1964b), eledoisin (Sandrin and Boissonnas, 1962), and physalaemin (Bernardi et al., 1964b), these peptides have become readily available for pharmacological investigations. In addition, large numbers of analogs of the active peptides have been prepared (Bernardi et al., 1964a; Schröder and Hempel, 1964), which have been valuable for studies of the effect of changes in chemical structure on biological activity.

The purpose of this paper is to describe a new approach to the synthesis of peptides – one which, it is hoped, will also contribute to such studies. The objective has been to develop a simple and rapid method of synthesis which would, of course, give good yields of pure products. If a sufficiently simplified procedure could be devised, it was expected that the entire process could eventually be automated. A procedure that would thus cut to a minimum the time and effort and materials required for peptide synthesis would be useful in problems requiring the preparation of large numbers of analogs as well as in the synthesis of longer-chain peptides.

The work was based on the idea that all the synthetic reactions required for the lengthening of a peptide chain could be carried out while the growing peptide was bound to an insoluble solid support. The chain would thus be immobilized and, there-

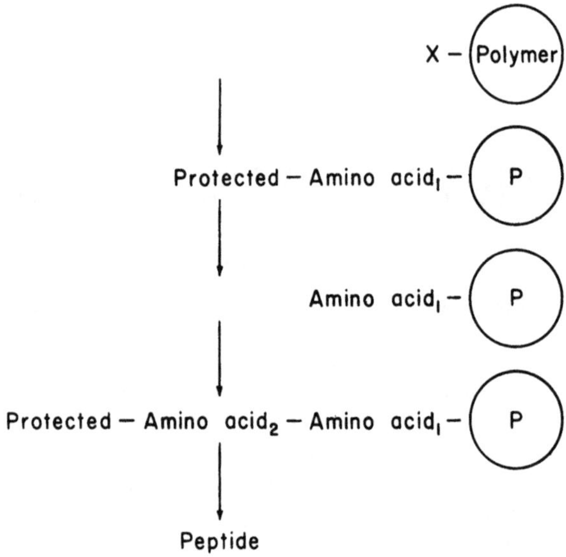

Fig. 1. The general scheme for solid phase peptide synthesis.

fore, would be subject to certain manipulations which would not otherwise be possible. The basic idea (Merrifield, 1963) is illustrated in Fig. 1. The synthesis begins with a solid particle containing a functional group, X. Through this group the first protected amino acid is attached. It is bound to the solid by a stable covalent bond, and will be the C-terminal residue of the proposed peptide chain. The protecting group is then removed, exposing the amino group which is coupled with the next protected amino acid of the peptide chain. The last two steps are repeated alternately until the desired peptide sequence is assembled. Finally, the completed peptide is cleaved from the solid support and is isolated and purified by standard procedures.

This process resembles the customary stepwise procedure in many ways, but differs from it in certain important respects. First, the peptide chain while bound to the solid support is totally insoluble in all the solvents that are used and, equally important, it is in a convenient physical form which permits rapid filtration and washing. Thus the peptide, after each step in the synthesis, can be filtered and thoroughly washed to remove excess reagents and by-products without danger of losses

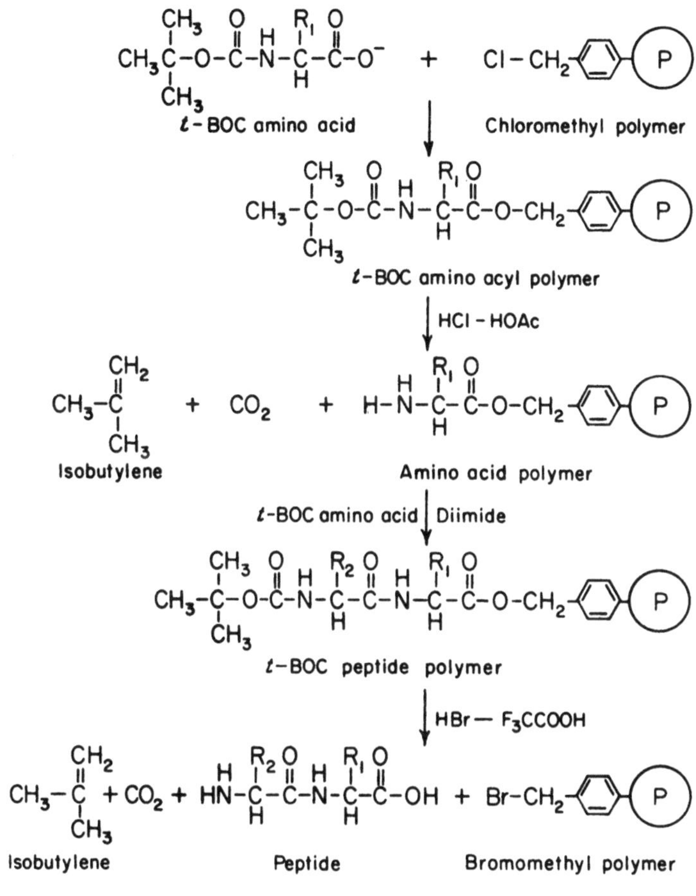

Fig. 2. The specific reactions now used for solid phase peptide synthesis.

of the peptide itself either through solubility or by mechanical means. Therefore, all the intermediates in the synthesis are purified, not by conventional recrystallization procedures but simply by thorough washing, and the usual tedious and wasteful workup of intermediates is avoided in this way. All the reactions are heterogeneous ones in which soluble reagents react with the peptide chain which is in the solid phase. For this reason the name solid phase peptide synthesis was introduced to describe the process.

The general scheme just described was eventually developed into a useful working method by employing the reactions shown in Fig. 2. (Merrifield, 1964a). The solid support was a copolymer of styrene and divinylbenzene which was partially chloromethylated to provide points of attachment for the peptide chains. The first amino acid was converted to the t-butyloxycarbonyl derivative (Carpino, 1957; Anderson and McGregor, 1957; McKay and Albertson, 1957) and condensed with the resin to form a stable benzyl ester derivative. Removal of the t-butyloxycarbonyl group was readily accomplished in N hydrochloric acid dissolved in glacial acetic acid. The resulting hydrochloride of the amino acyl resin was neutralized with triethyl amine, and the free amine was then coupled with the second amino acid to form the initial peptide bond. The activating agents have been dicyclohexylcarbodiimide (Sheehan and Hess, 1955) or active esters such as p-nitrophenyl esters (Bodanszky, 1955) or hydroxyphthalimide esters (Nefkins and Tesser, 1961). Cleavage of the completed peptide was carried out by passing a stream of hydrogen bromide through a suspension of the peptide-resin in trifluoroacetic acid. The free peptide, then in solution, was removed from the resin by filtration and was subjected to appropriate purification procedures.

The principal reason why this method was especially adapted to automation was the fact that all the chemical reac-

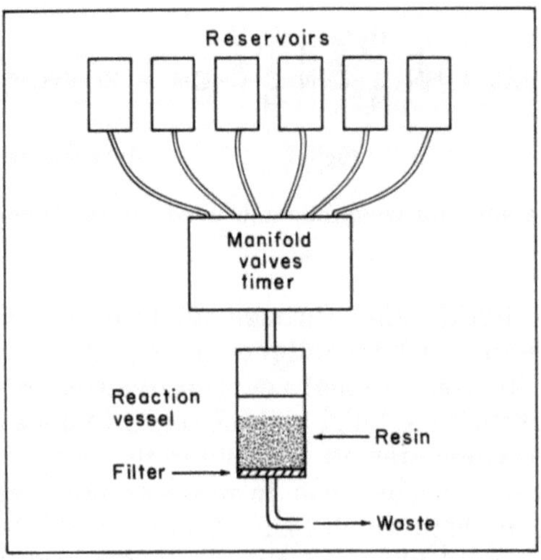

Fig. 3. Sketch of the basic requirements for automated peptide synthesis.

tions could be carried out in a single reaction vessel and, in addition, all the intermediate purification steps could also be carried out in the same vessel. Once the supporting resin containing the C-terminal amino acid was in place, it was never removed during the entire synthesis. Thus, the overall problem was simply to arrange to introduce and remove the proper solvents or reagents in the proper sequence at the proper times.

The process as originally envisioned (Merrifield, 1962) is sketched in Fig. 3. It seemed reasonable that the solvents could be pumped from a series of reservoirs through suitable manifolds and valves into the reaction vessel, and after an appropriate period of time could be removed from the vessel. The whole process could be controlled by a programmer.

A machine that will carry out a peptide synthesis in the manner just indicated has recently been constructed (Merrifield and Stewart, 1965). The apparatus can be divided into two main parts, the first being the reaction vessel with the components required to store and select reagents and to transfer them into and out of the vessel (Fig. 4), and the second being the programmer which automatically controls and orders the sequences of operation of the various components (Fig. 5).

The reaction vessel is similar to the one previously described (Merrifield, 1963) but modified to contain a stopper and small fritted disk at the top in addition to the large disk at the bottom. Reagents and solvents are stored in reservoirs which are connected to a motor-driven 12-port circular selector valve. The proper solution is then pumped into the bottom of the reaction vessel for a specified time by a metering pump while air is displaced at the top through a capillary tube. The latter is closed by a solenoid-operated valve, and the contents of the vessel are mixed by rocking the vessel through 180°. After a suitable reaction time the vessel is stopped in the vertical position and the solvent removed by filtration under vacuum through the fritted disk at the bottom while dry air is admitted at the top. The valve then rotates to select the next reagent, and the pumping, mixing, and filtering steps are repeated. During one complete cycle of the synthesis, the rotary valve makes one revolution and thus selects in sequence each of the reagents required for the lengthening of the peptide chain by one amino acid residue. The correct amino acid solution for each succeeding cycle is selected by a second rotary valve.

These events are controlled by the stepping drum programmer shown in Fig. 5. Properly placed pins on the drum activate

Fig. 4. The apparatus for automated peptide synthesis.

microswitches which in turn operate individual components in the apparatus. Each operation is controlled by a timer which also provides the signal for the drum to step to the next position. At the end of one cycle the drum automatically resets and begins the next cycle, which consists of exactly the same series of events except that a new amino acid is introduced. In this way six amino acids can be added to the peptide chain during a 24-hour period without manual attention. Two interchangeable

Fig. 5. The programmer for the automated peptide synthesis apparatus.

programmed drums are used, one for the carbodiimide-mediated coupling steps and one for the active ester coupling reactions.

An abbreviated program for the automated synthesis is illustrated in Fig. 6. For simplicity of presentation a few details of the actual program have been omitted. For one cycle of this program 83 steps were required, and the total time was 222 min.

The instrument was first tested by repeating the solid-phase synthesis of bradykinin by the same series of reactions as described earlier for the manual procedure in Fig. 2 (Merrifield, 1964a). The reaction vessel was loaded with 10 g of t-butyloxycarbonyl-nitro-L-arginine-resin. The coupling of the

Step	Reagent	Operation controlled by the program drum								
						Timers				
		Solvent selector	Pump	Shake	Filter	30 sec.	1 min.	10 min.	30 min.	2 hr.
1	HOAc*	X	X				X			
2				X		X				
3					X	X				
10	HCl-HOAc	X	X				X			
11				X					X	
12					X	X				
13	HOAc*	X	X				X			
14				X		X				
15					X	X				
22	EtOH*	X	X				X			
23				X		X				
24					X	X				
31	DMF*	X	X				X			
32				X		X				
33					X	X				
40	Et_3N-DMF	X	X				X			
41				X				X		
42					X	X				
43	DMF*	X	X				X			
44				X		X				
45					X	X				
52	CH_2Cl_2*	X	X				X			
53				X		X				
54					X	X				
61	Amino Acid CH_2Cl_2	X	X			X				
62				X				X		
63	Diimide/CH_2Cl_2	X	X			X				
64				X						X
65					X	X				
66	CH_2Cl_2*	X	X				X			
67				X		X				
68					X	X				
75	EtOH*	X	X				X			
76				X		X				
77					X	X				
84	Return to Step 1									

Fig. 6. An abbreviated program for the automated synthesis. For the reagents marked with an asterisk, the pump, shake, and filter operations are repeated in sequence two more times.

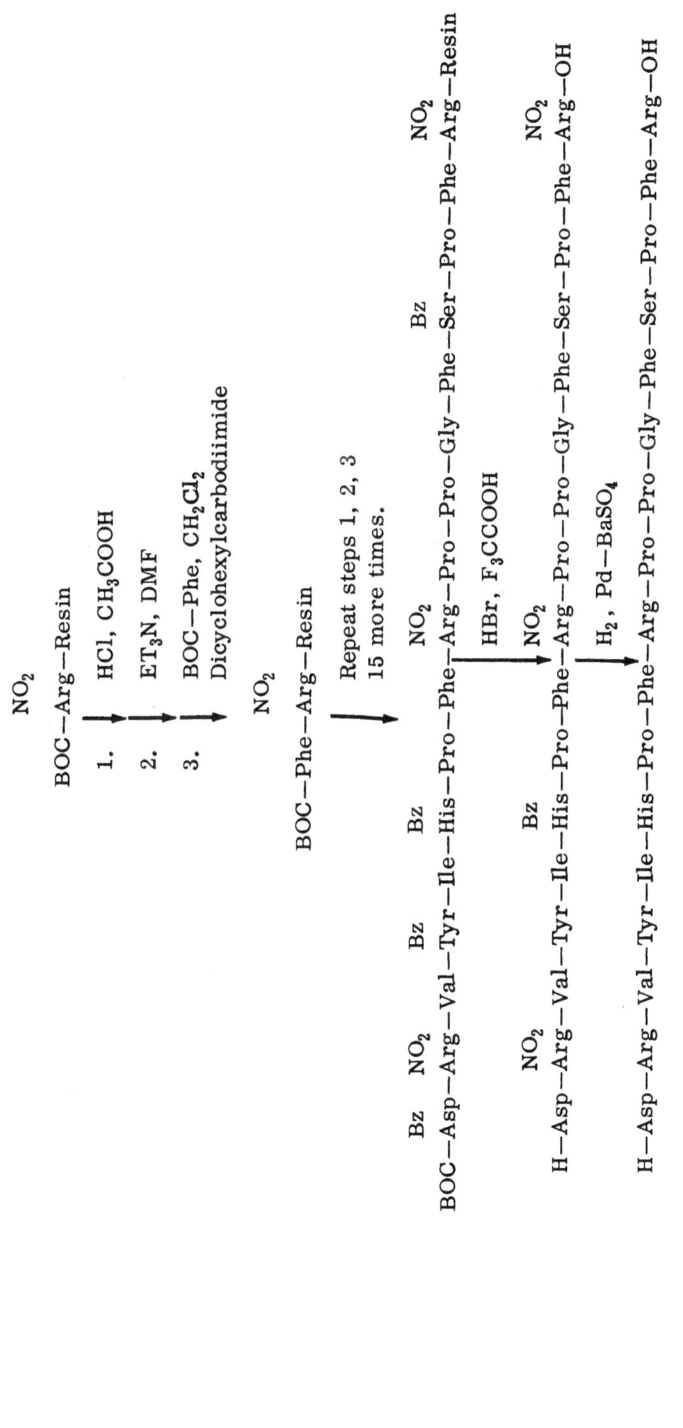

Fig. 7. The solid phase synthesis of angiotensinyl-bradykinin.

No.	Peptide
1.	Leu—Ala—Gly—Val
2.	Leu—Asp—Ala—Val
3.	Leu—Met—Asp(NH_2)—Thr
4.	Asp—Arg—Val—Tyr—Ile—His—Pro—Phe
5.	β—Asp—Arg—Val—Tyr—Ile—His—Pro—Phe
6.	Asp(NH_2)—Arg—Val—Tyr—Ile—His—Pro—Phe
7.	Arg—Pro—Pro—Gly—Phe—Ser—Pro—Phe—Arg
8.	Met—Lys—Arg—Pro—Pro—Gly—Phe—Ser—Pro—Phe—Arg
9.	Asp—Arg—Val—Tyr—Ile—His—Pro—Phe—Arg—Pro—Pro—Gly—Phe—Ser—Pro—Phe—Arg
10.	Arg—Pro—Pro—Gly—Phe—Ser—Pro—Phe—Arg—Arg—Pro—Pro—Gly—Phe—Ser—Pro—Phe—Arg

Fig. 8. Some peptides synthesized by the solid phase method.

eight additional amino acids to complete the protected nonapeptide chain on the resin was accomplished in 32 hours of continuous operation of the machine. The peptide was cleaved from the resin, hydrogenated, and purified by countercurrent distribution as previously described. In this particular synthesis the overall yield calculated from the first arginine residue was 52%. The chemical purity, based on amino acid analyses, paper chromatography, electrophoretic mobility, and countercurrent distribution was comparable with that reported previously. The biological activity of the product was identical with the standard bradykinin in the rat uterus assay. The saving in overall time and effort expended in the synthesis of this relatively small peptide was significant and is expected to become progressively more important as the length of the peptide chain is increased. This is illustrated by a recent synthesis of the heptadecapeptide angiotensinyl-bradykinin which is outlined in Fig. 7. In this instance the chain was assembled in 3 days and cleavage, hydrogenation, and countercurrent purification were completed after 4 more days.

Some of the peptides which have been prepared in this laboratory by the solid-phase method are shown in Fig. 8. Included are three representative short-chain peptides which were useful as model compounds. Also shown are angiotensin II (4), β-aspartic angiotensin II (5), 1 - asparagine angiotensin II (6), bradykinin (7), methionyl-lysyl-bradykinin (8), angiotensinyl-bradykinin (9), and bradykininyl-bradykinin (10). The results have been encouraging, and we are continuing to explore the potentialities of this approach to the synthesis of peptides.

REFERENCES

1. Anderson, G. W., and A.C. McGregor: t-Butyloxycarbonylamino acids and their use in peptide synthesis. J. Amer. chem. Soc. 79, 6180 (1957).
2. Bernardi, L., G. Bosisio, F. Chillemi, G. De Caro, R. De Castiglione, V. Erspamer, A. Glaesser, and O. Goffredo: Synthetic peptides related to eledoisin. Experientia (Basel) 20, 306 (1964a).
3. Bernardi, L., G. Bosisio, O. Goffredo, and R. De Castiglione: Synthesis of physalaemin. Experientia (Basel) 20, 490 (1964b).

4. Bodanszky, M: Synthesis of peptides by aminolysis of nitrophenyl esters. Nature (Lond.) 175, 685 (1955).
5. Boissonnas, R. A., St. Guttmann, and P.-A. Jaquenoud: Synthesis of L-arginyl-L-prolyl-L-prolyl-glycyl-L-phenylalanyl-L-seryl-L-prolyl-L-phenylalanyl-L-arginine, a nonapeptide with the properties of bradykinin. Helv. chim. Acta 43, 1349 (1960).
6. Carpino, L. A.: New protective groups on nitrogen. J. Amer. chem. Soc. 79, 98 (1957).
7. Elliott, D. F., G. P. Lewis, and E. W. Horton: The structure of bradykinin — a plasma kinin from ox blood. Biochem. biophys. Res. Commun. 3, 87 (1960).
8. Guttmann, St., J. Pless, and R. A. Boissonnas: A new synthesis of bradykinin. Helv. chim. Acta 45, 170 (1962).
9. McKay, F. C., and N. F. Albertson: New amine-masking groups for peptide synthesis. J. Amer. chem. Soc. 79, 4686 (1957).
10. Merrifield, R. B.: Peptide synthesis on a solid polymer. Fed. Proc. 21, 412 (1962).
11. Merrifield, R. B.: Solid phase peptide synthesis I. The synthesis of a tetrapeptide. J. Amer. chem. Soc. 85, 2149 (1963).
12. Merrifield, R. B.: Solid phase peptide synthesis III. An improved synthesis of bradykinin. Biochemistry 3, 1385 (1964a).
13. Merrifield, R. B.: Solid phase peptide synthesis IV. The synthesis of methionyl-lysyl-bradykinin. J. org. Chem. 29, 3100 (1964b).
14. Merrifield, R. B., and J. M. Stewart: Automated peptide synthesis. Nature (Lond.) 207, 522 (1965).
15. Nefkins, G. H. L., and G. I. Tesser: A novel activated ester in peptide synthesis. J. Amer. chem. Soc. 83, 1263 (1961).
16. Nicolaides, E. D., and H. A. De Wald: Studies on the synthesis of polypeptides. Bradykinin. J. org. Chem. 26, 3872 (1961).
17. Nicolaides, E. D., H. A. De Wald, and D. A. McCarthy: The synthesis of a biologically active decapeptide having the structure proposed for kallidin II. Biochem. biophys. Res. Commun. 6, 210 (1961).
18. Pless, J., E. Stürmer, St. Guttmann, and R. A. Boissonnas: Kallidin, synthesis and properties. Helv. chim. Acta 45, 394 (1962).
19. Sandrin, E., and R. A. Boissonnas: Synthesis of eledoisin. Experientia (Basel) 18, 59 (1962).

20. Schröder, E.: Synthesis of methionyl-lysyl bradykinin, a kinin from ox blood. Experientia (Basel) 20, 39 (1964).
21. Schröder, E., and R. Hempel: Bradykinin, kallidin, and their synthetic analogs. Experientia (Basel) 20, 529 (1964).
22. Sheehan, J. C., and G. P. Hess: A new method of forming peptide bonds. J. Amer chem. Soc. 77, 1067 (1955).

SYNTHESIS OF ALL-D-ISOMERS IN THE BRADYKININ SERIES

K. VOGLER and P. LANZ

F. Hoffmann-La Roche & Co. Ltd., Basle, Switzerland

Until recently no antipodes of polypeptide hormones have been synthesized, probably because some of the D-amino acids are not easily available (e.g. D-proline). This paper deals with the synthesis and biological properties of all-D-bradykinin and all-D-retro-bradykinin. The preparation of the former has already been announced by Stewart and Woolley (1965).

Although the definition of an all-D-polypeptide is familiar to most pharmacologists and chemists, the term "retropeptide" seems to be less well known and needs to be defined, as shown in Fig. 1, where A represents the natural product in its full structure as well as in the usual abbreviated form. All-L-retro-bradykinin is the isomer having the reversed amino acid sequence (← in B′), as compared with natural bradykinin A having the same orientation of the CO—NH— bonds marked by the arrows (→). If B is turned through 180° in order to get B′, the polar end groups and the orientation of the —CONH— skeleton are reversed, whereas the amino acid sequence again becomes the same as in the natural hormone. B and B′ represent therefore the same compound, differing only in orientation.

For understanding the following speculation it is necessary to specify the exact stereochemistry of these isomers by means of a special projection method outlined in Fig. 2.

Figure 2 shows the natural kinin hormone bradykinin (I); its antipode (II); the reversed structures all-L-retro-bradykinin (III); and its antipode all-D-retro-bradykinin (IV). These polypeptide chains are all depicted in a special projection form in which the asymmetric centers, the CO— and NH— groups of the peptide bonds, lie in the same plane perpendicular to the

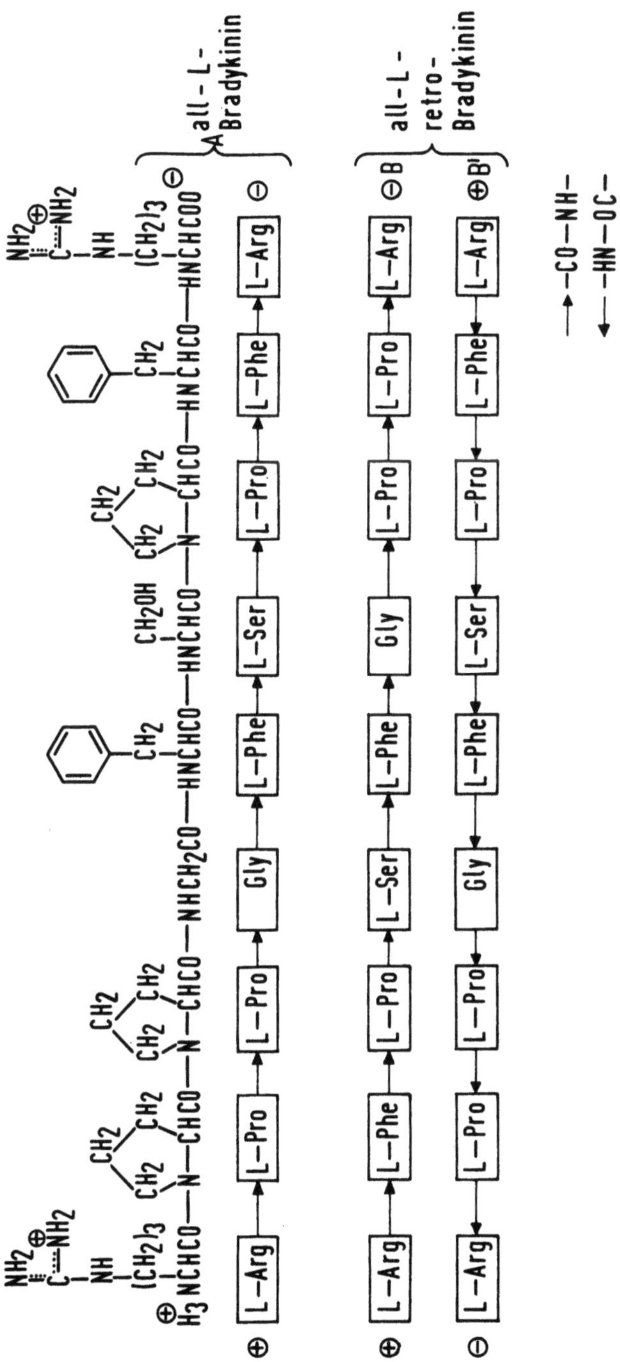

Fig. 1. Structure of all-L- and all-L-retro-bradykinin.

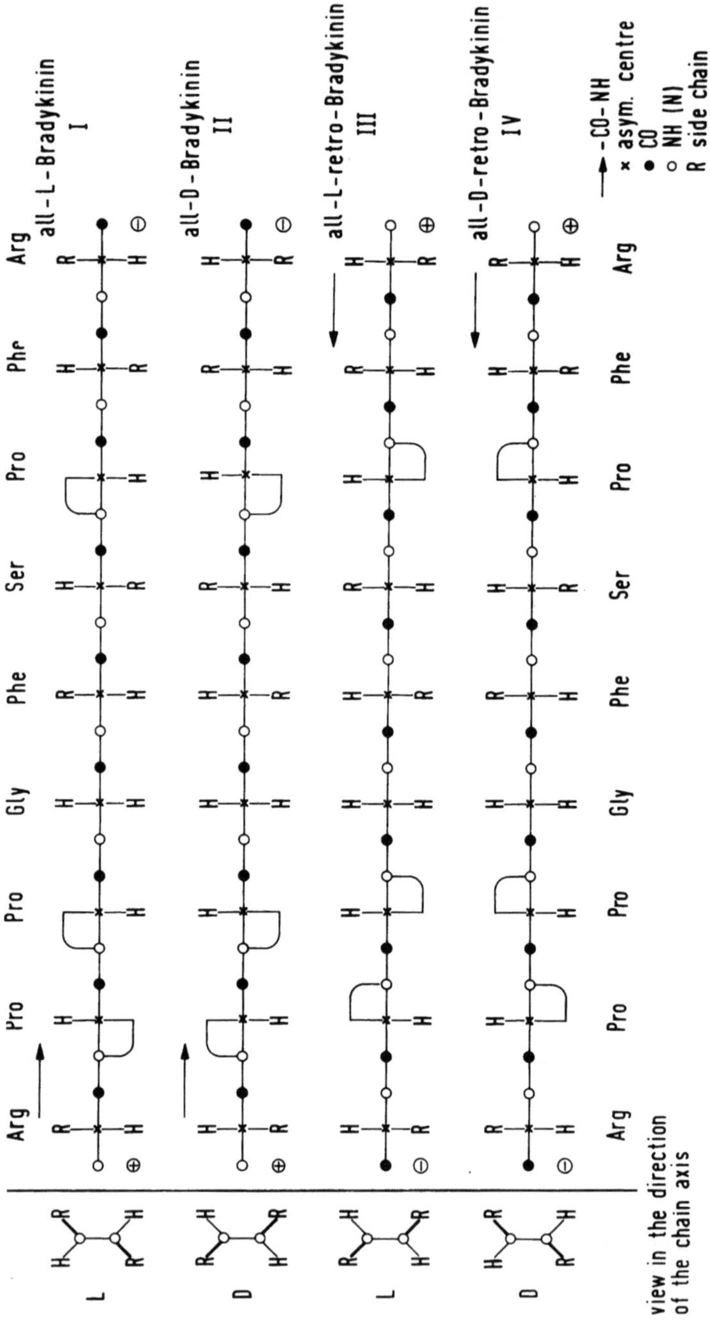

Fig. 2. Stereochemical relationship of some all-D-isomers of bradykinin.

plane of the screen and when seen from above, as here, appear in stretched form. If seen in side elevation, this line would be a zigzag. The amino acid side chains (R) are then arranged alternately to the right and left of this line. The figures on the left show the end view in the direction of the chain axis where the circles represent the two possible alternating positions of CO, NH, and the asymmetric center. In addition, these figures indicate the exact positions of the side chains R and the H atoms adjacent to the asymmetric centers. In I and II the direction of the peptide bond —CONH— is from left to right and in III and IV from right to left. I and II as well as III and IV are in the relationship of antipodes to each other, owing to the different arrangement of their side chains.

Our incentive to synthesize the antipode of bradykinin was not only the desire to fill a gap in peptide chemistry but also the feeling that antipodes of natural peptide hormones could have some antagonistic effect in biological application. This idea was catalyzed after the publication of some Japanese colleagues (Hano et al., 1964) came to our attention. These workers observed that D—His → D—Phe → D—Arg → D—Try → Gly, the antipode of an active fragment of α-MSH, showed inhibition against the corresponding all-L-pentapeptide. Another positive approach in this direction was made (Schröder et al., 1965) in the eledoisin series by means of the all-D-heptapeptide fragment derived from the C-end of the natural compound.

A second stereochemical aspect concentrates on the spatial arrangement of all-D-retro-bradykinin (IV) in comparison with the natural hormone I. All amino acid side chains except proline have the same arrangements in both compounds, whereas the direction of the —CO—NH— linkages is reversed. Proline is in fact an exception in peptide chemistry, since the side chain engages the nitrogen group in a ring system. In peptides containing no proline, the all-L-isomer and the all-D-retro-isomer have exactly the same spatial arrangements in the side chains. If the latter plays some important role in biological activity, one has an exact model for checking the concept that biological systems might be capable of "reading" a peptide chain in the reverse direction. Of course, there are the polar end groups to prevent such an accident.

We should like to explain briefly the synthesis of II and IV; III was obtained synthetically in 1962 by Lande and by the authors (Vogler et al., 1962).

All-D-bradykinin, which Stewart and Woolley (1965) pro-

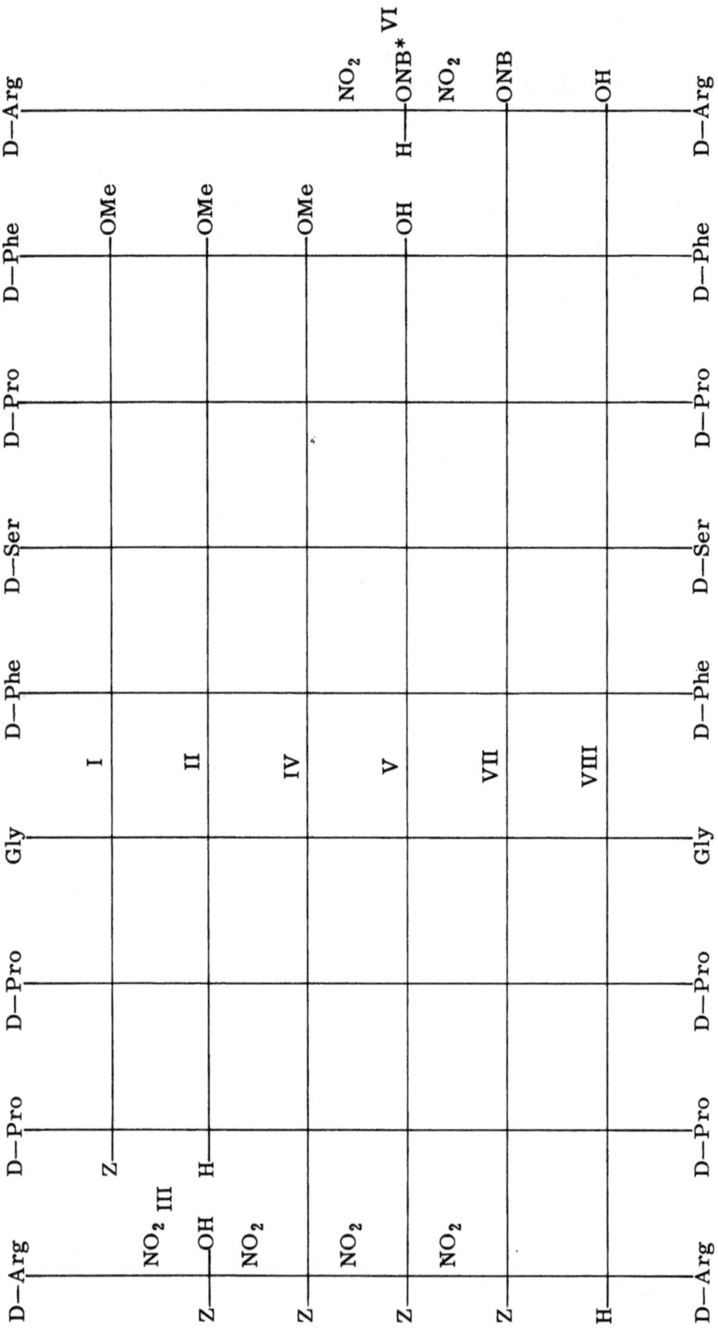

*ONB = p-nitrobenzyl ester; Z = benzyloxycarbonyl; OMe = methyl ester.

Fig. 3. Scheme of synthesis of all-D-bradykinin.

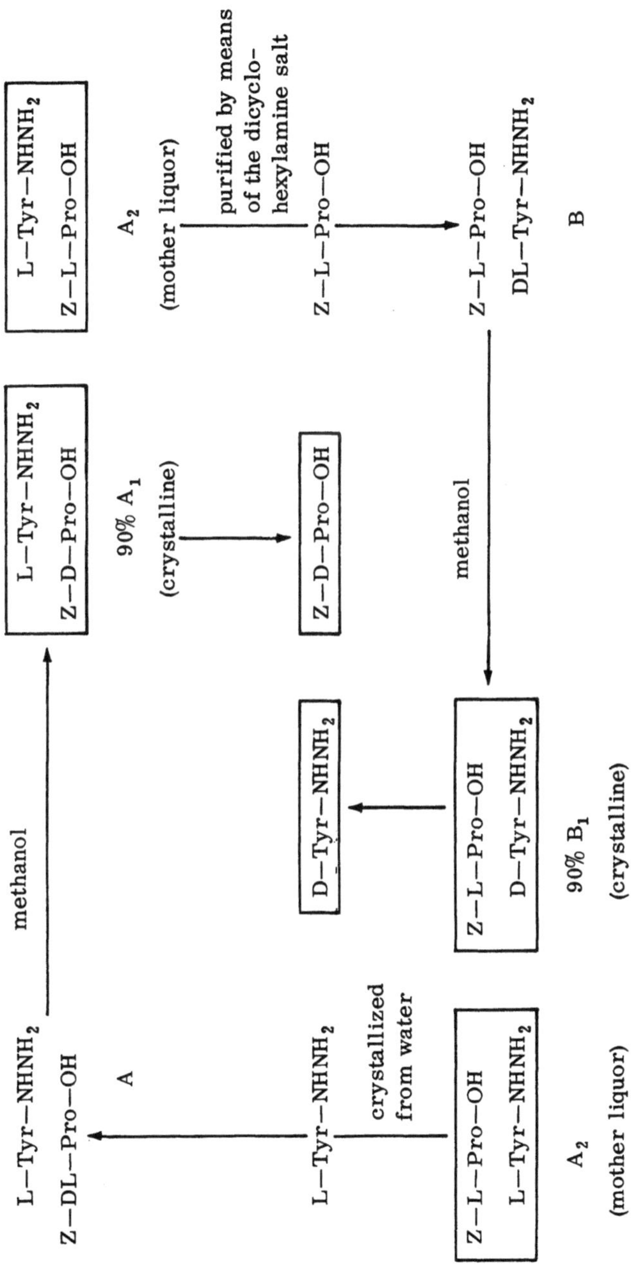

Fig. 4. Resolution scheme of DL-proline und DL-tyrosine.

duced by Merrifield's solid-phase technique (1964, a, b), was obtained by conventional methods according to this scheme (Fig. 2).

Our principle was to build up first a core heptapeptide lacking the arginine moieties at each end, which were added last. This procedure resulted from our preparation of D-arginine starting with D-glutamine, transforming this into D-ornithine, and introducing the guanido group according to the method of Zaoral and Rudinger (1959). This tedious approach carried out in the D-series justified the introduction of D-arginine at so late a step in our synthetic program.

The core heptapeptide I (Fig. 3), obtained in two ways (for details see Vogler and Lanz, 1966), was decarbobenzoxylated and condensed with III, yielding the octapeptide ester IV. After hydrolysis of the ester bond and coupling with VI to the fully protected nonapeptide VII, we obtained the end product VIII by catalytic hydrogenation.

All-D-retro-bradykinin, on the other hand, was obtained according to the scheme already used for the synthesis of all-L-retro-bradykinin (Vogler et al., 1962). Both end products were purified by countercurrent distribution and by precipitation from organic solvents and collected in the form of the trihydrochlorides in an analytically pure state. The two all-D-isomers are stable against chymotrypsin and carboxypeptidase (A+B), as is to be expected from their structures.

The biological effect of these stereoisomers (tests carried out by Dr. W. Haefely) was rather disappointing. The activity in vitro of all-D-bradykinin and all-D-retro-bradykinin was tested on the isolated guinea pig ileum. Both preparations were devoid of bradykinin-like smooth muscle-stimulating activity up to final bath concentrations 10^4 times higher than that necessary for a maximal effect of all-L-bradykinin. All-D-bradykinin had practically no antagonistic effect against contractions produced by all-L-bradykinin. In vivo, neither product had bradykinin-like hypotensive action in the anesthetized cat. In the same animal no antagonistic effect against all-L-bradykinin was observed.

After these unprofitable stereochemical games, we should like to present a positive one of interest to chemists — namely, the preparation of D-proline and D-tyrosine in a simple and most efficient way.

This new resolution procedure (Fig. 4) starts with the very easily available and inexpensive L-tyrosine hydrazide, which is combined in equimolar amounts with benzyloxycarbonyl (Z)-DL-proline (A) and crystallized from methanol. The more insoluble salt A_1, consisting of L-Tyr—$NHNH_2$ and Z-D-Pro—OH, is obtained in a yield of over 90%, from which Z-D-proline results almost quantitatively. This, as a suitable starting material for peptide synthesis, can be used as such or can be hydrogenolyzed to D-proline. The more soluble and optically impure salt A_2 contains L-Tyr—$NHNH_2$ and Z-L-Pro—OH. The latter after decomposition can be purified by means of the dicyclohexylamine salt in ethylacetate. It is obtained also in an optically pure state.

The same process can, of course, be applied for the resolution of DL-tyrosine. For that purpose Z-L-Pro—OH is combined with DL-Tyr—$NHNH_2$ into the salt B which again, after crystallization from methanol, yields the insoluble and crystallized salt B_1. This as the mirror image of A_1 shows the same physical constants, but the opposite sign of rotation. B_1 produces optically pure D-Tyr—$NHNH_2$ after decomposition. From the more soluble and optically impure salt A_2 in the mother liquor, L-Tyr—$NHNH_2$ is obtained after decomposition and can be purified by a simple crystallization from water, since the antipodes are less soluble than the corresponding racemate. And so we are again at the starting point of our cycling process.

REFERENCES

Hano, K., M. Koida, K. Kubo, and H. Yajima: Evaluation of the physiological properties of D-histidyl-D-phenylalanyl-D-arginyl-D-tryptophyl-glycine in frog melanocyte. Biochim. biophys. Acta 90, 201 (1964).

Lande, S.: Retrobradykinin: A peptide with the reverse sequence of bradykinin. J. org. Chem. 27, 4558 (1962).

Merrifield, R. B.: An improved synthesis of bradykinin. Solid phase peptide synthesis III. Biochemistry 3, 1385 (1964a).

Merrifield, R. B.: The synthesis of methionyl-lysyl-bradykinin. Solid phase peptide synthesis IV. J. org. Chem. 3, 3100 (1964b).

Schröder, E., K. Lübke, and R. Hempel: D-Aminosäurehaltige Analoga des Eledoisins. Experientia (Basel) 21, 70 (1965).

Stewart, J. M., and D. W. Woolley: All-D-bradykinin and the problem of peptide antimetabolites. Nature (Lond.) 206, 619 (1965).

Vogler, K., P. Lanz, and W. Lergier: Synthese des Bradykinin-analogen mit umgekehrter Aminosäuresequenz. Helv. chim. Acta 45, 561 (1962).

Vogler, K., and P. Lanz: Helv. chim. Acta, in press (1966).

Zaoral, M., and J. Rudinger: Dehydration of derivatives of asparagine and glutamine; A new route to α,γ-diaminobutyric acid, ornithine and arginine. Coll. Czech. chem. Commun. 24, 1993 (1959).

THE SEARCH FOR PEPTIDES WITH SPECIFIC ANTIBRADYKININ ACTIVITY

JOHN MORROW STEWART and D. W. WOOLLEY

The Rockefeller University, New York, N. Y., U.S.A.

Bradykinin and other related kinins present to the pharmacologist a complex and sometimes confusing spectrum of activities. The biochemical mechanisms by means of which these effects are produced are as yet unknown. The use of specific antimetabolites has in the past provided much insight into the mode of action of a number of biologically important substances. Work in this laboratory has been directed toward the goal of the design and synthesis of specific antimetabolites of bradykinin in the hope that these antimetabolites might be used to clarify some of the present uncertainties about the mode of action of that hormone. In addition, such specific antimetabolites might be useful in those clinical conditions that are characterized by the production of excessive amounts of bradykinin.

The synthesis of antimetabolites of bradykinin has been attempted by making analogs of bradykinin in which certain of the amino acid residues were replaced by different or abnormal amino acids. About seventy analogs of bradykinin have been synthesized in other laboratories, and have been reviewed by Schröder and Hempel (1964). No instances of compounds with antibradykinin activity have been reported among these analogs. In this laboratory over forty analogs of bradykinin have been prepared (Table 1). Among these compounds some have been found which possessed varying degrees of antibradykinin activity as well as bradykinin-like potency.

During the early part of this work classical methods of peptide synthesis were used in which the carbobenzoxy residue was used for protection of amino groups and in which carbodiimide and nitrophenyl ester coupling reactions were employed.

TABLE 1
Bradykinin Analogs

	Compound*	Biological activity† Pro	Anti‡
Bradykinin:	Arg—Pro—Pro—Gly—Phe—Ser—Pro—Phe—Arg	1	
1.	Arg—Pro—Pro—Gly—Phe—Thr—Pro—Phe—Arg	1/3	
2.	Pro—Pro—Gly—Phe—Thr—Pro—Phe—Arg	1/3000	
3.	Arg—Pro—Pro—Gly—Phe—Thr—Pro—Leu—Arg	1/100	
4.	Pro—Pro—Gly—Phe—Thr—Pro—Leu—Arg	1/150,000	100% at 50 μg/ml
5.	Arg—Pro—Pro—Gly—Leu—Thr—Pro—Leu—Arg	1/30,000	50% at 0.0001
6.	Pro—Pro—Gly—Leu—Thr—Pro—Leu—Arg	1/250,000	
7.	Arg—Pro—Pro—Gly—Phe—gly—Pro—Phe—Phe—Arg	1/5,000	
8.	Pro—Pro—Gly—Phe—gly—Pro—Phe—Phe—Arg	1/10,000	50% at 0.01
9.	Arg—Pro—Pro—Gly—Phe—Ser—Pro—Phe—Arg—Gly	1/150	
10.	Pro—Pro—Gly—Phe—Ser—Pro—Phe—Arg—Gly	1/10,000	
11.	Arg—Pro—Pro—Gly—Phe—Ser—Pro—Phe—Arg—Arg	1/25	
12.	Pro—Pro—Gly—Phe—Ser—Pro—Phe—Arg—Arg	1/200	
13.	Arg—Pro—Pro—Gly—OMT—Ser—Pro—OMT—Arg	1/3	90% at 1×10^{-6} **
14.	Pro—Pro—Gly—OMT—Ser—Pro—OMT—Arg	1/1000	
15.	Acetyl—Arg—Pro—Pro—Gly—OMT—Ser—Pro—OMT—Arg	1/4	50% at 1×10^{-6} **
16.	Arg—Pro—Pro—Gly—OMT—Ser—Pro—OMT—Arg—Phe	1/1500	
17.	Arg—Pro—Pro—Gly—OMT—Ser—Pro—OMT—Arg—Gly	1/5000	
18.	Pro—Pro—Gly—OMT—Ser—Pro—OMT—Arg—Gly	1/100,000	30% at 1×10^{-5}
19.	Arg—Pro—Pro—Gly—Phe—Phe—Pro—Phe—Arg	1/1000	
20.	Pro—Pro—Gly—Phe—Phe—Pro—Phe—Arg	1/3000	
21.	Arg—Pro—Pro—Gly—Phe—TLy—Pro—Phe—Arg	1/2000	
22.	Pro—Pro—Gly—Phe—TLy—Pro—Phe—Arg	1/20,000	
23.	Arg—Pro—Pro—Gly—Leu—Gly—Pro—Leu—Arg	1/50,000	

24.	Pro—Pro—Gly—*Leu*—*Gly*—Pro—*Leu*—Arg	no effect	
25.	Arg—Pro—Pro—Gly—Phe—*Gly*—Pro—*OMT*—Arg	1.2	
26.	*Gly*—Arg—Pro—Pro—Gly—Phe—*Gly*—Pro—*OMT*—Arg	1/5	
27.	*Acetyl*—Arg—Pro—Pro—Gly—Phe—*Gly*—Pro—*OMT*—Arg	1/3	30% at 0.001
28.	Pro—Pro—Gly—Phe—*Gly*—Pro—*OMT*—Arg	1/1000	
29.	Arg—Pro—Pro—Gly—*OMT*—*Gly*—Pro—*OMT*—Arg	1/15	50% at 1×10^{-6}
30.	Pro—Pro—Gly—*OMT*—*Gly*—Pro—*OMT*—Arg	1/4000	
31.	Arg—Pro—Pro—Gly—*OMT*—*AsN*—Pro—*OMT*—Arg	1/200	30% at 1×10^{-5}
32.	Pro—Pro—Gly—*OMT*—*AsN*—Pro—*OMT*—Arg	1/1000	
33.	*arg*—*pro*—*pro*—Gly—*phe*—*ser*—*pro*—*phe*—*arg*	1/200,000	
34.	*arg*—Pro—Pro—Gly—*phe*—*ser*—Pro—*phe*—*arg*	1/10,000	
35.	Arg—*pro*—*pro*—Gly—Phe—Ser—*pro*—Phe—Arg	1/200,000	85% at 20
36.	*pro*—*pro*—Gly—Phe—Ser—*pro*—Phe—Arg	1/100,000	
37.	Arg—Pro—Pro—Gly—Phe—Ser—Pro—Phe—*arg*	1/1000	
38.	Pro—Pro—Gly—Phe—Ser—Pro—Phe—*arg*	1/500,000	50% at 1×10^{-5}
39.	Arg—*pro*—Pro—Gly—Phe—Ser—Pro—Phe—Arg	1/500	
40.	*pro*—Pro—Gly—Phe—Ser—Pro—Phe—Arg	1/1000	
41.	Arg—Pro—*pro*—Gly—Phe—Ser—Pro—Phe—Arg	1/10,000	
42.	Pro—*pro*—Gly—Phe—Ser—Pro—Phe—Arg	1/100,000	
43.	Arg—Pro—Pro—Gly—Phe—Ser—*pro*—Phe—Arg	1/75	
44.	Pro—Pro—Gly—Phe—Ser—*pro*—Phe—Arg	1/3000	
45.	*Acetyl*—Arg—Pro—Pro—Gly—Phe—Ser—*pro*—Phe—Arg	1/1000	
46.	*Acetyl*—Arg—Pro—Pro—Gly—Phe—Ser—Pro—Phe—Arg	1/2	
47.	*arg*—*phe*—*pro*—*ser*—*phe*—Gly—*pro*—*pro*—*arg*	1/30,000	

**Italicized* residues are different from the normal bradykinin structure. OMT = tyrosine methyl ether; TLy = ϵ-tosyl lysine; AsN = asparagine. Uncapitalized residues have the D configuration.

†On isolated rat uterus.

‡Maximum inhibition obtained. The inhibition was highly variable (see text).

**Recovery of the normal response to bradykinin was slow after administration of either of these peptides.

These methods proved to be quite laborious, since the intermediates in the syntheses were not easily crystallizable compounds. Countercurrent distribution was used extensively for purification. However, most of the peptides to be described were synthesized by the solid-phase method of peptide synthesis developed by R. B. Merrifield of The Rockefeller University (see this volume). This method has proved to be especially convenient for the synthesis of bradykinin analogs, and with the recent perfection of the automatic peptide-synthesizing machine (Merrifield and Stewart, 1965), the syntheses can now be carried out with great ease and rapidity.

Very little information is available to guide the peptide chemist in the choice of structural modifications for the purpose of changing metabolites into antimetabolites. Until very recently there have been no reported syntheses of good antimetabolites of any peptide hormones. The recent work of Smyth (1966) on carbamyl oxytocin is very promising, since he seems to have found an antioxytocin with no oxytocin-like action. However, the earlier oxytocin analogs which were reported to have antioxytocin activity also had oxytocin activity. An important goal in the design of antimetabolites is to obtain an analog with a prolonged inhibitory action. This can frequently be accomplished by building into the analog molecule features that will make it resistant to the action of the enzymes which normally inactivate the metabolite. This goal might be realized in bradykinin analogs by modification of either the C-terminal arginine or the phenylalanine residues. These changes might be expected to make the molecule resistant to degradation by enzymes with carboxypeptidase or chymotryptic activity. For the initial syntheses, the serine of bradykinin was replaced by a threonine residue, in accord with the general principle of adding methyl groups for the purpose of converting a metabolite to an antimetabolite. Also, Merrifield (1958) had found in the strepogenin work that such a replacement was an effective method for making an antimetabolite.

The synthesis of these threonine analogs (compounds 1-6) by classical methods has already been described (Stewart and Woolley, 1964). Two of them (4 and 5) were found to have antibradykinin activity in the rat uterus assay, although this activity was not seen on all the animals tested. An interesting observation was that the O-acetyl threonine methyl ester derivative of compound 5 was able to cause complete, specific, and irreversible inhibition of the bradykinin response in uteri of galactosemic rats.

The remaining analogs were synthesized by the solid-phase method. For the synthesis of these compounds, the original method (Merrifield, 1964) was modified by the use of 4 N hydrogen chloride in dry dioxane for removal of t-butyloxycarbonyl (BOC) protecting groups, by use of 10% triethyl amine in chloroform for neutralization of the hydrochlorides, and by carrying out all the coupling reactions in dichloromethane except that of BOC-nitroarginine, which was done in dimethylformamide for reasons of solubility. By the use of short washing periods and efficient operation, the time necessary for addition of an amino acid to the peptide-resin was reduced to 3 hours. In this way three residues could be coupled in a working day. These peptides, obtained in high yield (generally 75-85%), were pure compounds having the correct amino acid analyses.

TABLE 2
Importance of the Carboxyl End of Bradykinin

Compound	Activity	Investigator
Bradykinin	1	
Lys-Bradykinin	1/2	Pless
Met-Lys-Bradykinin	1/3	Elliott
Gly-Bradykinin	3/2	Schröder
Phe-Bradykinin	1	Schröder
Acetyl-Bradykinin	1/2	Stewart
Bradykinin-NH_2	0	Erdös
Bradykinin-Arg	1/25	Stewart
Bradykinin-Gly	1/150	Stewart

Several analogs of bradykinin are known in which the peptide has been lengthened at the amino end. However, none has previously been described in which the chain has been lengthened at the carboxyl end. The great loss in biological activity caused by addition of glycine or arginine residues to the C-terminus (compounds 9-12) is in marked contrast to the high activity of analogs with added N-terminal residues. These results are summarized in Table 2. This finding is in accord with evidence in the literature on other peptides of biological importance (Stewart and Woolley, 1965b). Thus an intact carboxyl end seems to be extremely important in the case of angiotensin, gastrin, eledoisin, and insulin, as well as for the enzymes ribonuclease and aldolase. Evidently the carboxyl end of peptides and proteins is much more important for biological activity with receptors or substrates than is the amino end.

The observations of Law and du Vigneaud (1960) on an analog of oxytocin containing tyrosine methyl ether suggested that this amino acid might be useful in the synthesis of other peptide antimetabolites. Analogs containing this residue (compounds 13-15) were indeed found to have very high antibradykinin activity as well as high bradykinin-like potency. In an attempt to lower the bradykinin-like action while retaining the antibradykinin activity, compounds 16-18 were synthesized. These did indeed have lowered bradykinin potency, but the antagonism also suffered.

The early finding of some antagonistic threonine analogs suggested further modifications in the serine position of the bradykinin molecule. Other investigators (Bodanszky et al., 1963; Schröder and Hempel, 1964) had found that 6-glycine- and 6-alanine-bradykinin retained high potency. The present findings with 6-phenylalanine- and 6-epsilontosyl lysine-bradykinin (compounds 11-22) indicate that the size of the side chain in that position is critical for bradykinin activity. The synthesis of additional analogs with modifications in both the serine and aromatic residues (compounds 23-32) failed to yield any compounds with enhanced antibradykinin activity.

It has been found by Hano et al. (1964) that an all-D pentapeptide sequence from melanophore-stimulating hormone (MSH) has anti-MSH activity. This observation, and the fact that no peptide hormone had ever been synthesized in the all-D configuration, prompted the synthesis of all-D bradykinin (Stewart and Woolley, 1965a) and a number of peptides related to it (compounds 33-45). All-D bradykinin was found to have a vanishingly small amount of bradykinin-like potency, and no antibradykinin activity. More recently, Flouret and du Vigneaud (1965) have synthesized all-D oxytocin and found it to be inactive. Evidently optical inversion cannot be a general rule for the synthesis of peptide antimetabolites.

The extreme importance of proline residues in determining the conformation of peptides is well demonstrated by the fact that compound 35, which has just the three prolines in the D configuration, has as little bradykinin potency as the all-D

TABLE 3
Comparison of Alanine and D-Proline Analogs

Arg-Pro-Pro-Gly-Phe-Ser-Pro-Phe-Arg

1 2 3 4 5 6 7 8 9

Substitution	Position		
	2	3	7
D-Proline	1/500	1/10,000	1/75
L-Alanine*	1/300	1	1/80

*Schröder and Hempel, 1964.

analog. The opposite analog (34), which has all of the molecule in the D configuration except the prolines, is twenty times as active. The very low activity of 9-D-arginine bradykinin (37) further emphasizes the importance of the carboxyl end of bradykinin. This analog has also been synthesized by Nicolaides et al. (1965), who found that it had no prolonged action in vivo. The three analogs having only one proline in the D configuration (39, 41, 43) provide an interesting comparison of the relative importance of these three positions. Especially interesting is the comparison of these compounds with the corresponding alanine analogs (Table 3). It can be seen that, although replacements of proline by L-alanine or D-proline in the 2- and 7-positions have similar effects on the biological activity, the effect of such replacements in the 3-position is strikingly different. While 3-alanine-bradykinin has full potency, 3-D-proline bradykinin has very low activity.

Early in the bradykinin work, Vogler et al. (1962) and Bodanszky et al. (1963) reported the synthesis of "retrobradykinin," in which the sequence of the amino acids was reversed. They found that this analog was inactive. Sheppard (1964), in commenting on this result, suggested that perhaps the "retro" analog should have been made in the D configuration. Retro-D-bradykinin has now been synthesized in this laboratory (compound 47) and compared with a sample of retro-L-bradykinin, kindly supplied by Dr. Bodanszky. In that assay, retro-L-bradykinin was found to have 1/100,000 of the activity of bradykinin, whereas retro-D-bradykinin was three times as active. Retro-D-bradykinin was thus found to be six times as potent as D-bradykinin, although the activities of all these analogs were extremely low.

The most potent inhibitor of bradykinin action found was 5,8-di-O-methyl tyrosine bradykinin (13), which was effective at a very low concentration. However, this inhibitor, as well as all the others listed, suffered from two serious defects: (1) when the concentration of the analog was raised above that shown in the table, the antibradykinin action first disappeared, and then bradykinin-like action appeared at about 1,000 times the dose listed; (2) more serious, however, was the fact that none of the analogs exhibited a reliable antibradykinin action. The degree of inhibition and the effective dose varied widely from animal to animal and seemed to depend on factors which are as yet not understood.

While the ideal antibradykinin peptide has as yet not been found, some progress has been made, and there are indications that it may well be possible to make such a compound. Perhaps a slight further modification of the bradykinin structure, or a combination of modifications, will achieve the desired result.

REFERENCES

Bodanszky, M., M. A. Ondetti, J. T. Sheehan, and S. Lande: Synthetic peptides related to bradykinin. Ann. N.Y. Acad. Sci. 104, 24 (1963).

Flouret, G., and V. du Vigneaud: The Synthesis of D-oxytocin, the enantiomer of the posterior pituitary hormone, oxytocin. J. Amer. chem. Soc. 87, 3775 (1965).

Hano, K., M. Koida, K. Kubo, and H. Yajima: Evaluation of the physiological properties of D-histidyl-D-phenylalanyl-D-arginyl-D-tryptophyl-glycine in frog melanocytes. Biochim. biophys. Acta 90, 201 (1964).

Law, H. D., and V. du Vigneaud: Synthesis of 2-p-methoxyphenylalanine oxytocin and some observations on its pharmacological behavior. J. Amer. chem. Soc. 82, 4579 (1960).

Merrifield, R. B.: Competitive inhibition of a strepogenin-active peptide by related peptides. J. biol. Chem. 232, 43 (1958).

Merrifield, R. B.: Solid-phase peptide synthesis. III. An improved synthesis of bradykinin. Biochemistry 3, 1385 (1964).

Merrifield, R. B., and J. M. Stewart: Automated peptide synthesis. Nature (Lond.) 207, 522 (1965).
Nicolaides, E. D., D. A. McCarthy, and D. E. Potter: Bradykinin: Configuration of the arginine moieties and biological activity. Biochemistry 4, 190, 1965.
Schröder, E., and R. Hempel: Bradykinin, kallidin, and their synthetic analogues. Experientia (Basel) 20, 529 (1964).
Sheppard, R. C.: Amino acids and peptides, in Annual Reports on the Progress of Chemistry, 60, 460. London: the Chemical Society (1964).
Smyth, D. G.: Molecular mechanism of oxytocin action. J. biol. Chem., in press (1966).
Stewart, J. M., and D. W. Woolley: Threonine analogs of bradykinin designed as antimetabolites. Biochemistry 3, 700 (1964).
Stewart, J. M., and D. W. Woolley: All-D-bradykinin and the problem of peptide antimetabolites. Nature (Lond.) 206, 619 (1965a).
Stewart, J. M., and D. W. Woolley: Importance of the carboxyl end of bradykinin and other peptides. Nature (Lond.) 207, 1160 (1965b).
Vogler, K., P. Lanz, and W. Lergier: Synthesis of the bradykinin analog with reversed amino acid sequence. Helv. chim. Acta 45, 561 (1962).

DISCUSSION

T. J. Bardos
Department of Medicinal Chemistry, School of Pharmacy,
University of New York, Buffalo, N. Y., U.S.A.

During the past year, a considerable amount of preliminary work has been carried out by Dr. P. S. Venkateswaran in my laboratory toward the synthesis of modified peptides with longer duration of action. I take this opportunity to present for the first time our basic approach to this problem.

Previous reports in the literature relating to modified, biologically active peptides dealt with structural modifications which involve addition or omission of one or several amino acids, changes of amino sequence, replacement of single amino acids with other L-amino acids or with the corresponding D-amino acid, and, in a few cases, replacement with an amino acid analog (arginine with nitroarginine, phenylalanine with p-fluorophenylalanine). Such modifications usually decreased the biological activity of the peptide without increasing its duration of action and without conferring antagonistic activity.

It appears that most of the biologically active peptides are highly structure-specific; that is, they must fulfill a multiplicity of steric and functional requirements in order to "fit" their specific receptors and "bind" to them. Optimal fit and effective binding to the receptor sites are absolute conditions for the effectiveness of either long-acting or antagonistic peptide analogs. We were particularly impressed with the high degree of structure-specificity of bradykinin and, therefore, we decided to attempt the synthesis of bradykinin analogs in which:

(a) the original sequence and configuration of the amino acid residues and thus the size, general shape, and hydrophobic binding sites of the entire molecule) are maintained;

(b) all free functional groups of the original peptide which may participate in binding to the receptor (amino, guanidino, carboxyl) are left unchanged.

Thus, all possible modifications were limited to the amide linkages of bradykinin. It was also decided to replace only one of the amide linkages with isosteric groups, because the simultaneous replacement of several amide groups would be expected to alter the basicity and conformation of the molecule. In the case of bradykinin, the (carboxy-terminal) phenylalanine-arginine linkage appears to be the site most sensitive to the degradative action of bradykininage and we have been directing our efforts therefore, to the synthesis of modified peptides in which this particular —CO—NH— linkage is replaced by —CH_2—NH—, —CO—O—, or other isosteric moieties. Such peptides would contain a "dipeptide-analog" which is not a dipeptide at all but a secondary amine, an ester ("depsipeptide") etc. which, however, must have the same stereochemistry as the dipeptide. The preparation of peptides containing such units (as we found to our chagrin) multiplies the usual problems of peptide synthesis. (For example, the secondary amine group requires selective protection throughout the further course of peptide synthesis; the ester group must be protected from hydrolysis, etc.) However, we have been making some progress in solving these problems, and I hope to be able to present the results of our endeavors, at our next meeting, using this new type of approach in the preparation of biologically active peptide analogs.

Collier: Some years ago, in testing bradykinin analogs synthesized by Dr. Nicolaides and his colleagues, as antagonists of bradykinin we used two in-vivo methods (guinea pig hypotensin and bronchoconstriction) because in-vitro results do not always apply in vivo. Occasionally we got an inconsistent antagonism, which we may ascribe to catecholamine release rather than to a receptor antagonism. I am not sure whether Dr. Stewart used any in-vivo methods of testing his analogs, but I noticed he spoke of inconsistent antagonism. I would like to ask whether the complication of catecholamine release might not apply to any of his results.

Stewart: Since the assays were done in isolated uterus, which does not respond well to catecholamines, I do not believe that this could be the explanation of the variable antibradykinin results found in our work. We hope in the future to use in-vivo assays for our analogs.

Eisen: Would Dr. Stewart tell us about his experiences with isolated uterus from galectosemic rats? Does any of the bradykinin analogs with antibradykinin action, antagonize also any other pharmacologically active polypeptides (e.g. angiotensin, oxytocin)?

Stewart: In our laboratory Dr. Woolley has been studying the mechanism of action of serotonin. He has a hypothesis that serotonin acts by combining with a receptor in the smooth-muscle membrane and with calcium to make the calcium pass across the muscle membrane and thus cause the muscle to contract. He has isolated the serotonin receptor from tissue and found it to be a galactolipid — a ganglioside. He has also postulated that the hereditary disease of galactosemia creates its effects by causing a disturbance in the functioning of serotonin. He has found that tissues from rats raised on high galactose diets after weaning show a diminished response to serotonin. For that reason the uteri from galactosemic animals were available. It has also been found that the serotonin receptor can be destroyed in rat stomach by treatment of the tissue with neuraminidase and EDTA. This treatment does not destroy the response to bradykinin. Thus, while these data clearly show that the bradykinin and serotonin receptors are different, the results with our inhibitor show that galactose is involved. Perhaps the bradykinin receptor is a different galactolipid which is not sensitive to neuraminidase.

RECENT DEVELOPMENTS IN THE CHEMISTRY OF GASTRIN

J. BEACHAM, P. H. BENTLEY, R. A. GREGORY, G. W. KENNER
J. K. MacLEOD, and R. C. SHEPPARD

The Robert Robinson and Physiological Laboratories,
Liverpool University, England

The antral hormone gastrin has but a small depressant effect on the blood pressure of experimental animals, and it falls outside the group of powerful hypotensive agents with which this volume is mainly concerned. It is therefore hoped that our results with gastrin will be of interest despite their slight connection with the main theme of the symposium.

The principal physiological action of gastrin is of course the stimulation of gastric acid secretion. Various other actions on alimentary tract structures can be observed under appropriate experimental conditions. These include the inhibition of gastric acid secretion, the stimulation of pepsin and pancreatic secretion, and the effects on gastric and intestinal musculature. Which of this latter group may be regarded as true physiological properties of gastrin is as yet unknown, for the normal concentration of gastrin circulating in the blood stream after indigestion of food has not so far been determined.

Our chemical investigations followed the successful isolation of pure gastrin by our colleagues Professor Gregory and Dr. Tracy in the Physiological Laboratory of Liverpool University (Gregory and Tracy, 1964). Two hormones (G.I and G.II) were in fact isolated from hog antral mucosa, and these proved to be indistinguishable in physiological properties and isomeric in amino acid composition. Initially we chose to work with the more abundant gastrin II, and structural studies led eventually to the amino acid sequence (I) for the hormone (Gregory *et al.*, 1964).

Glu–Gly–Pro–Try–Met–Glu–Glu–Glu–Glu–Glu–Ala–Tyr–

–Gly–Try–Met–Asp–Phe–NH_2 (I)

We know now that although this amino acid sequence for porcine gastrin II is correct, it does not in itself represent the complete structure of the molecule. This became apparent only when our synthetic studies directed at this structure matured (Anderson et al., 1964). The synthetic heptadecapeptide amide (I), while possessing the full ability of gastrin II to stimulate gastric secretion, was not identical with gastrin II. It was in fact identical with gastrin I! Indeed we had been puzzled for some months previously about the precise structural relationship between the two hormones, for degradative studies on gastrin I concurrent with the synthetic work were strongly suggesting the same amino acid sequence that we had previously deduced for gastrin II, despite the clearly different electrophoretic behavior of the two substances. With the realization that (I) is genuinely the structure of gastrin I, its relationship with the more acidic gastrin II soon became clear. Gastrin II is in fact the phenolic sulfate ester of the single tyrosine residue of gastrin I, and this has now been confirmed by their interconversion (Gregory et al., 1964).

During the course of the synthetic work, numerous intermediate peptide derivatives were prepared, corresponding in amino acid sequence to various parts of the gastrin molecule. Our physiological colleagues volunteered to test these various derivatives for gastrin-like activity in the conscious dog, and they were soon rewarded when it was found that even quite small fragments corresponding to the C-terminus of gastrin were active in stimulating gastric secretion (Tracy and Gregory, 1964). The minimal requirements for biological activity are illustrated below:

Z–Asp–Phe–NH_2	–
Z–Met–Asp–Phe–NH_2	–
Z–Try–Met–Asp–Phe–NH_2	+
Z–Gly–Try–Met–Asp–Phe–NH_2	+
Z–Gly–Try–Met–Asp–Phe–OMe	–
Try–Met–Asp–Phe–NH_2	+

(Active compounds, +; inactive compounds, –)

Peptide fragments corresponding to the N-terminal or central parts of the molecule were quite inactive, but peptides containing as few as four residues in the sequence of the C-terminus possessed the full range of biological actions (though in reduced degree) of natural gastrin. The presence of a terminal acylating group was not essential, but replacement of the terminal amide function resulted in virtually complete loss of activity. It is remarkable that of the seventeen residues of the natural hormone, only the C-terminal tetrapeptide sequence and the terminal amide group are essential for biological activity and, as far as we are aware, this is the smallest fragment of a natural peptide that has been shown to contain all the biologically essential structural features of the parent. This aspect of the chemistry of gastrin has recently been further explored by a colleague (Morley et al., 1965).

In Liverpool our current work on gastrin has centered around the structure and synthesis of gastrins of other species. A special interest and fascination has always been attached to natural products of human origin, and recently we have had the opportunity of investigating the nature of human gastrin. Special problems are of course also associated with work on human hormones, particularly in obtaining sufficient material for structural studies. It is fortunate for us, however, that the source of gastrin, the antral region of the stomach, is often removed surgically during the treatment of certain gastric conditions. Our colleagues Professor Gregory and Dr. Tracy in Liverpool and Dr. Grossman in Los Angeles have collected a substantial number of antra from patients undergoing partial gastrectomy, and these have been extracted in Liverpool by a method closely akin to that used in the isolation of porcine

	H.I	H.II	G.I	G.II
Aspartic acid	1	1	1	1
Glutamic acid	5-6	5-6	6	6
Proline	1	1	1	1
Glycine	2	2	2	2
Alanine	1	1	1	1
Methionine	1	1	2	2
Leucine	1	1		
Tyrosine	1		1	
Tyrosine-O-sulfate		1		1
Phenylalanine	1	1	1	1
Tryptophan	2	2	2	2

gastrin. In this way very small amounts of two human gastrins (H.I and H.II) have been obtained (Gregory *et al.*, 1966).

The results of quantitative amino acid analysis of both human hormones are shown above, together with the known compositions of the porcine gastrins I and II. The very close relationship between the human and porcine series is immediately apparent, and indeed the hormones from the two species are indistinguishable in their physiological actions. However, the human hormones each contain one residue less of methionine and one residue more of leucine than do their porcine counterparts. Because of the very small quantities of material available for analysis (0.02-0.03 μmole, 0.04-0.06 mg) we were unable on the basis of these early results to decide conclusively between the presence of 5 or 6 residues of glutamic acid, although the latter was favored and supported by analogy with the porcine series. The relationship between H.I and H.II appears to be the same as between G.I and G.II; i.e. H.II is the sulfated form of H.I.

On the basis of the above amino acid analyses, we considered as a working hypothesis for synthetic studies four possible structures (II-V) for human gastrin I. These were related to the established structure (I) of porcine gastrin I by replacement of one or other of the methionine residues at positions 5 and 15 by leucine, and also by the possible deletion of one of the five central glutamyl residues.

Glu—Gly—Pro—Try—Leu—Glu —Ala—Tyr—Gly—Try—Met—Asp—
—Phe—NH_2

(II, *n* = 5; III, *n* = 4)

Glu—Gly—Pro—Try—Met—Glu —Ala—Tyr—Gly—Try—Leu—Asp—
—Phe—NH_2

(IV, *n* = 5; V, *n* = 4)

It was possible of course that, despite their impressive similarity in amino acid content, the human and porcine gastrins differed more extensively in amino acid sequence. However, the simplest and most attractive hypothesis was that of a single amino acid interchange between methionine and leucine residues, possibly coupled with the loss of one glutamyl residue. Differences of this type between proteins of different species have been observed previously, and the genetic codes for leucine and methionine (Nirenberg et al., 1965) are such that interchange of these residues could occur by a single mutation during the differentiation of the species. No experimental evidence was available favoring substitution at position 5 (i.e. II and III) over position 15 (i.e. IV and V), although intuitively it was felt that substitution in the latter position was less likely since this residue was part of the "essential" C-terminal sequence, the structure of which was expected to be fairly critical for full biological activity.

These four possible structures for human gastrin I were therefore made the objectives for synthetic studies, with special attention focused on the Leu-5 G.1 structure (II). However, while this synthetic work was in progress, a larger quantity of human gastrin I (0.6 mg) became available, and all of this was committed to a single degradative experiment in the hope of distinguishing among the four favored structures. In the event, this experiment clearly excluded the three structures (III, IV, and V) from further consideration, and provided very strong support for the favored alternative (II) (Bentley et al., 1966).

In parallel experiments, approximately equal quantities of H.I and G.I were degraded by the proteolytic enzyme papain. The resulting complex mixtures of peptides were separated by paper electrophoresis; corresponding bands in the H.I and G.I digests were eluted, compared, and checked for purity by thin-layer chromatography; and the most abundant peptides were analyzed quantitatively for amino acid content. The compositions of some of the peptides thus obtained are indicated below, together with the favored structure (II) for human gastrin I (Fig. 2).

Glu—Gly—Pro—Try—Leu—Glu_5—Ala—Tyr—Gly—Try—Met—Asp—

—Phe—NH_2

(Glu, Gly, Pro, Try)
(Glu, Gly, Pro, Try, Leu)
(Glu, Gly, Pro, Try, Leu, Glu)
(Glu_5, Ala, Tyr, Gly)
(Glu_4, Ala, Tyr, Gly)

Clearly the leucine residue is located at position 5 and there are five glutamyl residues in the central sequence. Structures (III, IV, and V) are therefore excluded. No peptides were obtained that were incompatible with structure (II), although the limited evidence from this single experiment could not of course completely eliminate other, less likely, alternatives.

Shortly afterward the syntheses of the heptadecapeptide amide (II) and hexadecapeptide amide (III) were completed (Beacham et al., 1966). The methods used for the final stages of these syntheses differed slightly from those in our original synthesis of the porcine hormone (Anderson et al., 1964). Thus in the original synthesis of porcine gastrin I (scheme I), peptide derivatives comprising residues 1-5 (VI) and 6-13 (VII) were coupled by the azide method to yield the fully protected peptide 1-13 (VIII). After saponification of the terminal ester function, further coupling with the tetrapeptide amide (IX) by the mixed anhydride procedure yielded the protected heptadecapeptide amide, which in turn afforded porcine gastrin I by removal of the protecting groups.

This synthetic scheme suffered from two main disadvantages. During the elaboration of the central octapeptide derivative (VII) by the stepwise addition of γ-t-butyl glutamyl residues, it was found that a considerable increase in steric hindrance of the reacting amino group occurred with increasing chain length. This was of relatively little importance during the synthesis of (VII) itself, since the active ester method was used for the formation of peptide bonds, and this is not normally prone to serious side reactions. In the union of (VII) with the pentapeptide (VI), however, the azide method was necessary in order to avoid any possibility of racemization of the methionine residue. Steric hindrance of this condensation was a much more serious problem, for the slow rate of reaction of the acyl azide gave greater opportunity for the well-known Curtius rearrangement to occur, leading eventually to urea derivatives rather than the desired peptide. Although the importance of this side reaction was minimized by careful selection of reaction conditions, further improvement was desirable. An additional disadvantage of this scheme was that the intermediate polyester (VIII) proved to be a very insoluble compound and not readily amenable to purification by column chromatography or countercurrent distribution.

Both of these disadvantages could be overcome in our synthesis of human gastrin I by a simple reversal of the order in which the peptide fragments were assembled into the main chain (Beacham et al., 1966). The only structural change required is replacement of the terminal benzyloxycarbonyl group in (VII)

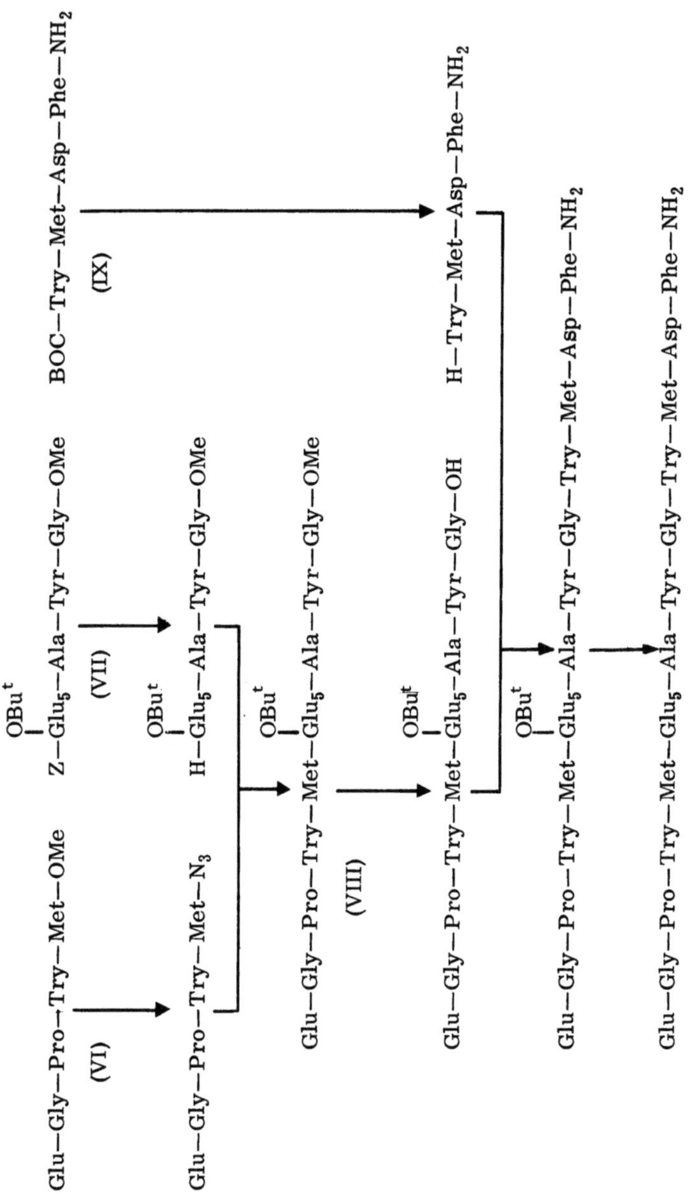

Porcine Gastrin I

Fig. 1. Scheme I; porcine gastrin I.

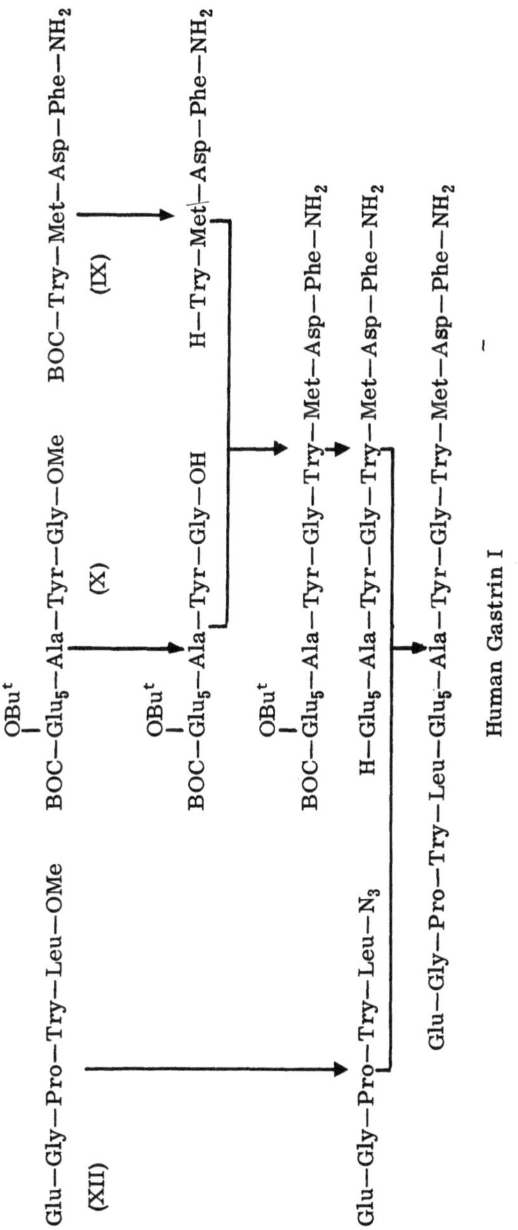

Fig. 2. Scheme II; human gastrin I.

by t-butyloxycarbonyl, as in (X) (scheme II). This was simply achieved by the use of γ-t-butyl N-t-butyloxycarbonylglutamate as its α-trichlorophenyl ester in the final step in the synthesis of (X). The terminal methyl ester of (X) was saponified and the free acid condensed with the C-terminal tetrapeptide amide (IX) by the mixed anhydride method. All the protecting groups were then cleaved from the product (XI), and the resulting water-soluble dodecapeptide purified by chromatography on Sephadex and aminoethyl cellulose. The final azide condensation with the N-terminal pentapeptide (XII) was no longer sterically hindered, and proceeded smoothly without appreciable urea formation. The synthetic heptadecapeptide amide (II) was then obtained directly. The des-glutamyl analog (III) was also prepared in a similar manner.

The synthetic product (II) has been carefully compared with natural human gastrin I, and we believe them to be identical. They have the same mobilities on paper electrophoresis (clearly different from that of synthetic III) and also on thin-layer chromatography. The natural and synthetic gastrins have been subjected to degradation by papain, and the resulting peptide mixtures examined by electrophoresis and thin-layer chromatography as in the structural studies already described. No evidence for any structural differences have been obtained. This synthetic work therefore confirms the structure of human gastrin I suggested by degradative studies, and we conclude that human gastrin differs from the porcine hormone only by replacement of the methionine residue at position 5 by leucine.

REFERENCES

Anderson, J. A., M. A. Barton, R. A. Gregory, P. M. Hardy, G. W. Kenner, J. K. MacLeod, J. Preston, R. C. Sheppard, and J. S. Morley: Synthesis of gastrin. Nature (Lond.) 204, 933 (1964).

Beacham, J., P. H. Bentley, R. A. Gregory, G. W. Kenner, J. K. MacLeod, and R. C. Sheppard: Synthesis of human gastrin I. Nature (Lond.), 209, 585 (1966).

Bentley, P. H., G. W. Kenner, and R. C. Sheppard: Structures of human gastrins I and II. Nature (Lond.), 209, 583 (1966).

Gregory, H., P. M. Hardy, D. S. Jones, G. W. Kenner, and R. C. Sheppard: Structure of gastrin. Nature (Lond.) 204, 931 (1964).

Gregory, R. A., and H. J. Tracy: The constitution and properties of two gastrins extracted from hog antral mucosa. Gut 5, 103 (1964).
Gregory, R. A., H. J. Tracy, and M. I. Grossman: Isolation of two gastrins from human antral mucosa. Nature (Lond.), 209, 583 (1966).
Morley, J. S., H. J. Tracy, and R. A. Gregory: Structure-function relationships in the active C-terminal tetrapeptide sequence of gastrin. Nature (Lond.) 207, 1356 (1965).
Nirenberg, M., P. Leder, M. Bernfield, R. Brimacombe, J. Trupin, F. Rottman, and C. O'Neal: RNA codewords and the general nature of the genetic code. Proc. nat. Acad. Sci. (Wash.) 53, 1161 (1965).
Tracy, H. J., and R. A. Gregory: Physiological properties of a series of synthetic peptides structurally related to gastrin I. Nature (Lond.) 204, 935 (1964).

DISCUSSION

Lembeck: It is quite remarkable that already a tetrapeptide has a gastrin-like action. Do you think that such a peptide could be derived from digested proteins by the action of pepsin during the digestion in the stomach?

Sheppard: It is, of course, possible that such a tetrapeptide could be formed in the stomach during digestion of a protein containing the appropriate amino acid sequence, but I think it unlikely that it would survive further cleavage long enough to be absorbed into the circulation. We have not studied the action of pepsin on gastrin at the C-terminal tetrapeptide amide, but other proteolytic enzymes (i.e. papain and subtilisin) degrade the C-terminal part of gastrin extensively.

Prusik: Do you know anything about the interaction of secretin with gastrin or with analogs of gastrin? Secretin, as you know inhibits the secretion of gastric acid.

Sheppard: Secretin has been shown to be capable of inhibiting gastric acid secretion stimulated by gastrin in conscious dogs provided with gastric pouches (Gillespie and Grossman, Gut 5, 342, 1964). However, the inhibition of gastric secretion by gastrin observed under certain conditions is not a consequence of the liberation of secretin. (Morley et al., Nature, Lond. 207, 1356, 1965).

Gladner: Is there any difference when one removes the sulfate from the tyrosine of the gastrin? Are there any changes in physiological activity?

Sheppard: No considerable differences.

Erdös: Is the hypotensive effect of gastrin due to the release of histamine? How quantitative is your assay when you compare the activities of the tetrapeptide with the heptadecapeptide?

Sheppard: The hypotensive effect is small and inconstant. It is not known whether this is due to release of histamine. The gastric secretory tests on conscious dogs are not easily interpreted in strictly quantitative terms, but the C-terminal tetrapeptide amide has about 20% of the activity of natural gastrin; i.e. a dose of about 5 times by weight elicits a similar response.

THE INFLUENCE OF SUBSTITUTION OR OMISSION OF AN AMINO GROUP ON THE HYPOTENSIVE ACTIVITY OF THE C-TERMINAL SEQUENCES OF ELEDOISIN

KLAUS LÜBKE, GEORGE ZÖLLNER, and EBERHARD SCHRÖDER

Schering AG, Hauptlaboratorium, Berlin, West Germany

Our knowledge about the relation between structure and activity in the peptide field originates almost completely from changes in the amino acid sequence or chain length of the native molecule. Only little is known about the influence of the substitution of functional groups by nonpeptidic residues. In some cases, however, the importance of unsubstituted functional groups is established, as well as the possibility of substitution or omission of some functional groups without reducing the activity to a large extent.

It is not surprising that in several cases the same substituent introduced in different peptides has a different influence on the activity (Table 1). For instance, the introduction of a nitro group in an arginine residue of the angiotensin molecule reduces the activity to 50% (Schwyzer and Turrian, 1960), whereas 1, 9-dinitro-bradykinin shows an activity of only 1% (Bodanszky et al., 1963). Such an influence is even more marked in the case of formylation of the lysine in melanocyte-stimulating hormone (MSH) and adrenocorticotropic hormone (ACTH) (Hofmann and Katsoyannis, 1963). Formylation of the lysine^{11} residue in MSH has no influence on the activity. In contrast, formylation of the lysine residues in positions 11, 15, 16, and 21 of α^{1-23}-ACTH leads to a nearly inactive product.

Of much more interest, however, are those examples in which different groups substitute the same functional group of one peptide. In the MSH field the difference in activity of the ϵ-formyl-lysine^{11} and the ϵ-tosyl-lysine^{11} derivative is tenfold. The same is also true for the partial sequence 6-13 (Hofmann and Katsoyannis, 1963).

TABLE 1

Substitution of the Basic ω-Function of Lysine and Arginine

	(%)
NH₂ NO₂ H—Asp—Arg—Val—Tyr—Val—His—Pro—Phe—OH (Asp with NH₂; Arg underlined, with NO₂)	50
H—Arg—Pro—Pro—Gly—Phe—Ser—Pro—Phe—Arg—OH (both Arg underlined, each with NO₂)	1
Ac—Ser—Tyr—Ser—Met—Glu—His—Phe—Arg—Try—Gly—Lys—Pro—Val—NH₂ (Glu with NH₂; Lys underlined, with Form)	100
Ac—Ser—Tyr—Ser—Met—Glu—His—Phe—Arg—Try—Gly—Lys—Pro—Val—NH₂ (Glu with NH₂; Lys underlined, with Tos)	10
H—His—Phe—Arg—Try—Gly—Lys—Pro—Val—NH₂ (Lys underlined, with Form)	100*
H—His—Phe—Arg—Try—Gly—Lys—Pro—Val—NH₂ (Lys underlined, with Tos)	6*
H—His—Phe—Arg—Try—Gly—Lys—Pro—Val—NH₂ (Arg with NO₂; Lys underlined, with Form)	6*
H—Ser—Tyr—Ser—Met—Glu—His—Phe—Arg—Try—Gly—Lys—Pro—Val—Gly—Lys— —Lys—Arg—Arg—Pro—Val—Lys—Val—Tyr—NH₂ (each Lys with Form; the first Lys of each line and the last Lys underlined)	inactive

*Related to the unsubstituted partial sequence = 100%.

In the case of angiotensin II (Arakawa et al. 1962) or oxytocin and vasopressin (Du Vigneaud et al., 1960) the omission of the α-amino group results in very highly active compounds (Table 2). In contrast to the omission of the N-terminal amino group, the substitution leads to nearly inactive compounds (Boissonnas et al., 1961; Cash and Smith, 1963; Rudinger, 1963; Studer and Cash, 1963). Furthermore, such changes can result in a differentiation of the various activities of one peptide. For example, the antidiuretic activity of the desamino-vasopressins is more distinct than the pressor activity.

TABLE 2
Substitution or Omission of the N-Terminal Amino Group

Desamino-angiotensin	100% = protracted effect			
	Rat uterus (%)	Chicken depressor	Rat pressor (%)	Anti-diuresis
Desamino-oxytocin	150	160	22	300
N-Acetyl-oxytocin	0.4	anti-oxytocin		
N-Methyl-oxytocin	0.3	0.2		1
Desamino-Lys^8-vasopressin	240	150	46	120
N-Acetyl-Lys^8-vasopressin		inactive		
Desamino-Arg^8-vasopressin	170	260	100	300
N-Acetyl-Arg^8-vasopressin		inactive		

Some facts are known about the phenolic hydroxyl function of tyrosine in position 4 of angiotensin II and in position 2 of oxytocin and vasopressin (Table 3). Omission of this group leads in all cases to rather active compounds (Schwyzer and Turrian, 1960; Boissonnas and Guttmann, 1960; Huguenin and Boissonnas, 1962; Jaquenoud and Boissonnas, 1959). It may be of some interest that the oxytocic activities of these compounds (uterotonic activity and depressor activity in the chicken) decreased more than the vasopressin-like activities. Substitution of the hydroxyl function with the methyl or the ethyl residue, however, results in nearly all cases in a strong decrease of the activity (Schröder and Hempel, 1965; Rudinger, 1963; Zhuze

TABLE 3

Substitution or Omission of the Hydroxyl Function of Tyrosine

	Rat uterus (%)	Chicken depressor (%)	Rat pressor (%)	Anti-diuresis (%)
H—Asp(NH_2)—Arg—Val—Tyr(Me)—Val—His—Pro—Phe—OH			0.2%	
H—Asp(NH_2)—Arg—Val—Phe—Val—His—Pro—Phe—OH			10%	
H—Cys—Tyr(Me)—Ileu—Glu(NH_2)—Asp(NH_2)—Cys—Pro—Leu—Gly—NH_2	1	2	100	1
H—Cys—Tyr(Et)—Ileu—Glu(NH_2)—Asp(NH_2)—Cys—Pro—Leu—Gly—NH_2	0.03		inactive	0.1
H—Cys—Phe—Ileu—Glu(NH_2)—Asp(NH_2)—Cys—Pro—Leu—Gly—NH_2	7	13	10	10
H—Cys—Tyr(Me)—Phe—Glu(NH_2)—Asp(NH_2)—Cys—Pro—Lys—Gly—NH_2			1	32
H—Cys—Tyr(Et)—Phe—Glu(NH_2)—Asp(NH_2)—Cys—Pro—Lys—Gly—NH_2			inhibition	2
H—Cys—Phe—Phe—Glu(NH_2)—Asp(NH_2)—Cys—Pro—Lys—Gly—NH_2	2	2	22	12
H—Cys—Phe—Phe—Glu(NH_2)—Asp(NH_2)—Cys—Pro—Arg—Gly—NH_2	1	2	32	80

et al., 1964; Zaoral et al., 1965). Only O-methyl-vasopressin is an exception, since it shows a significant antidiuretic activity.

Also some substituted derivatives are known in the eledoisin field, but most of them are intermediate compounds obtained during synthesis (Bernardi et al., 1964). To get more material we investigated the influence of different substituents on the hypotensive activity of the C-terminal hexapeptide sequence of eledoisin H—Ala—Phe—Ileu—Gly—Leu—Met—NH_2. Several reasons lead us to choose this compound. It is known that a substitution of the free amino group with the carbobenzoxy or the t-butyloxycarbonyl group does not decrease the activity of the hexapeptide (Bernardi et al., 1964). Furthermore, the increase in activity by the transition to the peptapeptide H—Asp—Ala—Phe—Ileu—Gly—Leu—Met—NH_2 is approximately threefold (Bernardi et al., 1964; Schröder and Lübke, 1964). This increase is not connected with the original aspartic acid residue at the N-terminal position. A great number of other amino acids also causes an increase in activity which is nearly independent from the kind of amino acid (Table 4). Aliphatic amino acids as well as basic, aromatic and hydroxy amino acids show the same effect.

To clarify the question of whether an extension of the hexapeptide by other acyl residues also produces and increase in

TABLE 4

C-Terminal Heptapeptide Analogs of Eledoisin

Amino-acyl—Ala—Phe—Ileu—Gly—Leu—Met—NH_2

Hypotensive activity (rabbit), eledoisin = 1.

Amino-acyl	Activity	Amino-acyl	Activity
H—Gly	0.9	H—Phe	1.2
H—Leu	1.3	H—Tyr	0.8
H—NLeu	1.0	H—Ser	2.0
H—Pro	0.9	H—Asp(NH_2)	1.0
H—Lys	1.1	H—Glu(NH_2)	0.9
H—Orn	3.0	H—Sar	0.9
H—His			

activity and whether there is an influence on the type of substituents, we synthesized a number of substituted C-terminal hexa- and heptapeptides of eledoisin.

For the synthesis of these compounds different routes were employed. In some cases the substituent was first introduced in the dipeptide ester H—Ala—Phe—OMe, and after hydrazinolysis the derivatives were coupled via the azide method with the C-terminal tetrapeptide H—Ileu—Gly—Leu—Met—NH_2. For the other compounds the C-terminal hexapeptide H—Ala—Phe—Ileu—Gly—Leu—Met—NH_2 was used as the starting material, and the substituent was introduced via the anhydride, the azide, or the p-nitrophenyl ester method. Poorer results were obtained via the acid chloride method. For the synthesis of the acetyl derivatives of the heptapeptides with lysine or ornithine in the N-terminal position, the necessary derivatives of these amino acids were coupled with the hexapeptide, and the protecting group was removed with hydrochloric acid in acetic acid. The acetyl-Ser-heptapeptide was obtained from the tripeptide H—Ser—Ala—Phe—OMe via selective N-acetylation, hydrazinolysis, and coupling with the C-terminal tetrapeptide.

The hypotensive activity of the individual compounds was determined on mongrel dogs (weight range, 16-24 kg), which were anesthesized with sodium pentobarbital. Arterial blood pressure was measured with a Statham strain-gauge transducer from the arteria carotis communis dexter and registered with a Hellige direct recorder. The substances were administered intravenously into the vena femoralis dexter. The doses were plotted against the reaction of the mean arterial blood pressure.

The heptapeptides H—Gly—Ala—Phe—Ileu—Gly—Leu—Met—NH_2, which could be obtained only in aqueous solution with the aid of dimethylformamide, and H—Lys—Ala—Phe—Ileu—Gly—Leu—Met—NH_2, which is easily soluble in water, were used as standards. Both peptides are typical representatives of C-terminal heptapeptides and show the same activity as eledoisin in the dog.

Substitution of the C-terminal hexapeptide with pure acyl residues leads to substances which are active, less active, or nearly inactive according to the chain length of the carbonic acid (Table 5). The activity of the peptide depends on the chain length (Fig. 1). The formyl derivative has the highest activity; the acetyl and the propionyl derivatives still have a significant activity; whereas the longer acyl residues lead to derivatives with only little activity.

TABLE 5
C-Terminal Acyl Hexapeptides of Eledoisin
C_nH_{2n+1}—CO—Ala—Phe—Ileu—Gly—Leu—Met—NH_2

Hypotensive activity (dog),
H—Gly—Ala—Phe—Ileu—Gly—Leu—Met—NH_2 = 1.

n = 0	Formyl	2.0	n = 5	Capronyl	0.2
n = 1	Acetyl	1.2	n = 9	Caprinoyl	0.08
n = 2	Propionyl	0.9	n = 15	Palmitinoyl	0.05

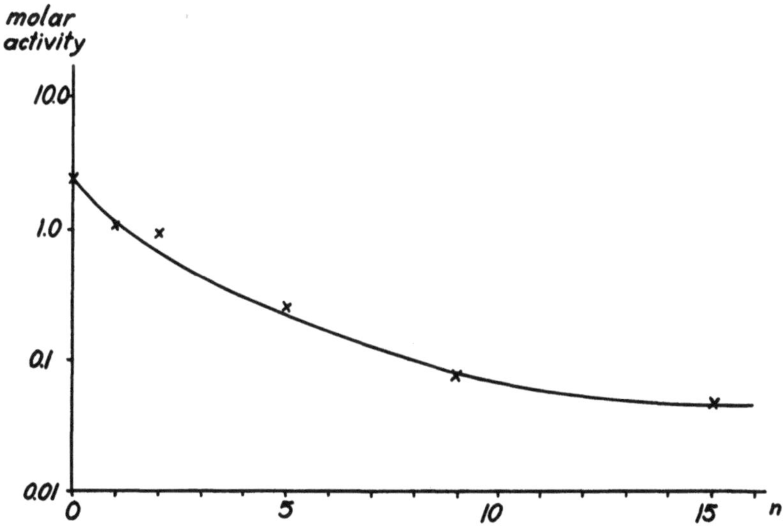

Figure 1.

It was also seen, however, that those derivatives having an acyl residue with an additional functional group show a very high activity, even higher than the activity of real heptapeptides (Table 6). The type of functional group seems to have little influence; hydroxyl function, as well as amino or chloro function, shows the same increase. Only the succinyl hexapeptide, which can be considered as the desamino derivative of the original C-terminal heptapeptide, has a lower activity. The high activities of the substituted hexapeptides cannot be explained by the impossibility of an enzymatic degradation by aminopeptidases, because the succinyl derivative and also some of the acyl de-

TABLE 6
C-Terminal Acyl Hexapeptides of Eledoisin
R—CO—Ala—Phe—Ileu—Gly—Leu—Met—NH_2

Hypotensive activity (dog),
H—Gly—Ala—Phe—Ileu—Gly—Leu—Met—NH_2 = 1.

R—CO	Name	Activity
$Cl-CH_2-CO$	Chloroacetyl	1.5
$CH(CH_3)_2$ $HO-CH-CO$	L-α-Hydroxy- isovaleryl	2.0
$H_2N-(CH_2)_5-CO$	ε-Aminocapronyl	2.5
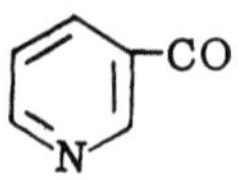	Nicotinoyl	1.5
CH_2-COOH CH_2-CO	Succinyl	0.7

rivatives are only slightly active. The substitution in the last two derivatives also prevented an enzymatic degradation.

Since most of the compounds are almost completely insoluble in water, the application was made in dimethylformamide or in a mixture of dimethylformamide and water. To obtain substances with a better solubility, we synthesized several salts from the succinyl hexapeptide. Salt formation, however, reduces the activity to a small extent (Table 7).

To establish these results we synthesized some substituted C-terminal heptapeptides besides the substituted hexapeptides (Table 8). It may be of some interest that the formylation,

TABLE 7
Salts of the C-Terminal Desamino Heptapeptide of Eledoisin

Hypotensive activity (dog),
H—Gly—Ala—Phe—Ileu—Gly—Leu—Met—NH_2 = 1.

Succinyl—Ala—Phe—Ileu—Gly—Leu—Met—NH_2	0.7		
Triethylammonium salt	0.7	Dicyclohexylammonium salt	0.3
Tryptammonium salt	0.5	Methylglucammonium salt	0.5

TABLE 8
C-Terminal Heptapeptide Derivatives of Eledoisin
R—Ala—Phe—Ileu—Gly—Leu—Met—NH_2

Hypotensive activity (dog),
H—Lys—Ala—Phe—Ileu—Gly—Leu—Met—NH_2 = 1.

R	Activity	R	Activity
H—Gly	1.0	Form—Gly	1.0
Ac Ac—Lys	1.2	Capronyl (α-ε-desamino-lysine)	0.3
Ac H—Lys	0.5	H—NLeu (ε-desamino-lysine)	1.0
Ac—Lys	3.5	ε-Aminocapronyl (α-desamino-lysine)	2.5
H—Ser	0.5	Ac—Ser	2.0
H—Orn	1.0	Ac—Orn	3.0

which results in hexapeptide with a strong increase of activity, has no influeence on the activity of the Gly-heptapeptide. The same is true for the diacetylated Lys-heptapeptide. In contrast, the monoacetylated derivatives have different activities. According to the results obtained with the substituted hexapeptide, the derivative with the unsubstituted ε-amino group shows the highest activity. The same increase in activity is found by the α-acetylated Orn- and Ser-heptapeptides, which also have a free funtional ω-group. In comparison, the desamino derivatives of the Lys-heptapeptide show an analogous effect. The bis-desaminopeptide, which is a pure acyl derivative, shows only a small activity, and the des-ε-amino peptide has the normal activity of a heptapeptide, whereas the des-α-amino peptide, which is an acyl hexapeptide with an additional functional ω-group in the acyl residue, has a very high activity.

To summarize the results we can say that the free amino group of the C-terminal hexapeptide and of the C-terminal heptapeptide is not essential for the biological activity. Dependent on the substituent, an increase in activity can be observed. The kind of substituent has a very important influence, whereby differences greater than tenfold are possible. Compounds with an additional functional ω-group show the highest activity. Although an enzymatic degradation by aminopeptidases was prevented by the substitution of the N-terminal amino group, none of the compounds shows a significantly prolonged effect.

REFERENCES

Arakawa, K., R. R. Smeby, and F. M. Bumpus, J. Amer. chem. Soc. 84, 1424 (1962).

Bernardi, L., G. Bosisio, F. Chillemi, G. de Caro, R. de Castiglione, V. Erspamer, A. Glaesser, and O. Goffredo, Experientia (Basel) 20, 303 (1964).

Bodanszky, M., M. A. Ondetti, J. T. Sheehan, and S. Lande, Ann. N.Y. Acad. Sci. 104, 24 (1963).

Boissonnas, R. A., and St. Guttmann, Helv. chim. Acta 43, 190 (1960).

Boissonnas, R. A., St. Guttmann, B. Berde, and H. Konzett, Experientia (Basel) 17, 377 (1961).

Cash, W. D., and B. L. Smith, J. biol. Chem. 278, 994 (1963).

Du Vigneaud, V., G. Winestock, V. V. S. Murti, D. B. Hope, and R. D. Kimbrough, Jr., J. biol. Chem. 235, PC64 (1960).

Hofmann, K., and P. G. Katsoyannis, in The Proteins (N. Neurath, ed.), Vol. I, p. 53. New York and London: Academic Press (1963).

Huguenin, R. L., and R. A. Boissonnas, Helv. chim. Acta 45, 1629 (1962).

Jaquenoud, P. A., and R. A. Boissonnas, Helv. chim. Acta 42, 788 (1959).

Rudinger, J., Acta biol. med. Germ. Suppl. II, 38 (1963).

Schröder, E., and R. Hempel, Justus Liebigs Ann. Chem. 648, 243 (1965).

Schröder E., and K. Lübke, Experientia (Basel) 20, 19 (1964).

Schwyzer, R., and H. Turrian, Vitam. and Horm. 18, 237 (1960).

Studer, R. O., and W. D. Cash, J. biol. Chem. 238, 657 (1963).

Zaoral, M., E. Kasafirek, J. Rudinger, and S. Sorm, Coll. Czech. chem. Commun. 30, 1819 (1965).

Zhuze, A. L., K. Jošt, E. Kasafirek, and J. Rudinger, Coll. Czech. chem. Commun. 29, 2618 (1964).

NOVEL TECHNIQUES FOR THE SYNTHESIS OF LINEAR AND CYCLIC PEPTIDES

EPHRAIM KATCHALSKI

Department of Biophysics, Weizmann
Institute of Science, Rehovoth, Israel

I shall describe in brief three novel techniques, worked out in our laboratory, for the preparation of linear peptides with a predetermined amino acid sequence, cyclic peptides, and peptides with a repeating amino acid sequence.

A. PEPTIDES WITH A PREDETERMINED AMINO ACID SEQUENCE

In the method to be described below insoluble high-molecular weight active esters of α-N-benzyloxycarbonyl amino acids were employed (Fridkin et al., 1966). These were allowed to react with the desired amino acid or peptide esters, according to the scheme in Fig. 1, and the newly formed terminally masked peptides separated from excess insoluble amino acid ester by filtration or centrifugation. Practically quantitative yields of the newly formed peptide were obtained in the presence of a marked excess of the insoluble reagent.

Removel of the N-terminal masking group by catalytic hydrogenation or with HBr in glacial acetic acid enables the repetition of the coupling reaction by making use of an insoluble active ester of a different amino acid.

The insoluble active esters used so far were derived from poly-4-hydroxy-3-nitrostyrene cross-linked with divinyl benzene. Insoluble chemical reagents containing up to 1 mmole bound amino acid per gram could be readily obtained.

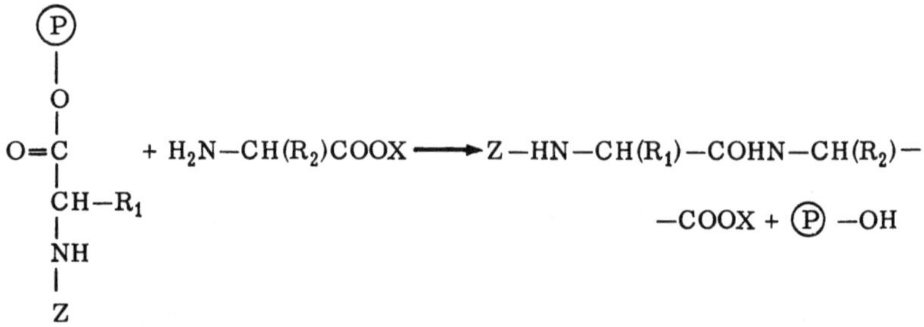

Fig. 1. Scheme describing the method employed for the synthesis of peptides. (P) – stands for insoluble polymer, Z for benzyloxycarbonyl.

Using different insoluble N-carbobenzyloxy-amino acid esters we were able to prepare in high yield peptides such as Cbz-Phe—Phe—Ala—PNB, benzoyl—Ala_3—PNB, and glutathione. The possible extension of the method described to the preparation of more complicated peptides is under investigation.

B. CYCLIC PEPTIDES

Cyclic peptides are usually prepared from linear peptides by intramolecular cyclization. The carboxylic end of the peptide is as a rule activated by the formation of active esters, anhydrides, azides, or chlorides, and the free terminal amino end allowed to react with the active terminal carbonyl group at high dilution. Because of intermolecular condensation occurring even under these conditions, linear oligopeptides are formed in addition to the desired cyclic peptide. The techniques available so far thus lead to reaction mixtures from which cyclic peptides are usually isolated only in relatively low yields. The use of peptide active esters of type II (see Fig. 2), in which the peptide is bound to a high-molecular weight polyalcohol-carrier as intermediate compounds in the synthesis of cyclic peptides, has been described recently by Fridkin et al. (1965). When insoluble esters of this type are employed, condensation between the activated peptide moieties is suppressed, and internal aminolysis leads to the formation of the desired cyclic peptide (IV) which is released from the insoluble polyhydroxy carrier (III). Intermolecular condensation might be expected to be markedly reduced even when soluble high-molecular weight peptide esters of type II are used.

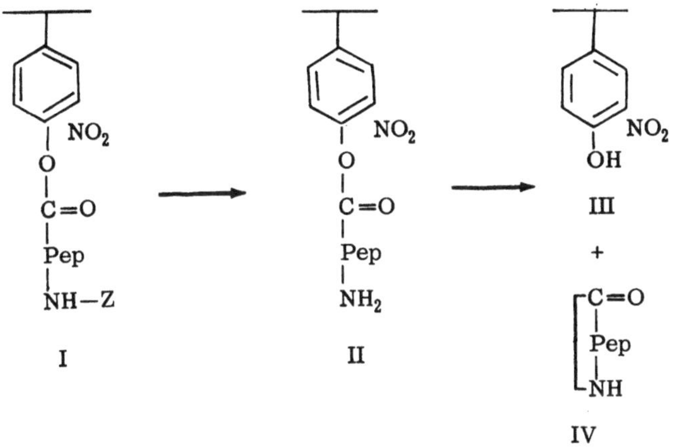

Fig. 2. Scheme describing the method employed for the synthesis of cyclic peptides. IIIa stands for poly-4-hydroxy-3-nitrostyrene; Ia and IIa are the corresponding N-benzyloxycarbonyl (Z) peptide and free amino peptide ester derivatives of IIIa. IIIb stands for a branched copolymer of DL-lysine and 3-nitro-L-tyrosine in which free amino groups have been blocked by acetylation; Ib and IIb are the corresponding N-benzyloxycarbonyl (Z) peptide and free amino peptide ester derivatives of IIIb.

Two high-molecular weight poly-(nitrophenol) derivatives have been used in the preparation of the peptide active esters: cross-linked poly-4-hydroxy-3-nitrostyrene (IIIa), and a branched copolymer of DL-lysine and 3-nitro-L-tyrosine (IIIb) in which free amino groups have been blocked by acetylation. The former has been prepared according to the literature (Packham, 1964) and is insoluble in the usual organic solvents. The latter has been prepared by total acetylation with acetic anhydride of a branched copolymer of DL-lysine and L-tyrosine (molar residue ratio 3:1) (Sela et al., 1962), removal of the O-acetyl groups in alkali, and nitration in concentrated nitric acid at 0° C; IIIb is insoluble in dioxane, ether, and acetone, but is soluble in dimethylformamide (DMF), ethanol, and water at pH values above 10. N-Benzyloxyczrbonyl derivatives of peptides were coupled with both poly-(nitrophenol) compounds in DMF with N, N′-dicyclohexylcarbodiimide as the coupling reagent. Highest yields of the N-benzyloxycarbonyl active esters (I) were obtained when the reaction mixture contained at least a threefold excess of the N-benzyloxycarbonyl peptide in comparison with the nitrophenol content of the polymer. Excess of unreacted peptide was removed from Ia by washing and from Ib by dialysis. The protecting groups of I were removed with anhydrous hydrogen bromide in glacial acetic acid (Ben-Ishai and Berger, 1952) and the high-molecular weight peptide ester hydrobromides isolated by filtration (esters derived from IIIa) or by precipitation with ether (esters derived from IIIb). The free amino groups of the peptide esters (II) were liberated from the corresponding hydrobromides by neutralization with triethylamine in DMF and the cyclization reaction allowed to take place at room temperature in the same solvent. The DMF-insoluble free peptide esters studied (IIa), derived from IIIa, yielded under our experimental conditions chromatographically pure cyclopeptides in 60 to 80% yield. The latter could readily be isolated from the DMF solution. The DMF-soluble peptide esters studied (IIb), derived from IIIb, yielded the corresponding cyclic peptides also in 60 to 80% yield. The cyclic peptides in this case, however, which were identified only by chromatography, were contaminated with some peptide oligomers.

The following cyclic peptides have been obtained so far by the above technique: cyclo (Gly—Gly); cyclo (Ala—Ala); cyclo (Phe—Gly); cyclo (tetra Ala) and cyclo (Ala—Gly—Ala—Ala). The last cyclic tetrapeptide was found identical in its chromatographic behavior with the cyclic peptide derived from Gly, Ala—Ala—Ala *p*-nitrophenyl ester.

C. POLYPEPTIDES WITH A REPEATING AMINO ACID SEQUENCE

Amino acid homopolymers and random copolymers have been used extensively as model compounds in the study of the physicochemical and biological properties of native peptides and proteins (Katchalski et al., 1964). Polymers with a repeating amino acid sequence might be even more useful in the elucidation of the factors determining protein conformation, of the behavior of high-molecular weight substrates toward proteolytic enzymes, and of the chemical basis of antigenicity. Several attempts have been made to synthesize polymers with a repeating amino acid sequence, but most of these resulted in products of rather low degrees of polymerization. Among the methods employed, the one that offers most promise is the active ester method. De Tar et al. (1963) investigated the use of *p*-nitrophenyl esters for the synthesis of polymers with repeating sequences and reported the formation of polymers of β—benzyl—aspartyl—glycyl—glycine with an average degree of polymerization of 25 (75 amino acid residues). Subsequently this approach was employed by Fraser and co-workers (1965) for the preparation of poly (Ala—Gly) from the *p*nitrophenyl ester of alanyl—glycine, and for the synthesis of polymers of γ—methyl—glutamyl—valyl—γ-methyl—glutamic acid from the corresponding *p*-nitrophenylester. Again, average chain lengths of 70 to 80 acid residues were estimated by end-group analysis. Recently, Kovacs and Kapoor (1965) reported the use of pentachlorophenyl esters in the synthesis of polypeptides with repeating amino acid sequences. The advantage of this method lies in the fact that pentachlorophenyl esters are stable to catalytic hydrogenolysis, thus permitting the selective removal of carbobenzoxy groups from the amino terminal prior to polymerization. Kovacs and Kapoor prepared the pentachlorophenyl ester of carbobenzoxy-γ-*t*—butyl—glutamyl—alanyl—γ-*t*—butyl—glutamic acid, decarbobenzoxylated the tripeptide active ester by catalytic hydrogenolysis, and polymerized the product. The polymer was deblocked by treatment with trifluoroacetic acid, and an average molecular weight of 20,000 was estimated from viscosity measurements.

Esters of peptides with N-hydroxysuccinimide have been employed in our laboratory for the preparation of polyamino acids with a repeating sequence by Dr. Ramachandran.* The utility of esters of protected amino acids with N-hydroxysuccinimide in peptide synthesis was demonstrated by Andersor. et al. (1964). These esters are highly reactive and can be used in aqueous media. The by-product, N-hydroxysuccinimide, is soluble in water and hence readily removed.

The synthesis of the polymer $[—Tyr—Ala—Glu]_n$ is illustrated in Fig. 3. The protected tripeptide was synthesized in a stepwise manner by carrying the various coupling reactions recorded with the N-hydroxysuccinimide esters of the corresponding amino acid derivatives. γ-Benzyl-glutamic acid was reacted with the N-hydroxysuccinimide ester of *t*-butyloxycarbonyl-alanine in aqueous dioxane to yield the crystalline dipeptide (V) in 85% yield. The *t*-butyloxycarbonyl group was selectively removed by treatment with 2 N HCl in dioxane, and the product was reacted with the N-hydroxysuccinimide ester of *t*-butyloxycarbonyl-O-benzyl-tyrosine. The crystalline tripeptide VI was obtained in 88% yield; VI was converted to

*J. Ramachandran, unpublished results.

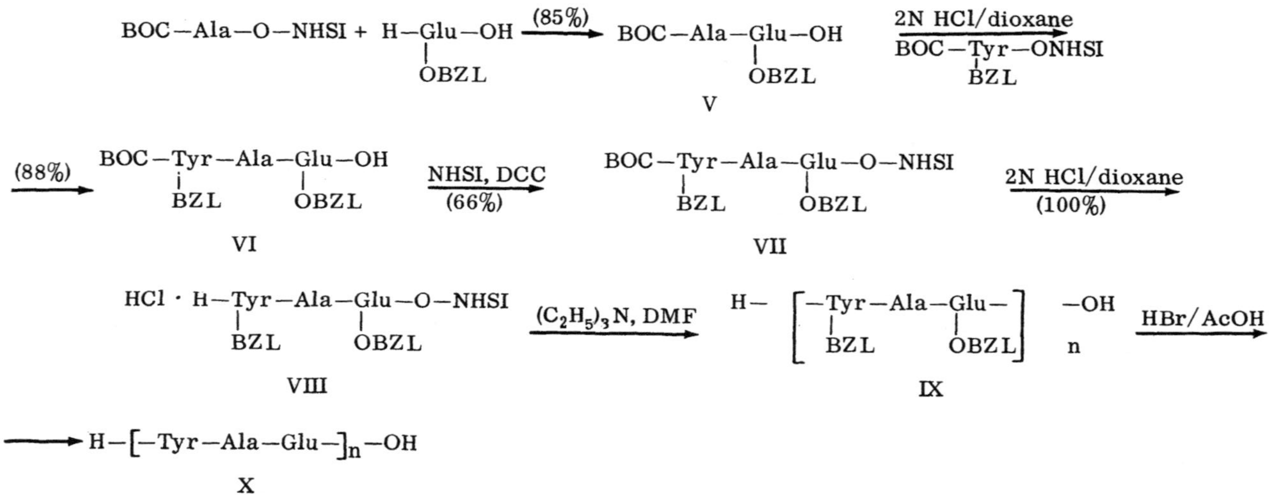

Fig. 3. Synthesis of —[Tyr—Ala—Glu—]$_n$—. NHSI stands for N-hydroxysuccinimide, BZL for benzyl, and BOC for *t*-butyloxycarbonyl.

the active ester VII by reaction with N-hydroxysuccinimide in the presence of dicyclohexylcarbodiimide. The active ester VII was purified by recrystallization. The *t*-butyloxycarbonyl group was again selectively removed from VII by treatment with 2 N HCl in dioxane. The hydrochloride of the active ester of the tripeptide VIII was dissolved in dimethylformamide (35% solution) and treated with one equivalent of triethylamine. Polymerization was allowed to proceed at room temperature for 8 days, and the product was isolated by precipitation into water. The polymer IX was then deblocked by treating with HBr in glacial acetic acid. The product $(Tyr—Ala—Glu)_n$ was found to have an average molecular weight of 20,000 from measurements of sedimentation and diffusion.

Another polymer $(Pro—Gly—Pro)_n$ has been recently prepared by us by the polycondensation of the tripeptide HPro—Gly—Pro—OM in pyridine, in the presence of tetraethylpyrophosphate (Engel et al., 1966). The polyamino acid obtained is readily soluble in water and has an average molecular weight of about 6,000. Fractionation by gel filtration yielded samples of average molecular weights between 1,000 and 12,000. The poly-$(Pro—Gly—Pro)_n$ fractions of high molecular weight showed remarkable resemblance to collagen in the solid state and in solution.

REFERENCES

Anderson, G. W., J. Zimmerman, and F. M. Callahan: J. Amer. chem. Soc. 35, 3039 (1963); 86, 1879 (1964).

Ben-Ishai, D., and A. Berger: J. org. Chem. 17, 1564 (1952).

DeTar, D. F., W. Honsberg, U. Honsberg, A. Wieland, M. Gouge, H. Bach, A. Tahora, W. S. Brinigar, and F. F. Rogers, Jr.: J. Amer. chem. Soc. 85, 2873 (1963).

Engel, J., J. Kurtz, E. Katchalski, and A. Berger: J. molec. Biol. in press (1966).

Fraser, R. D. B., T. P. MacRae, F. H. C. Stewart, and E. Suzuki: J. molec. Biol. 11, 706 (1965).

Fridkin, M., A. Patchornik, and E. Katchalski, J. Amer. chem. Soc. in press (1966).

Fridkin, M., A. Patchornik, and E. Katchalski: J. Amer. chem. Soc. 87, 4646 (1965).

Katchalski, E., M. Sela, H. I. Silman, and A. Berger: in The Proteins, H. Neurath, Ed., vol. 2, p. 405. New York: Academic Press (1964).

Kovacs, J., and A. Kapoor, J. Amer. chem. Soc. 87, 118 (1965).
Packham, D. I., J. chem. Soc. 2617 (1964).
Sela, M., S. Fuchs, and R. Arnon, Biochem. J. 85, 223 (1962).

POLYPEPTIDES ACTIVE ON PLAIN MUSCLE IN THE AMPHIBIAN SKIN

V. ERSPAMER and A. ANASTASI

Institute of Pharmacology of the University of Parma, and
Farmitalia Research Laboratories, Milan, Italy

This report describes the main findings obtained in the field of active polypeptides of the amphibian skin by the research groups of the Institute of Pharmacology of the University of Parma and of the Farmitalia Research Laboratories, Milan, which were joined by Dr. J. M. Cei, Mendoza, Argentina, and Dr. R. Endean, Brisbane, Australia, who were responsible for collection, classification, and occasional extraction of the amphibian material.

The fact that prompted us to undertake a systematic biological and chromatographic screening of skin extracts of the largest possible number of amphibian species, collected throughout the world, was the observation made a few years ago that crude methanol extracts of the skin of *Physalaemus fuscumaculatus* possessed a tremendously intense hypotensive action in the dog and a potent stimulant action on the large intestine of rabbit. (Erspamer et al., 1962).

It will be seen that the agent responsible for these effects, the endecapeptide physalaemin, was isolated in a pure form and, after elucidation of its amino acid composition and sequence, reproduced by synthesis.

MATERIALS AND METHODS

So far, more than 200 amphibian species have been subjected to our screening procedure. Generally, dried skins were examined, but the most important results were nearly always checked also with fresh skins.

Methanol (99% for fresh skins, 80% for dried skins) was the solvent of choice for the extraction of the peptides. First biological screening was carried out with crude methanol extracts. If results were positive, a more accurate pharmacological study was performed with semipurified or purified material. It is obvious that separation of single active constituents was imperative when dealing with mixtures of active compounds (Fig. 2).

The purification of crude extracts and isolation of the active principles was carried out with all the available methods: column chromatography using alumina, silicic acid, ion-exchange resins, Sephadex, and carboxymethylcellulose; paper chromatography; high-voltage electrophoresis; countercurrent distribution; dialysis; etc.

Test objects and systems used in the routine biological screening were as follows: dog and rabbit blood pressure, rabbit and rat large intestine, rat uterus, guinea-pig ileum, and diuresis of hydrated rats. Several other subsidiary test objects were employed in particular cases.

The use of these test objects was of decisive importance in the demonstration and discrimination of the polypeptides occurring in the amphibian skin. Other important criteria for the differential diagnosis and classification of the peptides were afforded by treatment with proteolytic enzymes, by some color reactions of particular amino acid residues, and finally by the behavior of the peptides in high-voltage electrophoresis and other purification procedures.

RESULTS

Active polypeptides so far traced in the amphibian skin may be divided, at least provisionally, into four groups characterized by distinctive features: eledoisin-like polypeptides, bradykinin-like polypeptides, caerulein-like polypeptides and, finally, miscellaneous polypeptides including also active compounds, the peptide nature of which has yet to be established unequivocally.

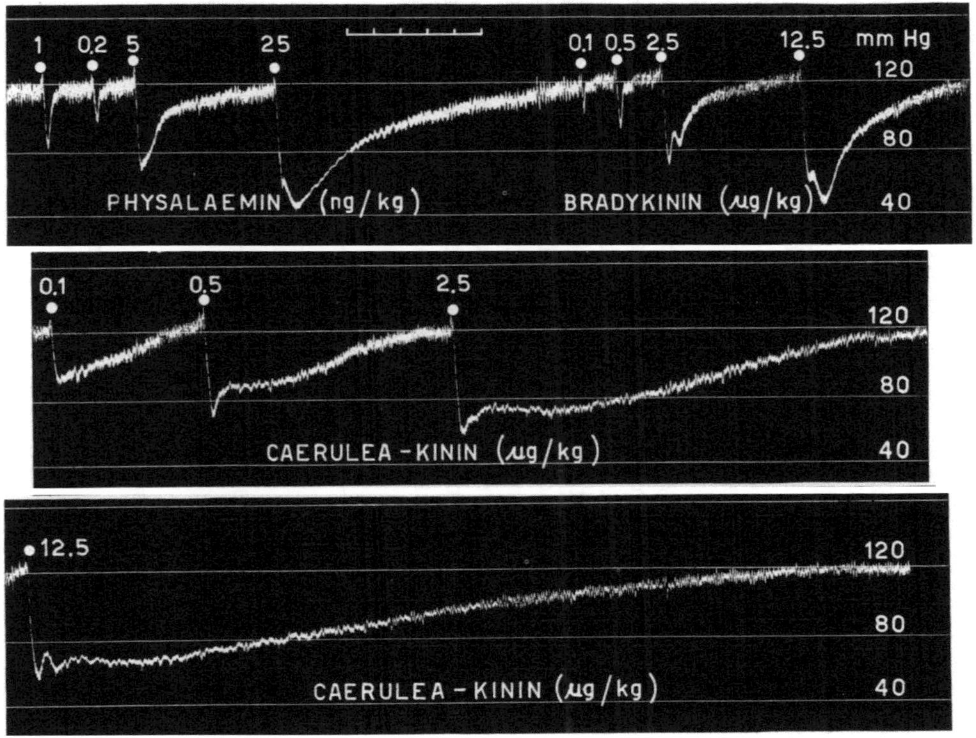

Fig. 1. Blood pressure of a dog weighing 13 kg, anesthetized with sodium pentobarbital (30 mg/kg, i.v.) after treatment with 0.2 mg atropine sulfate/kg, intramuscularly. Time marks 1 min. The pressure response to different, rapidly increasing intravenous doses of physalaemin, bradykinin, and caerulein is shown. Doses for physalaemin are in nanograms, for the other two polypeptides in micrograms per kilogram body weight. It may be noted in this experiment that physalaemin is at least 500 times more potent than bradykinin, the shape of the hypotensive response being similar. The pressure response to caerulein is strikingly different, as it lasts considerably longer. Increasing the dose of caerulein increases the duration of hypotension more than its intensity.

Eledoisin-like polypeptides

The most important representative of this polypeptide group in amphibian skin is physalaemin, which has been isolated in a pure form, reproduced by synthesis and submitted to a thorough pharmacological study (Anastasi et al., 1964; Bertaccini et al., 1965a, 1965b). Elucidation of its chemical structure (I) fully confirmed the strict resemblance to eledoisin (II), first suggested only on the basis of biological data (Erspamer et al., 1962).

Like eledoisin, physalaemin is characterized by a tremendously intense hypotensive action in the dog (threshold intravenous dose, 0.1 to 0.5 ng/kg) (Fig. 1), by a potent stimulant action of the large intestine of the rabbit (threshold, 0.2 to 1 ng/ml), by a lesser action on the guinea pig ileum, and by a poor action on the rat uterus and the rat colon. Moreover, the polypeptide displays a powerful stimulant action on the salivary, pancreatic, and lacrimal secretions in the rat and dog, and strikingly increases capillary permeability in the guinea pig, rat, and man.

(I) Pyr—Ala—Asp(OH)—Pro—Asp(NH_2)—Lys—Phe—Tyr—Gly—
—Leu—Met—NH_2

(II) Pyr—Pro—Ser—Lys—Asp(OH)—Ala—Phe—Ile—Gly—Leu
Met—NH_2

(III) Glu,Pro,Ser,Lys,Asp(OH),Ala,Phe,Ile,Gly,Leu,Arg,Val,Thr*

Pyr- = pyroglutamyl-

*Amino acid composition according to Haefely and Hürlimann, 1962

Physalaemin is completely inactivated not only by chymotrypsin but also by trypsin, whereas it withstands treatment with carboxypeptidase and leucine aminopeptidase. Owing to the presence of the tyrosyl residue, it is largely inactivated by treatment with diazonium salts in alkaline medium.

In its biological and chemical characteristics physalaemin is closely related not only to eledoisin but also to substance P (III) which we believe should be included in this polypeptide group. In fact, it may be seen from (II) and (III) that as many as ten of the eleven amino acid residues of eledoisin are common to substance P, the only amino acid characteristic of eledoisin being methioninamide; the three amino acids characteristic of substance P are arginine, valine, and threonine.

The possibility that the labile methioninamide residue has escaped the attention of investigators who have been studying substance P has been suggested in a previous paper (Bertaccini et al., 1965b). Meinardi and Craig, and Zuber have now demonstrated that methionine is a constituent of at least some samples of substance(s) P (see in this volume).

Physalaemin or physalaemin-like polypeptides are present in methanol extracts of the skin of other *Physalaemus* species, as well as in skin extracts of several *Phyllomedusa* species. Whereas it is likely for *Physalaemus* that the active polypeptide is authentic physalaemin, in the case of *Phyllomedusa* it would be prudent to speak of "physalaemin-like" polypeptides. It is certain, for example, that the active polypeptide of *Phyllomedusa hypochondrialis*, although biologically mimicking physalaemin, is different from physalaemin, as is shown by its behavior in high-voltage electrophoresis (Bertaccini et al., 1965a).

The content of physalaemin or physalaemin-like polypeptides may be considerable: 370 to 700 μg/g dry skin in *Physalaemus fuscumaculatus*, and 270-500 μg/g in *Phyllomedusa helenae*, *Ph. callidryas*, and *Ph. dacnicolor*.

Of interest and practical importance for the study of these polypeptides is the fact that they seem to withstand perfectly the drying process; i.e. they may be obtained without appreciable loss from dry skins.

Eledoisin, physalaemin, substance P, and related polypeptides are characterized by a prompt stimulant action on extravascular smooth muscle and a prompt hypotensive action. One could perhaps suggest for this polypeptide group the comprehensive denomination of tachykinins, as opposed to the group of slow-acting kinins, the true bradykinins.

Bradykinin and bradykinin-like polypeptides

Polypeptides of this group are characterized by a remarkable, but not exceptional, hypotensive action in the dog (Fig. 1), rabbit, and cat; by a potent stimulant action on the guinea pig ileum and the dog intestine; and by a formidable stimulant action on the rat uterus. They have a poor stimulant action on the rabbit and rat colon and an inhibitory action on the rat duodenum. Bradykinin-like polypeptides effectively increase capil-

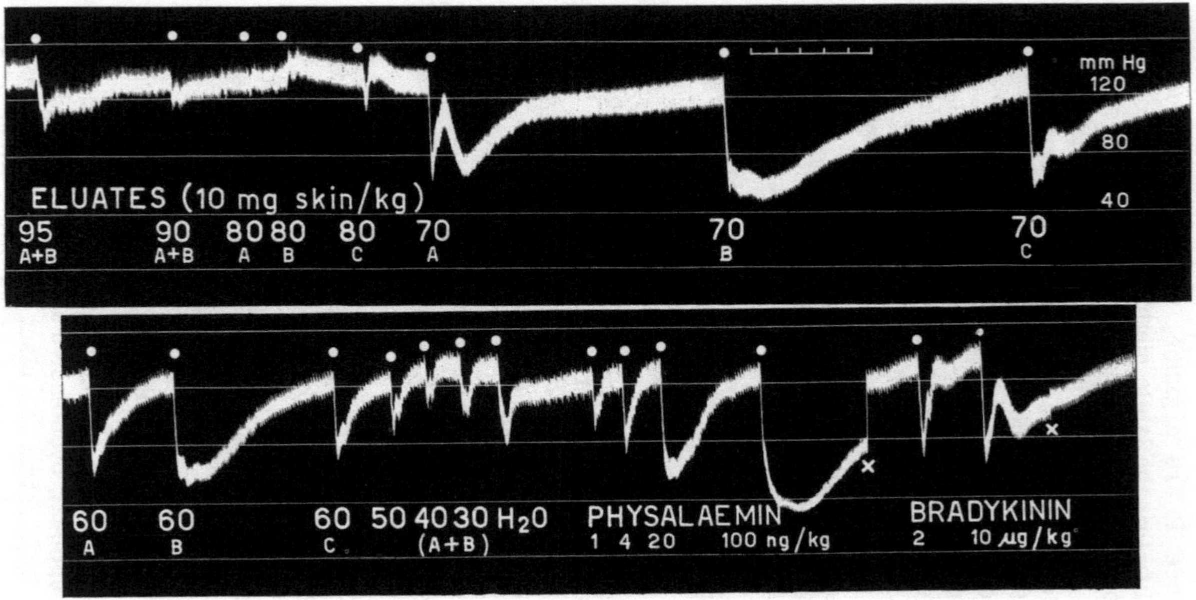

Fig. 2. Blood pressure of a dog anesthetized with pentobarbital after treatment with atropine. Time marks 1 min. The pressure response to the eluates obtained from an alumina column loaded with an extract of *Phyllomedusa rohdei* is shown. Elution was performed with descending concentrations of ethanol. Three peaks of hypotensive activity may be observed, caused by ethanol eluates 95 A + B, eluates 70 A + B + C, and eluate 60 B, respectively. The first peak is produced by a tryptophan-containing pentapeptide, the second by phyllokinin, the third by a physalaemin-like polypeptide. The effect of different doses of physalaemin and bradykinin is shown for comparison.

lary permeability in man and experimental animals.

In addition to authentic bradykinin (I), three natural bradykinin-like polypeptides have been so far described: kallidin or lysyl-bradykinin (II), methionyl-lysyl-bradykinin or methionyl-kallidin (III), and now glycyl-bradykinin (IV). All of them contain the nonapeptide bradykinin and one or two amino acid residues at the amino terminal of the bradykinin molecule.

(I) Arg—Pro—Pro—Gly—Phe—Ser—Pro—Phe—Arg

(II) Lys—Arg—Pro—Pro—Gly—Phe—Ser—Pro—Phe—Arg

(III) Met—Lys—Arg—Pro—Pro—Gly—Phe—Ser—Pro—Phe—Arg

(IV) Gly—Arg—Pro—Pro—Gly—Phe—Ser—Pro—Phe—Arg

(V) Arg—Pro—Pro—Gly—Phe—Ser—Pro—Phe—Arg
—Ile—Ileu—Tyr
|
SO_3H

Amphibian skin contains authentic bradykinin and at least one new highly active bradykinin-like polpeptide.

Authentic bradykinin has been isolated from the skin of the common European brown frog *Rana temporaria*, in which it is contained in amounts as high as 200 to 250 μg/g fresh tissue (Anastasi et al., 1965). This is 7 to 50 times the quantity that may be obtained from the bradykininogen present in 1 ml of mammalian blood plasma. Whether authentic bradykinin occurs in the skin of other *Rana* species, including our common *Rana esculenta* presumably will be solved by research in progress.

Although the occurrence of authentic bradykinin in amphibian skin is of considerable interest, the occurrence of new, hitherto unknown bradykinin-like polypeptides in *Phyllomedusa* and in *Ascaphus truei* is even more interesting.

Methanol extracts of the skin of *Phyllomedusa rohdei*, a Brazilian amphibian, contain at least one active bradykinin-like polpeptide in addition to other polpeptides (Fig. 2). This is nothing but bradykinin with two additional amino acid residues at the carboxyl terminal of the molecule. For the new polypeptide the name of phyllokinin is suggested. Anastasi et al. presents a detailed report on isolation and properties of phyllokinin in another communication in this volume.

Methanol extracts of the skin of *Ascaphus truei*, a rare North American amphibian, contain, in turn, a bradykinin-like polypeptide that may be easily distinguished in parallel assay from authentic bradykinin (Fig. 3). In fact, 1 g dry *A. truei* skin

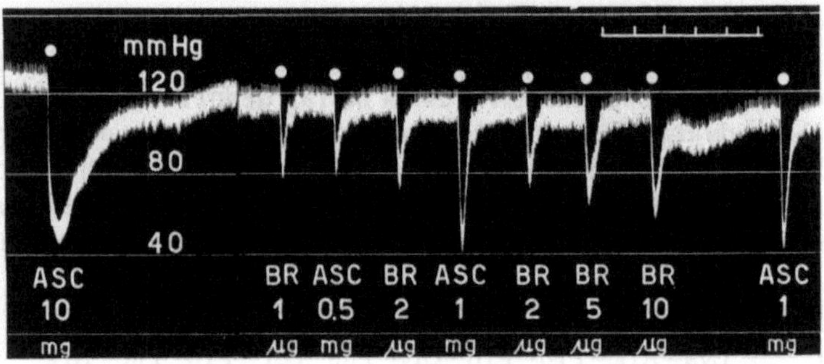

Fig. 3. Blood pressure of a dog weighing 10 kg, anesthetized with pentobarbital after treatment with atropine. Time marks, 1 min. The effect on blood pressure of different doses of bradykinin and of a crude methanol extract of Ascaphus truei skin (doses are referred to dry tissue) is shown. It may be noted that, whereas increasing the dose of bradykinin 5- to 10-fold produces only a relatively moderate increase in the hypotensive response. doubling the dose of A. truei extract causes a striking difference in the hypotensive response. This points to the non-identity of Ascaphus-kinin with bradykinin.

is equiactive to 1,400 to 2,500 µg bradykinin on the dog blood pressure, 500 to 600 µg bradykinin on the guinea pig ileum, as much as 8,000 µg bradykinin on the rabbit uterus, and as little as 150 to 300 µg bradykinin on the rat uterus. At present it is impossible to determine whether this polypeptide is identical with phyllokinin. The lack of material is a serious obstacle to its isolation and study; until now we have had at our disposal only 0.6 g of dry skin.

Caerulein and related polypeptides

The skin of *Hyla caerulea*, an Australian amphibian, contains, besides considerable amounts of 5-hydroxytryptamine and histamine, a very interesting polypeptide which is characterized, among other things, by a potent relatively long-lasting hypotensive action in the dog (Fig. 1) and the rabbit and by a spasmogenic action on the pylorus of the rat stomach. For this new polypeptide the name caerulein is suggested.

On the other smooth-muscle preparations usually employed in our screening, caerulein does not seem to be particularly active, which makes its discrimination from the other hypotensive polypeptides quite easy.

Drying the skin causes a 25 to 35% loss of kinin activity. The polypeptide is rapidly inactivated by chymotrypsin and by treatment with diazonium salts, since its molecule has radicals of aromatic amino acids capable of coupling with diazonium salts.

Caerulein has been obtained in a pure form, and elucidation of its amino acid composition and sequence is well advanced. Our studies show that caerulein differs strikingly from all other known polypeptides active on plain muscle not only in biological action but also in chemical constitution. It may well be that caerulein represents a prototype of a new group of hypotensive polypeptides. Actually, methanol extracts of the skin of some South American *Phyllomedusae* and of some *Leptodactylus* contain a hypotensive polypeptide that closely resembles caerulein. Work is already in progress to solve the problem of whether active polypeptides of the South American amphibians are identical with each other and with caerulein.

Miscellaneous polypeptides

Samples of active polypeptides in the amphibian skin are not exhausted with physalaemin, bradykinin, caerulein, and related polypeptides. Several other active compounds, of certain or probable polypeptide nature, have been traced in skin extracts of *Rana, Phyllomedusa,* and *Physalaemus.*

Generally, these polypeptides are characterized by a more or less pronounced stimulant action on the rat colon and the rat uterus, by a moderate stimulant effect on other intestinal preparations and, possibly, by a moderate hypotensive action in the dog.

In *Rana esculenta* the principle active on the rat colon may be separated from the bradykinin-like polypeptide by chromatography on alkaline alumina column. Upon elution of the column with descending concentrations of ethanol it appears in the eluate before bradykinin. A principle identical with or strictly related to that of *R. esculenta* is present also in skin extracts of other *Rana* species, sometimes in considerably higher concentrations. However, owing to the lack of sufficient material of these species, our present efforts are directed to the isolation of the active principle from semipurified material of *R. esculenta.*

The skin of *Physalaemus fuscumaculatus* contains two different polypeptides active on the uterus and colon of rat. They are completely inactivated by digestion with chymotrypsin and apparently only partially so by treatment with trypsin. As much as 90 to 95% of the activity displayed on the rat colon and uterus by crude skin extracts of *Physalaemus* is due to these kinins and barely 5 to 10% activity is due to physalaemin (Bertaccini et al., 1965a).

Phyllomedusa is more complex. Extracts of the skin of *Ph. rohdei* and *Ph. hypochondrialis* contain, besides bradykinin- and physalaemin-like polypeptides (Fig. 2), a number of polypeptides carrying in their molecule a tryptophan residue. A first gross separation of these peptides may be obtained by chromatography on alumina, and their subsequent purification is facilitated by their ability to give color reactions wi the NNCD reagent (a stable diazonium salt) and with *p*-dimethylaminobenzaldehyde.

So far, the majority of these tryptophan-containing peptides seem to be inactive on our test objects. However, one of them, possibly a pentapeptide, shows a remarkable stimulant action on the uterus and colon of rat (threshold concentrations 5-20 and 20-40 ng/ml respectively) and a moderate hypotensive action in the dog (Fig. 2).

The amino acid composition and sequence of some of these tryptophan-containing peptides have already been established, but completion of their synthesis seems necessary to confirm our analytical data.

It should be stressed that for some of the active compounds dealt with here the polypeptide nature, although probable, has not been unequivocally demonstrated, since chymotrypsin inactivation is not a mandatory criterion. For example, the possibility cannot be excluded that in some cases we have to do with biologically active lipid-soluble organic acids.

DISCUSSION

This communication is intended chiefly as a description of results so far obtained. A discussion of the results seems premature, since work on active polypeptides of the amphibian skin, although satisfactorily advanced, is far from being concluded. However, some conclusions can be advanced.

First, it is certain that the studies reported represent a considerable widening of our knowledge in the field of polypeptides. Not only was an unexpected localization of authentic bradykinin found outside the blood plasma, which hitherto had been the only known source of the polypeptide, and not only were new bradykinin- and eledoisin-like polypeptides identified, but completely new groups of polypeptides active on vascular and/or extravascular smooth muscle have been described for the first time (caerulein, small tryptophan-containing kinins).

Thus, amphibian skin has proved to be an enormous storehouse of aromatic biogenic amines and of highly active polypeptides. It is hoped that the progress of our studies will gradually bring to light the correlations in biosynthesis which exist among the different members of a single polypeptide group and among the different polypeptide groups.

The significance of the polypeptides in the amphibian skin is as obscure as that of biogenic amines. It may be tentatively suggested that they interfere in some basic function of the skin, for example in the regulation of water and electrolyte exchanges through the skin. It should be kept in mind in this respect, however, that often polypeptides seem to be mainly localized in the cutaneous glands, the secretion of which is believed to be only external.

It is obvious that the same function may be displayed in the different amphibian species by different polypeptides, and it is conceivable that polypeptides inactive in our screening systems are active on amphibian skin. Some work in this field is in progress.

And now a final consideration. We hope that the study of the spectra of biogenic amines and active polypeptides occurring in the skin of the largest possible number of amphibian species will be of interest not only from the viewpoint of comparative biochemistry and pharmacology but also from that of biochemical taxonomy.

Acknowledgment. This research was supported by grants from the Consiglio Nazionale delle Ricerche, Rome.

REFERENCES

Anastasi, A., V. Erspamer, and J. M. Cei: Isolation and amino acid sequence of physalaemin, the main active polypeptide of the skin of *Physalaemus fuscumaculatus*. Arch. Biochem. 108, 341 (1964).

Anastasi, A., V. Erspamer, and G. Bertaccini: Occurrence of bradykinin in the skin of *Rana temporaria*. Comp. Biochem. Physiol. 14, 43 (1965).

Bertaccini, G., J. M. Cei, and V. Erspamer: Occurrence of physalaemin in extracts of the skin of *Physalaemus fuscumaculatus* and its pharmacological actions on extravascular smooth muscle. Brit. J. Pharmacol. 25, 363 (1965a).

Bertaccini, G., J. M. Cei, and V. Erspamer: The action of physalaemin on the systemic arterial blood pressure of some experimental animals. Brit. J. Pharmacol. 25, 380 (1965b).

Erspamer, V., G. Bertaccini, and J. M. Cei: Occurrence of an eledoisin-like polypeptide (physalaemin) in skin extracts of *Physalaemus fuscumaculatus*. Experientia (Basel) 18, 562 (1962).

Haefely, W., and A. Hürlimann: Substance P, a highly active naturally occurring polypeptide. Experientia (Basel), 18, 297 (1962).

DISCUSSION

Goodwin: I should like to tell Professor Erspamer a story. Some years ago while working in East Africa, I found I had inherited a Dachshund with the house I was living in. He was very interested in a small hole in the ground near the house, which I found contained a small brown toad. One evening he caught up with the toad and touched it with his nose. He gave a yelp of pain. In about 30 seconds his nose had swollen to two or three times its normal size and his nose, eyes, and mouth streamed for about 3 hours after touching the toad. I suggest that this may be one possible function for the peptides in amphibian skin.

Erspamer: I completely agree with the illuminating hypothesis of Dr. Goodwin. It is well possible, in fact, that down from the Carboniferous period, for more than 100 million years, amphibians have slowly prepared themselves to face the arrogance of the Dachshund, especially that of Dr. Goodwin's exceedingly bold Dachshund.

Prado: We have been unable to find, in our preliminary experiments, any wasp kininogen or wasp kininogenases. I wonder if Professor Erspamer would care to comment on the possible origin of kinins in amphibian skin.

Erspamer: Concerning the origin of our kinins in the amphibian skin, we have, so far, made only fragmentary observations. Our purpose is to search for kininogens in *Rana temporaria* and *Rana esculenta* (the only living amphibians practically available), and to establish the polypeptide spectrum in the most important of our extracts. In fact, an active polypeptide may equally well originate by fission of a larger peptide or by condensation of smaller peptides.

A BRADYKININ-LIKE ENDECAPEPTIDE OF THE SKIN OF PHYLLOMEDUSA ROHDEI

A. ANASTASI, V. ERSPAMER, G. BERTACCINI, and J. M. CEI

Institute of Pharmacology of the University of Parma, and Farmitalia Research Institute, Milan, Italy; and Institute of Biology, National University of Cuyo, Mendoza, Argentina

In the preceding communication Erspamer and Anastasi reported that a number of peptides active on smooth muscle could be extracted from the skin of the South American amphibian *Phyllomedusa rohdei*. We will now report briefly on the isolation and determination of the structure of phyllokinin, a peptide which strikingly resembles bradykinin in its pharmacological actions.

By submitting crude methanol extracts of P. rohdei skin to chromatography on an alkaline alumina column (Anastasi and Eerspamer, 1963) and eluting with ethanol-water mixtures of decreasing ethanol concentration, phyllokinin appears in the 70% ethanol fraction together with four other peptides and several low-molecular weight contaminants, largly amino acids.

Figure 1 shows the spots of all peptide components of the 70% ethanol eluate on a high-voltage electrophoresis sheet performed at pH 1.9: phyllokinin, giving a positive Sakaguchi reaction of arginine and possessing bradykinin-like activity; other two Sakaguchi positive peptide, probably lacking pharmacological activity; finally, two more peptides positive to the p-dimethylaminobenzaldehyde (Ehrlich) reagent for tryptophan, also probably inactive. The study of the peptides accompanying phyllokinin is in progress.

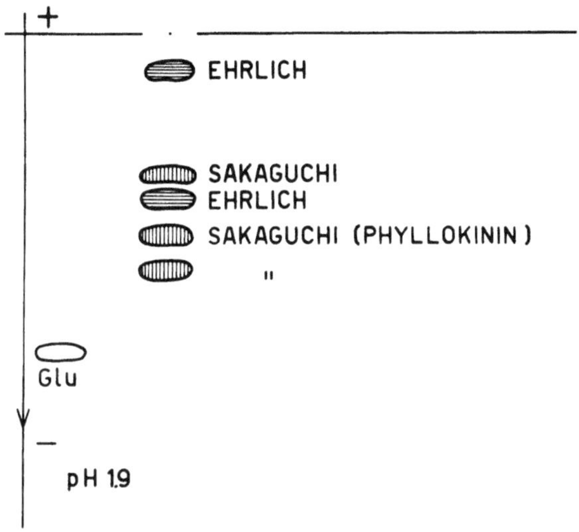

Fig. 1. High-voltage paper electropherogram of the 70% ethanol eluate from an alumina column loaded with an extract of *Phyllomedusa rohdei* skin.

Phyllokinin has been identified as an endecapeptide possessing the structure of a bradykinin molecule with two additional residues attached to its C-terminal arginine. The proposed structure has been recently confirmed by synthesis (Bernardi and co-workers, to be published).

Isolation of phyllokinin

The rather troublesome separation of phyllokinin from the other peptides contained in the 70% ethanol eluate was carried out with the usual procedures: filtration on Sephadex G-25, in which the peptides containing tryptophan are retarded with respect to the arginine peptides; countercurrent distribution; and ion-exchange chromatography on the carboxylic resin CG-50 with volatile buffers of increasing pH and concentration. Generally two or more recycles in these procedures had to be made to separate the partially overlapping bands. Occasionally, final purification was accomplished by high-voltage electrophoresis on thick paper.

On ion-exchange chromatography the peak of phyllokinin emerged at an intermediate position between the peaks of the two other arginine peptides. Upon countercurrent distribution in a system with a pH of 2 in the water phase, phyllokinin showed a distribution coefficient of about 0.65 — again intermediate value between those of the other two arginine peptides. Finally, on high-voltage electrophoresis at pH 1.9, phyllokinin had a migration rate of 0.65 relative to glutamic acid, and at pH 5.8 it had a migration rate of 0.3 relative to histidine, indicating a basic isoelectric point. The migration rates of bradykinin are known to be respectively $E_{1.9}$ 1.1 Glu, and $E_{5.8}$ 0.8 His (Boissonnas et al., 1960).

Phyllokinin could be spotted on paper with ninhydrin, giving a brown color turning slowly to purple, with chlorine and with the Sakaguchi reagent for arginine. It also gave a slowly appearing positive reaction with the α-nitroso-β-naphthol reagent for tyrosine.

Structure of phyllokinin

Pure peptide preparations gave, on total acid hydrolysis 3 moles proline, 2 moles of phenylalanine and arginine, and 1 mole each of glycine, serine, isoleucine, and tyrosine. This result was obtained with a semiquantitative method on paper (Blackburn and Lee, 1963), and was confirmed by sequential degradation.

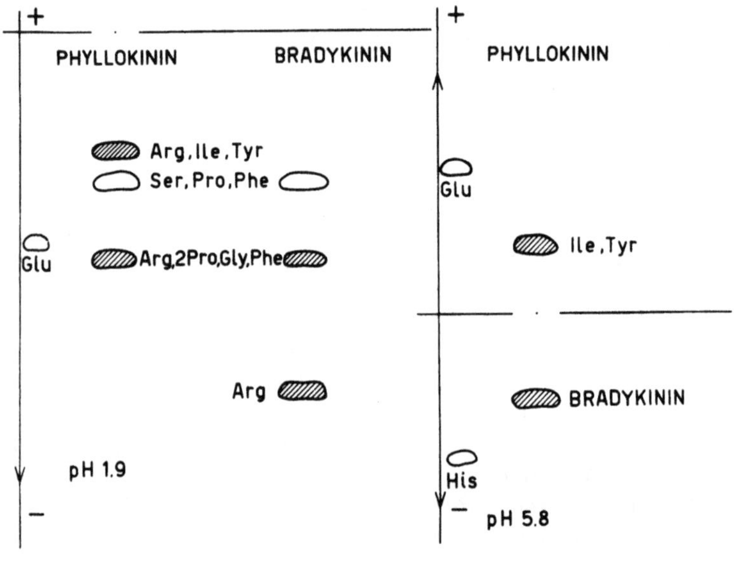

Fig. 2. Electrophoretic patterns of chymotryptic and tryptic digests of phyllokinin.

The sequence determination of phyllokinin rested mainly on the results of degradation by the enzymes chymotrypsin and trypsin. Carboxypeptidase and leucine aminopeptidase showed no action on the intact molecule.

As can be seen in Fig. 2, chymotrypsin splits the molecule into three fragments, two of which perfectly correspond on an acid electropherogram to two of the fragments resulting from chymotryptic degradation of bradykinin. The third fragment showed a position remarkably retarded, although it yields on acid hydrolysis only three residues among which is arginine, whose presence should have increased the migration rate of the tripeptide to a value higher than that of the N-terminal pentapeptide. Trypsin digestion on the other hand produced bradykinin and a strongly acidic dipeptide which on acid hydrolysis yielded isoleucine and tyrosine.

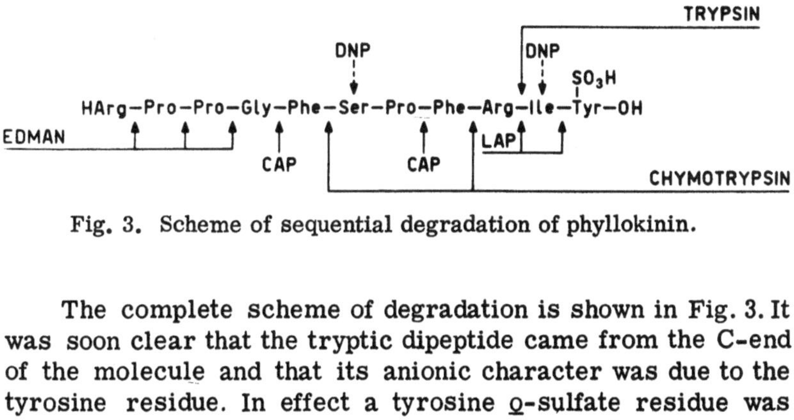

Fig. 3. Scheme of sequential degradation of phyllokinin.

The complete scheme of degradation is shown in Fig. 3. It was soon clear that the tryptic dipeptide came from the C-end of the molecule and that its anionic character was due to the tyrosine residue. In effect a tyrosine o-sulfate residue was readily obtained upon leucine aminopeptidase digestion of the chymotryptic C-terminal tripeptide. The amino acid was identified by direct comparison with a sample of the tyrosine o-sulfate ester after excluding the presence of phosphate with the specific molibdate reagent.

As shown in Fig. 3, the sequence of the bradykinin moiety has been checked by the well-known experimental approaches. The C-terminal sequence has been unequivocally confirmed by applying Sanger's procedure in the tryptic dipeptide.

Phyllokinin is thus a bradykinyl-isoleucyl-tyrosine o-sulfate.

$$\text{H—Arg—Pro—Pro—Gly—Phe—Ser—Pro—Phe—Arg—Ileu—}\underset{}{\overset{SO_3H}{\overset{|}{\text{Tyr}}}}\text{—OH}$$

There are two characteristic features in the structure of phyllokinin. The first is the presence of tyrosine sulfate. At the time the amino acid was identified in phyllokinin its presence in natural peptides was considered rather unusual since previously it had been found only in fibrinopeptides. But shortly thereafter, Gregory et al. (1964) reported its presence in gastrin, and very recently we have found that it is also contained in caerulein, a new active polypeptide occurring in the skin of *Hyla caerulea*. The second feature worthy of comment is that phyllokinin is the first natural bradykinin-like peptide with the bradykinin molecule lengthened from the carboxyl side, and this peculiarity is interesting especially with regard to its pharmacological activity.

Parallel biological assay. Natural and synthetic phyllokinin was tested in parallel assays and with synthetic bradykinin in five test systems: dog and rabbit blood pressure, guinea pig ileum, rat duodenum and rat uterus.

Synthetic phyllokinin was purified by countercurrent distribution. Synthetic bradykinin was the commonly used Sandoz product provided in buffered solution containing 100 μg of the peptide/ml. We have checked this by determining the proline content (Sarid et al., 1959) which was found to be exact. However, all our comparative data are valid only if we assume that the potency of the sample is also 100 per cent.

TABLE 1
Relative Potencies (per cent of bradykinin)
of Natural and Synthetic Phyllokinin

Test preparation	Bradykinin	Phyllokinin natural	Phyllokinin synthetic
Dog blood pressure	100	270-340	270-340
Rabbit blood pressure	100	65-95	70-90
Guinea pig ileum	100	25-40	35-47
Rat uterus	100	33-40	55-65
Rat duodenum	100	6.5-8.7	-

Table 1 gives a preliminary idea of the relative potencies of natural and synthetic phyllokinin (m.w.1,417) against synthetic bradykinin (m.w. 1,060), the activity of a mole of which was arbitrarily taken as 100. It may be seen that parallel assay, while satisfactorily proving the identity of natural and synthetic phyllokinin, shows a remarkable dissociation in potency between bradykinin and phyllokinin. For example, in its effect on the dog blood pressure phyllokinin is 2.7 to 3.4 times more potent on a molar basis than is bradykinin; on the rat duodenum it is 11 to 15 times less potent.

Quantitative and qualitative differences in the pharmacological actions of phyllokinin and bradykinin are shown in Figs. 4 to 6. It may be seen that in the dog and rabbit the blood pressure fall caused by phyllokinin lasts longer than that produced by bradykinin, and that the spasmogenic action of phyllokinin on the guinea pig ileum is less prompt and rapid than that of bradykinin.

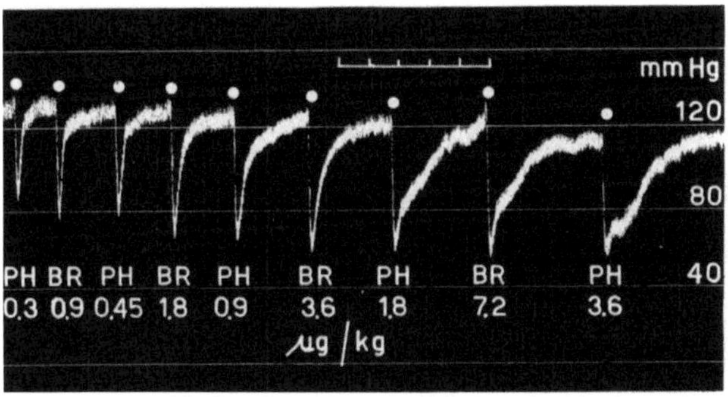

Fig. 4. Dog weighing 10 kg, anesthetized with sodium pentobarbital, 30 mg/kg, i.v. Time marks, 1 min. The action on blood pressure of different doses of synthetic bradykinin (BR) and natural phyllokinin (PH) is shown. In this experiment phyllokinin was, on a weight basis, about twice as active as bradykinin. Moreover, the pressure fall caused by phyllokinin lasted longer than that caused by bradykinin.

At this point the main problem that arises from our still incomplete pharmacological data is whether phyllokinin acts on vascular and extravascular smooth muscle as an intact molecule or through the release of bradykinin.

We have observed that phyllokinin rapidly loses its C-

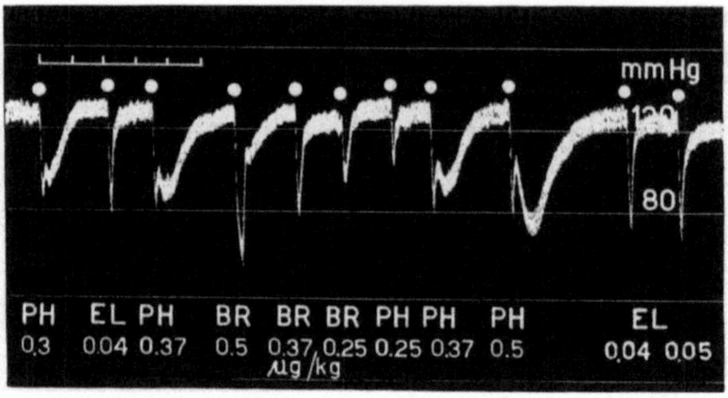

Fig. 5. Blood pressure of a rabbit anesthetized with urethan, 1.2 g/kg, i.v. Time marks, 1 min. The effects of different intravenous doses of phyllokinin (PH), bradykinin (BR), and eledoisin (EL) are shown. Whereas phyllokinin was, on a weight basis, slightly less potent than bradykinin, the pressure fall caused by phyllokinin lasted considerably longer than that caused by bradykinin.

terminal dipeptide when digested with trypsin, complete splitting occurring in about 5 minutes with an enzyme substrate weight ratio of 1:10 at both neutral and alkaline pH. Also, incubation with rabbit plasma causes the rapid removal of the same dipeptide, but we have no idea as yet about the rate of this process in vivo.

The problem is interesting because the conspicuous potency of phyllokinin seems to constitute an exception to the phenomenon shown by most of the active peptides – the marked fall in potency when modified at the carboxyl side.

Stewart and Woolley (1965) have recently discussed this topic, especially pointing out the great decrease in activity of bradykinin upon attachment at its C-end of even one amino acid, glycine or arginine.

In this regard we think it would be interesting to explore the behavior of these two peptides, bradykinyl-glycine and bradykinyl-arginine, upon digestion with enzymes. However, it does not seem unlikely to anticipate that the release of bradykinin by an enzyme behaving like trypsin should be considerably slower in both these decapeptides than in phyllokinin. On the other hand, in contrast with phyllokinin, bradykinyl-glycine and bradykinyl-arginine look as though they could be degraded at the C-end by an exopeptidase possessing carboxypeptidase-B-like activity, although at different rates. But this

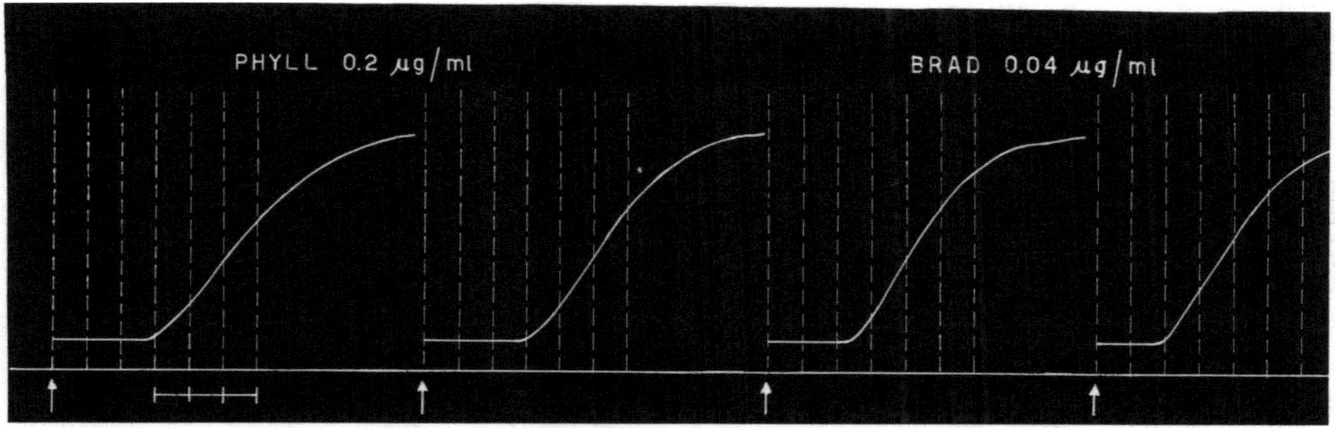

Fig. 6. Guinea pig ileum preparation suspended in 10 ml. of Krebs solution at 32°. Time marks, 3 sec. Spasmogenic response to approximately equiactive doses of phyllokinin (PHYLL) and bradykinin (BRAD). Drum speed was relatively high, and the distance between two consecutive broken vertical lines was run in 3 seconds. It may be seen that the latency period, i.e. the time elapsing between the introduction of the peptide into the bath (at arrows) and the start of contraction, was longer for phyllokinin than for bradykinin. Moreover, the ascent of the contraction curve was somewhat more rapid for bradykinin than for phyllokinin.

would cause, in both cases, the almost simultaneous splitting of the arginine residue in penultimate position and hence loss of activity.

Thus the question of whether the free C-terminal end of bradykinin is essential for activity remains, but we hope that a thorough study of phyllokinin will contribute to the solution of the problem.

Acknowledgments. This work was in part supported by grants from the Consiglio Nazionale delle Ricerche, Rome. We are indebted to, and wish to thank, the members of the Direction of the Proquifar S. A., Rio de Janeiro, for their valuable collaboration in the collection and preparation of the *Ph. rohdei* skins.

REFERENCES

Anastasi, A., and V. Erspamer: The isolation and amino acid sequence of eledoisin, the active endecapeptide of the posterior salivary glands of Eledone. Arch. Biochem. 101, 56 (1963).

Blackburn, S., and G. R. Lee: A micromethod for the estimation of amino acids. Biochem. J. 81, 1P (1963).

Boissonnas, R. A., S. T. Guttmann, and P. A. Jaquenoud: Synthèse de la L-arginyl-L-prolyl-L-prolyl-L-glycyl-L-phenylalanyl-L-seryl-L-prolyl-L-phenylalanyl-L-arginine, un nonapeptide présentant les proprietés de la bradykinine. Helv. chim. Acta 43, 1349 (1960).

Gregory, H., P. M. Hardy, D. S. Jones, G. W. Kenner, and R. C. Sheppard: The antral hormone gastrin. Nature (Lond.) 204, 931 (1964).

Sarid, S., A. Berger, and E. Katchalski: Proline iminopeptidase J. biol. Chem. 234, 1740 (1959).

Stewart, J. M., and D. W. Woolley: Importance of the carboxyl end of bradykinin and other peptides. Nature (Lond.) 207, 1160 (1965).

DISCUSSION

Erdös: We have tested an analog of bradykinin which has an additional arginine at the C-terminal end. The compound, which was synthesized by Dr. Stewart, was virtually inactive in our hands.

SYNTHETIC PEPTIDES RELATED TO PHYSALAEMIN AND ELEODOISIN

L. BERNARDI

Instituto Richerche Farmitalia, Milan, Italy

This paper will summarize briefly the results of the studies on eledoisin and physalaemin done at Farmitalia Research Institute under the direction of Professor Camerino.

In Fig. 1 the formulae of eledoisin (Erspamer and Anastasi, 1962; Sandrin and Boissonnas, 1962) and physalaemin (Erspamer et al., 1964; Bernardi et al., 1964a) are reported, and it can be seen that eledoisin and physalaemin are surprisingly similar in spite of the fact that they come from very different sources; the first, the seventh, and the last three amino acid residues are in fact identical. Since both peptides have similar biological activities, we expected that these properties depend on the presence of these amino acids.

ELEDOISIN

Pyr—Pro—Ser—Lys—Asp—Ala—Phe—Ile—Gly—Leu—Met—NH_2

PHYSALAEMIN NH_2 (on Asp)

Pyr—Ala—Asp—Pro—Asp—Lys—Phe—Tyr—Gly—Leu—Met—NH_2

Fig. 1. Structure of eledoisin and physalaemin.

In order to prove this point we synthesized (Bernardi et al., 1964b; 1966) the C-terminal shorter homologs of physalaemin and eledoisin and Ile^{8}-physalaemin. The compounds were tested by Professor Erspamer and co-workers, and the results are summarized in Figs. 2 and 3.

Figure 2 shows the activity on the dog blood pressure of the C-terminal homologs of eledoisin and physalaemin. The

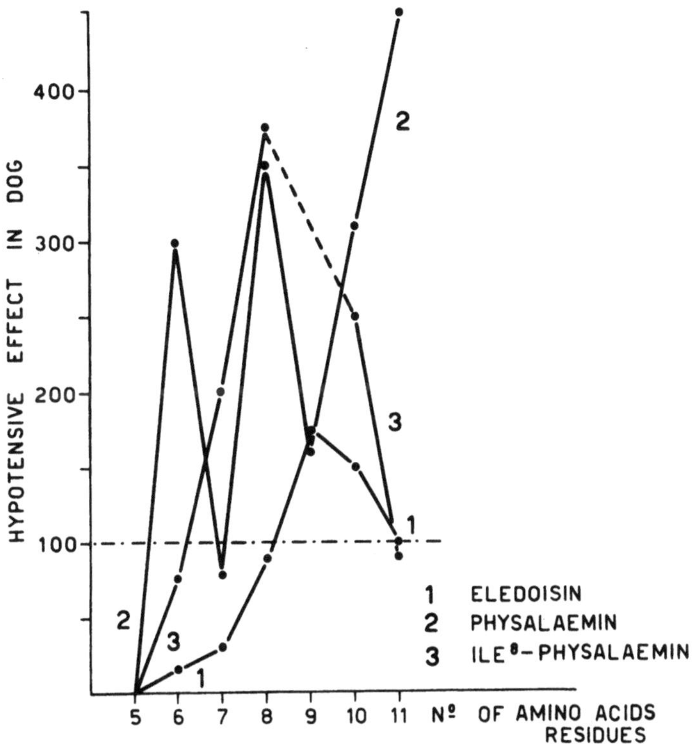

Fig. 2. Hypotensive effect in dog of the C-terminal inferior homologs of eledoisin, physalaemin, and Ile^8-physalaemin (eledoisin = 100).

number of the amino acid residues are shown on the abscissa. In all cases the C-terminal pentapeptide presents an appreciable activity, whereas the tetrapeptide is inactive. When the amino acid chain is lengthened, the activity increases, but the behavior of physalaemin is somewhat different from eledoisin. In the case of eledoisin the activity increases regularly and the nonapeptide is the most active. With physalaemin, the hexa- and octapeptides are the most active fragments, but they are not as active as physalaemin itself. The general trend in isoleucine^8-physalaemin is similar to that of eledoisin, but all the fragments are distinctly more active.

When a different test object is used the results can be different. In Fig. 3 we summarized the activity of the same pep-

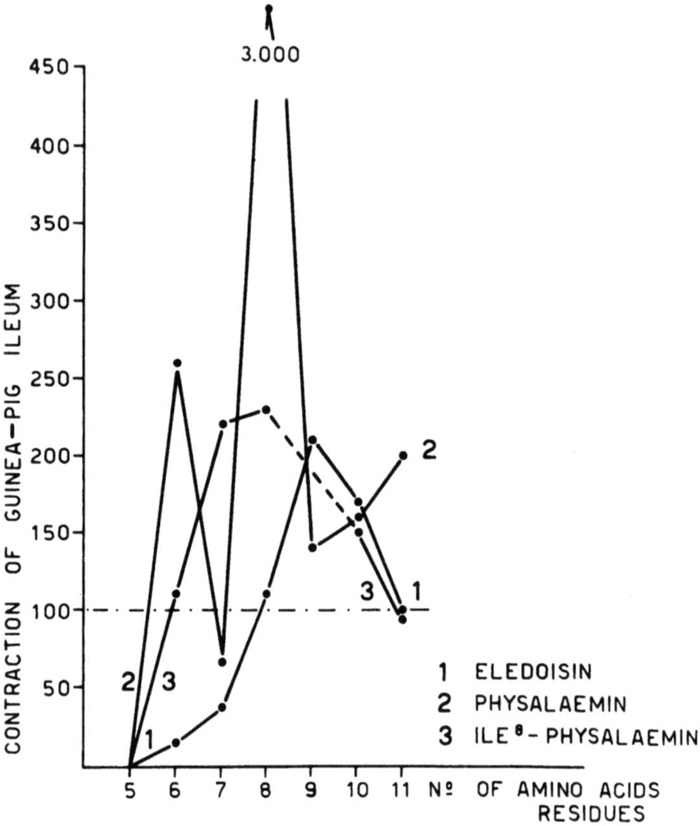

Fig. 3. Spasmogenic effect on the guinea pig ileum of the C-terminal inferior homologs of eledoisin, physalaemin, and Ile^8 = physalaemin (eledoisin = 100).

tides as measured by the contraction of the guinea pig ileum. In this case the physalaemin octapeptide is extremely active and both hexa- and octapeptide are more active than physalaemin itself.

Summing up, the reported results show that the pyroglutamyl residue is not essential for the activity and that the hexapeptides are already very active substances.

In order to find out whether any of the amino acids is essential for the biological activity, a number of analogs of the C-terminal hexapeptide of eledoisin and physalaemin were prepared; the results are graphically presented in Fig. 4, where the number of the black dots is roughly proportional to the

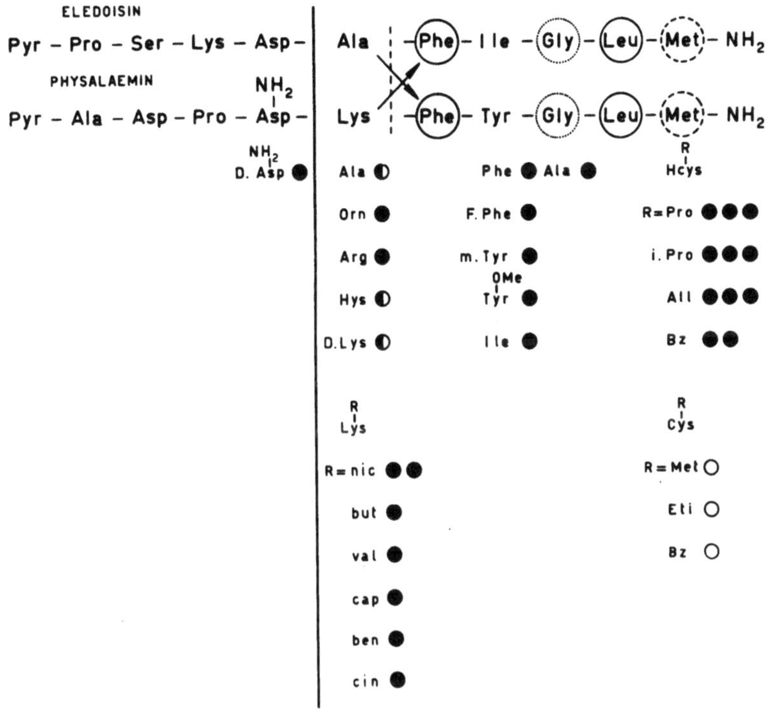

Fig. 4. Activity ratio of analogs of the C-terminal hexapeptides of eledoisin and physalaemin. One full black dot means that the analog has practically the same activity as the parent natural peptide.

Abbreviations. F.Phe = p-fluorophenylalanine; H cys = homocysteine; Pro = propyl; i.Pro = isopropyl; All = allyl; Bz = benzyl; Met = methyl; Eti = ethyl; nic = nicotinyl; but = butyryl; val = valeryl; cap = caproyl; ben = benzoyl; cin = cinnamyl.

activity ratio (Camerino et al., 1963; Bernardi et al., 1964b; 1965, 1966). Any variation of the phenylalanine and leucine residues (full circle) afforded inactive compounds, and these amino acids can therefore be considered essential. Of course not every variation has been tried, and our proof is a negative one; however, the results of the work of Boissonnas (Stürmer et al, 1964) and Schröder (Schröder and Lübke, 1964; Lübke et al., 1965) are in accord with our conclusions.

The importance of methionine is almost as great. However, we have found that it can be replaced with advantage by a group of unnatural amino acids. We have synthesized several hexapeptides containing alkylhomocysteinamides in place of methioninamide, and the activity was enhanced several fold. We can therefore conclude that from a chemical point of view methionine is not strictly essential. If more profound changes are made, for instance substituting alkylcycteinamides for methioninamide, the activity presents a 100-fold reduction.

One of the most interesting points in eledoisin and physalaemin is the role of isoleucine and tyrosine. We have synthesized a group of analogs of the C-terminal hexapeptides of eledoisin and physalaemin and have found that isoleucine and tyrosine can be interchanged without major change in activity. Moreover, tyrosine itself can be substituted with a group of aromatic amino acids (phenylalanine, m-tyrosine, etc.) without loss of activity; neither tyrosine nor isoleucine is therefore essential, but they do impart specific activities. For instance, peptides containing tyrosine or other aromatic amino acids are less active on the rat uterus but more active on the dog pressure.

Since the pentapeptides are already active, alanine and lysine were expected to be nonessential and, in fact, every substitution we have tried gave active hexapeptides; however, lysine-containing peptides are distinctly more active.

Substituting another basic amino acid (ornithine, arginine) for lysine gives highly active peptides, whereas substitution with neutral amino acids (alanine, serine) reduces the activity about 5-fold. At this point one could assume that the activity is in part due to the presence of a basic ω-group, but this would be a rash conclusion. In fact we have found that, at least in the case of the peptide lysyl-phenylalanyl-isoleucyl-glycyl-leucyl-methioninamide, acylation of the ω-amino group does not reduce the activity; in one case the activity was found to be increased.

In conclusion I will mention that while substitution of its D-enantiomer for an amino acid residue usually gives inactive compounds, we have found that this is not true in the case of physalaemin. Substitution of D-lysine for L-lysine, and of D-asparagine for L-asparagine yields active compounds (somewhat more active in the case of D-asparagine, somewhat less active in the case of D-lysine); however, substitution in the C-terminal pentapeptide resulted in inactive compounds. Similar conclusions in the field of eledoisin have already been reported by us (Bernardi et al., 1964b) and by Schröder et al. (1965).

Acknowledgments — The work I have reported is the result of the joint efforts of my co-workers, Drs. Bosisio, Chillemi, de Castiglione, and Goffredo; and of the pharmacological studies of Drs. Erspamer, De Caro, and Glaesser.

REFERENCES

Bernardi, L., G. Bosisio, R. De Castiglione, and O. Goffredo: Synthesis of physalaemin. Experientia (Basel) 20, 490 (1964a).

Bernardi, L., G. Bosisio, F. Chillemi, G. De Caro, R. De Castiglione, V. Erspamer, A. Glaesser, and O. Goffredo: Synthetic peptides related to eledoisin. Experientia (Basel) 20, 306 (1964b)

Bernardi, L., G. Bosisio, F. Chillemi, G. De Caro, R. De Castiglione, V. Erspamer, A. Glaesser, and O. Goffredo: Synthetic peptides related to eledoisin. Experientia (Basel) 21, 695 (1965).

Bernardi, L., G. Bosisio, F. Chillemi, G. De Caro, R. De Castiglione, V. Erspamer, and O. Goffredo: Synthetic peptides related to eledoisin. Physalaemin-like peptides. Experientia (Basel) 22, 29 (1966).

Camerino, B., G. De Caro, R. A. Boissonnas, Ed. Sandrin and E. Stürmer: Synthetic peptides related to eledoisin. Experientia (Basel) 19, 339 (1963).

Erspamer, V., and A. Anastasi: Structure and pharmacological actions of eledoisin, the active endecapeptide of the posterior salivary glands of Eledone, Experientia (Basel) 18, 58 (1962).

Erspamer, V., A. Anastasi, G. Bertaccini, and J. M. Cei: Structure and pharmacological actions of physalaemin, the main polypeptide of the skin of Physalaemus fuscumaculatus, Experientia (Basel) 20, 489 (1964).

Lübke, K., R. Hempel, and E. Schröder: Über den Einfluss der Kettenlänge auf die biologische Aktivität eledoisinanaloger Verbindungen. Experientia (Basel) 21, 84 (1965).

Sandrin, E., and R. A. Boissonnas: Synthesis of eledoisin. Experientia (Basel) 18, 59 (1962).

Schröder, E., and K. Lübke: Über Peptidsynthesen. C-Terminale Teilsequenzen des Eledoisins und eledoisinanaloger Verbindungen. Experientia 20, (Basel), 19 (1964).

Schröder, E., K. Lübke, and R. Hempel: D-Aminosäurehaltige Analoga des Eledoisins. Experientia (Basel) 21, 70 (1965).

Stürmer, E., E. Sandrin, and R. A. Boissonnas: Synthetic peptides related to eledoisin, Experientia (Basel) 20, 303 (1964).

CHARACTERIZATION OF KININS IN WASP VENOM

J. L. PRADO, *Z. TAMURA, E. FURANO, J. J. PISANO, and S. UDENFRIEND

Laboratory of Clinical Biochemistry, National Heart Institute, Bethesda, U.S.A.

In mammals, kinins are formed from inactive protein precursors by the action of specific proteases at or near the site of their action. Wasp and hornet venoms are in interesting contrast to the mammalian system in that they contain free kinins which act immediately upon injection (Jacques and Schachter, 1954; Schachter and Thain, 1954; Holdstock et al., 1957; Mathias and Schachter, 1958; and Bhoola et al., 1961). In *Vespa vulgaris* a single major and two minor components were resolved on columns of Amberlite (XE-64). All three kinin activities were completely destroyed by chymotrypsin and greatly reduced by trypsin. Bradykinin and kallidin are not inactivated by trypsin and could be distinguished chromatographically from the wasp kinins (Mathias and Schachter, 1958). Hornet venom contains a single, different kinin which is not inactivated by trypsin (Bhoola et al., 1961). Peptides from wasps and hornets have pharmacological actions very similar to bradykinin and kallidin; they contract isolated smooth muscle preparations, lower arterial blood pressure, increase capillary permeability, and cause pain when applied to an exposed blister base on human skin.

The present study was undertaken to characterize the potent venom kinins chemically. Their structural relationship to bradykinin and kallidin, if any, is of interest, since virtually any amino acid substitution in bradykinin causes an almost com-

---

*Fulbright-Hays visiting scholar on leave of absence from Escola Paulista de Medicina and Instituto Adolfo Lutz, Sao Paulo, Brazil.

†Visiting scholar from the faculty of Pharmaceutical Sciences, University of Tokyo, Japan.

plete loss of biological activity (Boissonnas et al., 1960; Bodanszky et al., 1963; Nicolaides et al., 1963; Schröder and Hempel, 1964). It would be of equal interest to find the bradykinin structure in such a highly specialized fluid as venom which also contains serotonin, histamine (Jaques and Schachter, 1954), acetylcholine (Bhoola et al., 1961), and pharmacologically active enzymes (Beard, 1963). The discovery of mammalian kinins in venom should contribute to a better understanding of them and strengthen the prevailing opinion that these peptides are uniquely active substances of physiological importance in man.

MATERIALS AND METHODS

Three species, *Polistes exclamans* Viereck, *P. annularis* Linnaeus, and *P. fuscatus pallipes* Lepeletier, were brought to the laboratory and their venom mixed. When not used immediately, wasps were kept alive 1-3 days at 3-5°C or for as long as 2 weeks at room temperature when fed fresh or canned fruit. Most of the wasps, about 6,000, were frozen at −10°C upon arrival at the laboratory. The abdomens were dissected and the lower halves, which contain the venom glands, sacs, ducts, and lancets were stored up to 6 months at −10°C.

Extraction of peptides and amines

For purpose of comparison, several methods of extraction were employed. The venom sac, duct, and lancet were removed intact from anesthetized insects by enlarging the abdominal opening slightly with forceps, grasping the lancet, and pulling the entire apparatus through the opening (Schachter and Thain, 1954). The apparatus was homogenized in glacial acetic acid and the peptides precipitated from the extracts with diethyl ether as previously described (Holdstock et al., 1957). Extracts were also prepared with 0.15 M saline instead of acetic acid. These were bioassayed directly for their amine content, and precipitation with diethyl ether was omitted. The electric shock technique (O'Connor et al., 1963) was used to obtain pure venom.

Bioassays

Rat uteri, guinea pig ilea, and rat duodena were suspended in a 10-ml muscle chamber and the assays performed in the usual manner. Rat blood pressure was recorded with a Statham strain gauge and a Sanborn recorder. Assays were based on responses to single and double doses of unknowns and standards. Chemical assays for histamine (Shore et al., 1959) and serotonin (Bogdanski et al., 1956) were also employed.

Chromatography and electrophoresis

Columns of CM (carboxymethyl)-cellulose, 0.8 x 23 cm, were eluted with ammonium formate essentially as described previously (Pierce and Webster, 1961). High-voltage paper electrophoresis* was performed on Whatman 3MM paper in 4% formic acid for 4 hours at 20°C with a potential of 5,000 V. Descending paper chromatograms were developed with 1-butanol:glacial acetic acid:water (60:15:25). Whatman 3MM paper was washed previously by the descending technique, with a solution of equal volumes of formic acid, ethanol, and water. After chromatography or electrophoresis, the papers were dried and either sprayed with a modified Sakaguchi reagent (F. Irreverre, personal communication) or cut into segments, eluted with Tyrode's buffer, and assayed directly for biological activity. Synthetic bradykinin and kallidin were used as standards.

RESULTS AND DISCUSSION

Yields of active peptides and amines were essentially the same from samples obtained by electric shock or by acetic acid extraction of the venom apparatuses. It was usually necessary to shock wasps 3 to 4 times before their venom supply was depleted. In some cases venom could be obtained daily for as long as 5 days; however, mortality was higher than 50%. Although the electric shock technique was efficient, it was time-consuming and impractical for handling large numbers of insects. One important difference was noted in the various prepa-

---

*Electrophorator model D, Gilson Medical Electronics Co., Middletown, Wis.

rations: i.e. kinin activity of venom obtained by electroshock was more stable than saline and acetic acid extracts of the venom apparatus. The latter slowly lost their activity on standing at room temperature. Activity of the extracts could be stabilized by heating in a boiling water bath as noted earlier (Schachter and Thain, 1954).

Active amines

The histamine content of several venom preparations was determined both by bioassay and chemical techniques. An average value of 0.8 μg per insect was obtained with the guinea pig ileum (after Code's treatment; Code, 1937) while 0.6 μg per insect was the value obtained fluorometrically. All the contractile activity of these preparations was blocked by diphenhydramine.

The fluorometric procedure for serotonin gave an average of 1.2 μg per insect. All biological activity on the rat uterus of butanol extracts from the above procedure was inhibited by lysergic acid diethylamide. Acetylcholine-like activity was not observed in any of the venom preparations. However, if acetylcholine were present in amounts less than 0.15 μg per insect, it would have been undetected in the bioassays. The European wasp (Jaques and Schachter, 1954) and hornet (Bhoola et al., 1961) also contain high levels of these amines.

Active peptides

The estrus rat uterus pretreated with lysergic acid diethylamide (1 μg/ml muscle chamber) was used to detect and quantify peptides. Antihistaminics were usually omitted because the rat uterus is relatively insensitive to the amine. Contraction of this tissue was of the slow type (Fig. 1) characteristic of bradykinin and kallidin and, like these peptides, activity was abolished by previous incubation with chymotrypsin but not trypsin. Assays of several preparations gave a maximal value of activity per insect equivalent to 1.5 μg of synthetic bradykinin. However, a sample assayed on uteri of different rats frequently gave varying values, suggesting the presence of one or more peptides whose activity on this tissue did not always parallel that of the bradykinin standard.

When various crude extracts were examined by CM-cellulose chromatography, five peaks of activity were observed.

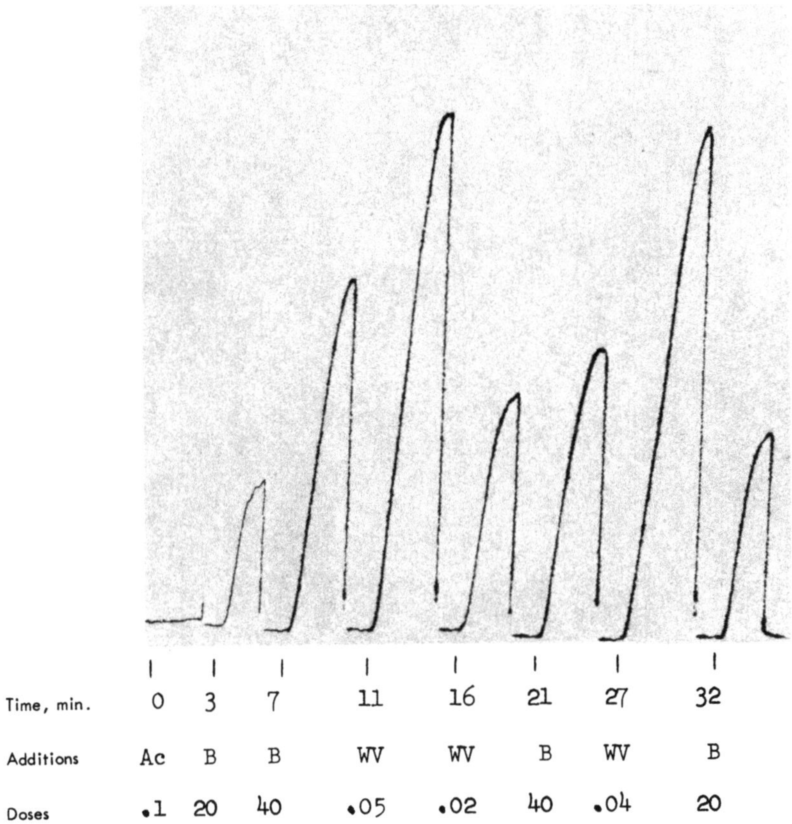

Fig. 1. The rat uterus was suspended in a 10-ml muscle chamber containing atropinized (0.010 μg/ml) de Jalon's solution maintained at 30°C. A 2-g load was applied to the muscle, and contractions (7 × magnification) were recorded for 60 sec. The drum was stopped when test solutions were washed out of the chamber. Acetylcholine, 0.10 μg, produced no contraction, and 0.020 ml of a 1:10 dilution of a wasp venom (WV) preparation (10 wasps/ml) gave a contraction equivalent to 0.03 μg bradykinin. Venom was obtained by the electric shock technique. Additions were: acetylcholine (Ac) in micrograms, bradykinin (B) in nanograms, and wasp venom solution in milliliters.

Peaks 1 and 2 were indistinguishable by this technique from bradykinin and kallidin. Further similarities in these peptides with regard to electrophoretic and chromatographic mobilities, susceptibility to inactivation by enzymes, and activity on various tissues were also noted (Table 1).

CM-Cellulose chromatography of several extracts always

TABLE 1
Properties of Bradykinin, Kallidin, and Wasp Kinins (Peaks 1, 2, and 3)

	Bradykinin	Peak 1	Kallidin	Peak 2	Peak 3
Active fraction numbers from CM-cellulose[a]	21-25	19-23	31-35	26-35	43-51
Paper electroph.[b]	47-51	40-47	55-58	45-51	44-54
Paper chromat.[c]	8-10	5-10	3-6	0-5	0-2.5
Chymotrypsin[d]	+	+	+	+	+
Trypsin, pepsin, collagenase[e]	–	–	–	–	–
LSD, atropine[e]	–	–	–	–	–
Guinea pig ileum and rat uterus	contr.	contr.	contr.	contr.	contr.
Rat duodenum	relax.	relax.	relax.	relax.	relax.
Rat blood pressure	hypot.	hypot.	hypot.	hypot.	hypot.

[a] Fractions of 2.4 ml were collected. Eluent was ammonium acetate (Pierce and Webster, 1961).
[b] Distance in centimeters from origin cathodically. The buffer was 4% formic acid.
[c] Distance in centimeters from the origin. The solvent front was 38 cm and the solvent 1 – butanol:acetic acid: water (60:15:25).
[d] The enzymes, 100 μg/ml, either inactivated (+) or did not inactivate (–) the kinins and the standards.
[e] Lysergic acid diethylamide (LSD), 1 μg/ml, and atropine, 0.01 μg/ml, did not affect (–) the activity on the uterus.

showed that peak 1 contained the least and peak 3 the most biological activity. The variability in assays of crude venom noted above was due largely to peak 3. Such variability was not noted with peaks 1 and 2. An example of the activity in each of the peaks is given in Table 2, which shows several different assays employed.

TABLE 2
Bioassay of Peaks 1, 2, and 3 from CM-Cellulose

Bioassay and standard	Peak 1	Peak 2	Peak 3
G × B	1	20	135
G × K	3	50	400
RU × B	3.5	52	48*
RU × K	4.5	100	56
BP × B		360	540
BP × B		270	540
D × B	1	40	20
D × K		40	20

Activities expressed as micrograms of either bradykinin (B) or kallidin (K). Activity (from 625 wasps) added to the column was equivalent to 900 μg bradykinin when assayed on a previous day with the rat uterus. G, guinea pig ileum; RU, rat uterus; BP, rat blood pressure; D, rat duodenum.

*Reassay of this material with several uteri gave higher results. Since no variability was ever noted in peaks 1 and 2, it is apparent most of the 900 μg of bradykinin activity assayed previously in the extract was due to peak 3.

Further characterization of peaks 1, 2, and 3

Peaks 1 and 3 from CM-cellulose chromatography also gave single peaks of kinin activity when chromatographed on columns of CM-Sephadex. However, peak 2 was resolved into two peaks. Amino acid analysis of peaks 1 and 2 revealed the presence of all twenty amino acids. These results were unexpected, since up to this time the wasp kinins had been indistinguishable from bradykinin and kallidin. It became apparent that pure peptides could not be obtained by column chromatography alone. Consequently, the fluorescent derivative technique was introduced with the expectation that it could be used as both an analytical and preparative procedure.

A dramatic demonstration of the heterogeneity of peaks 1 and 2 was obtained when aliquots were reacted with 1-dimethylamino-naphthalene-5-sulfonyl chloride (Gray and Hartley, 1963) and the fluorescent peptide derivatives examined by thin-layer chromatography. With silica gel H (without binder) and the solvent system 1-butanol:acetic acid:water (5:1:5), seven fluorescent bands were observed on the plate. Furthermore, when eluted from the plate, three of the bands had biological activity. The less basic component of peak 2 from the CM-Sephadex step, when similarly treated, gave six fluorescent bands, of which five were biologically active. It is interesting that the fluorescent derivatives themselves were biologically active. The 1-dimethylamino-naphthalene-5-sulfonamides of bradykinin and kallidin have 1-2% the potency of the free peptides in the rat uterus assay. Because of the minute quantities of the subfractions of peaks 1 and 2, it was not possible to perform quantitative amino acid analyses on the fluorescent wasp kinin derivatives eluted from the thin-layer plates. However, qualitative results indicated relatively high levels of arginine, proline, and phenylalanine.

Peak 3, the most abundant peptide, appeared homogeneous in thin-layer chromatography. The amount isolated from 3,000 wasps was 1.3 mg. A portion, 0.15 mg, was hydrolyzed with HCl, and the liberated amino acids were determined by an ion-exchange procedure (Spackman et al., 1958). Preliminary results indicate that peak 3 is composed of Arg_2, Asp, Gly_2, Glu, Leu, Lys_2, Phe_2, Pro_3, Ser, and Thr. No tryptophan was found in alkaline hydrolysates.

Treatment of peak 3 with trypsin changed its chromatographic properties on columns of CM-Sephadex and on thin-layer plates. However, the biological activity was increased slightly. (As indicated earlier activity is variable because rat uteri differ in their sensitivity to peak 3.) The biological activity in the trypsin digest was recovered as a single peak from a column of CM-Sephadex, and one fluorescent band was obtained on thin-layer chromatography of the 1-dimethylamino-naphthalene-5-sulfonamide. Amino acid analysis of the free peptide obtained from the column step showed that it consisted of

Arg_2, Gly_2, Pro_3, Phe_2, Ser. The fluorescent peptide derivative eluted from the thin-layer plate had the identical composition less one glycine residue which was identified as the 1-dimethylamino-naphthalene-5-sulfonamide. Hence glycine was N-terminal, and the active tryptic peptide was glycine[1]-kallidin (Fig. 2). Neither glycine nor any of the other constituent amino acids was N-terminal in the native peptide, suggesting the possibility that this position is occupied by pyroglutamic acid.

H–Gly–Arg–Pro–Pro–Gly–Phe–Ser–Pro–Phe–Arg–OH

Fig. 2. Structure of glycine[1]-kallidin (glycylbradykinin), the active peptide released from peak 3 by trypsin.

Currently under study is the sequence of the other amino acids (Asp, Glu, Leu, Lys_2, and Thr) in peak 3 as well as the identification of the many other kinins found in peaks 1, 2, 4, and 5 obtained from the CM-cellulose step.

If the biosynthesis of kinins in wasps is similar to the mammalian system, kininogens might also be present in these insects. This does not seem to be the case, however, since preliminary experiments indicate the absence of kininogens in venom and protein extracts of the entire wasp.

It is of interest that peptides containing the bradykinin structure are found in such a highly specialized fluid as wasp venom with histamine, serotonin and probably other pharmacologically active substances. Probably all these substances play a role in the pain, swelling, and inflammation caused by stings from these insects. Not only might these substances have individual roles but their combination could be particularly effective. For example, it has recently been shown that serotonin has a remarkable potentiating action on bradykinin-induced pain in man (Sicuteri et al., 1965).

From an evolutionary viewpoint there may be a correlation between the venom composition and the behavior of *Hymenopteri*. For example, the more primitive solitary wasp, *Sceliphron caementarium* (mud-dauber), preys primarily on insects. Its venom, which is less effective on mammals, lacks histamine, serotonin, acetylcholine, and presumably kinins (O'Connor and Rosenbrook, 1963). The more advanced social wasps and hornets, on the other hand, are not known to use their stings to capture insects; instead they sting only in defense of the colony, and their venom contains potent, pharmacologically active substances such as histamine, serotonin, kinins, phospholipase A, hyaluronidase, and histidine decarboxylase (Beard, 1963), which are native to mammals. Bees are very advanced in the evolutionary scheme; their venom is similar to that of wasps and hornets but lacks kinins. However, bee venom does contain a substance, presumably an enzyme, which produces a delayed, rapidly desensitized contraction in the guinea pig ileum (Schachter and Thain, 1954). This could be a kallikrein which liberates kinins from the host's substrate.

Regarding peak 3 isolated from wasps, it will be recalled that this is the most abundant kinin and that only after trypsin treatment and the formation of glycine[1]-kallidin was there uniformly high activity on the rat uterus. Conceivably, the native peptide is a kininogen which after injection is hydrolyzed by the host's enzymes to form glycine[1]-kallidin. This could provide a prolonged kinin action which, when considered with the synergistic action of serotonin on bradykinin, makes wasp venom a rather potent and exquisitely active toxin.

Acknowledgments — It is a pleasure to acknowledge the helpful suggestions and sustained interest of Drs. Jack V. Pierce, Marion Webster, and Nicholas Bachur of the National Institutes of Health. Thanks are also due to Dr. Karl V. Krombein and Mr. John A. Fluno of the United States Department of Agriculture, Agriculture Research Service, Entomology Research Division, for the identification of the wasps and many valuable discussions. Synthetic peptides were kindly supplied by E. D. Nicolaides, E. Schröder, and M. Bodanszky. Sandoz Ltd., Basle, Switzerland, generously donated lysergic acid diethylamide.

REFERENCES

Beard, R. L.: Insect toxins and venoms. Ann. Rev. Entomol. 8, 1 (1963).

Bhoola, K. D., J. Calle, and M. Schachter: The identification of acetylcholine, 5-hydroxytryptamine and other substances in hornet venom (*Vespa crabro*). J. Physiol. (Lond.) 159, 167 (1961).

Bodanszky, M., M. A. Ondetti, J. T. Sheehan, and S. Lande: Synthetic peptides related to bradykinin. Ann. N. Y. Acad. Sci. 104, 24 (1963).

Bogdanski, D. F., A. Pletscher, B. B. Brodie, and S. Udenfriend: Identification and assay of serotonin in brain. J. Pharmacol. exp. Ther. 117, 82 (1956).

Boissonnas, R. A., S. Guttmann, P.–A. Jaquenoud, H. Konzett, and E. Stürmer: Synthesis and biological activity of peptides related to bradykinin. Experientia (Basel) 16, 326 (1960).

Code, C. F.: The histamine activity of white blood cells. J. Physiol. (Lond.) 90, 485 (1937).

Gray, W. R., and B. S. Hartley: A fluorescent end-group reagent for proteins and peptides. Biochem. J. 89, 59P (1963).

Holdstock, D. J., A. P. Mathias, and M. Schachter: A comparative study of kinin, kallidin and bradykinin. Brit. J. Pharmacol. 12, 149 (1957).

Jaques, R., and M. Schachter: The presence of histamine, 5-hydroxytryptamine and a potent, slow-contracting substance in wasp venom. Brit. J. Pharmacol. 9, 53 (1954).

Mathias, A. P., and M. Schachter: The chromatographic behaviour of wasp venom kinin, kallidin and bradykinin. Brit. J. Pharmacol. 13, 326 (1958).

Nicolaides, E. D., H. A. Dewald, and M. K. Craft: The synthesis of kinin analogues. Ann. N. Y. Acad. Sci. 104, 15 (1963).

O'Connor, R., and W. Rosenbrook, Jr.: The venom of mud-dauber wasps I. *Sceliphron caementarium:* Preliminary separations and free amino acid content. Canad. J. Biochem. 41, 1943 (1963).

O'Connor, R., W. Rosenbrook, Jr., and R. Erickson: *Hymenoptera:* Pure venom from bees, wasps and hornets. Science 139, 420 (1963).

Pierce, J. V., and M. E. Webster: Human plasma kallidins: Isolation and chemical studies. Biochem. biophys. Res. Commun. 5, 353 (1961).

Schachter, M., and E. M. Thain: Chemical and pharmacological properties of the potent, slow contracting substance (kinin) in wasp venom. Brit. J. Pharmacol. 9, 352 (1954).

Schröder, E., and R. Hempel: Bradykinin, kallidin and their synthetic analogues. Experientia (Basel) 20, 529 (1964).

Shore, P. A., A. Burkhalter, and V. H. Cohn, Jr.: A method for the fluorometric assay of histamine in tissues. J. Pharmacol. exp. Ther. 127, 182 (1959).

Sicuteri, F., M. Fanciullacci, G. Franchi, and P. L. Del Bianco: Serotonin-bradykinin potentiation on the pain receptors in man. Life Sci. 4, 309 (1965).

Spackman, D. H., W. H. Stein, and S. Moore: Automatic recording apparatus for use in the chromatography of amino acids. Analyt. Chem. 30, 1190 (1958).

DISCUSSION

Austen: What is the potential sensitivity of chemical assay with the dansyl reagent?

Pisano: It has been reported (Gray and Hartley, 1963) that 10^{-10} to 10^{-10} mole of dansyl derivative can be seen on paper when irradiated with ultraviolet light. In solution we can detect 10^{-10} mole/ml with the spectrophotofluorimeter, and with microcells it should be possible to measure 10^{-11} mole in 0.1 ml of solution. This would be approximately 0.01 μg of bradykinin or 0.005 μg of kallidin, since 2 equivalents ot reagent react with this peptide (α and ϵ amino groups of lysine). In an actual application higher quantities would be required, since unavoidable losses would be encountered in the preparation and isolation of the dansyl derivative.

Schachter: There are many similarities in the results of the elegant chemistry of Drs. Prado and Pisano and their colleagues and our own simple chemistry carried out some years ago. Nevertheless, there are also some possibly significant differences. For example, none of our three kinin-like activities in wasp venom (or the one in hornet venom) corresponded to either kallidin or bradykinin, and all were inactivated by trypsin as well as by chymotrypsin. It may well be that there are species differences in the kinins in insect venoms and that the ones in *Vespa vulgaris*, *polistis* and *crabro* are different. If this is so, there is an abundance of work in this sphere for the organic and synthetic chemist.

Gladner: How reliable is the certainty of the amino acid analysis that relates to the molecular weight and purity?

Pisano: There is some question, since the analysis was not done by us.

Sicuteri: We have found a potentiation between bradykinin and 5-HT in pain receptors. This may be the reason for the association of these two substances in wasp venom.

Prado: We would agree with Dr. Sicuteri that this is quite possible. Serotonin and bradykinin in wasp venom have a potentiating effect on the pain produced by the sting.

STUDIES OF BOVINE PLASMA KININOGEN, ORNITHO-KALLIKREIN, AND ORNITHO-KININ

E. WERLE, K. HOCHSTRASSER, and I. TRAUTSCHOLD

Klinisch-Chemisches Institut an der Chirurgischen Klinik der Universität München, Munich, Germany

In order to find out how kinin is bound in the kininogen molecule, pure kininogen should be hydrolyzed by kininogenases and the free end-groups of the resulting products determined.

In our attempts to obtain pure kininogen from bovine plasma, we have encountered difficulties owing to the very high loss of kininogen during the purification procedures. The starting material was the 33-45% ammonium sulfate fraction of bovine plasma treated with acid. This kininogen was further fractionated on hydroxylapatite, P-cellulose, or DEAE-cellulose columns as other workers have done before. The purest preparations yielded up to 20 μg bradykinin from 1 mg kininogen. The ratio between the amounts of kallidin and bradykinin obtained was 1:3 throughout the purification procedure. Spontaneous kinin liberation occurred during the different stages of the fractionation procedures.

In the meantime we found (Werle and Hochstrasser, 1963) that pepsin was able to break down kininogen in such a way that kinins can be liberated by trypsin or kallikrein in high yields from the kininogen fragments. Thereafter we tried to isolate this kininogen derivative which we call, for the time being, prokinin. The intravenous injection of a product resulting from the digestion of plasma by pepsin produced a biphasic blood pressure reaction in the dog (Fig. 1). A few years ago Croxatto (1955) described the initial purification to some degree of three substances which he found in pepsin-digested plasma, namely pepsitensin, pepsitocin, and pepsanurin. We (Werle and Hoch-

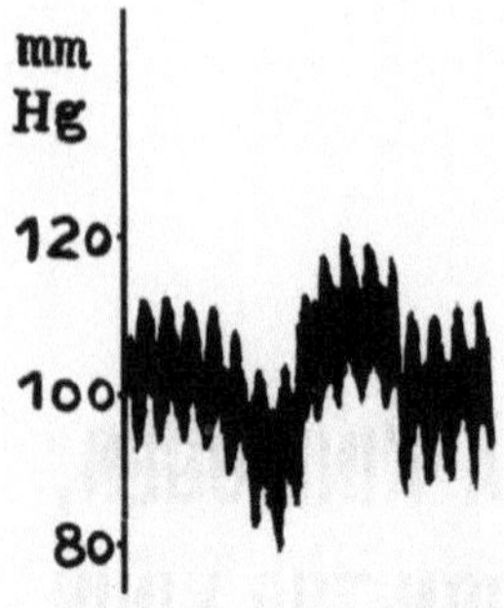

Fig. 1. Dog's blood pressure; pernocton; i.v. injection of 2 ml of pepsin digested plasma.

strasser, 1963) decided to isolate prokinin and a blood pressure-raising peptide, presumably pepsitensin. We started with Cohn fraction IV of plasma, treated it with perchloric acid to destroy some interfering enzymes, and dialyzed it against tap water. The pH of the suspension of the proteins was adjusted to 2.5 with hydrochloric acid and digested for 40 hours at 37° C with crystalline pepsin. We used 1 mg enzyme/1,000 mg protein. After 20 hours the same amount of pepsin was added again. The pH of the suspension was brought to 6.0 with NaOH, then centrifuged and the supernatant partially evaporated. This solution was separated into acid, neutral, and basic peptide-containing fractions by means of a Dowex 1 × 2 ion-exchange column according to Rudloff and Braunitzer (1961) (Fig. 2). Pepsitocin was in the basic, and pepsitensin in the neutral fraction. The acid

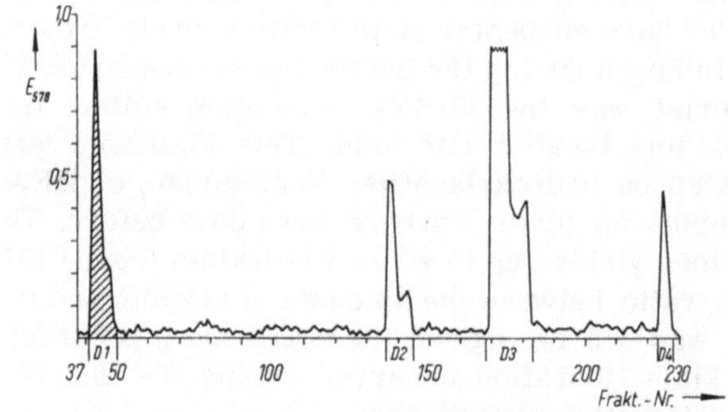

Fig. 2. Fractionation of the peptides on Dowex 1 × 2 ion exchange column after the peptic digestion of Cohn's fraction IV of bovine plasma.
The prokinins are located in the shaded area.

peptide fraction was without pharmacological activity. In the basic fraction kinin appeared after treatment with kallikrein or trypsin. Our initial attempts at further purification of prokinin were not very successful, probably because of a possible oxidation of a methionine group of the prokinin. This has to be assumed since, after oxidation by performic acid to methionine sulfon or after reduction of the methionine residue with a Raney nickel to β-hydroxybutyric acid, the prokinin no longer yielded kinins. After the incubation of prokinin with N-ethylmaleinimide the liberation of bradykinin was not abolished; therefore it seems unlikely that a free SH-group in the prokinin is necessary for the liberation of kinins by the corresponding enzymes.

Following this line of thought, we found that substances with thiol groups prevented, to some degree, the loss of the prokinin function; we therefore used thiodiglycol. Thioglycolic acid inhibits (Werle et al., 1964) the degradation of kinins by peptidases because the substance is an inhibitor of those enzymes, which can be present in highly purified kallikrein — or even in crystalline trypsin — preparations.

Table 1 shows the method of purification of prokinins in the presence of thiodiglycol (50 ml/10 liters). The loss in the yield of kinins during the first step was kept at a tolerable level.

The subsequent steps are without drastic loss of prokinin. Fractionation on P-cellulose resulted in two different prokinin fractions, a main fraction P_1 and a small fraction P_2; P_2 has a higher positive charge than P_1. The next step is the distribution chromatography on Sephadex G-25 columns with butanol:pyridine:acetic acid:water (50:20:6:24). The P_1- and P_2-containing fractions are separately subjected to thin-layer chromatography on silica gel with the solvent mentioned, because repeated paper chromatography was again very detrimental. Thin-layer chromatography of both prokinin fractions resulted in a very well-defined separation of the prokinins from inert material. Elution of the prokinins was performed with 1% trimethylamine solution without loss of yield. By using this method we are now able to isolate prokinins in amounts suitable for analysis. It is noteworthy that both P_1 and P_2 have a distinct kinin activity, equivalent to about 5% of the total kinin content, which can be liberated by trypsin or kallikrein. Therefore, the so-called pepsitocin of Croxatto (1955) is most likely identical with the mixture of at least two prokinins. The possibility that the activity of P_1 and/or P_2 originates from known kinins can be excluded because they do not run on paper in solvents where kinins have higher R_f values.

The molecular weights of P_1 and P_2 are not higher than 2,000, to judge by their behavior in Sephadex filtration. After

TABLE 1
Purification of Prokinin From Pepsin-Digested Cohn Fraction IV (cattle plasma)

Step	Treatment		(μg bradykinin/ mg peptides)	Total brady- kinin (μg)	Yield (%)	Purificat. (x-fold)
1	Cohn fraction IV perchloric acid; proteins dialyzed		0.3	360,000	100	
2	Pepsin digest. then pH 6; centrifuged, dialyzed		0.3	270,000	78	
3	Fractionation on Dowe x 1 x 2		0.6	190,000	54	4
4	CM-Sephadex		1.5	150,000	42	10
5	P-cellulose	P_1	1.5	127,000	36	10
		P_2	1.7	22,000	6	11
6	Distr.chrom. G-25	P_1	9	105,000	29	60
		P_2	12	20,400	6	80
7	Thin-layer chrom.	P_1	85	102,000	28	560
	(silica gel acc. to Stahl)	P_2	96	19,500	6	640

dinitrophenylation, which blocks the free NH_2-end group and the ε-amino group of lysine, liberation of kinins is no longer possible. Some of our findings are in agreement with those published in the meantime by Habermann and Blennemann (1964) and by Greenbaum et al. (1965).

We should like to mention briefly the isolation of pepsitensin, which was done mainly by fractionation on Dowex 1 × 2-Sephadex G-25, and P-cellulose columns and by four subsequent runs in paper chromatography with two different solvents. Finally we obtained one ninhydrinpositive zone. The purification was 1,000-fold. Val^5-angiotensin II served as a reference substance for standardization during the purification steps; 1.8

μg of the substance corresponded to the blood pressure activity of 1 μg of Val^5-angiotensin II but only to 9% and 11% of the activity in the discrimination tests on isolated rat uterus and guinea pig ileum (Figs. 3-5). The two peptides behave differently also in paper chromatography and somewhat in the shape of the blood pressure curve they cause (Fig. 6). Pepsitensin is an endecapeptide (Table 2). The free NH_2-end group is lysine.

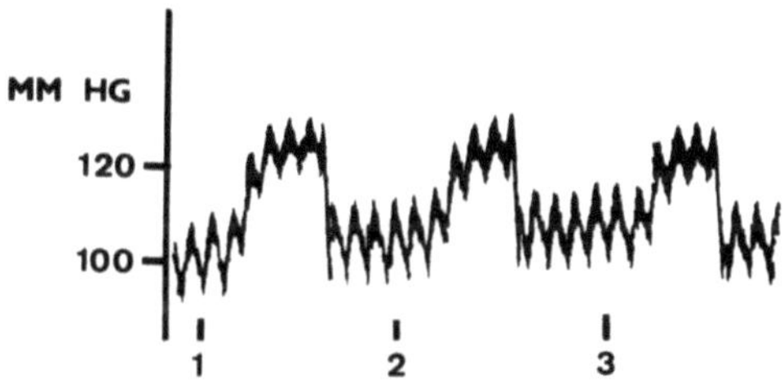

Fig. 3. Dog's (12 kg) blood pressure; pernocton; i.v. injection
1) 0,4 μg Val^5-angiotensin II;
2) 0,7 μg "pepsitensin"
3) 0,3 μg Val^5-angiotensin II.

The sequence is yet to be elucidated. Renin does not convert the substance into angiotensin I or II. It is very unlikely that pepsitensin has any relationship to angiotensin or to the minimal substrate of renin, the tetradecapeptide described by Skeggs

Fig. 4. Isolated rat's uterus; bath 20 ml 30°C
1) 0,03 μg Val^5-angiotensin II
2) 0,35 μg "pepsitensin"

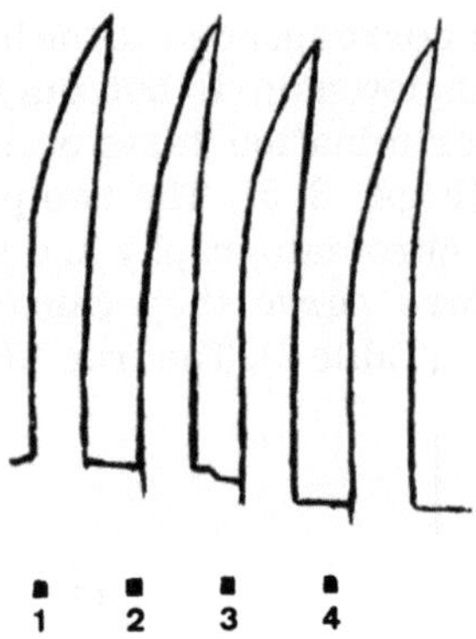

Fig. 5. Isolated guinea pig's ileum; bath 20 ml
1 and 3) 0.03 μg Val^5-angiotensin II
2 and 4) 0,28 μg "pepsitensin"

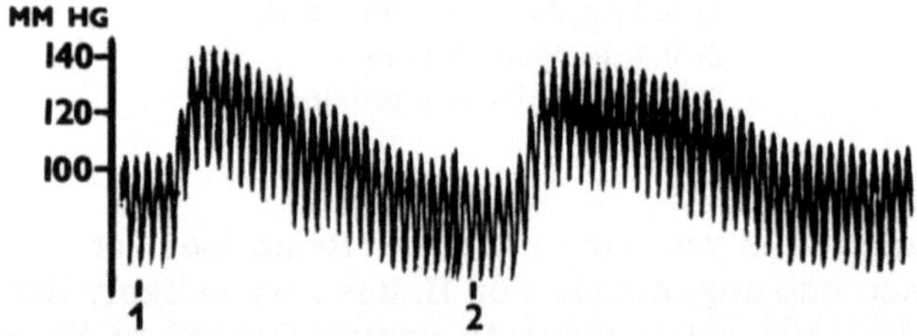

Fig. 6. Dog's (16 kg) blood pressure; pernocton; i.v. injection.
1) 0,4 μg Val^5-angiotensin II;
2) 0,8 μg "pepsitensin"

(1960), because of their greatly different amino acid compositions.

Birds such as hens, geese, pigeons, and ducks have

TABLE 2
Amino Acid Composition of "Pepsitensin"

Amino Acids	Lys,	His,	Asp,	Ser,	Glu,	Pro,	Gly,	Ala,	Leu,	Tyr,	Phe
	1.00	0.97	1.08	1.15	1.00	0.5	1.02	0.82	0.95	1.03	0.99
Number	1	1	1	1	1	1	1	1	1	1	1

pancreas kallikrein which lowers the blood pressure of birds but not of mammals (Werle and Hürter, 1936). Kallikreins of mammals do not influence the blood pressure of birds. This so-called ornitho-kallikrein splits BAEe and is blocked by hen and cattle plasma and Trasylol as well, but not by soybean trypsin inhibitor (SBTI). In the presence of SBTI, purified ornitho-kallikrein liberates a kinin from hen plasma (Fig. 7). This kinin,

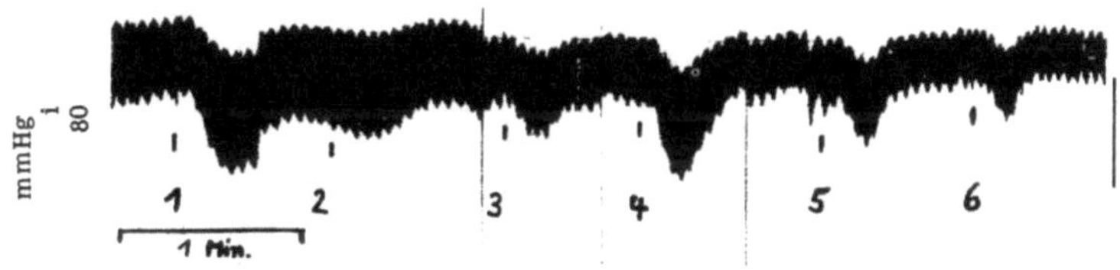

Fig. 7. Liberation of kinins from hen (H) serum by H-pancreas-kallikrein. Cock's (2 kg) blood pressure; pernocton (butallylonal narcosis): (1) H-pancreas-kallikrein; (2) H-pancreas-kallikrein boiled; (3) H-serum + H-pancreas-kallikrein boiled after 1 min of incubation; (4) as (3) but boiled after 20 sec; (5) H-serum + H-pancreas-kallikrein boiled after 5 min of incubation; (6) H-serum.

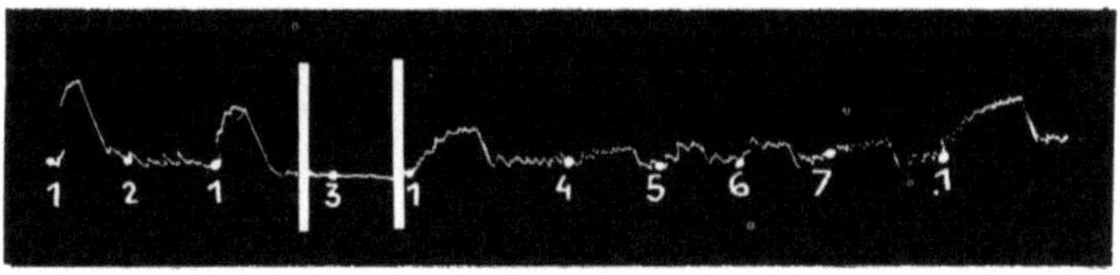

Fig. 8. Liberation of ornitho-kinin
Test: isolated hen gut
Isolated organ bath with 8 ml Tyrode's solution; (1) hen serum + hen pancreas kallikrein; (2) hen serum; (3) hen pancreas-kallikrein; (4) hen serum + pig pancreas-kallikrein; (5) bovine serum; (6) bovine serum + hen pancreas-kallikrein; (7) bovine serum + pig pancreas-kallikrein.

which we call ornitho-kinin, lowers the blood pressure of hens but not of dogs. It is rapidly inactivated by hen plasma. Ornitho-kinin contracts the hen ileum, whereas mammalian kinins are without influence on this tissue (Fig. 8). There is no doubt that this kinin is different from the plasma kinins of mammals. In this connection it is remarkable that (according to Anastasi,

1965) amphibians, which are lower phylogenetically, have bradykinin.

Unlike kallikrein, trypsin of bovine pancreas liberates ornitho-kinin because it is less substrate-specific than kallikrein. Untreated hen plasma may lower blood pressure in hens. This hypotensive factor is heat stable, and the activity is often augmented by treatment with weak acid. In contrast to ornitho-kinin, this substance is very slowly inactivated by plasma (Fig. 7). There is no competitive inhibition of the action of kallidin or bradykinin on rat uterus by ornitho-kinin.

It is of interest of course to know the structure of the ornitho-kinins. To prepare ornitho-kallidin we first purified (about 15-fold) ornitho-kallikrein from hen pancreas (Table 3).

TABLE 3
Purification (15-fold) of Ornitho-Kallikrein From 1 kg Hen Pancreas; Separation From Trypsin

Step	Treatment	Yield in (ornitho-kinin U)	Yield in (%)	(μg Protein/ 1 KU)*
1	Homogenization in 2 vol. water	639,000	100	500
2	Precipitation of impurities at pH 5	540,700	86	380
3	Fractionation with acetone	500,000	80	280
4	Fractionation on DEAE-cellulose column	63,000	10	50
5	Dialysis	40,000	6	33

*One unit (U) ornitho-kallikrein corresponds to 1 (pig) pancreas kallikrein unit on the basis of esterolytic activity(BAEe).

This preparation was free of trypsin, as suggested by its inability to split benzoylarginine-*p*-nitroanilide, a well-known substrate for trypsin. We tried to isolate not only ornitho-kallidin but also the spontaneously liberated kinin. Some purification steps are described in Tables 4 and 5.

TABLE 4
Purification of Ornitho-Kallidin Liberated By Ornitho Kallikrein

Step	Treatment	Yield in (kinin U)	Yield in (%)	(μg Protein/U)
1	24 liters plasma; acid treated in presence of thioglycolic acid			
2	24,000 O.K.U.* 30 min 20° added	960,000	100	1,700
3	Fractionation on XE-64	300,000	31	30
4	Fractionation on Sephadex G-25, lyophilization and a second Sephadex filtration	230,000	24	5
5	Fractionation on CM-cellulose	180,000	18	4

*One U ornitho-kallikrein corresponds to 1 (pig) pancreas kallikrein unit on the basis of esterolytic activity(BAEe).

TABLE 5
Ornitho-Kinin Liberated by Acid Treatment From 23.5 Liters Hen Plasma

Step	Treatment	Yield in (kinin U)	Yield in (%)	(μg/kinin U)*
1	Acid	705,000	100	2,200
2	Fractionation on XE-64	415,000	60	7
3	Desalting on Sephadex G-25	264,000	37	2.3
4	First fractionation on CM-cellulose	282,000	40	1.7
5	Second fractionation on CM-cellulose	280,000	40	0.8

*One unit (standard pool) lowers hen blood pressure to about 20 mm Hg.

As expected, after incubation of acid-treated hen plasma with ornitho-kallikrein we got a mixture of two active kinins. Therefore two active fractions appeared during the purification procedure. In a very preliminary test the hitherto purest preparations of the "spontaneously" liberated ornitho-kinin showed the following amino acid composition: Lys (2), His (1), Arg (2), Thr (1), Ser (1), Pro (2), Gly (1), Val (1), Leu (2), Tyr (1), Phe (1). The reaction with tryptophan was positive. Five of these twelve amino acids are not constituents of the known kinins. This also proves that kinins of birds and of mammals are chemically different substances.

REFERENCES

Anastasi, A., V. Erspamer, and G. Bertaccini: Occurrence of bradykinin in the skin of *Rana temporaria*. Comp. Biochem. Physiol. 14, 43 (1965).

Croxatto, H.: Active polypeptides formed by pepsin, in Polypeptides Which Stimulate Plain Muscle, H. Gaddum, Ed., p. 92. Livingstone, Edinburgh-London (1955).

Greenbaum, L. W., K. Yamafuji, and T. Hosoda: Studies of the peptic degradation of bovine bradykininogen. Biochem. Pharmacol. 14, 411 (1965).

Habermann, E., and G. Blennemann: Zum Mechanismus der Kininfreilegung aus der gereinigten Vorstufe und kininliefernden Teilsequenzen. Naunyn-Schmiedeberg's Arch. exp. Path. Pharmak. 247, 324 (1964).

Rudloff, V., and G. Braunitzer: Ein Verfahren zur präparativen Gewinnung natürlicher Peptide. Hoppe Seyler's Z. physiol. Chem. 323, 129 (1961).

Skeggs, L. T., Jr.: Polypeptides of the renal pressor system, in Polypeptides Which Affect Smooth Muscles and Blood Vessels, M. Schacter, Ed., p. 106. Pergamon Press, London (1960).

Werle, E., and J. Hürter: Über Ornitho-Kallikrein I. Biochem. Z. 285, 175 (1936).

Werle, E., and K. Hochstrasser: Versuche zur Isolierung eines niedermolekularen Kininogens. Hoppe Seyler's Z. physiol. Chem. 332, 322 (1963).

Werle, E., K. Hochstrasser, I. Trautschold, and G. Leysath: Wirkung von Thioglykolsäure auf Kininogen und Kininase. Hoppe Seyler's Z. physiol. Chem. 337, 286 (1964).

ENZYMATIC KININ RELEASE FROM KININOGEN AND FROM LOW-MOLECULAR COMPOUNDS

E. HABERMANN

Institut für Pharmakologie und Toxikologie der Universität, Würzburg, Germany

During the 1962 Symposium in New York, I reported the isolation and described some of the properties of beef serum kininogen and hog pancreatic kallikreins, components of the last step of kinin formation (Habermann, 1963). Since that time my efforts have been directed toward a better understanding of this part of the kinin system.

Experiments with kininogen

The principal aim of this series of experiments was the identification of the kinins formed by various enzymes, thus to determine the specificity of the substrate as well as that of the enzymes. Although the results have already been published (Habermann and Blennemann, 1964a,b; Habermann and Müller, 1966), they may be summarized briefly, since the experiments to be mentioned were based on them.

Enzymes can be divided into two separate groups with respect to the kinins formed. *Crotalus adamanteus* venom, trypsin, and serum kallikrein (pig) form kinin-9. In addition, trypsin releases a small amount of kinin-10 (a few per cent of the total kinin formed). We call these K-9 enzymes. On the other hand, pancreatic kallikrein (pig) produces quite predominantly kinin-10 enzymes. Our results are in accord with those of Webster and Pierce (1963).

The second aim of the experiments deals with the substrate function of beef serum kininogen. Do the active peptides originate from the same kinin-yielding group of our protein preparation? To answer that, we incubated kininogen first with *C. adamanteus* venom or trypsin and then with pancreatic kallikrein, and vice versa. If we treated the substrate first with K-9 enzymes, it was exhausted for the K-10 enzyme (pancreatic kallikrein) applied subsequently. When we reversed the sequence, we found about 10% of kininogen to be resistant against the K-10 enzyme used in the first place; trypsin or crotalus venom produced additional amounts of kinin-9. Up to this time we have failed to find an explanation for this somewhat puzzling observation.

The next step was to characterize the kinin formed by activation of human plasma. There was little hope that sufficient amounts for chemical identification could be obtained, because the blood has to be withdrawn in the laboratory immediately before use. We chose glass activation of EDTA-plasma prepared in siliconized vessels. This may be the closest approach to physiological or pathophysiological phenomena. The purification procedure was similar to that used previously (Habermann and Okon, 1961) with two exceptions: between ether precipitation and column chromatography we included gel filtration on Sephadex G-50; and the final chromatography on carboxymethylcellulose was continued into high ionic strength to avoid losses of bradykinin derivatives of stronger basicity (Table 1). Only one

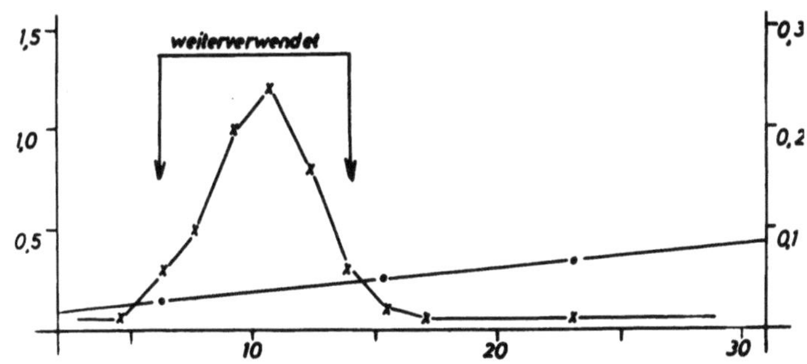

Fig. 1. First chromatography of contact kinin on carboxymethylcellulose. Conditions: Step 6 of Table 1; chromatography according to microcolumn device of Habermann and Blenneman (1964b).

TABLE 1
Purification of Kinin-like Material from Glass-activated Human Plasma

Step		μg kinin-9/ml starting plasma recovered
1.	Activated plasma	
2.	Ethanol 80% (supernatant)	0.33
3.	Drying, extraction with glacial acetic acid (supernatant)	0.48
4.	Addition of 5 vol. diethyl ether (precipitate)	0.35
5.	Eluate from sephadex G-50	0.42
6.	Eluate from the first CM-cellulose column	0.16
7.	Eluates from the second CM cellulose column	0.076

Step 1: Fresh EDTA-plasma (113 ml), 1:1 diluted with 0.85% NaCl, rotated for 10 min at room temperature with 20 mg glass beads/ml plasma. Step 2: Mixture poured into 4 vol. 96% boiling ethanol, kept boiling for 5 min, and filtered. Step 3: Supernatant dried in the rotary evaporator, stirred twice intensively with glacial acetic acid and centrifuged. Step 4: Supernatant mixed with 5 vol. water-free ether, precipitate spun down and washed with ether. Step 5: Dried precipitate dissolved into 2 ml 0.1 M NH_4-formate of pH 4.8, then filtered through a 1 × 90 cm Sephadex G-50 column in the same buffer. The smooth muscle active fractions eluted immediately before chloride were combined, twice lyophilized (100.5 mg dry substance) and dissolved in water. Step 6: Chromatography in 2 portions of 12.5 ml (pH 4.8) on 0.4 × 9 cm carboxymethylcellulose (microcolumn according to Habermann and Blennemann, 1964b). Step 7: The active eluates twice lyophilized and rechromatographed as in step 6.

peak of kinin-like activity has been detected by testing the eluates on the isolated guinea pig ileum. It could not be distinguished from kinin-9 added before chromatography, but it was different from kinin-10 (Figs. 1-4). Kinin-11 has been excluded as well. It appears in the chromatogram not far from kinin-10. Furthermore, in contrast to kinin-11, the activity of the kinin

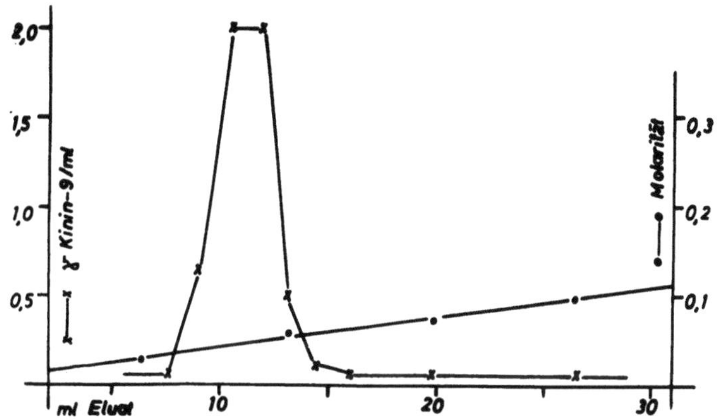

Fig. 2. Rechromatography of contact kinin on carboxymethylcellulose. Conditions: Step 7 of Table 1.

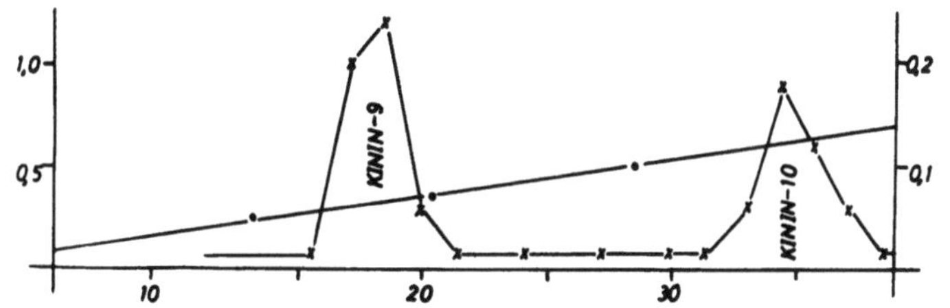

Fig. 3. Standardization of microcolumn of carboxymethylcellulose with a mixture of 4 μg kinin-9 and 5 μg kinin-10.

fraction from blood plasma is not increased by incubation with *C. adamanteus* venom. Third, the ratio of activity between the material formed in blood and synthetic kinin-9 is the same whether tested on rabbit blood pressure or guinea pig ileum (Table 2). The conclusion that the glass-activated factor is indistinguishable from kinin-9 is justified and is in accordance with our finding mentioned above that pig serum kallikrein releases kinin-9 from kininogen. It is in agreement with our earlier reports describing bradykinin as the main product of

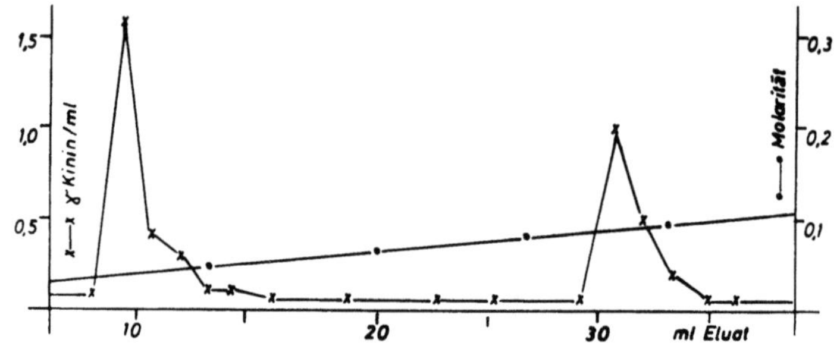

Fig. 4. Accelerated elution of kinins by addition of partially purified contact kinin. Because of the quicker elution of contact kinin (Figs. 1 and 2) when compared with standards (Fig. 3), 4 μg kinin-9 was mixed with 5 μg kinin-10 and contact kinin (step 7) equivalent to 6.6 μg kinin-9. Other conditions as in Fig. 1-3.

TABLE 2
Comparison of Glass-Induced Kinin and Kinin-10 with Bradykinin on Guinea Pig Ileum and Rabbit Blood Pressure

	Kinin-9 equivalents (μg/ml)		Discrimination index A/B
	Ileum (A)	Blood pressure (B)	
Kinin from donor H	5	5	1.0
Kinin from donor K	1	0.8	1.25
Kinin-10	0.33-0.5	2	0.17-0.25

acid activation of beef serum globulins (Habermann and Okon, 1961) and finally also with the observations of Webster and Pierce (1963). We found no kinin-11 in glass-activated plasma. Nevertheless, the possibility exists that it is formed as an intermediate. As will be mentioned below, serum kallikrein itself, and perhaps other enzymes of this kind, split kinin-11 into kinin-9 and methionyl-lysine. If such enzymes are present in excess, the transient occurrence of kinin-11 may escape detection.

Peptic fragments that yield kinin

During the course of our investigation of kininogen, we observed that peptic hydrolysates of this protein still can serve as substrates of kinin-releasing enzymes (Habermann and Schuck, 1964). This observation, in accordance with others (Greenbaum et al., 1965; Werle and Hochstrasser, 1963), stimulated the search for a relatively short sequence of amino acids containing the kinin-yielding group, i.e. a "low-molecular kininogen." Furthermore, peptic hydrolysates of kininogen acted similar to "pepsitocin" on blood pressure, guinea pig ileum, and rat uterus. This kinin-like activity was, however, very low (about 1%) when contrasted with the pharmacological potency that could be developed by *C. adamanteus* venom. The content of pepsitocin precursor of plasma fractions run parallel with kininogen content; there is no need to postulate the existence of any other substrate yielding pepsitocin.

Incubation of kininogen with pepsin leads to no significant loss of sensitivity to trypsin or *C. adamanteus* venom with respect to kinin yielded. The release of activity by pancreatic kallikrein is sharply decreased. This fact will be discussed later. The purification of the kinin-yielding material has been achieved by chromatography on carboxymethylcellulose; in this step (1), most of the kinin-yielding and pepsitocic materials appear in practically the same relatively basic fraction. A few per cent of both activities are less strongly adsorbed. The main fractions were then subjected to paper chromatography (2), which allowed the separation of two kinin-yielding peptides (PKFL and PKFS) from the bulk of impurities. Both were rechromatographed separately on phosphorylated cellulose (3). This step gave evidence for the existence of a distinct, kinin-10-like principle accompanying PKFL. All three factors (PKFL, PKFS, and "pepsitocic" principle) were associated with low kinin-like activity. PKFL and PKFS served as substrates for all kinin-releasing enzymes tested and – interestingly enough – for carboxypeptidase A (Habermann and Blennemann, 1964a; Habermann, 1965; Habermann et al., 1966).

This last observation led to the assumption (Habermann and Blennemann, 1964a) that the kinin-yielding peptides might be composed of two parts: the kinin itself and a C-terminal arrangement which might be different for PKFL or PKFS. Amino acid analysis demonstrated the presence of the acyls of kinin-11 (Met-Lys-bradykinin) in both of these peptides. In addition,

PKFL contained one serine, one valine, and one glutamic acid residue per mole. Analysis of carboxypeptidase hydrolysates identified the latter as being glutamine in the native peptide. PKFS contains the amino acids mentioned plus one valine plus an unidentified amount of methionine. Edman degradation revealed the N-terminal of both peptides as Met—Lys—Arg—Pro—Pro—Gly—. After treatment with pancreatic kallikrein, a neutral peptide has been isolated with the sequence Ser—Val—$GluNH_2$. The N-terminal amino acids before and after kallikrein (pancreas) digestion of PKFL were also identified as DNP derivatives of performic acid-treated reaction mixtures. In the former solution DNP-methionine sulfone, in the latter solution DNP-methionine sulfone and DNP-serine were found. The structure of the kinin-yielding group of kininogen thus can be written (Habermann, 1966) as

Met—Lys—Arg—Pro—Pro—Gly—Phe—Ser—Pro—Phe—Arg—Ser—

NH_2 (on Glu)

—Val—GluOH

This relatively simple compound helped in elucidating the action mechanism of kinin-releasing enzymes (Habermann and Müller, 1966). By our column chromatographic procedure, we are able to distinguish among kinin-9, kinin-10, and kinin-11. I therefore incubated PKFL with trypsin and found kinin-11 and kinin-9 when using low doses of enzyme, kinin-9 only when using high doses. *C. adamanteus* venom and serum kallikrein developed kinin-11 besides kinin-9. With pancreatic kallikrein, only kinin-11 has been found. This is quite in contrast to the behavior of intact kininogen with these enzymes. With the protein substrate, kinin-11 as reaction product has never been observed.

Schröder and his colleagues have synthetized PKFL as well as kinin-11—Ser—Val—$GluNH_2$—Val and kinin-11—Ser—Val—GluOH. Whereas the latter peptide is not a product of peptic degradation of kininogen, kinin-11—Ser—Val—$GluNH_2$—Val has been assumed to be one of the components of PKFS (Habermann, 1966). Since PKFL is the main kinin-yielding peptide originating from kininogen, the natural and synthetic materials were compared. As shown in Table 3, both the synthetic and the natural peptides are substrates for all kinin-releasing enzymes investigated; the yield of active material is, however, higher with the natural than with the synthetic product. Three points

TABLE 3
Comparison Between Natural and Synthetic PKFL

Paper electrophoresis	Identical	
Paper chromatography	" *	
Fingerprinting		
Intact peptide	"	
Kallikrein-treated peptide	"	
Used as substrate of	Activity equivalents (μg kinin-9/mg peptide) formed from:†	
	Natural PKFL	Synthetic PKFL
Trypsin	350	160
Crotalus venom	300	80
Serum kallikrein	200	50
Pancreatic kallikrein	80	60
Carboxypeptidase A	90	30
"Direct" activity		
Uterus	2	
Guinea pig ileum	2	1
Rabbit blood pressure	20	5

*Synthetic kinin contains about 20% of a slower compound susceptible to C. adamanteus venom also.
†Maximum activity obtained.

are to be considered for explanation. As Dr. Schröder informed us, methionine is oxidized partially to sulfoxide; furthermore, some racemization of the C-terminal arginine of the kinin-9 sequence cannot be excluded. Finally, the synthetic product is, in contrast to the natural, not quite homogeneous when subjected to paper chromatography in butanol-acetic acid-water. Nevertheless, the major component is identical with our natural PKFL. A mixture of materials from both sources migrates as one single main spot on fingerprinting. When digested by pancreatic kallikrein, the mixture yields only one electroneutral spot consisting of the tripeptide Ser—Val—GluNH$_2$, and one basic spot due to kinin-11 (Fig. 5). The latter spot has been

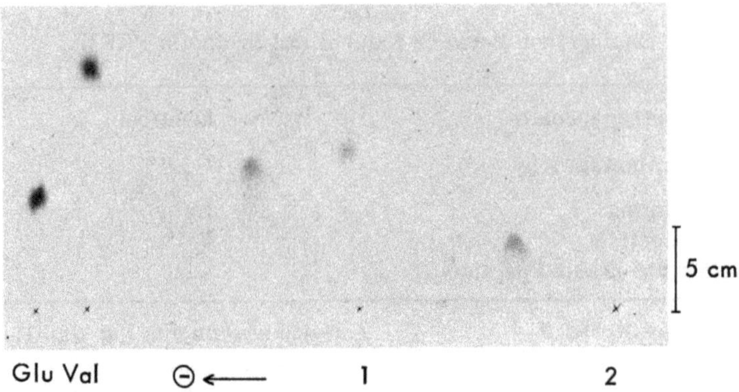

Fig. 5. "Mix-fingerprint" of natural and synthetic kinin-yielding peptide PKFL before and after kallikrein digestion. Glu = 2.5 μg glutamic acid; Val = 2.5 μg valine; 1 = 0.05 ml synthetic PKFL 1:2,000 + 0.05 ml natural PKFL 1:2,000, both in water, + 0.1 ml pancreatic kallikrein (pig) 1:50,000 in 0.05 M NH_4HCO_3, pH 8.0; 30 min at 37°C; 2 = 1 but kallikrein omitted. First dimension: 30 min electrophoresis (70 V/cm) in pyridine acetate, pH 5.9. Second dimension (glutamic acid and valine applied to the paper): descending chromatography in n-butanol:acetic acid:water (4:1:5). Ninnydrin staining.

identified by "mix-fingerprinting" of kinin-11 and kallikrein-digested natural or synthetic peptide.

PKFS is not a homogeneous peptide, as can be concluded from amino acid analysis and from fingerprinting of the kallikrein-digested material. Kinin-11—Ser—Val—$GluNH_2$ —Val probably corresponds to one of the components, as has been deduced from the chromatographic picture of the kallikrein digest. The other component may consist of kinin-11—Ser—Val—$GluNH_2$—Val—Met. Unfortunately, complete analysis of PKFS failed because of lack of material. It may be remembered that the main product of peptic hydrolysis of bovine kininogen is PKFL, the structure of which is now confirmed, thanks to the synthesis by Dr. Schröder.

N-terminal structural requirements of kinin-releasing enzymes

The "anomalous" behavior of PKFL when contrasted to kininogen could be predicted from previous experiments with kinin-11 and kinin-10. As in PKFL, bond (a) in kinin-11 could not be split by pancreatic kallikrein. One must conclude that kinin-11 is not an intermediate in the kinin formation by this enzyme. Bond (b) is susceptible to the kinin-9 forming enzymes, to *C. adamanteus* venom, trypsin, and serum kallikrein. This process takes place, however, at a slower rate with kinin-11 than with kininogen. For this reason, kinin-11 can not function as a major intermediate in the formation of kinin from kininogen by the enzymes mentioned. It may, however, act as such when PKFL is used as substrate. The transformation kinin-10 ⟶ kinin-9 is quite slow under the influence of trypsin or crotalus venom. This is in accordance with the structural requirements of trypsin-like enzymes and contradicts any intermediary function of kinin-10 in the formation of kinin-9 (Habermann and Blenneman, 1964a,b).

C-terminal structural requirements of kinin-releasing enzymes

The accessibility of peptic kinin-yielding fragments of kininogen to all kinin-releasing enzymes investigated shows that relatively short chains can be split off by them. The most simple model for a kinin-yielding peptide, bradykinylserine, is no longer a universal substrate. As is to be expected, its C-terminal serine is removed quite easily by trypsin and carboxypeptidase A, as well as by plasmin. It is, however, relatively resistant against *C. adamanteus* venom and quite resistant against pancreatic kallikrein. The latter enzymes are, therefore, endopeptidases (Habermann and Müller, 1966; Habermann, 1966).

DISCUSSION AND SUMMARY

The results presented in this paper bear on the reaction between high and low molecular kininogens with kinin-releasing enzymes, i.e. the second stage of kinin formation. A summary of the interaction of the various substances mentioned above is given in Table 4.

TABLE 4

Dependence of Kinin Formation on Enzymes and Substrates

Enzyme Type	Enzyme Name	Bovine kininogen	Kinin-yielding peptic peptide*	Kinin-11	Kinin-9 serine
K-9	Trypsin	Kinin-9 Kinin-10 (little)	Kinin-9 Kinin-11	Kinin-9	Kinin-9
K-9	C. adamanteus venom	Kinin-9	Kinin-9 Kinin-11	Kinin-9	Kinin-9 (little)
K-9	Pig serum kallikrein	Kinin-9	Kinin-9 Kinin-11	Kinin-9	Kinin-9
K-10	Pig pancreatic kallikrein	Kinin-10	Kinin-11	No degradation	No degradation
	Carboxypeptidase A	No kinin	Kinin-11	No degradation	Kinin-9

*PKFL, natural and synthetic.

From this compilation the following rules can be deduced. In the molecule

$$\mathrm{Met} \overset{}{\underset{a}{|}} \mathrm{Lys} \underset{b}{|} \mathrm{kinin{-}9} \underset{c}{|} \mathrm{Ser} \underset{d}{|} \mathrm{Val} \underset{e}{|} \mathrm{GluNH_2}$$

the bond (c) is more susceptible to K-9- enzymes than is bond (b), whereas (a) is nearly completely resistant. In the intact kininogen, however, splitting at (a) (K-10-enzymes) or (b) (K-9-enzymes) takes place more easily than, or as easily as, at (c). Carboxypeptidase A attacks successively at (e), (d), and (c).

The kind of kinin formed depends not only on the enzyme but also on the substrate. Pancreatic kallikrein can be regarded as a true endopeptidase.

Finally, after all the biochemical work for which pharmacology has only aided in testing the active compounds, it seems necessary to establish some pharmacological classification of kininogen. This protein is a depot, but, like angiotensinogen, a mobile one, transported by the blood to every part of the body. It can be contrasted to the fixed depots of biogenic amines and heparin localized intracellularly in tissues (Table 5). For local efficiency of kinins, quick formation and quick destruction are important, otherwise undesirable general actions might occur. Substances released from fixed depots can be eliminated by reabsorption into the stores or by enzymatic modification; substances released from mobile depots must be destroyed. In our view, the kinin-yielding and -destroying system is a well-balanced biochemical arrangement to enable use of a circulating depot. Vasoactive amines originating from fixed depots and vasoactive peptides from circulating depots are convergent in most of their effects; thus the animal body goes — to use an anthropomorphic parable — on two feet toward one goal.

TABLE 5
Comparison Between Circulating and Fixed Depots

	Circulating	Fixed
Carrier	Glycoproteins	Grana
Content	Peptides	Amines
Formation	Liver	Locally
Binding	Peptidic	Saltlike
Release	Enzymatic	Cellular damage
Elimination	Enzymatic	Reabsorption
		Enzymatic

REFERENCES

Greenbaum, L. M., K. Yamafuji, and T. Hosoda: Studies on the peptic degradation of bovine bradykininogen. Biochem. Pharmacol. 14, 411 (1965).

Habermann, E.: Snake venoms, kallikreins, and kininogen. Ann. N. Y. Acad. Sci. 104, 130 (1963).

Habermann, E.: Isolierung, pharmakologische Eigenschaften, Strukturaufklärung und Spaltung peptischer kininliefernder Fragmente (PKF) aus Rinderserum-Kininogen. Naunyn-Schmiedebergs Arch. exp. Path. Pharmak. 251, 187 (1965).

Habermann, E.: Strukturaufklärung zweier kininliefernder Peptide aus Rinderserum-Kininogen. Naunyn-Schmiedebergs Arch. exp. Path. Pharmak., 253, 474 (1966).

Habermann, E., and G. Blennemann: Zum Mechanismus der Kininfreilegung aus der gereinigten Vorstufe und kininliefernden Teilsequenzen. Naunyn-Schmiedebergs Arch. exp. Path. Pharmak. 247, 324 (1964a).

Habermann, E., and G. Blennemann: Über Substrate und Reaktionsprodukte der kininbildenden Enzyme Trypsin, Serum- und Pankreaskallikrein sowie von Crotalusgift. Naunyn-Schmiedebergs Arch. exp. Path. Pharmak. 249, 357 (1964b).

Habermann, E., B. Müller, and G. Blennemann: Charakterisierung und Reinigung peptischer kininliefernder Fragmente (PKF) sowie von "Pepsitocin" aus Rinderserum-Kininogen. Naunyn-Schmiedebergs Arch. exp. Path. Pharmak. 253, 444 (1966).

Habermann, E., and B. Müller: Zur enzymatischen Spaltung peptischer kininliefernder Fragmente (PKF) sowie von Rinderserum-Kininogen. Naunyn-Schmiedebergs Arch. exp. Path. Pharmak. 253, 464 (1966).

Habermann, E., and M. E. Okon: Identifizierung eines "spontan" aus Serumglobulin entstehenden Peptids mit Bradykinin. Z. physiol. Chem. 326, 171 (1961).

Habermann, E., and S. Schuck: Einige Beziehungen zwischen "Pepsitocin" und Plasmakininen. Naturwissenschaften 51, 61 (1964).

Webster, M. E., and J. V. Pierce: The nature of the kallidins released from human plasma by kallikreins and other enzymes. Ann. N. Y. Acad. Sci. 104, 91 (1963).

Werle, E., and K. Hochstrasser: Versuche zur Isolierung eines niedermolekularen Kininogens. Z. physiol. Chem. 332, 322 (1963).

DISCUSSION

Greenbaum: Professor Habermann demonstrated that kallikreins split a polypeptide that does not have an ester bond. This is particularly important, since an ester bond has been proposed for many years as being the sensitive bond split by kallikreins.

Habermannn: Although we described the relationship between esterase and kinin-releasing properties of kallikrein and snake venoms, we regarded esterolysis merely as an aid in testing these enzymes. We found no evidence for ester-like bonds in our peptides.

Back: Have you studied the action of the fibrinolytic enzyme plasmin on the low molecular PKF and related compounds?

Habermann: The peptic kinin-yielding fragments have not yet been tested. Bradykinyl-serine, however, proved to be a good substrate for plasmin.

Prado: Horse urinary kallikrein has a proteolytic effect on salmine which parallels both its esterolytic and kininogenic effect. It probably splits peptide bonds involving the carboxyl group of arginine but not necessarily of serine.

Habermann: We used pig pancreatic kallikrein. Any proteolytic effect except that on the kinin-yielding groups could not be attributed to that enzyme.

Trautschold: What is the ratio of liberated kinins, with either trypsin or pancreatic kallikrein as kininogenase, to the peptic kininogen split product?

Habermann: As shown in the text, about 350 μg kinin-9/mg PKFL is the highest yield obtained. Since our material consists of about 68% of amino acids (32% is salts and impurities deriving from columns), the recovery in kinin-9 with respect to "true" kinin-yielding peptide is very close (90%) to the theoretical amount. Since kinin-11 has a much lower activity on the guinea pig ileum, the yield in activity is lower when pancreatic kallikrein is used than with kinin-9-liberating enzymes. No direct comparisons with kinin-11 were made.

THE PURIFICATION AND SOME PROPERTIES OF TWO DIFFERENT KALLIDINOGENS FROM HUMAN PLASMA

JACK V. PIERCE and MARION E. WEBSTER

Laboratory of Metabolism and Laboratory of Cardiovascular Physiology, National Heart Institute, Bethesda, U.S.A.

Although bovine kininogen has been highly purified (Habermann et al., 1963; Suzuki et al., 1965), very little has been published on the purification of human plasma kallidinogen (Webster and Pierce, 1963). For the isolation of native kininogens, mild conditions of separation should be used. Denaturing treatments such as heat and low pH should be avoided. In addition, the use of a glandular kallikrein would seem preferable to the use of trypsin in following kininogen activity. The former enzyme is more likely to require undenatured substrates since it has a higher substrate specificity than trypsin, whereas the latter enzyme generally works best on denatured or fragmented kininogens.

This paper describes the isolation from human plasma of two different forms of kallidinogen, I and II, and their preliminary characterization as glycoproteins containing hexose, sialic acid, and glucosamine, and having a molecular weight of about 50,000. During the isolation process only aqueous solvents of around pH 6 and generally a temperature of 2° C were used. Kallidinogen content of the fractions was determined by adding the solutions directly to the uterine bath and releasing kallidin with an excess of human urinary kallikrein.

PURIFICATION AND SEPARATION OF HUMAN KALLIDINOGENS

Table 1 summarizes the purification procedure. STEP 1. Human plasma (5.8 liters) was adsorbed to DEAE (diethylamino-

TABLE 1
Summary of Steps for Purification and Separation of Human Plasma Kallidinogens

Step	Purification procedure	Yield of activity (%)	Specific activity (units/E_{280})	Purification factor
0	Human plasma (ACD)	(100)	0.06	(1.0)
1	DEAE-Cellulose chromatography (phosphate stepwise)	59	0.55	9.2
2	DEAE-Cellulose chromatography (phosphate gradient)	53	0.81	13.5
3	Hydroxylapatite chromatography (phosphate gradient)			
	a. First active peak	9.4	2.8	47
	b. Second active peak	19.7	8.3	138
	Total activity recovered	35	2.6	43
4a	Sephadex G-200 gel filtration of 3a			
	(1) Main peak (K_d = 0.46)	3.3	4.8	80
	(2) Main peak (K_d = 0.46)	1.2	1.0	17
	(3) Minor peak (K_d = 0.22)	1.3	1.4	24
	Total activity recovered	6.9	2.4	40
4b	Sephadex G-200 gel filtration of 3b			
	(1) Main peak (K_d = 0.45)	6.7	17.2	290
	(2) Main peak (K_d = 0.45)	4.1	12.1	200
	(3) Main peak (K_d = 0.45)	2.5	6.3	105
	Total activity recovered	19.4	9.6	160
5	Hydroxylapatite chromatography (phosphate gradient in 1 M sodium chloride)			
	a. First active peak: Kallidinogen I	1.2	13.2	220
	b. Second active peak: Kallidinogen II	0.97	20.7	345

ethyl)-cellulose and eluted by increasing stepwise the concentration of phosphate buffers. STEP 2. The eluates (0.125 through 0.175 M) which contained all the recovered activity were pooled and rechromatographed on DEAE-cellulose, eluting with a linear gradient of 0.05 to 0.20 M phosphate. Two peaks of activity were obtained but were combined, since a similar result could be obtained with hydroxylapatite chromatography. STEP 3. Adsorption of the combined active fractions from Step 2 to a hydroxylapatite column and elution with a linear gradient of 0.002 to 0.20 M phosphate gave two active peaks, containing 32% and 68%, respectively, of the recovered activity. STEP 4a. Filtration of the first active peak (fraction 3a) through a column of Sephadex G-200 gave a small, broad peak of activity with a K_d of about 0.22 [fraction 4a (3)] and a large, sharp peak of activity with a K_d of 0.46; the latter peak was divided into two fractions, 4a(1) and 4a(2), on the basis of purity. These fractions have not been further investigated, with one exception. Bio-Gel P-200 gel filtration of fractions 4a(3) and 4b(2) (see below) in 8 M urea gave the same elution volumes for the activity peaks. STEP 4b. Sephadex G-200 gel filtration of the second active peak from Step 3 (fraction 3b) gave only a trace of activity with a K_d of 0.22 and a large peak of activity with a K_d of 0.45. Three pools of decreasing purity were made and called fractions 4b (1), 4b(2), and 4b(3), respectively. STEP 5. Hydroxylapatite chromatography of one half of fraction 4b(2) in the presence of 1 M sodium chloride (Ada, 1963), with a linear gradient of 0.01 to 0.15 M phosphate, gave two peaks of activity. The purest fractions from each peak were separately combined to give 9.7 mg of kallidinogen I and 4.9 mg of kallidinogen II.

PROPERTIES OF HUMAN KALLIDINOGENS I AND II

Molecular weight. High-speed equilibrium centrifugation (Yphantis, 1964) of fraction 4b(1) gave a weight average molecular weight of 50,000 ± 5,000, assuming a $\bar{v}$ of 0.70. Bio-Gel P-200 gel filtration experiments on fraction 4b(2), bovine plasma albumin, and ovalbumin gave a value of 52,000 ± 5,000 (Andrews, 1964).

Hexose determination. Preliminary hexose analysis by the method of Dubois et al. (1956) of I and II gave 5.4% and 5.0%, respectively, or about 15 moles hexose per mole kallidinogen.

Amino acid and hexosamine analyses. Analysis of hydrolysates (6 N HCl, 110°, 24 hours) by the method of Moore et al. (1958) showed no marked differences between I and II in amino acid content. All the usual amino acids, except tryptophan which is destroyed, were found. Both kallidinogens had a high content of glutamic and aspartic acids (about 50 and 38 residues/mole respectively), a considerable half-cystine content (10 to 15 residues/mole), and a low content of histidine (3 to 4 residues/mole) and of methionine (2 to 3 residues/mole). In the same analyses, peaks for glucosamine on both acidic and basic columns were found, but none for galactosamine. From the data obtained, I contains about 1.8% and II, about 1.2% of glucosamine.

Carboxypeptidase B treatment. Incubation of kallidinogens I and II with carboxypeptidase B, followed by the addition of 1,10-phenanthroline to inhibit the enzyme and digestion with kallikrein, showed that kallidinogen I no longer contained a biologically active peptide, whereas kallidinogen II retained ful activity.

Cyanogen bromide cleavage (Gross and Witkop, 1962). Treatment of kallidinogens I and II with cyanogen bromide, a reagent that selectively cleaves methionyl peptide bonds under mild conditions, resulted in the liberation of much more kinin activity from I than from II.

Disc electrophoresis: Treatment of kallidinogens I and II with human urinary kallikrein and neuraminidase. Kallidinogens I and II were incubated with kallikrein; neuraminidase (*Vibrio cholerae*); kallikrein followed by neuraminidase; and neuraminidase followed by kallikrein. It was found that the ability of kallikrein to release kallidin from kallidinogens was not significantly altered by the removal of some or all of their sialic acid residues. Disc polyacrylamide gel electrophoresis was done on I and II and on the four enzyme digests. In the experiments using both enzymes, the order in which they were added did not alter the gel patterns. These patterns were compared for all the experiments, and the results are summarized in Table 2.

Immunodiffusion experiments. Sheep antiserum to fraction 4b(1) was prepared. Immunodiffusion experiments by a modification of the method of Ouchterlony (1949) revealed, besides obvious precipitin lines with impurities, a single line of identity

TABLE 2
Mobility Changes of Kallidinogens I and II in Disc Electrophoresis After Treatment with Human Urinary Kallikrein (HUK) and *Vibrio Cholerae* Neuraminidase (NA)

Kallidinogen	Distance of control from origin* (cm)	Mobility change in cm. after		
		HUK	NA	HUK + NA or NA + HUK
I	2.75	+0.30	−0.75	−0.55
II	3.15	0.0	−0.75	−0.55

*Cathode end.

for human plasma and the two kallidinogens, as well as for the residual protein after losing its kallidin moiety by treatment with human urinary kallikrein.

DISCUSSION

The fast-moving fraction 4a(3) would appear to be a kallidinogen dimer or higher aggregate which can be dissociated in 8 M urea. This aggregate may be similar to the kininogen II described by Habermann (1963).

The molecular weight found for the human plasma kallidinogens is consistent with results obtained with highly purified bovine kininogen by Habermann (1963) and by Suzuki et al. (1965). The former author calculated values of 48,000 from sedimentation and diffusion constants and 53,000 ± 10% from the yield of kinin-9. The latter authors reported their preparation to have an S_{20} value of 2.98. Habermann (1963) has reported that a highly purified preparation of bovine kininogen contains 6.1% N-acetylneuraminic acid, 8.5% hexosamine, and 5.5% hexose. These values are roughly comparable to those so far determined for kallidinogens I and II.

Our results with cyanogen bromide cleavage furnish good evidence that human plasma kallidinogen has a methionine residue adjacent to the N-terminal lysine of the kallidin -10 sequence, as has been found for bovine kininogen (Elliott et al., 1963; Elliott and Lewis, 1965; Habermann, 1965).

The neuraminidase experiments with the two kallidinogens provide strong presumptive evidence for the presence of sialic acid (N-acetylneuraminic acid) residues occupying terminal positions on the assumed carbohydrate side chains.

The immunodiffusion experiment in which I and II were compared separately with their products from the action of human urinary kallikrein, with sheep antiserum to kallidinogen, indicates that the major antigenic sites are still present in one large fragment. Had this not been so, spurs would have been observed. Thus, if kallidinogen yields two fragments besides kallidin-10, one of them must represent a major part of the native molecule, and the other must be too small to have more than one antigenic site.

These considerations and results suggest the following models for kallidinogens I and II:

A_1-K	A_2-K-B
I	II

where A_1 and A_2 are large (mol. wt. about 45,000) N-terminal fragments with C-terminal methionine residues, K is the kallidin-10 sequence, and B is a small C-terminal fragment of II. It is further supposed that A_1, A_2, and B carry sialic acid residues and become A_1', A_2', and B', respectively, after treatment with neuraminidase.

Thus, carboxypeptidase B treatment of I removes the C-terminal arginine residue, so that the peptide formed by kallikrein will be inactive. In II, this arginine residue is protected by B; hence, the peptide released by kallikrein retains full activity.

Cyanogen bromide cleavage of I gives $A_1 + K$, whereas cleavage of II gives $A_2 + K - B$. This explains the significantly higher yield of kallidin activity from I than from II, since $K - B$ would probably have a low specific activity compared with K, as reported by Stewart and Woolley (1965) for kinins with additional amino acid residues on the C-terminus.

Treatment of I and II with kallidrein gives $A_1 + K$ and $A_2 + K + B$, respectively. A_1 has a higher electrophoretic mobility than I because it has lost three positive charges with K. A_2 has the same mobility as II, indicating the presence of three acidic groups in B which balance the three positive charges of K. Neuraminidase treatment of I and II gives $A_1' - K$ and

$A_2' - K - B'$, respectively, both of which now have lower mobilities than their precursors. Treatment of these with kallikrein gives, respectively, $A_1' + K$ and $A_2' + K + B'$, where both A_1' and A_2' have slightly higher mobilities than their immediate precursors, since now the loss of K is not balanced by the loss of B', which means that the three negative charges ascribed above to B must be due to terminal sialic acid residues.

A comparison of the mobility changes found after neuraminidase treatment with those found after a subsequent kallikrein treatment enables us to make a rough estimate of about ten sialic acid residues occupying terminal positions of the presumed carbohydrate side chains. The existence of such side chains is made more plausible by finding hexose, glucosamine, and a high aspartic and glutamic acid content for kallidinogens I and II.

The properties of kallidinogens I and II which we have described agree with the observations and proposals made by Margolis and Bishop (1963) and to some extent with those of Vogt (1965). For example, the former authors found, by comparing untreated human plasma with plasma deficient in Component B, that B is a better substrate for trypsin than is the other postulated kininogen (as we found for kallidinogens II and I, respectively) and that the two kininogens, like kallidinogens I and II, are about equally good substrates for human urinary kallikrein. We can offer further evidence for thus identifying kallidinogen II with Component B. Starting with a plasma of only 0.8 unit (instead of 3.2 units) per ml, we obtained a highly purified material which was identical with kallidinogen I by disc electrophoresis; there was no trace of a band for kallidinogen II.

Since Habermann (1965) has obtained from peptic digests of bovine kininogen kinin-containing peptides with amino acid residues on the C-terminal end of the kinin-10 sequence, it will be of some interest to compare purified bovine kininogen with kallidinogens I and II by disc electrophoresis and other methods and perhaps thereby to learn whether bovine plasma has also two distinct kininogens.

SUMMARY

Two kallidinogens (I and II) have been obtained in highly purified form entirely by chromatographic methods and under mild conditions. These preparations were shown to be immunologically identical, but clearly different by several physical, chemical, and biological criteria. The kallidin-10 moiety has been placed near (for II) or at (for I) the C-terminus of kallidinogen on the basis of results obtained from the following experiments: (1) immunodiffusion of I and II against sheep antibody to kallidinogen before and after treatment with human urinary kallikrein; (2) treatment with carboxypeptidase B; (3) cyanogen bromide cleavage; (4) disc electrophoresis of I and II before and after treatment with human urinary kallikrein, neuraminidase, and a combination of the two enzymes; (5) amino acid, glucosamine, and hexose analyses; and (6) molecular weight determination by ultracentrifugation and gel filtration.

The properties of kallidinogens I and II fit to some extent the data and hypotheses of other authors who postulated the existence of two different kininogens in human plasma. Kallidinogen II may be identical with Margolis' Component B and has some of the properties ascribed by Vogt to his kininogen II.

ACKNOWLEDGMENTS

The authors are especially grateful to Dr. Parker A. Small, Jr., for performing the ultracentrifugation analysis and doing the major part of preparing the sheep antisera; to Dr. Erhard Gross for the amino acid and glucosamine analysis; to Dr. Rolf Axen for the cyanogen bromide experiments; and to all of these for their kind and helpful advice and encouragement.

REFERENCES

Ada, G. L.: Purification and properties of avian neuraminidase. Biochim. biophys. Acta 73, 276 (1963).

Andrews, P.: Estimation of molecular weights of proteins by Sephadex gel filtration. Biochem. J. 91, 222 (1964).

Dubois, M., K. A. Gilles, J. K. Hamilton, P. A. Rebers, and F. Smith: Colorimetric method for determination of sugars and related substances. Analyt. Chem. 28, 350 (1956).

Elliott, D. F., G. P. Lewis, and D. G. Smyth: A new kinin from ox blood. Biochem. J. 87, 21P (1963).

Elliott, D. F., and G. P. Lewis: Methionyl-lysyl-bradykinin, a new kinin from ox blood. Biochem. J. 95, 437 (1965).

Gross, E., and B. Witkop: Nonenzymatic cleavage of peptide bonds: The methionine residues in bovine pancreatic ribonuclease. J. biol. Chem. 237, 1856 (1962).

Habermann, E.: Über pH-bedingte Modifikationen des kininliefernden α-Globulins (Kininogen) aus Rinderserum und das Molekulargewicht von Kininogen I. Biochem. Z. 337, 440 (1963).

Habermann, E.: Isolierung, pharmakologische Eigenschaften, Strukturaufklärung und Spaltung peptischer kininliefernder Fragmente (PKF) aus Rinderserum-Kininogen. Naunyn-Schmiedeberg's Arch. exp. Path. Pharmak. 251, 187 (1965).

Habermann, E., W. Klett, and G. Rosenbusch: Partielle Reinigung und einige Eigenschaften eines Kininogens aus Rinderblut. Hoppe-Seylers Z. physiol. Chem. 332, 121 (1963).

Margolis, J., and E. A. Bishop: Studies on plasma kinins I. The composition of kininogen complex. Aust. J. exp. Biol. med. Sci. 41, 293 (1963).

Moore, S., D. H. Spackman, and W. H. Stein: Chromatography of amino acids on sulfonated polystyrene resins. Analyt. Chem. 30, 1185 (1958).

Ouchterlony, O.: Antigen-antibody reactions in gels. Acta path. microbiol. scand. 26, 507 (1949).

Stewart, J. M., and D. W. Woolley: Importance of the carboxyl end of bradykinin and other peptides. Nature 207, 1160, (1965).

Suzuki, T., Y. Mizushima, T. Sato, and S. Iwanaga: Purification of bovine bradykininogen. J. Biochem. (Tokyo) 57, 14 (1965).

Vogt, W.: Ein komplettes zweites kininbildendes System in menschlichem Plasma. Naunyn-Schmiedeberg's Arch. exp. Path. Pharmak. 251, 186, (1965).

Webster, M. E., and J. V. Pierce: The nature of the kallidins released from human plasma by kallikreins and other enzymes. Ann. N.Y. Acad. Sci. 104, 91 (1963).

Yphantis, D. A.: Equilibrium ultracentrifugation in dilute solutions. Biochemistry 3, 297 (1964).

DISCUSSION

Armstrong: I should like to say that the kinins E, F, and S, which the human plasma kinin-forming enzyme KFS of Armstrong Mills produces from human plasma, should also be considered both in relation to Sicuteri's context in vivo and in relation to the Pierce and Webster kallidinogen.

PHYSIOLOGICAL INFLUENCE ON THE LIBERATION OF HUMAN PLASMA KININ AT LOW TEMPERATURES

DESIRÉE ARMSTRONG and G. L. MILLS
The Middlesex Hospital Medical School,
University of London, England;

and

F. SICUTERI
Centro Cefalee, Clinica Medica dell'Università
di Firenze, Florence, Italy

The condition of severe and abnormal failure to produce kinin on induction, which was noticed by Armstrong and colleagues in samples of human parturition plasma, was traced, by means of analyses involving whole plasma, to a state of kininogen deficiency (Armstrong et al., 1960). These findings were supported by the works of Martinez and Sicuteri and their respective collaborators (Martinez et al., 1962; Periti et al., 1962; Centaro et al., 1963; and Sicuteri et al., 1963), who estimated the kininogen by the Diniz technique (Diniz et al., 1961; see also Diniz and Carvalho, 1963).

However, it was soon found that *freshly prepared* parturition plasmas could make kinin by all the induction techniques which had failed on older plasmas. Moreover, when the fresh plasma was cooled, e.g. to 0°C, plasma kinin, equivalent to several micrograms of synthetic oxytocin (Syntocinon, Sandoz) or synthetic bradykinin (BRS640, Sandoz) per milliliter plasma, rapidly appeared. By contrast, in normal plasmas this reaction was relatively slow (Armstrong and Stewart, 1960a; Centaro et al., 1959; Armstrong and Mills, 1965).

Plasma which has thus formed kinin very rapidly spontaneously, equally rapidly becomes kininogen depleted. This

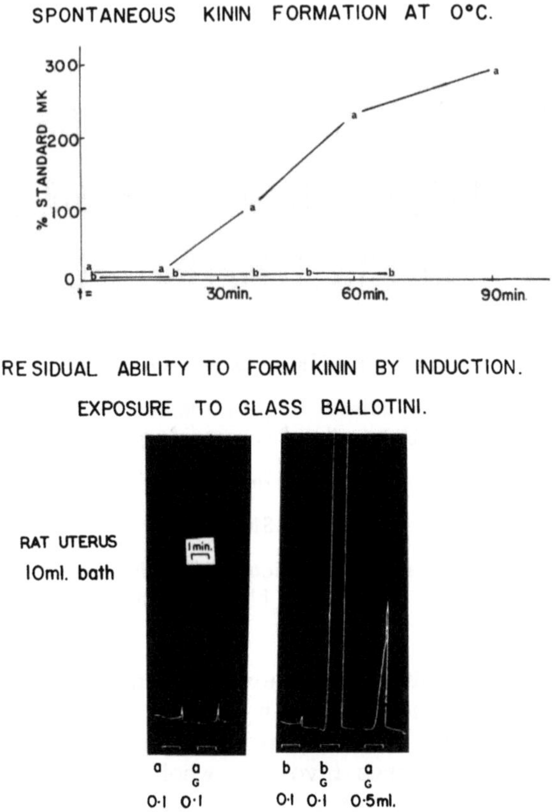

Fig. 1. Upper record. Relative spontaneous formation of plasma kinin at 0°C in two samples of fresh human plasma, collected from the same subject. Sample a was collected at term and sample b at four days post partum. Ordinate = scale of kinin concentration relative to an arbitrary standard, prepared by alcohol extraction of kinin from cold maternal parturition plasma, and called "Mk." Abscissa = duration in minutes of exposure of the fresh plasmas to the cold.

Lower record. The subsequent kinin-forming abilities of these plasmas measured by their response to exposure to glass. Isolated rat uterus contractions.

After the storage at 0°C, which lasted 2 hours, these plasmas were allowed to warm to 20°C for half an hour. G indicates that an aliquot has then been exposed to glass ballotini for 3 min; 0.1, 0.5, etc. indicate the volume (ml) of plasma administered to the organbath (10-ml capacity).

can be shown by stimulating the plasma by means of the induction techniques to produce kinin from the residual kininogen (Armstrong and Stewart, 1960a; Armstrong et al., 1960; Lewis, 1960).

In Fig. 1, upper record, the spontaneous kinin production in two samples of plasma from the same donor, both kept at 0°C, are shown; in the lower record is shown their subsequent ability to form kinin when exposed to glass. It can be seen that the intensities of the consecutive kinin generations, the one spontaneous (upper record) and the other induced (lower record) are in inverse relationship to each other.

Our recent work has concerned an investigation into the mechanism of this spontaneous kinin formation, which occurs in greater or lesser degree in all human plasma. Using, in particular, the dilution technique of Schachter (1956) to measure residual kininogen, we have traced the progression of the deple-

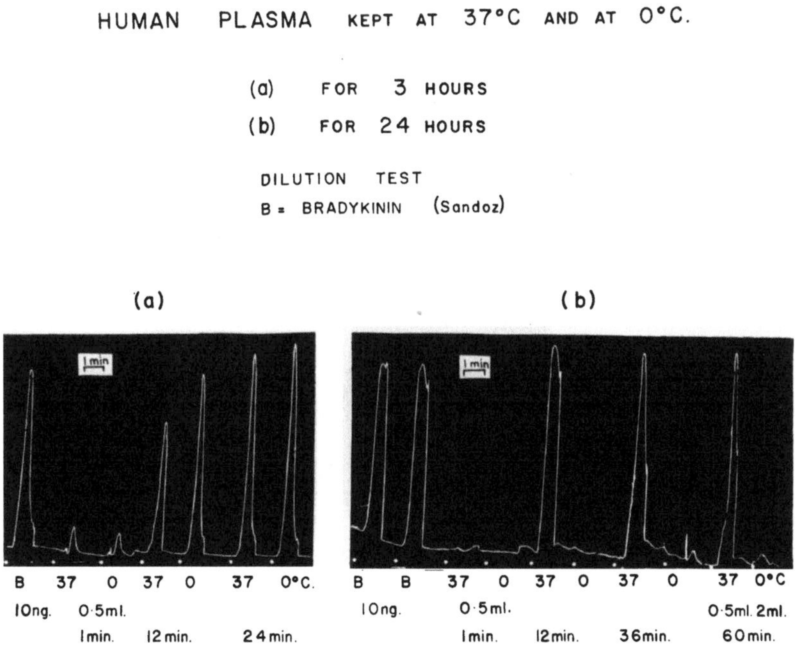

Fig. 2. Maintenance of kinin-forming ability in human plasma stored at 37°C, but loss of potential in the aliquots, of the same normal human plasma sample, that were stored at 0° C.

Kinin formation induced by 20-fold dilution of the plasma with Locke's solution, at 20°C. Min = time in minutes after dilution.

tion occurring in the plasmas during the course of their storage at various temperatures between 37° and 0°C.

Figure 2 shows the results of one such experiment. The freshly collected normal human plasma was divided into aliquots which were stored in closed polythene containers, some at 0°C and some at 37°C, and then tested concurrently. The two groups of aliquots behaved differently according to their storage temperature. We were astonished to observe the anomalous result that the aliquots stored in the *cold* showed the most rapid loss of kininogen. The plasma stored in the warm, at 37°C, maintained its induction kinin-forming ability during 24 hours (a) and (b); this ability, which persisted in the "cold-stored" plasma at 3 hours, (a), was lost during the 24-hour storage period; see (b). In this experiment the loss of production exceeded 75%.

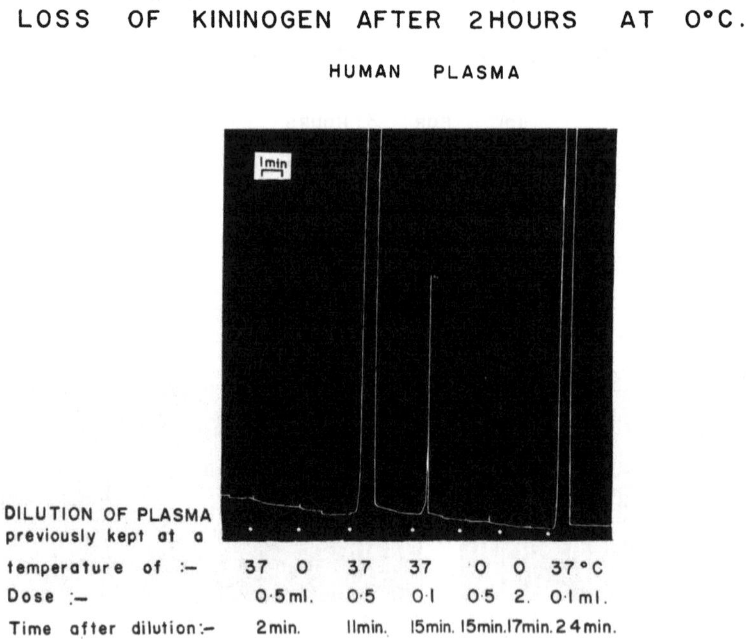

Fig. 3. The rapid deterioration of ability to form kinin by induction in a parturition plasma sample stored for only 2 hours at 0°C. The kinin-forming potential is, however, maintained in the aliquot of the same plasma stored at 37°C.

Kinin formation induced by 20-fold dilution in Locke's solution, at 20°C. Contractions of isolated rat uterus; 10-ml bath.

Not all plasmas lose their kininogen at the same rate. Figure 3 shows the very rapid loss that can occur. Here the behavior of two aliquots of a sample of parturition plasma is demonstrated. Within 2 hours the parturition plasma's ability to form kinin by induction has decreased by more than 95% at 0°C. Nevertheless the "warm-stored" plasma has *maintained* its kinin-forming potential.

Whatever the nature of the reaction that is induced by the cold, it does not produce a permanent change in the plasma even when temperatures of −20° or −50°C are used provided that immediately on melting, the plasma is at once taken back to 37°C and kept there (cf. Diniz and Carvalho, 1963).

The essential features of the relationship of available kininogen to the plasma's storage history are summarized below.

1. At 37°C the plasma maintains its kininogen level during many hours or for several days, and this is the temperature of maximum stability.

2. (a) Storage of the plasma at a given temperature between 37°C and 0°C results in the progressive loss of ability to produce kinin. During the spontaneous degradation of the plasma at 0°C the transient presence of kinin can be detected. In the presence of kininase-inhibiting concentrations of ϵ-amino caproic acid or disodium edetate, kinin can be detected at higher temperatures below 37°C. Little or no kinin is formed at 37°C (Armstrong and Stewart, 1960b; Erdös et al., 1963; Armstrong et al., 1955).

(b) The rate of loss of kininogen increases as the temperature falls below 37°C.

(c) This kininogen-depletion process can be started or stopped, at will, by merely lowering or raising the plasma temperature. In Fig. 4, upper records are shown the effects respectively on kininogen level in each of four aliquots of the same plasma, each of which received different temperature storage treatment. Their histories, at either 37°C of 0°C, are indicated in the corresponding lower records. The kininogen level, estimated by dilution kinin production, was measured at the end of each temperature-exposure period as shown. It can be seen that only during the cold exposure does kininogen depletion progress and that return of plasma temperature to 37°C arrests the process. It thus appears that the cold-induced effect leading to the kininogen depletion is *thermally reversible*.

Since the kinin-forming enzyme and kininogen cannot coexist in the plasma, and since there is evidence for the presence

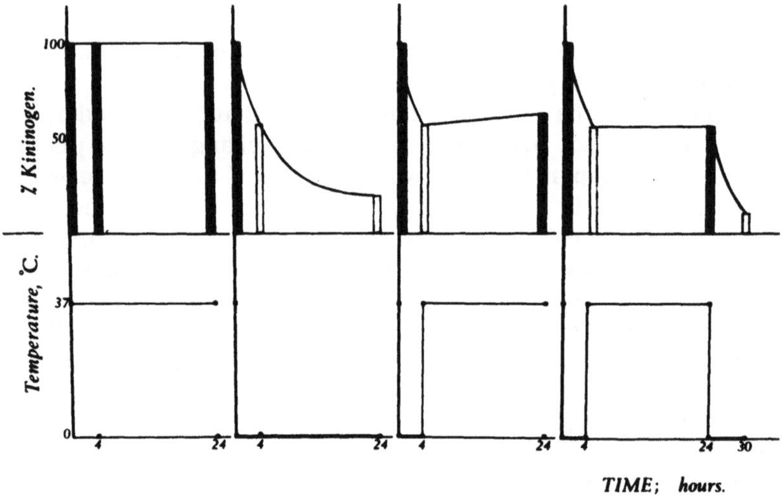

Fig. 4. The thermally induced digestion of human plasma kininogen.

Upper record. Kininogen levels. Ordinate = the peak kinin concentration in diluted plasmas (see below) expressed as a percentage of original production (i.e. of that at $\underline{t}$ = 0), using synthetic bradykinin as standard. Abscissa, the same for both records, indicates the time, absolute, in hours from the start of temperature treatment.

Lower record. Temperature histories. Ordinate = temperature, °C. Abscissa, as above.

The kininogen levels were measured by the kinin formation induced by 20-fold dilution in Locke's solution at 20°C, in samplings from four aliquots of the same sample of fresh normal human plasma, and made at the end of each period during the different temperature exposure histories, which are shown respectively, below each estimation.

of kininogen in plasma at 37°C (suggested by kinin generation on the addition of solutions of crystalline typsin or human KFS; Armstrong and Mills, 1964), it appears that the cold brings about the activation of the kininogenase.

A state of reversible activation may be depicted, as follows, where, for simplicity, the reaction is presumed to involve one molecule only:

$$\underset{\text{(inactive)}}{\text{kininogenase precursor}} \underset{37^{\circ}C}{\overset{0^{\circ}C}{\rightleftharpoons}} \text{kininogenase}$$

An inactive kininogenase precursor is presumed to be in dynamic equilibrium with the active enzyme, the equilibrium at 37°C being greatly in favor of the inactive form. Lowering the plasma temperature, e.g. to 0°C, increases the concentration of the active kininogenase, and kinin generation occurs.

Molecular rearrangement or disaggregation occurring on cooling has already been reported, e.g. in the tobacco mosaic virus protein and sickle cell hemoglobin molecules (Lauffer et al., 1958; Murayama, 1956), and these effects, which are reversible, take place between 40° and 0°C. This type of behavior has been attributed to a structure which is stabilized, at the higher temperatures, by intermolecular hydrophobic bonds. These bonds have a positive enthalpy of formation and are thus weakened by a fall in temperature, so leading to opening of the molecule (Kauzmann, 1959; Némethy and Scheraga, 1962). We thus postulate that the human kininogenase molecule is normally inactive at 37°C and is maintained in that condition by the integrity of hydrophobic bonds which may be inter- or intramolecular.

This work is now being extended to an investigation into the physiological relevance of nonthermal methods of activating the human kininogenase at 37°C.

REFERENCES

Armstrong, D., J. B. Jepson, C. A. Keele, and J. W. Stewart: Activation by glass of pharmacologically active agents in blood of various species. J. Physiol. (Lond.) 129, 80P (1955).

Armstrong, D., C. A. Keele, and J. W. Stewart: Observations on the formation of kinins in human plasma collected during labour. J. Physiol. (Lond.) 150, 20P (1960).

Armstrong, D., and G. L. Mills: A highly purified kinin-forming enzyme from human plasma. Biochem. Pharmacol. 13, 1393 (1964).

Armstrong, D., and G. L. Mills: The reversible cold-induced activation of the human plasma kinin-forming system between 37° and 0°C. J. Physiol. (Lond.) 179, 89P (1965).

Armstrong, D., and J. W. Stewart: Spontaneous plasma kinin formation in human plasma collected during labour. Nature (Lond.) 188, 1193 (1960a).

Armstrong, D., and J. W. Stewart: Formation of plasma kinin in human plasma by ε-amino caproic acid and certain related compounds. J. Physiol. (Lond.) 154, 19P (1960b).

Centaro, A., G. De Laurentis, and P. Periti: Ossitocina nel sangue di donna in travaglio di parto. Boll. Soc. ital. Biol. sper. 35, 1454 (1959).

Centaro, A., P. Periti, and F. Sicuteri: Bradykininogen and bradykininase in blood of women during labour. Settim. med. 51, 1st Suppl., 70 (1963).

Diniz, C. R., I. F. Carvalho, J. Ryan, and M. Rocha e Silva: A micromethod for the determination of bradykininogen in blood plasma. Nature (Lond.) 192, 1194 (1961).

Diniz, C. R., and I. F. Carvalho: A micromethod for determination of bradykininogen under several conditions. Ann. N. Y. Acad. Sci. 104, 77 (1963).

Erdös, E. G., A. G. Renfrew, E. M. Sloane, and J. R. Wohler: Enzymatic studies on bradykinin and similar peptides. Ann. N. Y. Acad. Sci. 104, 222 (1963).

Kauzmann, W.: Some factors in the interpretation of protein denaturation. Advanc. Protein Chem. 14, 1 (1959).

Lauffer, M. A., A. T. Ansevin, T. E. Cartwright, and C. C. Brinton: Polymerization—depolymerization of tobacco mosaic virus protein. Nature (Lond.) 181, 1338 (1958).

Lewis, G. P.: Active polypeptides derived from plasma proteins. Physiol. Rev. 40, 647 (1960).

Martinez, A. R., I. F. De Carvalho, and C. R. Diniz: The bradykininogen content of plasma from women in labour. J. Obstet. Gynaec. Brit. Cwlth. 69, 1011 (1962).

Murayama, M.: Titratable -SH groups of sickle-cell hemoglobin at 0° and at 38°C. Fed. Proc. 15, 318 (1956).

Némethy, G., and H. A. Scheraga: The structure of water and hydrophobic bonding in proteins. III. The thermodynamic properties of hydrophobic bonds in proteins. J. phys. Chem. 66, 1773 (1962).

Periti, P., A. Centaro, F. Sicuteri, and B. Leocani: Bradykininogeno durante il parto. Boll. Soc. ital. Biol. sper. 38, 672 (1962).

Schachter, M.: A delayed slow contracting effect of serum and plasma due to the release of a substance resembling kallidin and bradykinin. Brit. J. Pharmacol. 2, 11, 111 (1956).

Sicuteri, F., G. Franchi, and P. Periti: Physiopathological aspects of bradykinin and its precursor in man. Settim. med. 51, 1st Suppl. 53 (1963).

DISCUSSION

Trautschold: Can trypsin produce kinin from cold-depleted plasma and do you think that cooling impedes the kininogen determination with trypsin, done according to Diniz?

Armstrong: It is possible to get very severely cold-depleted plasmas (e.g. 95% and more depleted) by all my techniques, including the glass-exposure test, which give a normal response to the addition of trypsin. It thus appears that the specific substrate attacked is not identical with kininogen. Occasionally a cold-depleted plasma shows also a diminished response to trypsin. Perhaps this is the result of a nonspecific digestion in plasma with a very high "cold-activated kininogenase" activity. It appears that when the plasmas are not highly active in terms of the cold-activatable kininogenase the Diniz technique should give an accurate and useful measure of bradykininogen.

Webster: Did you try SBTI and changes in pH?

Armstrong: Yes, we have used both techniques. The results, which we have published in the Journal of Physiology, are as follows. SBTI: This inhibits the plasma kinin formation, which occurs in the cold, not only in the normal plasmas but also in the hyperactive parturition plasma. pH: When the plasma is acidified to a pH around 5.5 and stored in the cold for many hours the kininogen is not digested, whereas that in the unacidified aliquots is digested. The actions of SBTI and pH indicate that we are dealing with an enzymatic phenomenon.

Prado: Considering the experiments of Antonio and Rocha e Silva on thermal edema of rat paw, I wonder if you tried temperatures higher than 37°C?

Armstrong: We have also examined the plasma at temperatures between 37° and 57°C. In these circumstances the kininogen is also digested, and the rate increases with rise in temperature. At 37 the plasma is in its most stable condition and may be kept at that temperature during many hours or days without significant loss of kininogen. It is likely that at temperatures above 37° the hydrophobic bonds, opening of which activates the kininogenase, are split by chemical means. This all seems to support the work of Antonio and Rocha e Silva on the rat paw thermal edema, as I believe they employed temperatures around 43-45° C. At this temperature kininogen certainly disappears in my system but in a "stable" human plasma, outside the body, this would take a few hours to become easily detectable by the methods I have so far used.

Procacci: Is it possible that kinin formation at low temperatures would happen not only in vitro but also in vivo? In this case the phenomenon would be important in some physiological or pathological conditions.

Armstrong: Yes, when the body or parts of it are exposed to temperatures below 37° C, the hydrophobic bonds may be expected to open and kinin formation to occur. At temperatures near 37°C the kinin would normally be rapidly removed by kininase. At lower temperatures kininase becomes inhibited. This "cold-kinin" formation may well participate in the manifestations of frostbite. Though the hydrophobic bonds may be broken at low temperature by chemical means also, the cleavage at higher temperatures probably proceeds entirely by the chemical method. By whatever methods the bonds are ruptured, the in vivo manifestations would be aggravated by local stasis in circulatory arrest.

PURIFICATION AND PROPERTIES OF BRADYKININOGEN AND OF THE BRADYKININ-RELEASING AND -DESTROYING ENZYMES IN SNAKE VENOM

TOMOJI SUZUKI, SADAAKI IWANAGA,
SHIGEHARU NAGASAWA and TADASHI SATO

The Institute for Protein Research, Osaka University, Osaka, Japan

The arginine ester hydrolase in the venom of *Agkistrodon halys blomhoffii* ("Mamushi" in Japanese) was separated into three enzymatic entities: the "bradykinin-releasing," "clotting," and "capillary permeability-increasing" enzymes. The latter two enzymes were purified to physicochemically homogeneous states (Sato et al., 1965), and the bradykinin-releasing enzyme described in this paper contained no "clotting" or "capillary permeability-increasing" enzyme, and no bradykinin-destroying enzyme. The properties of the bradykinin-releasing enzyme of the venom were similar to those of pancreatic kallikrein (Werle and Trautschold, 1963). The authors also purified bovine bradykininogen, and from the results of Sanger's DNP method, exopeptidase digestion methods, and the release of bradykinin by the venom bradykinin-releasing enzyme, it was concluded that bradykinin is not located at either end of the polypeptide chain of bradykininogen.

It is well known that the *Crotalidae* venoms which release bradykinin from serum globulin also destroy bradykinin (Deutsch and Diniz, 1955; Rocha e Silva et al., 1949). The authors separated two enzymes with bradykinin-destroying activity from the venom of *A. halys blomhoffii*, and the sites of action of these enzymes on bradykinin are discussed in this paper.

Purification and properties of the bradykinin-releasing enzyme

A typical diagram of the elution of the venom of *A. halys blomhoffii* from a DEAE-cellulose column is given in Fig. 1.

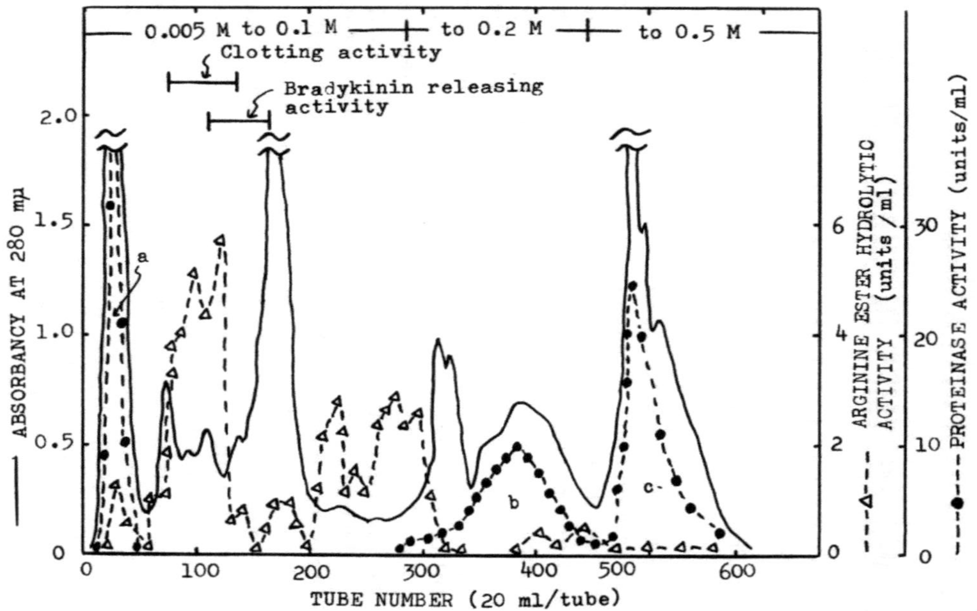

Fig. 1. Chromatography of the venom of *A. halys blomhoffii* on a DEAE-cellulose column. Six g. of venom was applied to a column (5.0 x 35 cm) of DEAE-cellulose (0.68 mEq/g). Concave gradient elution was carried out at 4°C with NaOAc buffer at pH 7.0. One unit of arginine ester hydrolytic activity was defined as the amount of enzyme that hydrolyzed 1 μg TAMe/min. The unit of proteinase activity was defined in a previous paper (Iwanaga et al., 1965).

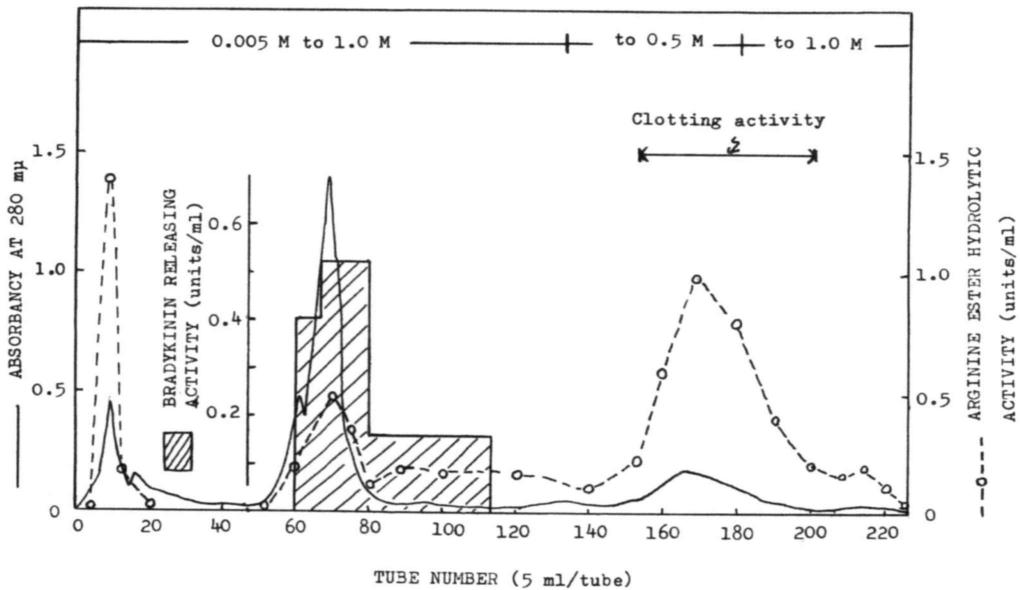

Fig. 2. Purification of the bradykinin-releasing enzyme on a CM-cellulose column. The bradykinin-releasing enzyme fraction (total absorbancy at 280 mμ = 134) was applied to a column (1.5 × 32.5 cm) of CM-cellulose (0.79 mEq/g). Concave gradient elution with NaOAc buffer at pH 6.0 was carried out at 4°C. One unit of bradykinin-releasing enzyme was defined tentatively as the amount that released bradykinin equivalent to 1 μg synthetic bradykinin under the conditions described previously (Sato et al., 1965).

The "clotting" (fibrinogen to fibrin conversion) and "bradykinin-releasing" enzymes were mainly in tubes 60 to 120, and 105 to 175, respectively. An enzyme with capillary permeability increasing and arginine ester hydrolytic activities was eluted in tubes 180 to 300. This enzyme showed no bradykinin-releasing activity. The fraction in tubes 60 to 175 was pooled, concentrated by lyophilization, and desalted on a Sephadex G-25 column. After rechromatography on a DEAE-cellulose column, it was purified further with CM-cellulose. In this way the bradykinin-releasing activity was completely separated from the clotting activity (Fig. 2). Arginine ester hydrolytic activity in the resulting purified bradykinin-releasing enzyme was less than 5% of its initial activity in the venom. Experiments on

TABLE 1
Inhibitory Effects of Trasylol and Soybean and Egg-white Trypsin Inhibitors on the Venom Bradykinin-Releasing Enzyme, Pancreatic Kallikrein and Trypsin

Inhibitor			Bradykinin released (μg)		
			Venom enzyme	Pancreatic kallikrein	Trypsin
None			0.75	0.75	0.75
Trasylol	5	Kallikrein inhibitor units	0.6	0.15	0
"	10	"	0.3	0	0
"	50	"	0.1		
"	100	"	0.08		
SBTI	5	μg			0
"	10	"	0.75	0.75	0
EWTI	5	"			0
"	10	"	0.75	0.75	0

Each mixture contained 34 μg venom enzyme, 5.8 μg pancreatic kallikrein, or 10 μg trypsin in 0.0066 M phosphate buffer (pH 8.0) and the indicated amount of inhibitor in a total volume of 0.7 ml. The mixtures were kept at 37°C for 5 min, and 260 μg bovine bradykininogen in 0.3 ml of water was then added. After 3 min the bradykinin was assayed on guinea pig ileum.

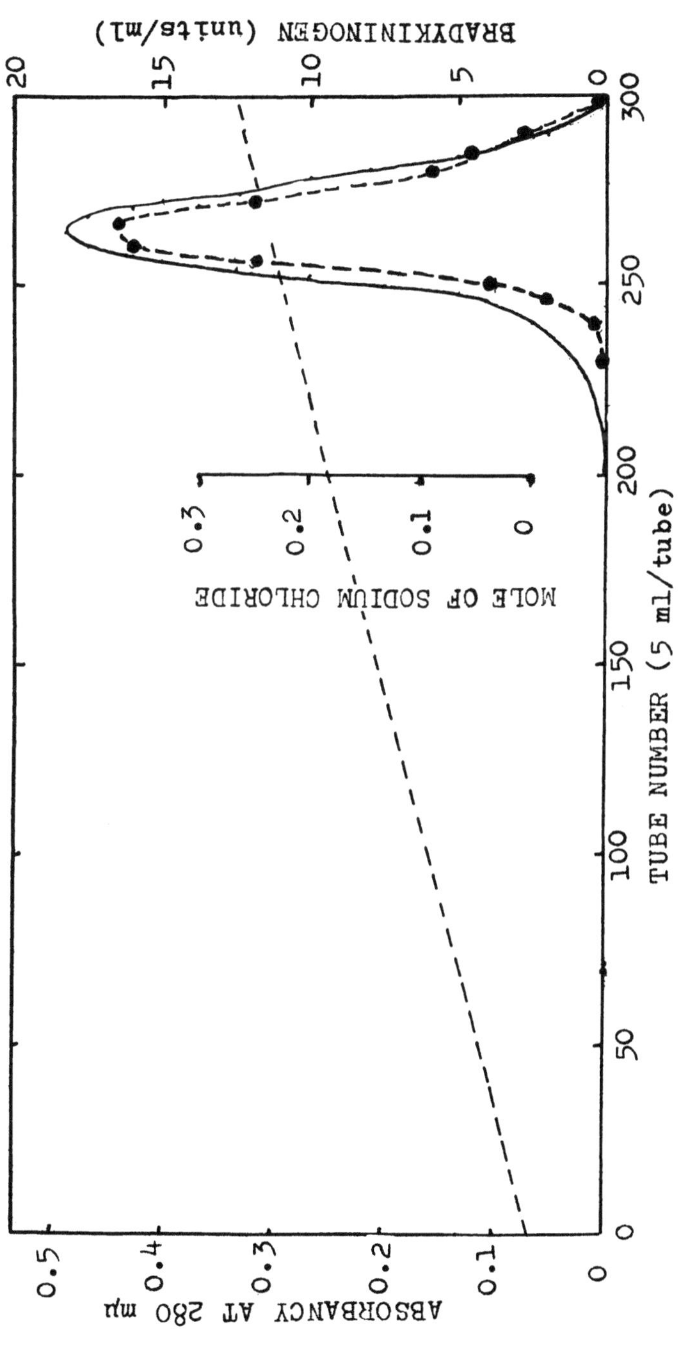

Fig. 3. Elution pattern of purified bradykininogen on DEAE-Sephadex A-50. Eighty-five mg of the preparation was applied to the column (2.0 × 25 cm), and linear gradient elution with 1,000 ml 0.05 M sodium phosphate buffer (pH 8.0) and 0.35 M sodium chloride (reservoir) was performed. One unit of bradykininogen is defined as the amount of bradykininogen that yields bradykinin equivalent to 1 μg synthetic bradykinin with an excess amount of the venom enzyme.

DFP-inhibition, heat-inactivation, and benzoylarginine ethylester-inhibition of the enzyme activity showed that the bradykinin-releasing activity depended completely upon the arginine ester hydrolytic activity. Trasylol, which inhibited the activity of the bradykinin-releasing enzyme (Table 1), had no effect on the arginine ester hydrolytic activities of the "clotting" and "capillary permeability-increasing" enzymes of the venom. Soybean trypsin inhibitor, which has been found to have an inhibitory effect on plasma kallikrein (Webster and Pierce, 1961), had no effect on the bradykinin-releasing enzyme, even when excess trypsin inhibitors were added to the purified enzyme.

Action of purified venom bradykinin-releasing enzyme on purified bovine bradykininogen

Procedures for the purification of bovine bradykininogen have been reported by various workers (Greenbaum and Hosoda, 1962; Habermann et al., 1963). The purification procedure described previously consisted of fractionation with ammonium sulfate, heat treatment, pH treatment at pH 4.9, and column chromatography on DEAE-cellulose and on CM-cellulose. The bradykininogen thus obtained was judged to be about 90% pure (Suzuki et al., 1965). In the present work, additional purification was accomplished by chromatography on DEAE-Sephadex A-50 and gel-filtration on Sephadex G-200 (or Sephadex G-100). The resulting purified bradykininogen was chromatographically homogeneous (Fig. 3), and gave a single peak on ultracentrifugation ($S^{1\%}_{20,w} = 2.65$) and a single band on electrophoresis, with polyacrylamide gel at various pH values. The amino acid and carbohydrate contents of this purified bradykininogen are shown in Table 2. The apparent molecular weight of Habermann's preparation was reported to be 48,000, based on the Svedberg equation (Habermann et al., 1963), whereas the average molecular weight of our preparation, calculated from amino acid analyses, was about 49,500.

DNP-Ser (0.4 mole/mole bradykininogen) and a small amount of DNP-Leu (DNP-Ileu) were found in the acid hydrolysate of dinitrophenylated bradykininogen. After incubation with leucine aminopeptidase for 1 to 3 min, the main product was serine, but there was also a little alanine and glycine. Thus, the N-terminal amino acid of bradykininogen must be serine. The C-terminal amino acid was determined by digestion with carboxypeptidases A and B. During incubation for 3 to 180

TABLE 2
Amino Acid and Carbohydrate Composition of Bradykininogen

Per cent of amino acids residue				Per cent of carbohydrate residue	
Lys	5.47	Ala	3.82	Galactose	4.4
His	2.29	Cys/2	2.44	Mannose	2.3
NH_3	1.58	Val	5.91		
Arg	3.07	Met	0.71	Glucosamine	6.0
Asp	8.14	Ileu	3.37	Glucosamine	6.0
Thr	4.58	Leu	5.89		
Ser	4.80	Tyr	3.59	Sialic acid	5.3
Glu	10.93	Phe	3.70		
Pro	4.94	Try	1.22		
Gly	2.15				
Total amino acid residues 78.60				Total sugar residues 18.0	

Average or extrapolated values after 24, 48, and 72 hours of hydrolysis. Values for Arg, Thr, Ser, and Tyr were obtained by extrapolation to the value after zero-time hydrolysis.

min, the liberation of leucine was always faster than that of alanine, serine, and valine. After 180-min digestion with carboxypeptidase A, about 1 mole leucine/mole bradykininogen was liberated. Thus, the C-terminal amino acid was leucine. Even after digestion of bradykininogen with carboxypeptidase A, leucine aminopeptidase, or a mixture of carboxypeptidases A and B, bradykininogen retained full biological activity.

The residual protein derived from bradykininogen by the action of venom enzyme was then prepared according to the experimental conditions shown in the legend of Fig. 4. The substance with ileum-contracting activity in tubes 23 to 31 in Fig. 4 was identified as bradykinin from its amino acid composition, and no free amino acid or peptide except bradykinin was detected in the eluate. The substance that was eluted in tubes 30 to 40 in Fig. 4 had an absorbancy at 280 mμ, but gave no color with ninhydrin reagent. A part of the residual protein (tentatively called "dekininated" protein) in tubes 10 to 20 in Fig. 4 was dinitrophenylated. After 2, 4, and 20 hours' acid hydrolyses,

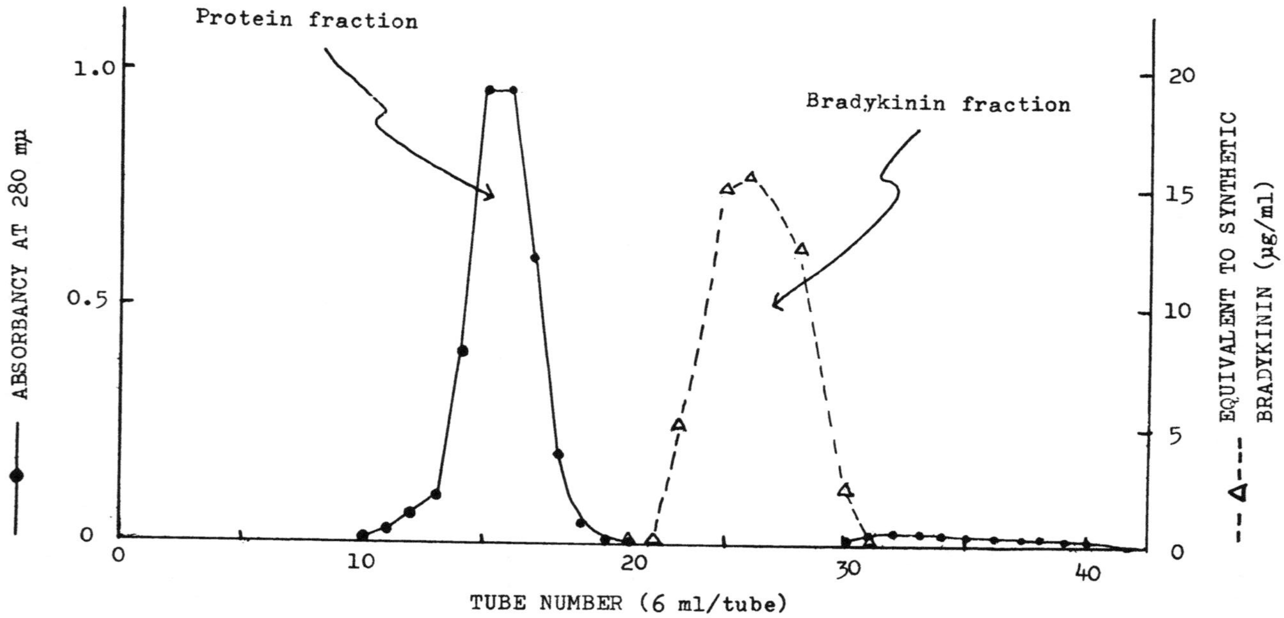

Fig. 4. Isolation of the residual protein derived from bradykininogen by the action of the purified venom bradykinin-releasing enzyme. Thirty mg of purified bradykininogen was incubated with 600 μg bradykinin-releasing enzyme at 30°C for 60 min in 10.9 ml of 0.004 M disodium phosphate. The reaction mixture was lyophilized. The residue was applied on a column of Sephadex G-25 (2.5 × 45 cm) and eluted with distilled water at 4°C.

the main substance found was DNP-Ser, while traces of DNP-Leu were detected in each hydrolysate by two-dimensional thin-layer chromatography with solvent systems toluene:pyridine: ethylenechlorohydrine:0.8 N ammonia (100:30:60:60; v/v) and chloroform:benzyl alcohol:glacial acetic acid (70:30:3; v/v). The N-terminal amino acid found in the dekininated protein was the same as that of the intact bradykininogen, and the amount of DNP-Ser did not increase even after removal of bradykinin from the bradykininogen. When the dekininated protein was gel-filtered through a column of Sephadex G-75 or G-100, only a single protein peak was obtained, and the molecular weight of this protein was considered to be almost the same as that of the intact bradykininogen. Thus, the bradykinin moiety is probably located in the peptide chain between the two cysteine molecules that form the disulfide linkage of bradykininogen.

Bradykinin-destroying enzymes in snake venoms and their sites of action on bradykinin

The proteinase activities of the venom of *A. halys blomhoffii* have been separated into three peaks on a DEAE-cellulose column (Satake et al., 1963), and a similar result was reported (Pfleiderer and Sumyk, 1961) with the venom of *Crotalus atrox*. The bradykinin-destroying activity in the venom of *A. halys blomhoffii* was found mainly in the fraction designated as the proteinase b fraction in Fig. 1. From this fraction, proteinase b was purified by rechromatography on a DEAE-cellulose column and gel-filtration on Sephadex G-100. The yield of physicochemically homogeneous proteinase b protein from lyophilized crude venom was 7.5%. Five ml of a solution containing 2 μ moles synthetic bradykinin (2.48 mg) and 195 μg proteinase b were incubated at pH 8.5 at 25°C. After 80 min, the hydrolysate was subjected to paper chromatography with a solvent system consisting of n-butanol:glacial acetic acid:water (63:10:27) and to electrophoresis in 2 N acetic acid at pH 3.0. In this way two peptides were isolated. The N-terminal amino acids of the two peptides were found by Sanger's method to be DNP-Arg and DNP-Phe. The C-terminal amino acids of these peptides analyzed by the hydrazinolysis method were glycine and arginine respectively. Thus, it was concluded that proteinase b hydrolyzed the glycylphenylalanyl linkage of bradykinin.

On the other hand, when purified bradykininogen was incubated with partially purified bradykinin-releasing enzyme at pH 8.5 at 37°C, it was found that the liberated bradykinin was slowly destroyed. To examine the site on bradykinin which is hydrolyzed by the enzyme contaminating this partially purified bradykinin-releasing enzyme, 1 μ mole synthetic bradykinin and 510 μg of the partially purified bradykinin-releasing enzyme were incubated at pH 8.5 at 37°C for 3 hours. By applying Sanger's method to the hydrolysate, DNP-Ser was identified as the newly formed N-terminus. By the hydrazinolysis method, phenylalanine was identified as the C-terminus. Thus it was concluded that an enzyme that hydrolyzes the phenylalanylseryl linkage of bradykinin is present in the venom.

The distribution of the bradykinin-destroying enzymes in various *Crotalidae* venoms was studied. It was found that the venoms of *C. atrox*, *C. adamanteus*, and *Trimerusurus flavoviridis* split the two peptide bonds of the bradykinin molecule, whereas the venoms of *A. contortrix contortrix*, *Bothrops jara-*

TABLE 3
Distribution of the Bradykinin-Destroying Enzyme and Some Physiological Activities of Snake Venoms

Snake venom	Newly formed amino acids		Hemorrhage	Bradykinin-releasing enzyme (unit*/mg)
	C-Terminus	N-Terminus		
A. halys blomhoffii	Gly > Phe	Phe > Ser	+ +	27.3
C. atrox	Gly > Phe	Phe > Ser	+	42.0
C. adamanteus	Gly > Phe	Phe > Ser	+ + +	63.0
T. flavoviridis	Gly > Phe	Phe > Ser	+	0
A. contortrix contortrix	Gly	Phe	–	52.5
B. jararaca	Gly	Phe	±	52.3
V. russellii	Gly	Phe	–	0

One μmole bradykinin and 300 μg venom were incubated for 3 hours. Newly formed N- and C-terminal amino acids were analyzed by the dinitrophenylation and hydrazinolysis methods respectively.

*See legend to Fig. 2.

raca, and *Vipera russellii* split only one peptide bond. All these venoms hydrolyzed the glycylphenylalanyl linkage of bradykinin (Table 3). In the venom of *A. halys blomhoffii*, the glycylphenylalanyl linkage of bradykinin was hydrolyzed by proteinase b in the venom. The proteinase b was found to have hemorrhagic activity (Iwanaga-et al., 1965). The venom of *A. contortrix contortrix* showed no hemorrhagic activity, and the venom of *B. jararaca* and *V. russellii* had very little, although all three hydrolyzed the same peptide bond of bradykinin as the venom of *A. halys blomhoffii*. These results indicate that different enzyme proteins in the venoms of poisonous snakes may hydrolyze the same peptide bond of bradykinin.

REFERENCES

Deutsch, H. F., and C. R. Diniz: Some proteolytic activities of snake venoms. J. biol. Chem. 216, 17 (1955).

Greenbaum, L. M., and T. Hosoda: Studies on the isolation of bradykininogen. Biochem. Pharmacol. 12, 325 (1962).

Habermann, E., W. Klett, and G. Rosenbusch: Partielle Reinigung und einige Eigenschaften eines Kininogens aus Rinderblut. Hoppe-Seyler's Z. physiol. Chem. 322, 121 (1963).

Iwanaga, S., T. Omori, G. Oshima, and T. Suzuki: Demonstration of a proteinase with hemorrhagic activity in the venom of *A. halys blomhoffii*. J. Biochem. 57, 392 (1965).

Pfleiderer, G., and G. Sumyk: Separation of rattlesnake venom proteinases by cellulose ion-exchange chromatography. Biochim. biophys. Acta 51, 482 (1961).

Rocha e Silva, M., W. T. Beraldo, and G. Rosenfeld: Bradykinin, a hypotensive and smooth muscle stimulating factor released from plasma globulin by snake venoms and by trypsin. Amer. J. Physiol. 156, 261 (1949).

Satake, M., Y. Murata, and T. Suzuki: Chromatographic separation and properties of three proteinases from *A. halys blomhoffii* venom. J. Biochem. 53, 438 (1963).

Sato, T., S. Iwanaga, Y. Mizushima, and T. Suzuki: Separation of arginine ester hydrolase of *A. halys blomhoffii* venom into three enzymatic entities. J. Biochem. 57, 380 (1965).

Suzuki, T., Y. Mizushima, T. Sato, and S. Iwanaga: Purification of bovine bradykininogen. J. Biochem. 57, 14 (1965).

Webster, M. E., and J. V. Pierce: Action of the kallikreins on synthetic ester substrates. Proc. Soc. exp. Biol. (N.Y.) 107, 186 (1961).

Werle, E., and I. Trautschold: Kallikrein, kallidin, kallikrein inhibitors. Ann. N. Y. Acad. Sci. 104, 117 (1963).

DISCUSSION

Habermann: As we pointed out some years ago, permeability-increasing potency of snake venoms would be different from esterolytic-proteolytic enzymes. However, a portion of activity in the whole venom is associated with them. Since all other kinin-releasing enzymes increase the capillary permeability as well, I wonder whether your kinin-releasing enzyme is completely free from such activity. The concordance of our results with bovine kininogen is striking indeed. We also have a series of observations leading to the conclusion that the kinin-yielding sequence resp. the peptic kinin-yielding peptides described by us stem from the inside, not from one of the ends of the molecule.

Suzuki: Kinin-releasing enzymes of the venom used in these experiments were completely free from "clotting" and "capillary permeability-increasing" enzymes. Physicochemically homogeneous capillary permeability-increasing enzyme of the venom, which was prepared in our laboratory, did not lower the blood pressure. Also when the capillary permeability-increasing enzyme was incubated with the purified bovine bradykininogen, no release of bradykinin was observed. The capillary permeability-increasing enzyme hydrolyzes arginine esters but does not hydrolyze casein.

HUMAN SALIVARY KALLIKREIN AND LIBERATION OF COLOSTROKININ

HIROSHI MORIYA, CHIAKI MORIWAKI, KEIKŌ YAMAZAKI, SETSUKO AKIMOTO, and HIROKO FUKUSHIMA

Laboratory of Physiological Chemistry, Tokyo College of Science, Tokyo, Japan

Kallikreins from various sources have been studied by many investigators including Moriya et al. (1958; 1959 a, b, c, d, e, f, g; 1963), to elucidate the physiological and pathological significance of the various kallikreins and of the vasodilator polypeptide, kallidin (or bradykinin), liberated by kallikreins. The present investigation describes the purification of a kallikrein from human mixed saliva and the liberation of colostrokinin from bovine colostrum by kallikrein. Some properties of colostrokinin will be presented and discussed.

MATERIALS AND METHODS

Human mixed saliva was collected from healthy students and members of our university. The material was stored in a freezer at −20°C. The period of collection (at room temperature) and storage of the saliva in the refrigerator overnight comprised less than 48 hours altogether.

Vasodilator activity was determined by measuring the increase in arterial blood flow after the injection of both standard and purified kallikrein in 0.2-0.4 ml of saline. Male and female dogs (7-10 kg) were anesthetized with 160 mg phenobarbital/kg given with food the night before the experiment. The electromagnetic flowmeter and autorecorder (Nihon Kohden and Co.,

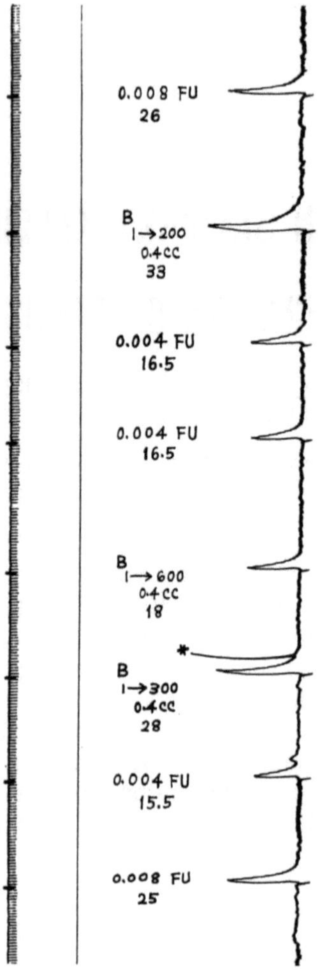

Fig. 1 The increase of arterial blood flow after the injections of both standard and our kallikrein preparations. Dog, 10.0 kg, male; the electromagnetic flowmeter and autorecorder (Nihon Kohden and Co., Tokyo). FU means standard. Arrow points to dilution of sample; 0.4 ml was injected at every 5-min interval; numbers show the height of response.

*Calibration: maximum obtained with the instrument.

Tokyo) were recently obtained by our laboratory. The bioassay was done according to Sarnoff et al. (1958); Fig. 1 shows some of our results obtained with this equipment.

Cellulose ion exchangers. Diethylaminoethyl (DEAE)-cellulose (Brown and Co., U.S.A., 0.9 mEq/g of exchange capacity) and carboxymethyl (CM)-cellulose (Serva and Co., Germany, 0.71 mEq/g) were treated to remove fine particles with decantation, then washed with hydrochloric acid, sodium hydroxide, water, and ethanol, and screened to give a 100-200 mesh fraction.

Cyanogum gel electrophoresis and esterase activity. These were done according to Moriya et al. (1963).

Sephadex gel-filtration. The columns (1.2 × 27 cm) were prepared with Sephadex G-75 or G-100, supplied in bead form, 40-120μ of particle size (Pharmacia, Uppsala, Sweden), and equilibrated with 0.05 M ammonium formate, pH 6.0. The active fraction in 2 ml was applied to the surface of this gel bed, then eluted with the same buffer. A Uvicord ultraviolet absorptiometer recorded automatically and semiquantitatively the protein absorption; a Miniflow micropump regulated the constant flow rate at 12.5 ml/hour. Each fraction (1.5-2.5 ml) was collected and each volume determination was based on the weight of the liquid. The protein was quantitatively measured in the spectrophotometer at 280 mμ.

Substrate in colostrum (colostrokininogen). Bovine colostrum obtained just after delivery was kindly supplied by the Zootechnical Institute, Ministry of Agriculture of the Japanese Government. Yellowish emulsive colostrum was stored frozen. Thawed colostrum was centrifuged, the supernatant adjusted to pH 4.6 with hydrochloric acid, and recentrifuged to remove impurities. This supernatant itself, or fractions obtained by salting out with ammonium sulfate (0.3, 0.5, and fully saturated) were treated at pH 3.0, 37°C for 5 min, then adjusted to pH 5.5 and dialyzed. Nondialyzable protein (crude colostrokininogen) was employed as substrate.

Colostrokinin and its activity. The crude colostrokininogen was added to salivary kallikrein (1-2 FU/mg, 2-4 FU/1 ml equivalent to original colostrum), then incubated while stirring at pH 7.5 for 2 hours. The pH of the reaction mixture was adjusted to 4.5 with acetic acid. Later, 2 volumes of ethanol were added, and the mixture was boiled for 10 min. After cooling, the mixture was centrifuged and the supernatant was concentrated and used as the crude colostrokinin fraction. Its activity was assayed by measuring the vasodilatation in dogs and the contractions of rat uterus and ileum. This activity was also compared with that of synthetic bradykinin.

I^{131}-labeled human salivary kallikrein. Commercial KI-I^{131} (5 mc, Dainabott Radioisotope Lab. Ltd.) was converted to I_2-I^{131} with $NaNO_2$ and H_2SO_4, extracted into $CHCl_3$-layer, washed with acidic water three times, then extracted to 1 ml of alkaline water, pH 8. Fifty mg per 0.5 ml of human salivary kallikrein (3.8-12.6 FU/mg) was added to 1 ml of I_2-I^{131} solution, incubated at 37°C for 1.5 hours, and stored overnight in a refrigerator; the reaction mixture was passed through a Sephadex G-25 column. The main protein fraction was lyophilized to afford 26-46 mg of I^{131}-labeled human salivary kallikrein bound to 18-19% of the original I^{131}. Crystalline KI (analytical grade, Tokyo Kasei Ltd.) was employed as carrier throughout the experiments.

The effect of colostrokinin on the absorption of I^{131}-labeled human albumin in rats. Experiments were done on 3 to 8 animals in each group of newborn (3 and 7 days old) and adult rats (body weight 120-150 g, used for control). Three hours after I^{131}-labeled human serum albumin (Dainabott Radioisotope Lab. Ltd.) was directly given in the closed intestinal sac of anesthetized rats with or without colostrokinin, the radioactivity in the tissues was checked. The residual activity in the intestine and the distribution in liver, kidney, spleen, and serum were assayed, and the protein-bound I^{131} in serum was separated by dialysis, gel filtration, and paper chromatography. The animals which received albumin with colostrokinin got 2 doses of the peptide (corresponding to 0.025 and 0.0025 μg bradykinin in bioassay in dog) given both at the same time and 30 min before the administration of I^{131}-labeled human albumin. All experiments were carried out by comparing KI-I^{131} given by the same method and I^{131}-albumin given by intravenous injection. Crystalline human serum albumin (Tokyo Kasei Ltd.) was employed as the carrier.

RESULTS

Purification of human salivary kallikrein

The kallikrein content of human mixed saliva was found to be about 1-2 Frey units/ml in many different lots. Table 1 summarizes the results of the main purification procedure. As starting material for purification, acetone dried powder of saliva was prepared by adding 3 volumes of acetone at pH 6.0 in an ice bath. This was the most suitable technique, among the methods examined, for preparing crude material for storage. The dissolved crude acetone powder was adsorbed on DEAE-cellulose at pH 7.0, while stirring it for 1.5 hours at 25°C; the adsorbed material was then eluted in a funnel with 0.6 M ammonium formate, pH 6.0, to yield almost the total activity of the starting acetone powder. Purification by adsorption on CM-cellulose and by acrinol precipitation was also tested; here a

TABLE 1
Purification of Human Mixed Salivary Kallikrein

Fractions or procedures	(FU/ml)	Purity index (FU/mg or FU/E_{280})	Purification factor
Human mixed saliva	0.92-2.25*	0.156	1
Crude acetone powder	1.71	1.42	9.1
DEAE-C adsorption, pH 6.8; eluate 0.6 M ammonium formate, pH 6.0	1.31	1.72	11
Acetone fractionation 55-60% (v/v)	0.6	22	141
Gel filtration			
1. Sephadex G-75	0.39	35-64	218-465
2. Sephadex G-100	0.2	365†	2,300

These values are averages obtained from more than 3 experiments except final gel filtration.

*More than 2 or even 3 FU/ml was found in few exceptions.

†This fraction still has 2 or 3 components by electrophoretic analysis on cyanogum 41 gelling agent.

somewhat higher purification was obtained. However, the yield of recovery varied and was inconsistent. The DEAE-cellulose eluate was purified further by acetone fractionation as described previously (Moriya et al., 1958; 1963). As shown in Table 1, about 50% of the kallikrein was recovered with a 15- to 26-fold increase in purity in the 55-60% acetone fraction, with a few exceptions.

Two Sephadex gel filtrations were carried out; the first with Sephadex G-75 with the material obtained by acetone fractionation, the second with Sephadex G-100 with the fractions pooled after the first filtration. Figure 2 shows the distribution of the protein and of the activity after the second filtration; the

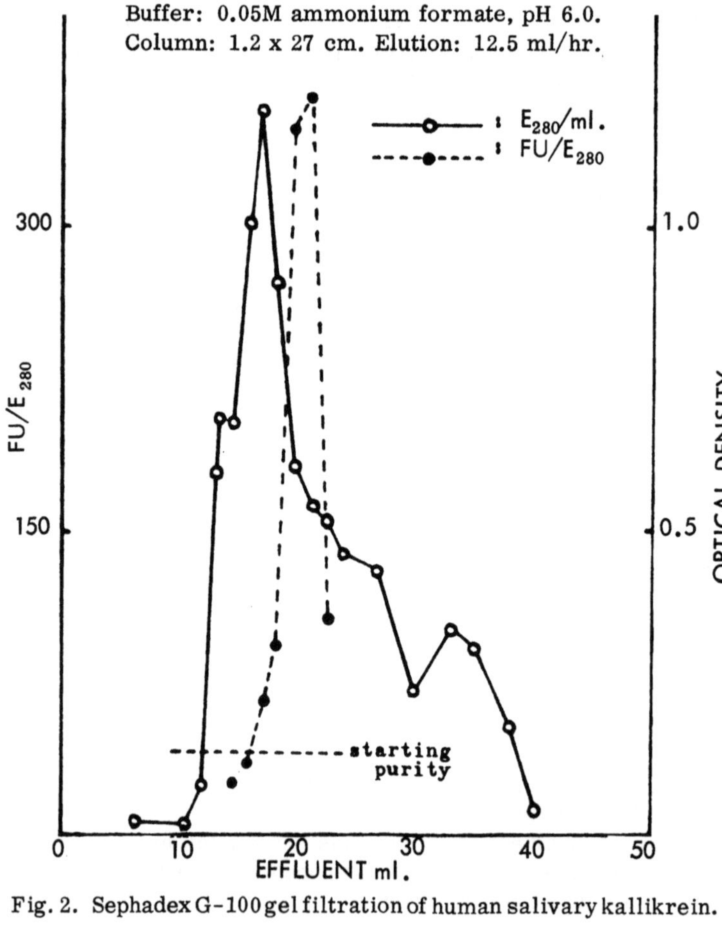

Fig. 2. Sephadex G-100 gel filtration of human salivary kallikrein.

TABLE 2
Liberation of Vasodilator Activity from Bovine Colostrokininogen

	Incubation with			Vasodilator
	Bovine crude colostrokininogen (mlEq original colostrum)	Same fraction from normal milk (mlEq normal milk	Human salivary kallikrein (total FU)	Activity (corresponds to μg syn. B/1 mlEq original saliva
Colostrokininogen + Human Salivary kallikrein	100		233	0.53
Control groups				
Colostrokininogen	100			0.0538*
Human salivary kallikrein only			233	0.0091*
Normal milk		60	233	0.0363*
Normal milk only		60		0.0054

Incubation at 37°C, 2 hours, pH 7.4–7.5.

*Small responses were observed at high concentration of sample, because of the quite high sensitivity of the instrument.

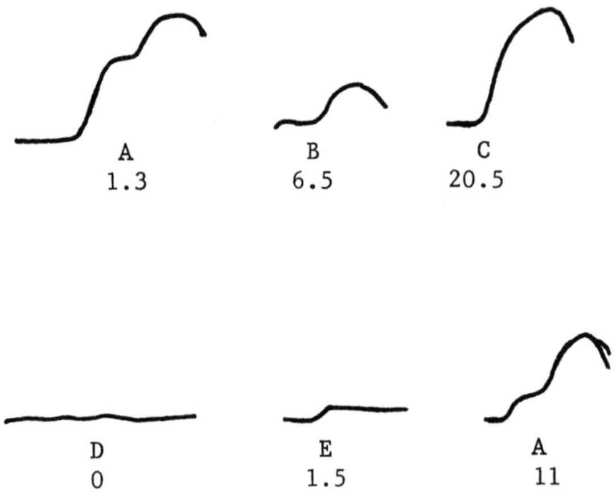

Fig. 3. Contraction of rat uterus by bovine colostrokinin.

Bath: 50 ml, at 24°C, modified Locke-Ringer's solution.

A: Syn. bradykinin 0.05 μg

B: Syn. bradykinin 0.025 μg

C: Colostrokininogen 0.625 ml. The equivalent of original colostrum was incubated with kallikrein, 1.45 FU.

D: Kallikrein, 7.05 FU incubated alone.

E: Colostrokininogen alone 2.0 ml Eq of original colostrum.

Each number shows the height of response.

material had 365 FR/E_{280} activity and a 9-fold-increased purification factor. The peaks of protein concentration and of activity were found in different fractions; such a pattern was quite reproducible within our experimental conditions. However, as indicated in Fig. 2, this highly purified fraction was found to be not yet homogeneous in cyanogum electrophoretic analysis.

Liberation of colostrokinin

As shown in Table 2 and Fig. 3, potent colostrokinin activity was found only when colostrum substrate was incubated with human salivary kallikrein. It was not found with milk substrate or when colostrum was incubated without kallikrein. To judge from these results, the kinin-liberating system seemed to be quite specific in the colostrum. This was also suggested by the electrophoretic patterns of normal and colostric proteins. A fraction precipitated from the treated colostrum (0-0.3 saturated with ammonium sulfate) showed a pattern relatively different from the normal substrate; and this fraction yielded almost all the kinin activity of the original colostrum. Therefore it can be concluded that the substrate colostrokininogen which is in the globulin fraction is a relatively specific colostric protein.

TABLE 3
Ratio of Responses to Colostrokinin and Bradykinin in Three Assays

Bioassays	Vasodilatation, dog		Contraction: Rat uterus	Contraction: Rat ileum	
Doses used, (mlEq original colostrum)	0.04	0.08	6	3	4
Responses	7	15	29	18	22
Corresponds to μg bradykinin	0.010	0.017	0.049	0.575	0.695
Corresponds to μg bradykinin 1 mlEq original colostrum	0.248	0.210	0.008	0.192	0.176
	0.229 (mean)			0.184 (mean)	

Comparison of colostrokinin (crude) with bradykinin

Stabilities of both colostrokinin and bradykinin in acid or alkali were examined, and it was found by measuring vasodilatation and by using the rat uterus method that colostrokinin was much more stable than bradykinin. With trypsin, both were fairly stable, but colostrokinin was much more unstable than bradykinin was with chymotrypsin in the same type of incubation. It was also seen that the activity of bradykinin on the rat uterus was almost lost after 30-min incubation with human serum, but a little of that of colostrokinin still remained after 1-hour incubation under the same conditions. The different ratios of responses in the various bioassays were also observed; Table 3 summarizes the comparison of three bioassays. As shown in Table 3, ratio of colostrokinin to bradykinin activity on the isolated rat uterus was much different from the ratios obtained with dog vasodilator and rat ileum methods. According to these results, colostrokinin seems to be somewhat different from bradykinin itself.

I^{131}-labeled human salivary kallikrein

This was prepared to study the physiological role and behavior of human salivary kallikrein in the body. It was established that radioactivity was bound to the protein molecule, as shown by gel filtration and paper chromatography, and that the

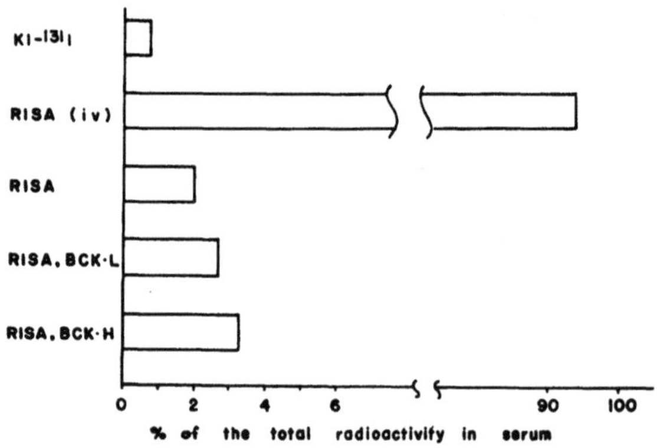

Serum was obtained 3 hrs. after RISA administration into intestinal sac.

RISA : ^{131}I-labelled human serum albumin (Dainabot Radioisotope Lab. Ltd.)

RISA (iv): Intravenous injection of RISA, serum was obtained 0.5 hrs. after RISA injection.

BCK : Bovine colostrokinin given with RISA.

L : Low dose, corresponds to 0.0025 μg bradykinin / head.

H : High dose, corresponds to 0.025 μg bradykinin / head.

Rats : Adult female, body weight 120 g, KI was given preliminarily, 4 – 8 rats for each group.

Fig. 4. Nondialyzable radioactivity in rat serum.

vasodilator activity of kallikrein in the protein fraction was quantitatively recovered. In preliminary experiments the following parameters were measured: instability against normal gastrointestinal juice *in vitro*; adsorption rate through intestine after administration in the intestinal sac of rats; distribution of radioactivity in liver, kidney, spleen, and serum; behavior of protein-bound radioactivity in serum in paper chromatography, gel filtration, and dialysis. Salivary kallikrein was not of special significance in these experiments. This might be due mainly to its instability in the gastrointestinal (GI) tract. However, the newborn animals usually have a different chemical milieu in the GI system, therefore the effect of colostrokinin on the absorption of I^{131}-labeled human albumin in rats was tested. Colostrokinin somewhat increased the absorption of protein-bound I^{131}, especially in newborn groups. Figure 4 shows the nondialyzable radioactivity in serum after absorption. Comparison of these values with those obtained without colostrokinin shows some increase in the absorption at the higher dose level of colostrokinin. However, more detailed experiments will be carried out in the future.

DISCUSSION

The kallikrein content in human mixed saliva varied somewhat in the different lots. Various proteins were also observed in the electrophoresis of crude acetone powder of saliva. During purification process, the reproducibility of methods was not definite sometimes, especially in the case of acrinol precipitation, CM-cellulose adsorption, and acetone fractionation. This must be due to the above-mentioned phenomena. To stabilize the kallikrein activity during purification, heating the crude kallikrein solution at 55°C for 5 min at pH 6.0 was also useful in destroying the "kallikreinase" present. The procedure did not have much effect on kallikrein itself.

As demonstrated by Webster and Pierce (1961), kallikrein has the ability to split TAMe esterase, thus the biological (FU) and esterase (EU) activities of the human salivary kallikrein were compared as well. While the ratio of EU/FU of crude acetone powder was a little less than 1, it usually decreased when the purified, fractionated form of DEAE-cellulose was used. This means that the ratio of activity between human salivary and urinary kallikreins is different.

It was observed that some of the proteins in colostrum originate from the plasma proteins of the mother (Askonas et al., 1954; Larson and Gillespie, 1957). When the colostrum is incubated with kallikrein, the resulting active peptide is different from bradykinin (Werle and Trautschold, 1960; Guth, 1959). The substance was named colostrokinin by Werle (1960). We had similar results when colostrokinin was liberated by partially purified human salivary kallikrein. Only crude colostrokinin was employed in our experiments for the differentiation of colostrokinin and bradykinin; therefore it is impossible to reach definite conclusions at the present stage. Additional studies of the purification of both colostrokininogen and colostrokinin and human salivary kallikrein will be carried out.

We became interested in clarifying the physiological role of salivary kallikrein, because kallikrein was found in the submaxillary gland first (Werle and Roden, 1936) and we wanted to isolate it from human material. In our present investigation, human salivary kallikrein was actually unstable in the milieu of the GI tract; however, this does not always hold true for newborn animals. Moreover, it is known that even such an unstable proteinase as the amylase can still have physiological significance after passing through the stomach. Johnson and Pierce (1959) showed the increased immune lactoglobulin content of newborn bovine serum by both electrophoretic and ultracentrifugal analysis, after adding colostrum. Our present study examined the possibility of whether the active polypeptide colostrokinin has an effect on the absorption of proteins through intestinal wall. Some observations were made about the absorption of albumin. An increased effect on the absorption was tentatively established by our immunological studies besides those described here. However, albumin does not seem to be a specific substance. In order to clarify the details of absorption mechanism, it seem to be necessary to study the absorption of substances such as globulin by using graded doses of colostrokinin, and to check the absorbed radioactivity more frequently after administration, etc. These problems will be studied in our laboratory.

ACKNOWLEDGMENTS

The authors are grateful to Professor E. Werle for his helpful suggestion on this work. The technical assistance of Miss E. Sugisawa and Mr. S. Maeda is gratefully acknowledged. This investigation was supported in part by Public Health Service Research Grant RIHE 007641 and -02 (1963 and 1964) from the National Heart Institute, U.S.A.

REFERENCES

Askonas, B. A., P. N. Campbell, J. H. Humphrey, and T. S. Work: The source of antibody globulin in rabbit milk and goat colostrum. Biochem. J. 56, 597 (1954).

Guth, P. S.: Kinins produced from bovine colostrum by kallikrein and saliva. Brit. J. Pharmacol. 14, 549 (1959).

Johnson, P., and A. E. Pierce: Ultracentrifugal and electrophoretic studies on neonatal calf sera and maternal colosstrum. J. Hyg. 57, 309 (1959).

Larson, B. L., and D. C. Gillespie: Origin of the major specific proteins in milk. J. biol. Chem. 227, 565 (1957).

Moriya, H., M. Deguchi, and H. Asano: Chemical studies on kallikrein. I. The amount of kallikrein contained in various organs and its isolation and purification. J. Pharm. Soc. Japan 78, 346 (1958).

Moriya, H.: II, Purification with acrinol. Ibid. 79, 32 (1959a).

Moriya, H., and E. Shimazawa: III. Purification by zone electrophoresis. Ibid. 79, 374 (1959b); IV. The isoelectric point of kallikrein. Ibid. 79, 688 (1959c).

Moriya, H.: V. Stability of water solution of kallikrein. Ibid. 79, 1216 (1959d); VI. Amino acid composition and terminal amino acids. Ibid. 79, 1390 (1959e); VII. Structural sugar of kallikrein. Ibid. 79, 1394 (1959f); VIII. Molecular weight of kallikrein and the supplement. Ibid. 79, 1451 (1959g).

Moriya, H., J. V. Pierce, and M. E. Webster: Purification and some properties of three kallikreins. Ann. N.Y. Acad. Sci. 104, 172 (1963).

Sarnoff, L. C., K. E. Sussman, and S. J. Sarnoff: Observations on the vasodilator properties of urine. II. The problem of reproducibility in blood flow bioassay technics for vasoactive substances: Studies with human urine. Circulat. Res. 6, 526 (1958).

Webster, M. E., and J. V. Pierce: Action of the kallikreins on synthetic ester substrates. Proc. Soc. exp. Biol. (N.Y.) 107, 186 (1961).

Werle, E., and P. von Roden: Über das Vorkommen von Kallikrein in den Speicheldrüsen und in Mundspeichel. Biochem. Z. 286, 213 (1936).

Werle, E.: Kallikrein, kallidin and related substances in Polypeptides Which Affect Smooth Muscle and Blood Vessels, M. Schachter, Ed., p. 199. Pergamon Press (1960).

Werle, E., and I. Trautschold: Colostrokinin und seine Bildung durch Kallikreine. Z. biol. 112, 170 (1960).

DISCUSSION

Guth: We reported five years ago the production of bovine colostrokinin using bovine saliva and bovine colostrum and suggested then that colostrokinin might function to enhance the intestinal absorption of α-globulin in the newborn. This mechanism would be of interest only in those species in which immunity is conferred via the colostrum rather than transplacentally. Does the newborn rat receive immunity transplacentally or via the colostrum? It might also be of importance to produce colostrokinins from homologous saliva and colostrum.

Moriya: I think species specificity must be important for such a study. We did not reach a final conclusion regarding immunity. However, a little effect of the absorption of high molecular protein was observed in rats.

Back: Was the enzymatic activity of kallikrein altered by the iodination procedure? To what extent was the enzyme iodinated?

Moriya: We have made sure that no loss of kallikrein activity occurred (vasodilatation). Because of very small amounts, we could not test it with TAMe or BAEe; 2% or 3% of the enzyme was iodinated.

Trautschold: Different colostrokinins are liberated by the various kallikreins from pancreas, submaxillary gland, urine, and serum as shown by comparing them in blood pressure tests. (See meeting in New York.) The kininase activity in colostrum is somewhat lower than in serum, but during the incubation time of 2 hours at 37°C. colostrokinin would certainly be destroyed if kininases are not inhibited.

Moriya: We used only salivary kallikrein as colostrokinin-liberating enzyme.

STUDIES OF THE SPECIFICITY OF KININ-FORMING ENZYMES

The Inhibition of the Kininogenic Activity of Trypsin and Kallikreins By Model Compounds

CARLOS R. DINIZ, ANIBAL A. PEREIRA, JADER BARROSO, and MARCOS MARES-GUIA

Departamento de Bioquimica, Faculdade de Medicina, Universidade de Minas Gerais, Belo Horizonte, Brazil

Enzymes of diverse origin form kinin when incubated with plasma globulins. Trypsin has been used as a model for kinin-releasing enzymes since the discovery of bradykinin by Rocha e Silva et al. (1949).

The application of small-molecular, exactly defined chemical compounds, such as amidines and guanidines as competitive inhibitors of trypsin (Mares-Guia and Shaw, 1965), or a specific reagent for the active center of trypsin such as TLCK (Shaw et al., 1965; Mares-Guia and Shaw, 1963), led us to study the effects of these substances upon kinin-releasing enzymes. Our aim was to better determine their relationship to trypsin in terms of specificity and mechanism of action.

MATERIALS AND METHODS

Trypsin was a twice-crystallized, lyophilized, salt-free preparation from Worthington Biochemical Corp. (lot 6223). Benzamidine—HCl·$2H_2O$ was a product from Aldrich Chemical Co. TLCK, L-(1-chloro-3-tosylamido-7-amino-2 heptanone) was kindly supplied by Dr. Elliott Shaw. N-α-Benzoyl-L-arginine ethyl ester hydrochloride (BAEe) was purchased from Mann Research Laboratories. Horse urinary kallikrein was a purified preparation kindly supplied by Dr. Eline Prado. Rat urine, extensively dialyzed against deionized water in the cold, was used as source of kallikrein. Human urinary kallikrein was purified according to Moriya et al. (1963).

The esterase activity of the enzymes, as measured with BAEe, was determined by the color reaction described by Hestrin (1949). To 0.20 ml of a 0.01 M stock BAEe solution in a test tube was added 0.70 ml of 0.05 M Tris buffer, pH 8.0. After 5 min allowed for thermal equilibration at 37°C, 0.10 ml of the enzyme solution was added with mixing, and a timer was started. After a suitable time interval the reaction was interrupted by the addition of 0.10 ml of 50% TCA, and the amount of remaining ester was determined. During the interval of incubation the kinetics followed were zero order. The spontaneous ester hydrolysis was assayed in control experiments. The results are expressed in $M.sec^{-1}$ (mg protein/ml)$^{-1}$. The kinin-releasing activity was measured with a partially purified bradykininogen (sp. act. = 8.1 units/mg) (BKGN) prepared from bovine plasma as described by Habermann et al. (1963). In a typical run, 2 ml of a BKGN solution in 0.9% (w/v) NaCl was added to 6 ml of 0.2 M Tris, pH 8.0, containing the enzyme, or the enzyme plus inhibitor, and a timer was started. At 2, 4, and 6 min, or as otherwise indicated, a 2.0-ml sample of the incubation mixture was pipetted into a centrifuge tube containing 5 ml of 95% ethanol at 70°C, and the tube left at this temperature for 10 min. After centrifugation, the supernatant was transferred to a round-bottom flask and the solvent removed in a rotary evaporator. The residue was dissolved in 2 ml of 0.9% NaCl and the amount of bradykinin determined with the guinea pig ileum preparation described by Rocha e Silva (1952). The amount of bradykinin released in the incubation mixture was then plotted as a function of time, and a tangent to the curve yielded the initial rates of kinin release. In the experiments with TLCK, the method used was similar to that described by Shaw et al. (1965). To 2.0 ml of enzyme solution in 0.05 M barbital buffer of pH 8.0, 0.20 ml of a TLCK solution in the same buffer was added, and a timer was started. At selected time intervals, 0.10-ml aliquots of the incubation mixture were assayed for the esterase activity, and at the end of the incubation the kinin-releasing activity was also determined. Protein determinations were carried out by absorption at 2,800 Å or by the method of Lowry et al. (1951), with serum albumin as standard.

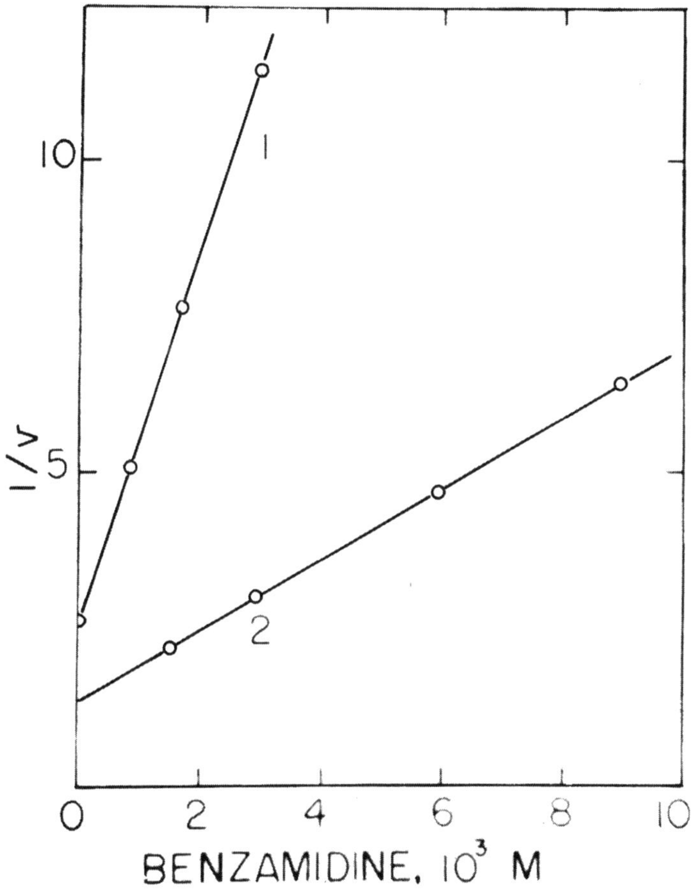

Fig. 1. The effect of benzamidine-HCl on the BAEe-esterase activity of rat and horse urinary kallikreins. Ordinate: reciprocal values of initial rate, in liters per mole × 10^{-5} sec. Abscissa: molar concentration of benzamidine-HCl times factor. Curve 1: horse urine kallikrein, 0.020 mg protein/ml. Curve 2: rat urine kallikrein, 0.09 mg protein/ml. Tris, 0.05 M, pH 8.0, 37°C, 5-min incubation. Data plotted according to Dixon (1953).

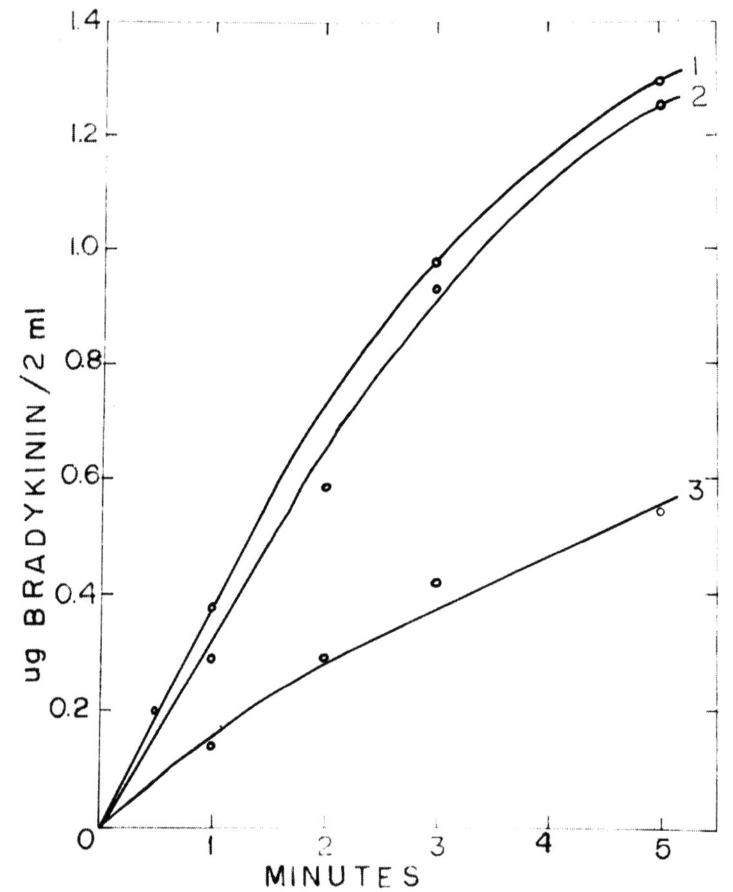

Fig. 2. The kinin-releasing activity of rat urine kallikrein: effect of benzamidine-HC1 and TLCK. Curve 1: enzyme preincubated with TLCK, 1.0×10^{-3} M. Curve 2: control. Curve 3: enzyme plus benzamidine-HC1, 8.85×10^{-4} M. Incubation with kininogen in Tris, 0.2 M, pH 8.0, 37°C.

RESULTS

The BAEe-esterase activity of rat and horse kallikreins was inhibited by benzamidine-HCl, as shown in Fig. 1. As al-

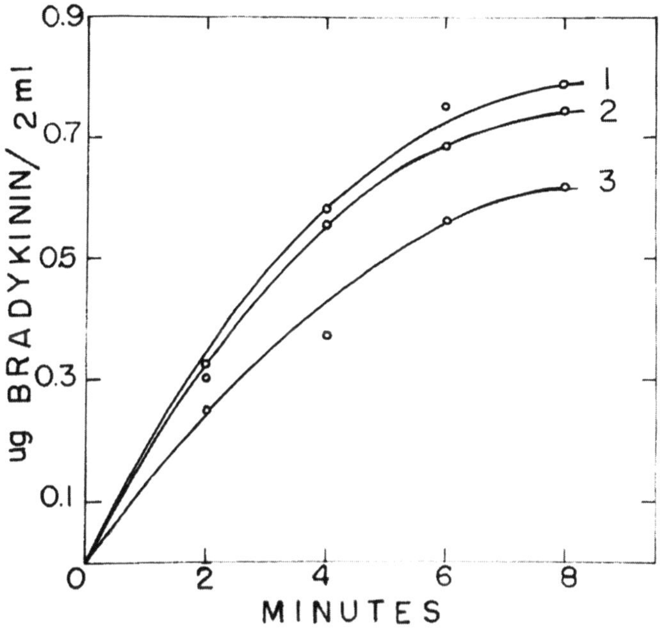

Fig. 3. The kinin-releasing activity of horse urine kallikrein: effect of benzamidine-HC1 and TLCK. Curve 1: control. Curve 2: enzyme preincubated with TLCK, 0.5×10^{-3} M. Curve 3: enzyme plus benzamidine-HC1, 8.85×10^{-4} M. Incubations with kininogen as in Fig. 2.

ready indicated (Diniz et al., 1965), the higher inhibition observed with the horse urine enzyme might be due to lower affinity of the enzyme for BAEe. When these enzymes were incubated with bradykininogen, their kinin-releasing activities were also inhibited by benzamidine-HCl, as can be seen from Figs. 2 and 3. The release of kinin by human urinary kallikrein was not affected by benzamidine-HCl, (Fig. 4), but the BAEe-esterase activity in the preparations was inhibited by benzamidine-HC1, as pointed out previously (Diniz et al., 1965). Our experiments in which rat and horse urinary kallikreins were preincubated with TLCK demonstrated that the reagent did not affect the BAEE-esterase activity of the kinin-releasing activity, as shown in Fig. 2 and 3 respectively.

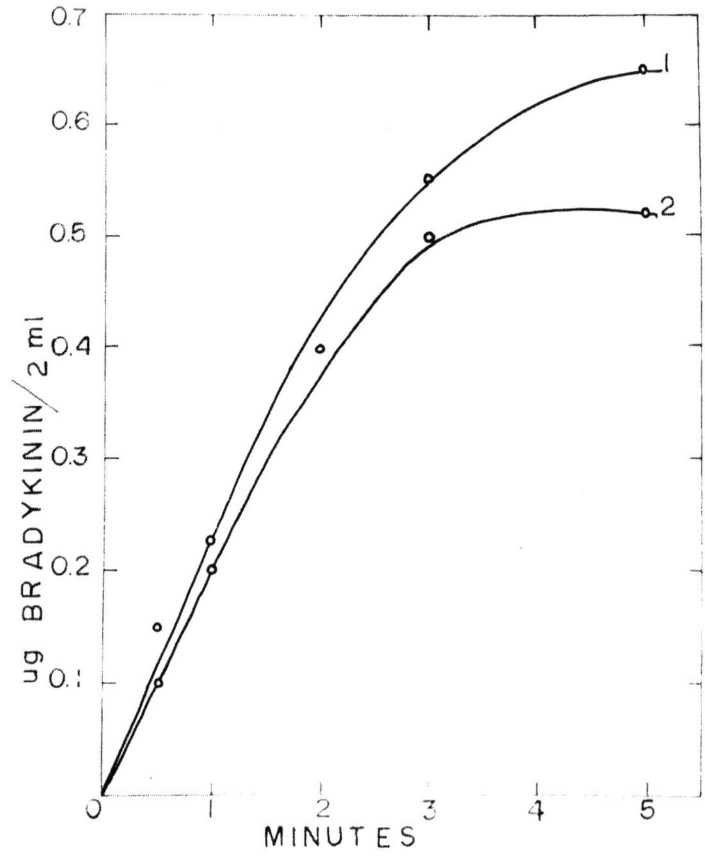

Fig. 4. The kinin-releasing activity of human urinary kallikrein: effect of benzamidine-HCl. Curve 1: enzyme plus benzamidine-HCl, 8.85 $\times 10^{-4}$ M. Curve 2: control. Incubation with kininogen as in Fig. 2.

Previously reported experiments (Diniz et al., 1965) demonstrated that the kinin-releasing activity of trypsin is affected by both benzamidine-HCl and TLCK.

DISCUSSION

Several guanidines and amidines, including benzamidine HCl, are good competitive inhibitors of trypsin (Mares-Guia and Shaw, 1965) when tested with a substrate of the amide type. As expected, benzamidine is also able to inhibit the proteolytic and esterolytic activities of trypsin when measured with casein and BAEE respectively. The kinin-releasing enzymes reported here have been known for some time to catalyze the hydrolysis of esters of N-α-acyl-L-arginine, as does trypsin. This places them in the group of trypsin-like enzymes, at least as far as specificity requirements go. The present work demonstrates that, although rat and horse urinary kallikreins are able to hydrolyze BAEe, a typical trypsin substrate, and are inhibited by benzamidine-HCl, a competitive inhibitor of trypsin, they differ from trypsin in one important respect. Whereas trypsin is rapidly inactivated by TLCK, a specific reagent for its active center (Shaw et al., 1965; Mares-Guia and Shaw, 1963), the kallikreins are not affected by the compound. Since TLCK was developed with the specificity requirements of trypsin as guide lines (Shaw et al., 1965), it is concluded that the trypsin-like kallikreins differ from trypsin in some aspects of their specificity requirements and/or in mechanism of action. It certainly cannot be said at the present time that the kallikreins reported here do not possess a histidyl residue important in their catalytic mechanism, for the lack of reaction with TLCK might as well be due to an unfavorable fitting into the enzyme, which would therefore preclude alkylation of the active center of the kallikreins and the consequent inactivation. Further study of this field is clearly required but, at least with some enzymes, it will have to wait until better purification procedures are found.

Although the results presented here show that rat and horse urinary kallikreins are different from trypsin, they confirm previous findings which allow the classification of these enzymes as trypsin-like, at least as far as the ability to hydrolyze N-α-acyl-L-arginine esters (Prado et al., 1961; I. Carvalho and C. R. Diniz, unpublished results) is concerned. To this, inhibition by competitive inhibitors of trypsin of well-defined chemical structure is added. These inhibitors are useful tools for obtaining much more information on the enzyme active center.

The data support the view that rat and horse urinary kallikreins, when acting on the kininogen, liberate bradykinin and not kallidin (lysyl-bradykinin), since the release of the latter would require the hydrolysis of a methionyl-lysyl peptide bond. This would be in contrast to their described specificity requirements. It is assumed here that the kininogen has a single kinin-containing sequence, which includes the partial sequence methionyl-lysyl-bradykinin (Elliott et al., 1963). This view is supported by the results of Habermann et al. (1963), who found that exhaustive reaction of the kininogen with one kinin-forming enzyme prevented kinin release by another enzyme.

On the other hand, our human urinary kallikrein preparation probably liberates lysyl-bradykinin, but not bradykinin, from the precursor. This view is supported by our finding that a kinin-releasing activity can be extracted from human urine free from BAEe-esterase activity (Diniz et al., 1965) and that the action of the enzyme on kininogen is not affected by benzamidine-HCl (Fig. 4). As a consequence, one is tempted to speculate that the bradykinin sequence is C-terminal in the primary structure of the precursor, for otherwise the latter type of kinin-releasing enzyme would have to catalyze hydrolysis of peptide bonds of the carboxyl groups of methionine and arginine, a conclusion hard to reconcile with known specificity properties of proteolytic enzymes.

The fact that rat and horse urinary kallikreins do not react with TLCK is taken as basis for distinction between trypsin-like kallikreins and trypsin. In addition, it demonstrates that enzymes with qualitatively identical specificity requirements might exhibit important quantitative differences or even different mechanisms of catalysis. A clear understanding of the lack of reaction of the kallikreins with TLCK has not yet been reached, but further work is being carried out on the subject.

ACKNOWLEDGMENTS

This investigation was supported in part by United States Public Health Service Research Grant TW-00127-1, from the National Institutes of Health. We are grateful to Dr. Elliott Shaw of Tulane University for a sample of TLCK, and to Drs. E. and J. L. Prado of the Escola Paulista de Medicina, São Paulo, for a sample of purified horse urinary kallikrein.

REFERENCES

Diniz, C. R., A. A. Pereira, J. Barroso, and M. Mares-Guia: On the specificity of urinary kallicreins. Biochem. biophys. Res. Commun., 21:448 (1965).

Dixon, M.: The determination of enzyme inhibitor constants. Biochem. J. 55, 170 (1953).

Elliott, D. F., G. P. Lewis, and D. C. Smyth: A new kinin from ox blood. Biochem. J. 87, 21P (1963).

Habermann, E., W. Klett, and G. Rosenbusch: Partielle Reinigung und einige Eigenschaften eines Kininogens aus Rinderblut. Hoppe-Seylers Z. physiol. Chem. 332, 121 (1963).

Hestrin, S.: The reaction of acetylcholine and other carboxylic acid derivatives with hydroxylamine, and its analytical application. J. biol. Chem. 180, 249 (1949).

Lowry, O. H., N. J. Rosebrough, A. L. Farr, and R. T. Randall: Protein measurement with the Folin-phenol reagent. J. biol. Chem. 193, 265 (1951).

Mares-Guia, M., and E. Shaw: The irreversible inactivation of trypsin by the chloromethyl ketone derived from N-α-tosyl-L-lysine. Fed. Proc. 22, 528 (1963).

Mares-Guia, M., and E. Shaw: Studies on the active center of trypsin. The binding of amidines and guanidines as models of the substrate side chain. J. biol. Chem. 240, 1579 (1965).

Moriya, H., J. V. Pierce, and M. E. Webster: Purification and some properties of three kallikreins. Ann. N. Y. Acad. Sci. 104, 172 (1963).

Prado, E. S., and J. L. Prado: Partial purification of plasma kinin-forming enzyme from horse urine. Experientia (Basel) 17, 31 (1961).

Rocha e Silva, M., W. T. Beraldo, and G. Rosenfeld: Bradykinin, a hypotensive and smooth muscle stimulating factor released from plasma globulin by snake venoms and by trypsin. Am. J. Physiol. 156, 261 (1949).

Rocha e Silva, M.: Method of assay of bradykinin upon the guinea pig gut. Acta physiol. lat.-amer. 2, 238 (1952).

Shaw, E., M. Mares-Guia, and W. Cohen: Evidence for an active center histidine in trypsin through the use of a specific reagent, TLCK, the chloromethyl ketone derived from N-α-tosyl-L-lysine. Biochemistry, in press (1965).

DISCUSSION

Prado: We have, indeed, observed that the proteolytic effect of horse urinary kallikrein on salmine is less marked than that of trypsin. Also the effect of Co^{++} is different on the two enzymes.

Diniz: Your preparation of kallikrein worked fine with benzamidine.

Back: Have you studied the action of plasmin in your system?

Diniz: We have not studied it yet, but I think Dr. E. Shaw is doing this.

DEMONSTRATION OF THE PRESENCE OF TWO SEPARATE KININ-FORMING SYSTEMS IN HUMAN AND OTHER PLASMA

W. VOGT

Department of Pharmacology, Max-Planck Institut für Experimentelle Medizin, Göttingen, Germany

Although the formation of plasma kinins by the various kallikreins in purified systems has been extensively studied and is well understood, their formation by endogenous mechanisms in whole plasma still poses unsolved problems which have incited a re-evaluation of the processes going on during activation of human plasma by contact with glass. The results have led to the conclusion that in human plasma two separate kinin-forming systems exist, each consisting of an enzyme and a substrate.

When human plasma is brought into contact with glass beads (0.1 mm diameter, 170 mg/ml, 5 min 37°C) it develops kinin activity in a few minutes and, unless precautions are made, the kinin is destroyed subsequently (Armstrong et al., 1957). A second glass activation will not produce kinin again (Margolis, 1958). This is due neither to a general depletion of (1) substrate nor (2) of enzyme, for the following reasons.

(1) When trypsin is added to glass-activated plasma, strong kinin activity develops. This was first demonstrated by Keele (1960). Eisen (1963), in an attempt to find an explanation, suggested that an incomplete loss of kininogen after the first glass contact might be sufficient to slow down the kinin production during the second activation, to an extent that no measurable level could develop, because of the simultaneously proceeding continuous kinin destruction. This explanation was rendered unlikely by the recent investigations of John and Vogt (1965), who found that plasma loses after glass contact less than 50% of its kininogen.

The kininogen was estimated as kinin liberated after treatment with pancreas kallikrein, in the presence of O-phenanthroline. Before incubation the plasma samples were kept for 30 min at pH 2 to 3, at 37°C. When the kininogen was measured by the method of Diniz et al. (1961), the loss of kininogen after glass contact was still smaller.

Moreover, when any destruction of possibly formed kinins is completely prevented by addition of O-phenanthroline-HCl (1 mg/ml) (Erdös and Sloane, 1962; John, to be published), again no kinin formation is seen upon a second glass contact in human plasma, within a reasonable time (30 min at 37°) (Table 1).

TABLE 1
Kinin Formation and Kininogen Content in Human Plasma

Plasma, treatment	Kininogen content*	Kinin formation: On 2nd glass contact	Kinin formation: "Spontaneous"
Fresh	6.2	+	
After glass contact	3.3	–	
After glass contact, kininogen added (2.1 μg)	5.4†	–	
Acid-treated plasma, (diluted 1:2.2)			0.9
Acid-treated, diluted, kininogen added (2.1 μg)			2.4

*The kininogen content is given in bradykinin equivalents (μg/ml) liberated after incubation of acid-treated, phenanthroline-containing samples with pig serum kallikrein. Spontaneous kinin formation was measured as μg bradykinin/ml diluted solution, formed after 2-hour incubation at 37°C.

†This figure was calculated from that above and that to the left.

Even when the previously glass-activated plasma is enriched with pure kininogen from ox serum (preparation kindly obtained from Prof. Habermann, Würzburg) no kinin activity appears after a second glass contact. On the other hand, this kininogen preparation is utilized by acid-treated human plasma for kinin production (Table 1). Its failure to serve as a substrate in glass-treated plasma is, therefore, not due to its foreign nature. These findings demonstrate that a general loss of kininogen cannot account for the lack of kinin formation after a second activation of human plasma.

(2) That a kinin-forming enzyme is still present in glass-activated plasma has been shown by Margolis (1958), who found that such plasma induces kinin liberation when transferred to fresh human plasma not brought into contact with glass.

Thus an enzyme as well as a substrate can be demonstrated in glass-activated plasma, but nevertheless no kinin is formed. These seemingly contradictory facts led Margolis and Bishop (1962) to postulate a third compound essential for kinin formation in addition to enzyme and substrate — component B. It was assumed to be an accelerator. Recently Margolis and Bishop (1963) recognized it as a special kind of substrate that becomes exhausted during glass activation. The latter view is in accord with our own findings, but it still leaves unsolved problems concerning the events going on in plasma during contact activation.

First, if purified bovine kininogen and purified pig serum kallikrein (both preparations kindly provided by Prof. Habermann) are mixed and incubated, they liberate kinin in a quick reaction that needs no cofactor or special substrate. Second, when serum kallikrein is added to glass-activated human plasma, it does liberate kinin despite the lack of component B. This indicates that serum kallikrein does not need component B for kinin formation and that in glass-activated plasma no plasma kallikrein is present. Then the question arises — what else does the kinin-forming activity of glass-activated plasma consist of? It is not merely a soluble fraction of active Hageman factor. This is shown by the fact that glass-activated human plasma induces kinin formation in nonactivated plasma that has been heated to 61°C for 30 min (Eisen, 1963, and our own experiments). Since, in the heated specimen, no precursor of activatable serum kallikrein is left, any Hageman factor contained in the glass-treated sample could not induce kinin formation. Consequently, glass-activated plasma contains a kinin-forming enzyme active by itself, but this is not plasma kallikrein.

This conclusion entails another one. Apparently, the other enzyme does not act on the kininogen that is still present in plasma after glass activation, nor on the purified bovine kininogen. It needs another substrate, present in fresh human plasma but not in glass-treated samples. Thus human plasma contains, in addition to plasma kallikrein and kininogen, a second kinin-forming system. A suggested term for the other enzyme is kininogenase II and its substrate kininogen II. Correspondingly, one might call the plasma kallikrein kininogenase I and its substrate kininogen I. The use of the term kallikrein would then be restricted to the glandular kinin-forming enzymes, whose properties differ from serum kallikrein in any case.

Kininogenase II can be demonstrated in an active state in glass-activated plasma. The existence of the substrate kininogen II is evident indirectly, so far, by its absence from glass-activated plasma. One can, however, also demonstrate its existence directly. To human plasma one unit pancreas kallikrein/ml was added five times, and each addition was followed by incubation for 10 min at 37°C. This treatment led to a complete depletion of kininogen I. After standing for 1 hour at 37°, after the last incubation all kinin activity formed had disappeared. Then *O*-phenanthroline (1 mg/ml) was added to prevent any destruction of kinin to be formed during subsequent procedures. New addition of pancreas kallikrein to this plasma did not release kinin. The plasma did, however, still liberate kinin when activated by contact with glass. This demonstrates that kininogen II and activatable kininogenase II were still present, and that pancreas kallikrein does not liberate kinin from kininogen II.

From these findings the events in human plasma during activation by contact with glass can be explained as follows. With the aid of Hageman factor (Margolis, 1960) the kininogenase II is activated; it converts all kininogen II into a kinin and remains active afterward for some time. More of it can be activated by a second glass contact. Kininogenase I is also activated but is subsequently destroyed so quickly that only less than 50% of kininogen I is split before this enzyme is entirely lost. Consequently, glass-activated human plasma contains a kinin-forming enzyme, kininogenase II, and a kininogen, kininogen I. Since these two are not components of the same system, no kinin is formed.

It is surprising that kininogenase I should be so quickly inactivated in human plasma. In fact, this occurs only shortly after plasma has been activated by glass contact (for the first or second time). When serum kallikrein is added immediately afterward (1 unit/ml), it will not liberate kinin from kininogen I which is still present. If, however, 1 hour is allowed to pass between glass activation and addition of serum kallikrein, the kallikrein remains stable enough to produce kinin activity. It is tempting to assume that it is kininogenase II itself that inactivates kininogenase I, but this assumption is unproved.

Possible interactions of the two kinin-forming systems have been studied by experiments in which rat or dog plasma was used in addition to human plasma. The conclusions drawn are regarded as valid under the following premises: Pancreas kallikrein attacks only kininogen I, kininogenase II attacks only kininogen II. Active kininogenase II but not I is present in human glass-activated plasma ("B-depleted" plasma prepared according to Margolis, 1958). Differences in the kinin-forming systems of various species are due to differences in their enzyme and substrate composition rather than to individual specificities. Corresponding enzymes and substrates are assumed to be interchangeable among species.

Under these premises one can explain some peculiar features of dog plasma which throw some light on the interactions of the two systems. Dog plasma forms kinin when incubated with pancreas kallikrein, whether or not the plasma is fresh or heated. Thus it contains kininogen I. Glass-activated human plasma does not liberate kinin or only traces from heated dog plasma (61°C, 30 min). Consequently, dog plasma is a poor source of kininogen II. In regard to enzymes, dog plasma contains kininogenase I: it is activated by acid, after neutralization forms kinin "spontaneously" from the kininogen I present. When glass-activated human plasma is allowed to act on fresh dog plasma it will form kinin, in contrast to its ineffectiveness in heated preparations. Apparently, kininogenase II of human glass-treated plasma activates the kininogenase I of dog plasma and by this indirect mechanism liberates kinin.

It is a well-known peculiarity of dog plasma that it does not form kinin when brought into contact with glass. Even glass beads coated with Hageman factor from human plasma (Margolis, 1960) are without effect, at least when the globulin fraction of dog plasma is used (Vogt, 1964) which contains the complete system I. This means that kininogenase I cannot be activated directly by glass or Hageman factor. What is activated in human plasma, primarily, must then be kininogenase II. This enzyme is apparently absent from dog plasma, which consequently contains only the complete system I. The lack of kininogenase II explains why dog plasma is not activated by contact with glass.

In rat plasma only system II is complete. After glass contact, rat plasma liberates large amounts of kinin. On the other hand, it does not form kinin when incubated with pancreas kallikrein (Fasciolo and Halvorsen, 1964; Jensen et al, 1965; and our own observation). Thus there is no kininogen I present in rat plasma. However, it possibly contains some kininogenase I, because after treatment with acid, it is able to liberate kinin when mixed with heated dog plasma. This result is not yet definite, as some of the kinin formed in that mixture may have originated from the sample of rat plasma. Human, dog, and rat plasmas are suitable tools for the study of kinin formation, since they represent solutions containing both or only one or the other of the two complete kinin-forming systems.

Trypsin in a concentration of 10^{-4} forms kinin from fresh rat plasma, but not when the plasma had been heated to 61°. From this we concluded that trypsin activates kininogenase II and does not act directly on kininogen II (Vogt, 1965). However, John (unpublished) recently found that at a concentration of 10^{-3}, trypsin does form kinin from heated rat plasma. Probably the ineffectiveness of the lower concentration of trypsin can be explained by the presence of trypsin inhibitors which persist in heated rat plasma. The life span of trypsin 10^{-4} then would be just sufficient to activate kininogenase II in fresh rat plasma but not to convert the major portion of kininogen II itself. In heated rat plasma, there is no kininogenase to be activated, thus trypsin (10^{-4}) seems inactive. The enzyme inhibition can be overcome by increasing its concentration to 10^{-3}.

The possible interaction of kininogenase I with system II is not yet clear. From fresh or heated rat plasma, kininogenase I liberates only traces or no kinin, when added in amounts of 1 unit/ml. On the other hand if human plasma is incubated several times with kininogenase I, it cannot afterward be induced to form kinin on contact with glass. It may be that kininogenase I slowly attacks kininogen II or that it activates kininogenase II. Instead of a biochemical incompatibility existing between enzyme and substrate, it may be that kallikrein inhibitors in rat plasma prevent it from acting specifically on kininogen II. On the other hand, the slight effect of pig serum kallikrein preparations on System II may be due to contamination with kininogenase II.

The scheme presented below shows the two kinin-forming systems and their possible interactions as they appear at present.

Plasma contains kininogenase I as an inactive proenzyme which can be activated by acid but not by glass. Its substrate,

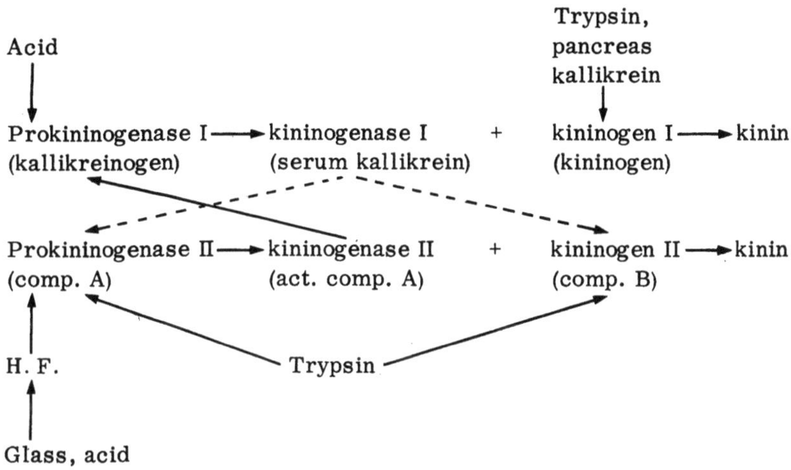

kininogen I, is also used by pancreas kallikrein and trypsin. Prokininogenase II, identical with component A (Margolis, 1958) is activated by glass, acid, or trypsin. The enzyme liberates kinin from kininogen II which is synonymous with component B. This enzyme also activates and possibly inactivates kininogenase I. Trypsin can, besides activating kininogenase II, liberate kinin directly from kininogen II.

REFERENCES

Armstrong, D., J. B. Jepson, C. A. Keele, and J. W. Stewart: Pain-producing substances in human inflammatory exudates and plasma. J. Physiol. (Lond.) 135, 350 (1957).

Diniz, C. R., I. F. Carvalho, J. Ryan, and M. Rocha e Silva: A micromethod for the determination of bradykininogen in blood plasma. Nature (Lond.) 192, 1194 (1961).

Eisen, V.: Observation on intrinsic kinin-forming factors in human plasma: The effect of acid, acetone, chloroform, heat and euglobulin separation. J. Physiol. (Lond.) 166, 496 (1963).

Erdös, E. G., and E. M. Sloane: An enzyme in human blood plasma that inactivates bradykinin and kallidins. Biochem. Pharmacol. 11, 585 (1962).

Fasciolo, J. C., and K. Halvorsen: Specificity of mammalian kallidinogen. Amer. J. Physiol. 207, 901 (1964).

Jensen, K. B., S. F. Rinvik, and A. M. Venneröd: Experiments on the in vitro inactivation of plasma kinins by carboxypeptidase B, plasmakininase or erythrocyte kininase in the presence of Disulfiram. Acta pharmacol. (Kbh.) 22, 187 (1965).

John, G., and W. Vogt: Kininbildung in frischem und glasaktiviertem menschlichen Plasma. Naunyn-Schmiedeberg's Arch. exp. Path. Pharmak. 251, 186 (1965).

Keele, C. A.: Formation of pain-producing polypeptide (PPS) from plasma, in: Polypeptides Which Affect Smooth Muscles and Blood Vessels, M. Schachter, Ed. p. 253 Oxford: Pergamon Press (1960).

Margolis, J.: Activation of plasma by contact with glass: Evidence for a common reaction which releases plasma kinin and initiates coagulation. J. Physiol. (Lond.) 144, 1 (1958).

Margolis, J.: The mode of action of Hageman factor in the release of plasma kinin. J. Physiol. (Lond.) 151, 238 (1960).

Margolis, J., and E. A. Bishop: Interrelation between different mechanisms of release of biologically active peptides from blood plasma. Nature (Lond.) 194, 749 (1962).

Margolis, J., and E. A. Bishop: Studies on plasma kinins. I. The composition of kininogen complex. Austr. J. exp. Biol. med Sci. 41, 293 (1963).

Vogt, W.: Kinin formation by plasmin, an indirect process mediated by activation of kallikrein. J. Physiol. (Lond.) 170, 153 (1964).

Vogt, W.: Ein komplettes zweites kininbildendes System in menschlichem Plasma. Naunyn-Schmiedeberg's Arch. exp. Path. Pharmak. 251, 186 (1965).

DISCUSSION

Erdös: Where does the permeability factor fit in your scheme?

Vogt: This is not yet known. As it activates serum kallikrein it may well contain kininogenase II; it seems also to contain Hageman factor-like material.

Armstrong: Human plasma does indeed contain another kinin-forming enzyme which can be distinguished from both plasma kallikrein and from plasmin. My colleague Dr. G. L. Mills has made a highly purified preparation of this enzyme and we have reported several observations which are in accord with those of Dr. Vogt.

The substrate attacked is different from that attacked by human plasma kallikrein because on addition of our human kinin-forming substance to fresh, intact human plasma it does not produce bradykinin or kallidin but a kinin distinguishable from both of these and which we have called "kinin E."

As I have reported at this symposium, the human plasma kinin-forming enzyme which occurs, normally, mainly in its precursor form, and which can be inhibited by the SBTI, specifically attacks a kininogen which is different from the trypsin substrate. I wonder if this corresponds to Dr. Vogt's kininogenase II. Like Dr. Vogt's enzyme and that which is activated by glass, the cold-activated kininogenase appears to be Hageman factor dependent at some stage in its operation, since the very fresh Hageman-deficient plasma we examined did not produce kinin in the cold.

Rat plasma which is an excellent substrate for the effects of glass also responds to cold as does human plasma (i.e. the normal rat plasma slowly digests its kininogen and that of the parturient rat does so rapidly). Thus again there is resemblance to Dr. Vogt's number II enzyme.

On the other hand, the dog plasma, which does not respond to glass, does not respond to our purified enzyme. It appears likely that the dog plasma, which lacks Dr. Vogt's enzyme II, is also lacking in the substrate for his enzyme. He has told me that when he has looked for dog kininogen II it was almost absent from the plasma, and when my colleagues and I have measured the activity of dog plasma in comparison with human plasma as a substrate for our enzyme we have found it very low indeed – around 5-7%.

Habermann: Which of the two systems mentioned is present in pig plasma? This would be of interest since the kallikrein used is prepared from pig serum or pancreas.

Vogt: From the behavior of pig plasma in various procedures of kinin formation, I would assume that it contains both systems.

Margolis: Does the acid activation of pro-kininogenase I depend on the presence of Hageman factor? In our experience HF-deficient plasma could not be activated by treatment with acids.

Vogt: We have not tried HF-deficient plasma.

Edery: I would like to ask if the two kinins formed from different precursors are the same or not.

Vogt: My colleague Dr. John was able to separate the kinin formed from kininogen II, from kallidin and from the hendecapeptide in paper chromatography. It was not separated from bradykinin. The experiments of Habermann do not show any difference from bradykinin. This would mean that both kinins are identical, and that formed by system I is also bradykinin.

V. EISEN

Department of Pharmacology
The Middlesex Hospital Medical School
London, England

Professor Vogt's admirable study has made it difficult to avoid the conclusion that there is more than one kinin-forming system in plasma —however hard one tries to follow Occam's Entia praeter necessitatem non multiplicanda sunt. I should like to mention two findings which relate to the relationship between the two systems: (1) The question has been raised whether factor XII is involved in kinin formation after acid treatment of plasma, and we believe that this is so (Eisen and Keele, 1961). (2) In plasma which has been depleted of kininogen II (component B) by glass, renewed exposure to glass will accelerate kinin formation induced by other means. Samples of previously "B-depleted" plasma were activated by mixing with glass powder for 2-3 min, and immediately treated with pH 2 for 15 min. After re-neutralization, they formed kinin faster than did control samples of B-depleted plasma which had been acid-treated without the activating glass contact.

Now that the number of kininogenases in plasma is growing, it may be useful to reconsider whether plasmin also acts on kininogen, and compare it with plasma kallikrein (Fig. 1). Purified human plasma

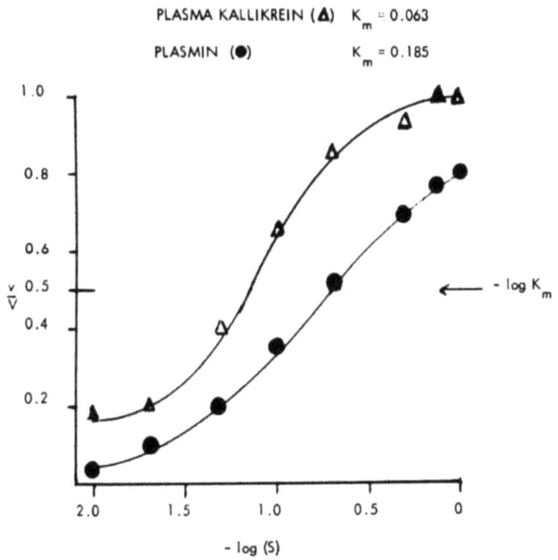

Figure 1

kallikrein was kindly given by Webster and Pierce (1961), and human plasmin by the American Red Cross who prepared it according to

Sgouris et al. (1960). This method includes prolonged exposure to acidity at 0° and 60°C, which is presumed to destroy plasma kallikrein.

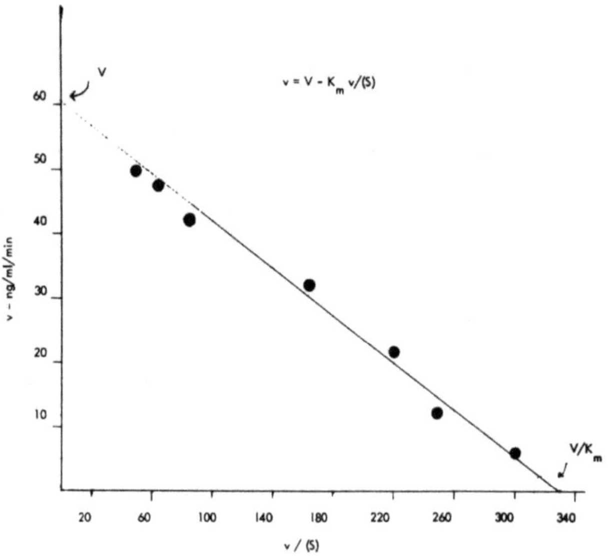

Figure 2

The substrate was human plasma in which kinin-forming and -destroying factors had been abolished by heat (61°C/60 min), acidity (pH 2/15 min), and ϵ-amino caproic acid. The substrate concentration (S) in undiluted heated plasma was denoted as 1, and the negative logarithm of (S) plotted on the abscissa against the ratio of the measured rate of kinin formation (V) over maximal rate of formation (v) on the ordinate. At higher (S), plasma kallikrein approached V and therefore zero-order kinetics; for plasma, V had to be obtained by extrapolation by means of Eadie's method (Fig. 2) in which v is plotted against v/(S) according to the equation $v = V - K_m \cdot v/(S)$. The Michaelis constants (K_m) of the two enzymes could then be assessed, and showed that plasmin has a lower affinity for kininogen than plasma kallikrein. Basic kinetic analyses of this type will help to establish and differentiate kininogenases old and new.

REFERENCES

Eisen, V., and C. A. Keele: Intrinsic plasma kinin-forming factors in human plasma. J. Physiol. (Lond.) 154, 20P (1960).

Sgouris, J. T., J. K. Inman, K. B. McCall, L. A. Hyndman, and H. D. Anderson: The preparation of human fibrinolysis (plasmin). Vox Sang. (Basel) 5, 357 (1960).

Webster, M. E., and J. V. Pierce: Action of the kallikreins on synthetic ester substrates. Proc. Soc. exp. Biol. (N. Y.) 107, 186 (1961).

QUANTITATIVE STUDIES OF KININ-RELEASING ENZYMES IN PLASMA

J. MARGOLIS

Children's Medical Research Foundation, Royal Alexandra Hospital for Children, Camperdown, New South Wales, Australia

Intact human plasma exposed to glass or treated with a preparation of plasma kallikrein (pKkr) releases rapidly 1-1.5 μg of bradykinin equivalent (BK-eq) and loses 20-40% kininogen which appears to be specifically susceptible to pKkr and probably corresponds to component B (Margolis and Bishop, 1963). This will be referred to as comp. B or [B]. Plasma depleted of [B] ($\bar{B}$ plasma) still reacts with trypsin, venoms, or glandular kallikrein but not with pKkr. As shown in Fig. 1, the initial velocity of kinin formation depends on the concentration of releasing enzymes (and of comp. B) and may be used for assays of activated Hageman factor (ac-HF) and other components of the pKkr system, in particular component A, (comp. A or [A]) which is necessary for the formation of pKkr and is exhausted by exposure to 100-200 cm^2 glass or kaolin per ml plasma (A and B depleted, or $\overline{AB}$ plasma). On the basis of qualitative data, [A] behaves as if it were plasma kallikreinogen (pKkrgn) which is activated to pKkr by ac-HF (Margolis, 1960; 1963).

MATERIALS AND METHODS

Preparation of intact, $\bar{B}$ and $\overline{AB}$ plasma and assays of bradykinin on rat uterus have been described elsewhere (Margolis and Bishop, 1963).

To measure kinin-releasing activity, 1 vol. (usually 0.05 ml)

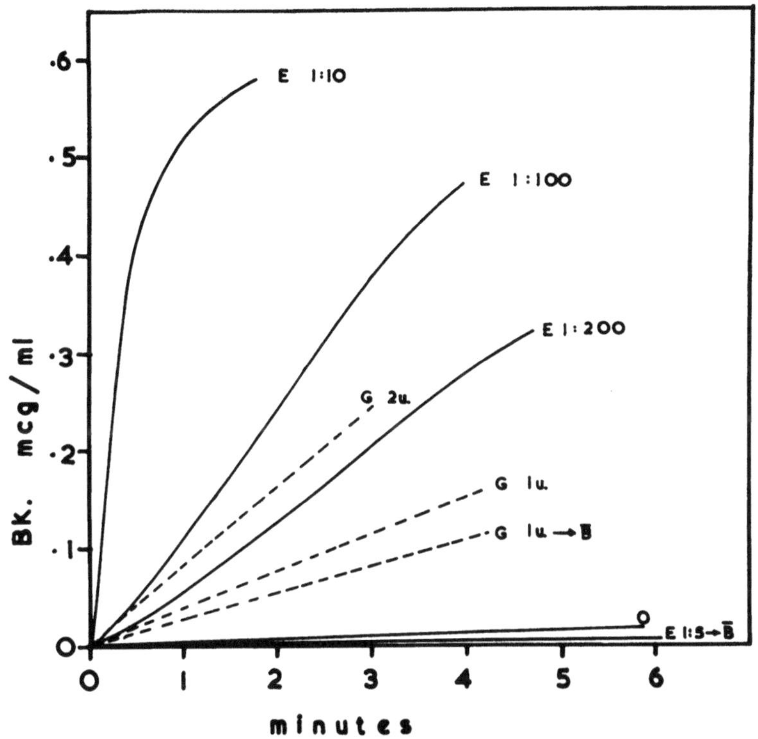

Fig. 1. Progress curves of kinin release by eluates of ac-HF (E), or "Glumorin" (G) from intact, or $\overline{B}$ plasma. Incubation mixture: test solution (E or G), 1 vol.; 0.04 M K_2 EDTA, 1 vol.; substrate plasma 4 vols. (see methods). Note that (E) has no effect on $\overline{B}$ plasma, which is still a suitable substrate for glandular kallikrein (G).

of test solution, 1 vol. of K_2-EDTA, and 4 vols of intact substrate plasma were incubated in 1-ml wells of Perspex agglutination plates at room temperature for 1-5 (usually 2) min. The mixture was assayed on rat uterus against synthetic bradykinin. The enzyme activity was expressed in μg BK-eq/min per ml or per mg of the material. Further details are given in text and Table 1.

Eluates of ac-HF were made from expired blood bank plasma, first exhausted for 1 hour with 1.5 mg kaolin/ml or 0.5-1 g glass ballotini/ml (0.1 mm). After separation and aging for 2-3 hours, the $\overline{AB}$ plasma was heated to 56°C for 10-20 min, centrifuged, and frozen until used. Adsorption was done with 10 mg kaolin/ml for 5 min, followed by 2 to 3 washes and elution for 5-15 min into an arbitrary volume of N/10 glycine or

TABLE 1

A System for Assay of Kinin-releasing Activity

Rat uterus in 10-ml oxygenated organ bath at 30° C. Tests 1-3 are direct applications of A and not of incubation mixtures B. Explanation of symbols in text.

(A) Enzyme preparation			(B) Reaction with substrate*			(C) Assay (rat uterus)		(D) Calculation of enzyme activity†				
(a)	(b)	(c)	(d)	(e)	(f)	(g)	(h)	(i)	(j)	(k)	(l)	(m)‡
Test material	Dilution	Treatment	Dilution of (A)	Incubation (min)	Test dose (ml)	Latent period (sec)	BK-eq. in (f) (ng)	Bk-eq. in (ng/ml) (h) : (f)	Bk-eq. in ng/min (i) : (e)	(j) Blank (= activity)	Activity/ml of (a) (k) × (d) × (b)	Conversion to standard units 1 × 10/1,000 ug/min/ml
1 BK (100 ng/ml)					0.05	10	5.0					
2 Plasma substrate	5	G, V, × 10 min			1/100	20	2.4	240				
3 BK					0.02	26	2.0					
4 Saline blank				5	1/5	50	1.0	5	1			
5 $\overline{B}$ plasma	12	G 2 × 2 min	1	2	1/20	30	1.8	36	18	17	204	2.04
6		6 min	1	2	1/30	18	3.2	96	48	47	564	5.64
7 El ($\overline{AB}^{56*}$) 1			20	2	1/30	23	3.2	96	48	47	940	9.40
8 2			10	1	1/25	5	10 +	250+				
9			50	1	1/30	22	3.0	90	90	89	4450	44.50
10 3			100	1	1/25	60+	1					
11			100	3	1/20	35	1.2	24	8	5	(?) 500	(?) 5.00
12			10	1	1/25	30	1.6	40	40	37	370	3.70
13 Saline blank				5	1/5	22	3.0	15	3			
14 "Glumorin" lu/ml			1	2	1/20	15	4.3	86	43	40	40	0.40

*One vol. A; 1 vol. 0.04 M K_2 EDTA; 4 vol. substrate plasma.

†Under defined conditions: e.g. substrate kininogen = 1μg/ml BK-eq, T = 22° C.

‡In round figures (l) × 10 (or 10/1,000 in μg). This allows for arbitrary dilution of enzyme (1/6) and substrate (2/3) in (B) and for anticoagulant in plasma.

phosphate buffer. If required, the sediment was eluted for the second and third times. After centrifuging at 12,000 *g* until the solution became clear, the pH was checked with a glass-calomel electrode. The eluates were then neutralized with 1 N HCl and usually stored at −20°C. When desired, they were concentrated by ultrafiltration and some were then freeze-dried. Protein content was estimated by the biuret method.

For activation of plasma, glass ballotini (0.1 mm, i.e. 200 cm^2/g or 300-350 cm^2/ml were used.

RESULTS

Measurement of kinin-releasing activity

Table 1 shows the method for assay and the steps in calculation. In test 2, intact plasma substrate, activated with optimal glass surface (30-50 cm^2/ml plasma) and K_2-versenate for 10 min (G,V, x 10 min), was tested directly to determine the available kininogen [B]. The standard procedure was: ballotini, 0.02 ml (= 6 cm); saline 0.7 ml; 0.04 M K_2-EDTA, 0.1 ml; plasma 0.2 ml; rotated 8-14 min. In this instance the result was 240 ng, i.e. 1.2 μg per ml of undiluted plasma. In tests 4-14, kinin-releasing activity was measured indirectly, as described above. In 5 and 6, 0.1 ml $\overline{B}$ plasma + 1.1 ml saline was activated with 0.2 ml ballotini (G2) for 2 and 6 min. In 7-12, El ($\overline{AB}^{56°}$) were preparations of eluted ac-HF (see methods). The dilution of enzyme, incubation time, and dose are adjusted to fall in the range of 1.5-5 ng BK-eq per 10-ml bath (h) and to consume no more than 15% of available substrate kininogen (j). Some tests in Table 1 (8, 10, and 11) had to be repeated to fall into this range. After allowing for dilutions of enzyme and substrate and corrections for blank, temperature, and kininogen, the activity is expressed in objective units as shown in the last column. A unit of activity is defined as 1 μg BK-eq released/min at 22°C in undiluted system from a substrate, containing 1 μg BK-eq/ml of available kininogen [B]. In the range 18° - 38°C the temperature coefficient (Q_{10}) is approximately 2.0. For convenience, 22°C was chosen as the standard temperature. For small deviations (within ± 3°) a correction of 7% for each 1°C is sufficiently accurate. Decay of the enzymes during actual incubation was not corrected for, but if the time is kept constant, e.g. 2 min, errors from this source are minimized. With overall reproducibility of ± 10-20% the typical examples of activity are as follows: 20% acetone-activated plasma, 4-6 μg/min/ml; guinea pig permeability globulin (G_2),* 6 μg/min/mg protein; ac-HF eluate (see above), up to 100 μg/min/mg protein, the same as purified ac-HF.† "Glumorin" (FBA) gives a value of 0.4 μg/min/unit, but being of glandular origin it utilizes

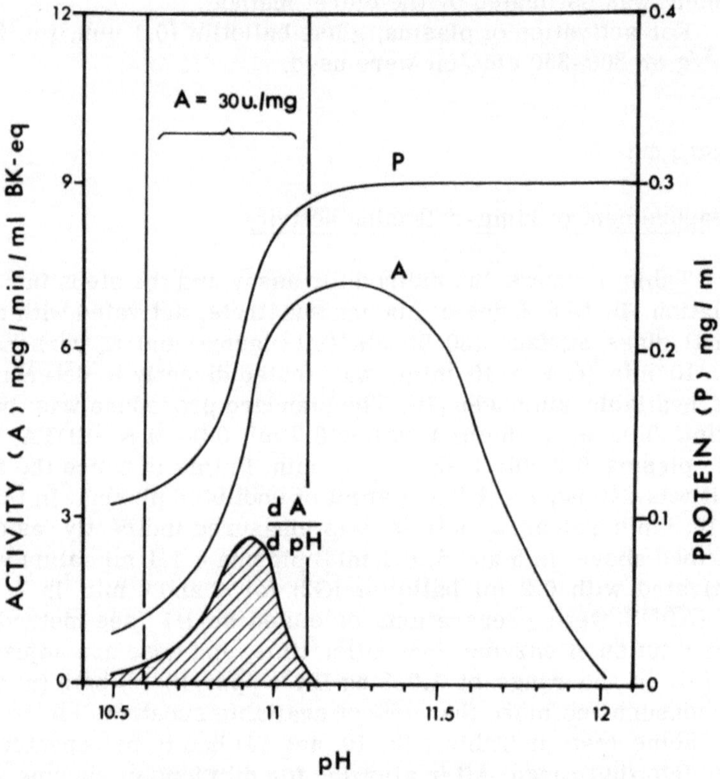

Fig. 2. Recovery of ac-HF in eluates from kaolin treated with AB[56]° plasma (see methods). Replicate samples were eluted in parallel at different pH values. Activity (A) was assayed and calculated as described. Protein (P) was estimated by the biuret method. Increment curve dA/dpH was constructed for increments of 0.1 pH unit. Between the vertical lines the specific activity per mg protein is practically uniform (A = 30 units/mg).

*Kindly supplied by D. L. Wilhelm.
†Kindly supplied by O. D. Ratnoff.

a different part of the substrate and is not strictly comparable (cf. Fig. 1).

Activated HF (ac-HF)

Although ac-HF probably acts indirectly (see below), it may be assayed empirically as in Table 1. Since the kinetics of the reaction are still unknown, it is advisable to keep the incubation time constant (e.g. 2 min). Figure 2 represents the recovery of activity and protein in ac-HF eluted at different pH values. The cross-hatched area under the differential curve shows that most of the activity is recovered between pH 10.6 and 11.2. Above pH 11.6 the enzyme is irreversibly damaged. The best preparations were obtained from $\overline{AB}$ plasma heated to 56°C, discarding the first eluate at pH 10.6 and retaining the second at pH 11.3. Their activity was comparable to highly purified HF preparations. Without preliminary exhaustion of comp. A, the yield was 50% or less (see below).

Measurement of comp. A

This was first assumed to be the precursor of pKkr. Attempts were made to measure the total [A] by constructing progress curves (1) of activity for plasma rotated with ballotini (cf. tests 5 and 6 in Table 1). Small corrections for direct transfer of kinin were introduced when necessary. Decay was determined separately by interrupting contact with glass. The rate of decay was then computed for a series of points on (1) and a second curve constructed. The area under this represents the total amount decayed (2). The sum of (1) + (2) gives the combined activity which tends to reach a constant level when all the precursor has been used. This may take 3 or 60 min, depending on dilution, temperature, and surface area, but we would expect the final result to be the same. Such, indeed, was the case at different dilutions of plasma as long as the ratio of glass surface to plasma (undiluted) was kept constant. However, when the surface was increased, not only was the reaction faster but also the total yield was higher. Since this was not in accordance with all the other observations, the validity of the method was repeatedly questioned and tested, but the unexpected results could not be explained away and had to be accepted. The maximal activity (15-20 units/ml) is obtained with 1,500-3,000 cm^2 glass surface per ml, which is sufficient to adsorb all the HF in plasma (Margolis, 1960; 1963).

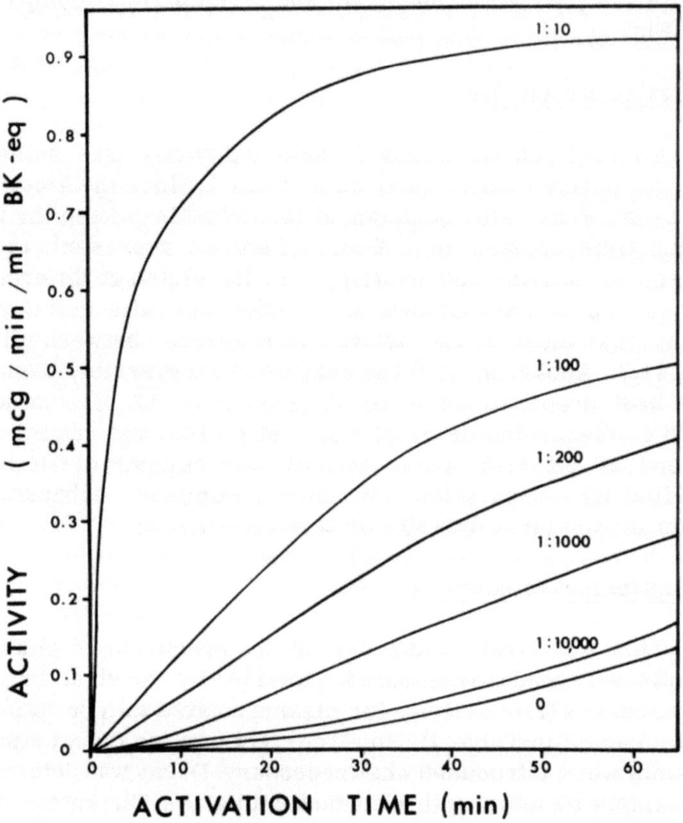

Fig. 3. Activation of diluted plasma with ballotini coated with ac-HF. Glass ballotini (0.3 ml = 100 cm^2) were treated for 2 min with 0.2 ml of $\overline{AB}^{56°}$ plasma and washed four times. They were then rotated with 1 ml of plasma (heated to 56° for 15 min), diluted 10^{-1} to 10^{-4} with buffered neutral saline. Activity calculated per ml of diluted plasma. See text.

In order to test the effect of still larger surfaces, the beads were precoated with ac-HF by brief exposure to $\overline{AB}$ plasma, followed by washing. Increasing dilutions of plasma were then activated with a constant volume of coated beads (Fig. 3). In plasma heated to 56°C and diluted 1:100 or more, decay is slow and may be ignored. It may be seen that the levels of activity (subtracting blank) multiplied by dilution do not reach ceiling values even in the range 10^{-3} to 10^{-4}, that is with surface areas 10^5 to 10^6 cm^2 per ml plasma. Since comp. A is easily exhausted by as little as 200 cm^2/ml, it is unlikely that this apparently limitless increase reflects the amount of comp. A in the plasma. However, if instead of the yield we look at the initial rates of the product formation, we find they are proportional to the concentration of plasma (cf. 1:100 and 1:200 curves). This was found also in experiments where $\overline{AB}$ plasma (which in this system develops 1-2% activity) was used as the diluent instead of saline. On the other hand, when plasma dilution was kept constant, but the (coated) surface was varied, the yield was dependent on the surface area. This could be interpreted to mean that ac-HF is the actual kinin-releasing enzyme and comp. A simply removes it from the surface. In support of this view are the earlier observations that the yield of the ac-HF was higher in kaolin eluates prepared from $\overline{AB}$ plasma than in those made from intact plasma. The following findings, however, speak against it: (1) beads coated with ac-HF can be used for activating successive samples of $\overline{HF}$ plasma (Margolis, 1960); (2) $\overline{AB}$ plasma, which contains at least 80% HF, is not activated by acetone (Margolis, 1963); (3) ac-HF and pKkr apparently act differently on substrates heated to 60°C (Eisen, 1963).

An alternative model (Miles, 1964), suggested by experiments with permeability factor (PF), introduces an intermediate precursor, which is activated by ac-HF and then, in turn, activates pKkrgn. Such an intermediate, if unstable, could account for the present results, because the yield of pKkr would then depend on the rate of activation. In support of an additional factor is also the following observation: when serial dilutions of intact plasma in $\overline{AB}$ plasma are incubated with 20% acetone, the early (1-4 hours) activity is proportional to the square of plasma concentration, as though two components were being diluted. After 8 hours the activity settles at a level directly proportional to intact plasma in the mixture (Fig. 4).

Comp. A in eluates

Glass or kaolin, when "coated" with plasma which had not been exhausted of comp. A, activates $\overline{AB}$ plasma to some extent (Margolis, 1960; Eisen, 1963). This could be attributed to a combination of some part of comp. A with the ac-HF fixed to the surface. However, this material in "pre-active" form is also adsorbed from HF-deficient plasma and may be eluted from kaolin at pH 11.0. Such eluate has no initial kinin-releasing activity but develops it when exposed to beads coated with

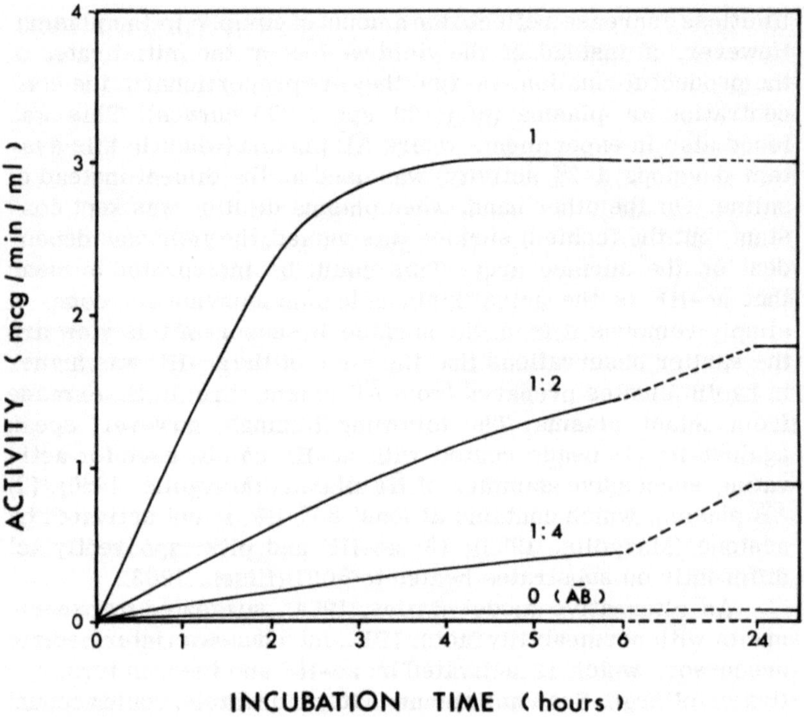

Fig. 4. Acetone activation. Acetone 0.2 ml, intact plasma diluted in AB plasma as indicated, 0.8 ml. Activity per ml of mixture (including acetone).

ac-HF, as in Fig. 3. When tested for clot-promoting properties, it behaves like isolated F Xl (unpublished work).

These observations, when followed up and correlated with physical studies (cf. Adams and Vroman, 1965), may help to unravel the composition of comp. A and also reveal some still unsuspected mechanisms in the reaction of plasma with solid surfaces.

CONCLUSION

The measurement of kinin-releasing activity in objective units does not appear to present difficulties except for the need to introduce a number of definable corrections to increase the accuracy (blank, temperature, substrate kininogen [B], kinin in the activator and, possibly, decay of enzyme during incubation).

However, when it is applied to the measurement of the precursor of kininogenase, the very fact that the information is more precise forces us to recognize inconsistencies which were glossed over in rough qualitative studies. Thus, the simplest explanation of the inability of $\overline{AB}$ plasma to produce pKkr would be a deficiency of a single factor, comp. A, which was not HF and could be identical with pKkrgn. Now, along with Miles (1964) we may be forced to postulate that at least two factors are exhausted (this is quite apart from comp. B, which is fully accounted for). If this is so, then, indeed, one of them may be identical with pKkrgn and the other with PF (Miles, 1964). On the other hand, as suggested by Lewis (1964), some of these entities may prove to be not enzymes, but accelerators of a slow underlying enzyme-substrate reaction. Lewis (1958), Margolis and Bishop (1963), and Vogt (1965) recognized the existence of a rapid and a slow phase in kinin release, but offered different explanations for them.

In the "spontaneous" plasma kinin system, whether there are one or more phases, HF is clearly indispensable. Comp. A may become resolved into two or more factors (which are also essential, at least to the rapid phase) with the aid of differential substrates and inhibitors. The term "comp. A" will then have served its purpose and will be replaced by more specific designations. In the meantime, this factor, or complex, may be measured empirically, by using $\overline{AB}$ plasma as diluent and glass or acetone as the activating agent. At present the results may be expressed only in per cent of normal, much in the same manner that "prothrombin activity" is still used as an index of the combined effect of five known clotting factors.

While it may be hoped that the kinin system will prove less intricate than blood coagulation, some of the lessons learned in the study of clotting may be relevant in the present context. First, most advances in the understanding of coagulation followed on the discovery of naturally occurring defects and this, in turn, had to await adequate test methods. Second, there is always some conflict of priorities: whether it is better to strive first for internal consistency within a well-defined but narrow frame of reference, as has been attempted here, or to extend this frame in the hope of finding shortcuts and connections with more firmly established disciplines.

ACKNOWLEDGMENT

I wish to thank Miss Sally Bruce for assistance in this investigation.

REFERENCES

Adams, A. L. and L. Vroman: Observations of factor XII and factor XII-deficient plasma at an interface. Thrombos. Diathes. haemorrh. Stuttg. 13, 387 (1965).

Eisen, V.: Observations on intrinsic kinin-forming factors in human plasma: The effect of acid, acetone, chloroform, heat and euglobulin separation. J. Physiol. (Lond.) 166, 496 (1963).

Lewis, G. P.: Formation of plasma kinins by plasmin. J. Physiol. (Lond.) 140, 285 (1958).

Lewis, G. P.: The role of peptides in the first stages of inflammation, in Injury, inflammation and immunity p. 242. L. Thomas, J. W. Uhr and L. Grant Eds. Williams and Wilkins Co. (1964)

Margolis, J.: The mode of action of Hageman factor in the release of plasma kinin. J. Physiol. (Lond.) 151, 238 (1960).

Margolis, J.: The interrelationship of coagulation of plasma and release of peptides. Ann. N.Y. Acad. Sci. 104, 133 (1963).

Margolis, J., and E. Bishop: Studies on plasma kinins. 1. The composition of kininogen complex. Aust. J. exp. Biol. med. Sci. 41, 293 (1963).

Miles, A. A.: Large molecular substances as mediators of the inflammatory reaction. Ann. N.Y. Acad. Sci. 116, 855 (1964).

Vogt, W.: Ein komplettes zweites kininbildendes System in menschlichem Plasma. Naunyn-Schmiedeberg's Arch. exp. Path. Pharmak. 251, 187 (1965).

DISCUSSION

Vogt: Is EDTA efficient enough to block carboxypeptidase activity in human plasma? It is not in rat plasma.

Margolis: In human plasma there is a loss of some 30-50% in 30 min. Since the usual incubation time of the enzyme with the plasma substrate is 2 min, the loss is negligible.

Erdös: EDTA is a good inhibitor of carboxypeptidase in human plasma but not in rat or other animal plasma as we have shown. The enzyme which inactivates bradykinin is inhibited better in the human plasma by 1.10^{-3} phenonthroline or 2,2 dipyridyl.

Margolis: While EDTA is not an ideal inhibitor of plasma kininase, it is sufficiently effective to prevent significant losses of kinin during the short periods of incubation used for measurement of kininogenase activity.

Eisen: What is the evidence that component B- depletion leaves 80% of component A in plasma? The fall in arginine esterase activity inducible by kaolin is certainly greater. Has Dr. Margolis excluded the possibility that 0.04 M EDTA depresses component A (kallikrein) activity?

Margolis: The kinin-releasing activity of B plasma was measured directly on numerous occasions. The degree of concomitant exhaustion of component A depends on the glass surface and activatiion time. With 20 cm^2 activation for 5 min, the loss of component A is no more than 20% as compared with intact plasma. The final concentration of EDTA in the mixture is less than 0.01 M. Qualitative comparison with the rate of kinin formation in the absence of EDTA does not suggest an inhibitory effect. I have no quantitative data on this subject because an essential part of the method is the inhibition of kininase by EDTA.

Trautschold: Did you try to get kinin liberation if you incubate the activated Hageman factor with purified kininogen preparation according to your proposed reaction mechanism No. 2?

Margolis: This is what we have to do next.

Habermann: You measured kinin activity on the uterus only, and you can see quantitative differences only. Would it not be possible that various enzymes or conditions of activation would lead to different kinins varying in activity spectrum?

Margolis: All assays were done on rat uterus only. If it were shown that at different stages of activation of the "plasma kallikrein" system the enzymes release different peptides, this could explain some present inconsistencies. However, based on the present evidence, contact activation can release kinin-9 only.

IMMUNOLOGIC STUDIES OF COMPONENTS OF THE KALLIKREIN-KININ SYSTEM*

C. W. AUNGST, N. BACK, B. BARLOW, and G. A. TSUKADA
From the Division of Medicine, Roswell Park Memorial Institute, and the State University of New York at Buffalo, School of Pharmacy, Buffalo, N. Y., U.S.A.

The components of the kallikrein-kinin system previously studied by immunologic techniques include pancreatic and urinary kallikrein (Webster et al., 1963) and bradykinin (Goodfriend et al., 1964). The purpose of this report is to present certain immunologic aspects of bradykinin, human bradykininogen, bovine kallikrein inhibitor (Trasylol), human pancreatic kallikrein, and human urinary kallikrein.

MATERIALS AND METHODS

Antigens. Synthetic bradykinin† was obtained in concentrations of 0.1 mg/ml and 1.0 mg/ml. Bovine kallikrein inhibitor (Trasylol‡) was prepared commercially from lung tissue and contained 10,000 kallikrein inhibitor units (KIU) per ml. Purified

*This investigation was supported by Public Health Service Research Grants HE 08633 and HE 03784, from the National Heart Institute and the Heart Association of Erie County, Inc., Buffalo, N. Y.

†Synthetic bradykinin was supplied by Dr. R. Bircher, Sandoz Pharmaceuticals, Hanover, N. J.

‡Trasylol and human pancreatic kallikrein was obtained from Dr. R. Goesswald, FBA, Division of Metachem, Inc., New York, N. Y.

bradykininogen* was the acetone-dried preparation from human plasma, prepared by the method of Greenbaum and Hosoda (1963), and contained 0.5 bradykininogen unit/mg protein. Crude human urinary kallikrein was prepared in our laboratories by the method of Gaddum and Guth (1960) and contained 10 mg protein/ml; 0.1 mg of this preparation activated all the kininogen from 1×10^{-6} g plasma. Human lyophilized pancreatic kallikrein contained 1 kallikrein unit/5.5 μg biuret protein.

Preparation of antiserum. Except for bradykinin the immunization procedure was the same for each antigen. The antigen solutions were mixed with equal volumes of Freund's complete adjuvant and injected subcutaneously into rabbits. Three injections were given during the first week, followed by a fourth injection after 7 days. Blood was drawn from the rabbits 1 week after the last injection and the serum separated. Booster injections of the same mixture of adjuvant and antigen were given at 2-month intervals and blood was obtained 1 week after each injection. The bradykinin-adjuvant mixture was given three times weekly for 3 weeks and the procedure as with the other antigens followed.

The dose of each antigen given at each injection was as follows: synthetic bradykinin, 0.1 mg; human bradykininogen, 1 mg; bovine kallikrein inhibitor, 10,000 KIU; human pancreatic kallikrein, 5 mg; and human urinary kallikrein, 1 ml.

In most studies a standard Ouchterlony (1949) technique was used, employing six equally spaced peripheral wells 10 mm from a central well. A 3-mm layer of 1%-2% agar in phosphate-buffered saline, pH 7.4, was used. The reactant wells had a volume of approximately 0.1 ml. In some studies a modification of the Ouchterlony technique with constant-feed reservoirs was employed (Aungst, in press). Immunoelectrophoresis was performed by the method of Scheiddeggar (1955).

RESULTS

Bradykinin. The antisera to bradykinin, developed in two rabbits (B-264 and B-114), demonstrated marked differences

*Purified human bradykininogen was prepared by Dr. L. M. Greenbaum, College of Physicians and Surgeons of Columbia University, New York, N. Y.

between the immunologic reactions obtained. The antiserum which was obtained from rabbit B-264, both at the time of the initial bleeding and 5 months later (after a booster dose of bradykinin), produced definite precipitin bands with bradykinin on the constant-feed-type Ouchterlony plates. Exactly 1 ml antiserum was placed in the central well with 1 ml bradykinin in concentrations of 80, 40, 20, and 10 μg/ml in the peripheral wells. Precipitin bands were obtained at all concentrations. No precipitin bands were seen when the conventional type of gel diffusion plate was employed. The precipitin bands were sharp nearest the antigen wells but became diffuse and cloudy near the antibody well. When normal human serum was used as antigen, two precipitin bands were formed. No precipitin bands were found on immunoelectrophoresis when bradykinin was used as antigen, but a precipitin band appeared near the albumin region when normal human serum was the antigen (Fig. 1). Ouchterlony plates, with dilutions of human serum albumin from 5% to 0.62% as antigen and rabbit antibradykinin in the central well, failed to reveal precipitin bands. Unfortunately, this

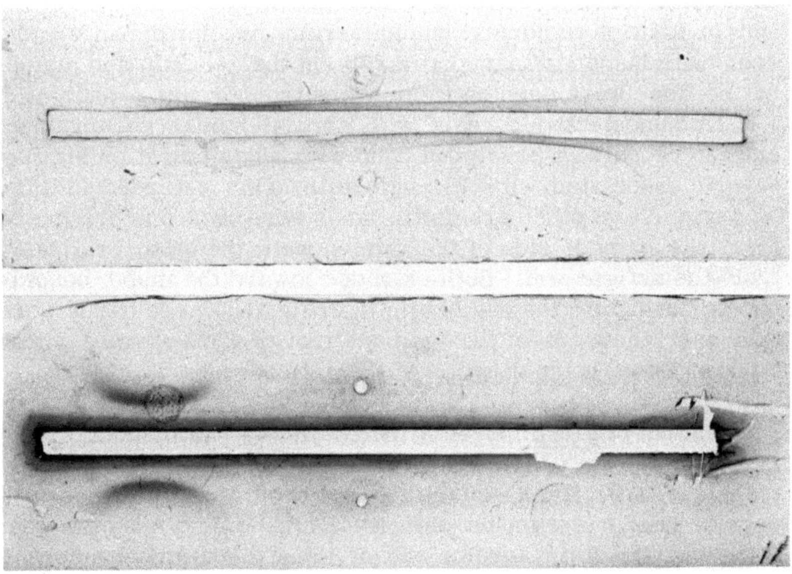

Fig. 1. Immunoelectrophoretic pattern obtained with rabbit antibradykinin. Antiserum from rabbit B-114 is in the trough and normal human serum in the antigen well (top). Antiserum from rabbit B-264 is in the trough and normal human serum in the antigen well (bottom).

rabbit died of trauma, and the supply of antibody was exhausted before further studies could be performed.

Serum from rabbit B-114 obtained 5 and 14 weeks after the start of immunization demonstrated no antibodies. The serum obtained at 24 weeks produced no precipitin bands on gel diffusion plates with bradykinin in concentrations varying from 2 to 100 μg/ml. However, when normal human serum was used as antigen, two definite precipitin bands were formed. Precipitin bands were present at a dilution of 1:8 when serial dilutions of human serum in 0.9% NaCl were placed in the peripheral wells with undiluted antiserum in the center well. Within 48 hours these bands became broadened and diffuse in the direction of the antibody well. When the antiserum was diluted 1:1 and the human serum diluted 1:4 with normal saline, no definite precipitin bands were formed. However, a haziness in the agar was evident near the antibody well. When the serial dilutions of the human serum were made with a solution of bradykinin containing 10 μg/ml, no differences were noted in the position of density of the precipitin bands. Absorbtion of the antiserum with bradykinin was unsuccessful, since dilution with saline alone resulted in a lack of activity.

In addition to normal human serum, precipitin bands were seen with human urinary kallikrein on the gel diffusion plates, but no lines were seen on immunoelectrophoresis.

Immunoelectrophoresis with normal human serum as an antigen revealed a precipitin band with a configuration similar to that associated with 7S γ-globulin. This extended into the α_2 region. Two other precipitin bands were seen: one originated from the cathode side of the antigen well; the other originated from the antigen well. Both extended toward the anode, becoming confluent with the γ-globulin line (Fig. 1). These bands were thin near the edge of the antibody trough and appeared within 2 hours after the application of the antibody.

Human bradykininogen (HBKG). Rabbit antihuman bradykininogen serum (RAHBKG) produced three bands (Ouchterlony technique) with HBKG as antigen and two precipitin bands with normal human serum as antigen. By the use of specific goat antihuman albumin serum, one of these precipitin bands was identified as albumin. Although only three bands were seen on immunoelectrophoresis when HBKG in a concentration of 20 mg/ml was electrophoresed, seven bands were seen with normal human serum. These consisted of a prealbumin, albumin, two

α_1-globulins and α_2-globulin, a β-globulin, and a γ-globulin. Since there were similarities between the electrophoretic patterns of HBKG and HUKK, suggesting the possibility of common contaminants, the HBKG antibody was absorbed with two parts HUKK (10 mg/ml) to three parts RAHBKG. With this as antibody, immunoelectrophoresis revealed a prealbumin, an α_2-globulin, and a large precipitin band in the α_1-region, with normal human serum of HBKG as antigen (lower part of Fig. 2). In an effort to determine which band represented HBKG, one part HUKK was incubated at room temperature with four parts normal human serum and electrophoresed as antigen with the absorbed RAHBKG in the center trough. This resulted in a diminution in the density of the precipitin band in the prealbumin and α_2-globulin regions, which was due to splitting of the bands. There was no apparent effect upon the α_1-globulin (upper part of Fig. 2).

Human urinary kallikrein (HUKK). Ouchterlony analysis of

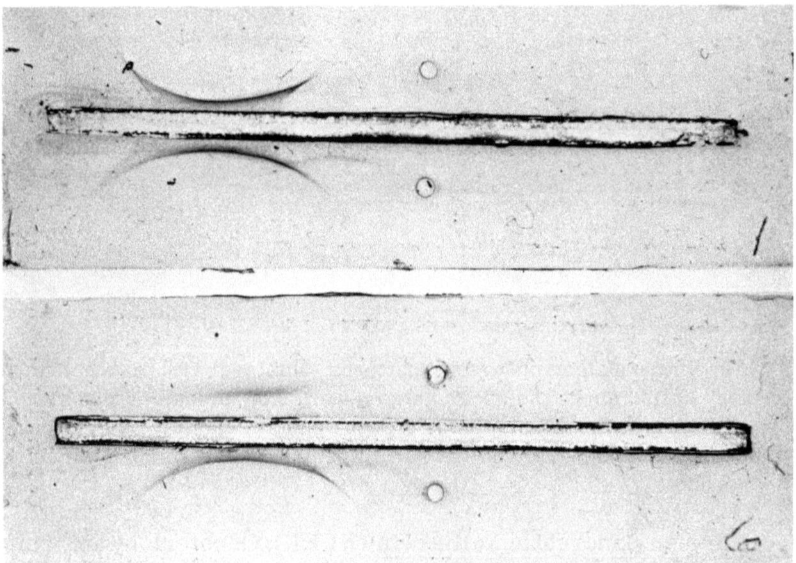

Fig. 2. Immunoelectrophoretic patterns obtained with rabbit antihuman bradykininogen absorbed with human urinary kallikrein in the antibody troughs. The lower antigen wells contain normal human serum. The upper antigen wells contain normal human serum plus human urinary kallikrein (top) and human bradykininogen (bottom).

the crude HUKK revealed five precipitin bands with the rabbit antihuman urinary kallikrein (RAHUKK). Albumin and 7S γ-globulin were identified by using specific antisera for these proteins. Immunoelectrophoresis revealed at least seven precipitin bands consisting of two α_1 -globulins, a faint albumin band, two α_2-globulins, a β-globulin, and 7S γ -globulin when HUKK or normal human serum was electrophoresed (upper part of Fig. 3). With HUKK, the latter band had a configuration similar to that of β_2 -A globulin (bottom part of Fig. 3). No bands were developed with normal human serum.

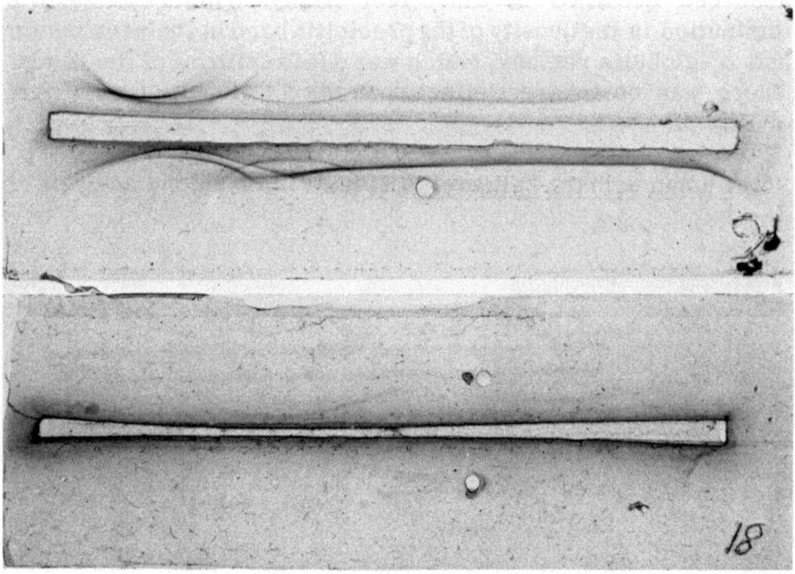

Fig. 3. Immunoelectrophoretic patterns obtained with crude human urinary kallikrein in the upper antigen wells and normal human serum in the lower wells. The antibody troughs contain undiluted rabbit antihuman kallikreins (top) and RAHUKK absorbed 1:1 with normal human serum (bottom).

Human pancreatic kallikrein (HPKK). Ouchterlony analysis of the HPKK revealed a single precipitin band. On immunoelectrophoresis a small band in the α_2 region was seen (upper part of Fig. 4). No cross-reaction was seen with HUKK but a single thin band which migrated slightly faster toward the anode was seen with normal human serum.

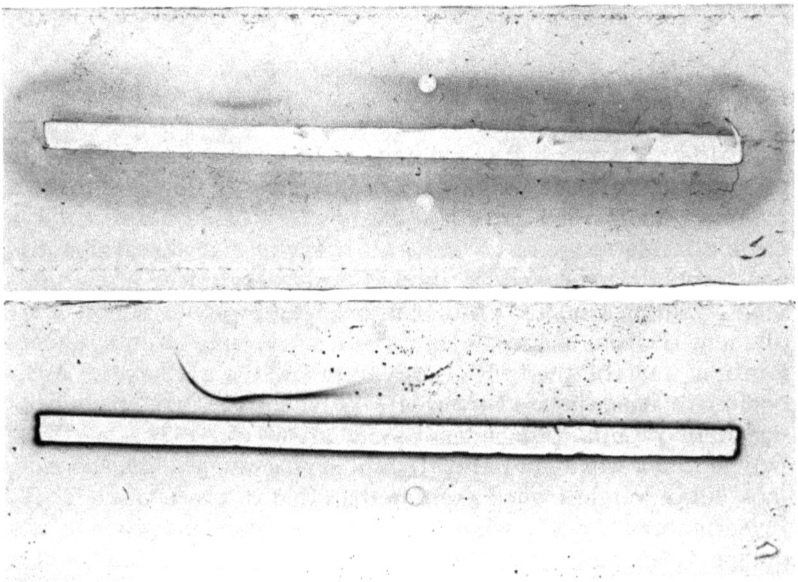

Fig. 4. The antigen wells of the upper slide contain purified human pancreatic kallikrein (top) and crude human urinary kallikrein (bottom) with rabbit antihuman pancreatic kallikrein in the antibody trough. The antigen wells of the lower slide contains Trasylol (top) and normal human serum (bottom) with rabbit anti-Trasylol antibodies in the antibody trough.

Bovine kallikrein inhibitor (Trasylol). Single precipitin bands were formed on both types of Ouchterlony plates. Immunoelectrophoresis also demonstrated a single band which originated near the antigen well and extended into the α_2 area in the shape of a hook. Normal human serum showed no precipitin bands with this antisera (bottom part of Fig. 4).

Inhibition studies. Inhibition of the hypotensive activity of HPKK and HUKK in the dog was demonstrated when these enzyme antigens were incubated in vitro with their respective antisera. Appropriate controls with normal serum and the antigens confirmed the specificity of the inhibition. No inhibition of bradykinin by RABK was obtained with this method. Inhibition studies using the isolated rat uterus (Back, et al., 1963) revealed no antagonism of the smooth-muscle-stimulating effect of bradykinin, nor were the kinin-forming capacities of HUKK and HPKK inhibited by their respective antisera. HBKG antiserum also did not interfere with the activation of HBKG.

DISCUSSION AND SUMMARY

Most of the previous studies of the immunological aspects of bradykinin have employed conjugation of bradykinin to a protein. Goodfriend and co-workers (1964) used water-soluble carbodiimide reagents to conjugate bradykinin to rabbit albumin, and antibody produced in rabbits had a high titer of complement-fixing antibodies. Although bradykinin would inhibit complement fixation of the conjugate with antiserum, no complement fixation was obtained with bradykinin and the antiserum. Antibodies to unconjugated synthetic polypeptides have been produced in rabbits (Buchanan-Davidson et al., 1959), but these polypeptides were not naturally occurring compounds, and their molecular weights were greater than that of bradykinin. These investigators found cross-reactions between antisera to the polypeptides and naturally occurring proteins such as bovine and human serum albumin, and variable reactions of the polypeptides with homologous antisera obtained from different rabbits.

In our study, immunization with bradykinin mixed with Freund's adjuvant produced antibodies in one rabbit which formed precipitin bands with bradykinin and an unidentified serum protein. The antibody developed in the other rabbit had no precipitins specific for bradykinin but did cross-react with at least three proteins in normal human serum. A possible explanation for this finding may be that the highly charged bradykinin reacted with normal rabbit protein in such a manner as to produce antigenic sites "foreign" to the rabbit but normally present in certain human serum proteins.

The HBKG used in the present study demonstrated a homogeneous peak in the α_2 region, on zone electrophoresis (Greenbaum and Hosoda 1963). The development of seven precipitin bands on immunoelectrophoresis of normal human serum as opposed to the three precipitin bands with HBKG probably relates to the development of antibodies to minor contaminants of the bradykininogen preparation caused by the method of immunization with Freund's adjuvant. The splitting of the precipitin bands in the prealbumin and α_2 regions after incubation of HBKG with HUKK may indicate that these bands rather than the large α_1 precipitin band represent the substrate for kallikrein.

Webster et al. (1963) developed antibodies to crude and partially purified HUKK and HPKK which formed multiple precipitin bands with their homologous antigens and single precipitin bands with highly purified antigens. These antisera inhibited the ability of kallikrein to increase blood flow after injection into the femoral arteries of dogs. They found cross-reactions between human urinary and pancreatic kallikrein but not between kallikrein of various animal species, when these were tested for pharmacological activity and by Ouchterlony analysis. RAHPKK from one rabbit formed strong precipitin bands but did not inhibit vasodilator activity.

In our studies we found no cross-reactions between HUKK and HPKK on Ouchterlony analysis or immunoelectrophoresis, but there were cross-reactions when inhibition of pharmacological activity was tested. Differences in these results could be explained if the precipitin bands found in the absorbed RAHUKK-HUKK system did not represent the active kallikrein. We are unable to make a definite statement regarding the identity of one or both of these bands as kallikrein at the present time. Further studies have been undertaken with starch block electrophoresis to separate the antigenic components of a more highly purified HUKK preparation.

The human pancreatic kallikrein was a highly purified preparation which produced a single precipitin band in the α_2 region on immunoelectrophoresis. A precipitin band also was formed with a protein in normal human serum, which has a slightly faster electrophoretic mobility in agar. This precipitin band can be eliminated by absorbtion of the antiserum with normal human serum.

Bovine kallikrein inhibitor is a polypeptide with a molecular weight of 6,500 (Anderer and Hörnle, 1965). This material has been administered intravenously to humans in intermittent courses of therapy over a period of months without causing overt manifestations of sensitization (Nardi, 1963). Despite the low molecular weight of the polypeptide and lack of sensitivity reactions in humans, precipitating antibodies were produced in rabbits when the material was injected with Freund's adjuvant. Antigenicity with doses of 2,000 to 10,000 KIU/kg in the guinea pig also has been demonstrated (Dr. R. Goesswald, personal communication). The use of Trasylol in experimental and clinical conditions associated with shock and fibrinolytic hemorrhage has been reviewed recently (Back, 1966; Back and Ambrus, in press).

ACKNOWLEDGMENTS

Appreciation for their assistance in this study is extended to Miss Joan Czarnecki, research assistant, and Miss Heidi Hansen, participant in the Research Participation Program of the Roswell Park Memorial Institute; supported by Grant GE-7396, National Science Foundation.

REFERENCES

Anderer, F. A., and S. Hörnle: Strukturunter suchungen am Kallikrein-Inaktivator aus Rinderlunge. 1. Molekulargewicht, Endgruppenanalyse und Aminosäure-Zusammensetzung. Z. Naturforsch. 20b, 457 (1965).

Aungst, C. W.: A specific and sensitive method for the detection of ferritin in body fluids. J. Lab. clin. Med. In press.

Back, N., and J. L. Ambrus: Fibrinolytic hemorrhage, in Surgical Bleeding. A Handbook for Surgery and Specialties, A. W. Ulin and S. Gollub, Eds. New York: McGraw-Hill, (in press).

Back, N.: Fibrinolysin system and vasoactive kinins. Fed. Proc. 25, 77 (1966)

Back, N., P. S. Guth, and A. E. Munson: On the relationship between plasmin and kinin. Ann. N. Y. Acad. Sci. 104, 53 (1963).

Buchanan-Davidson, D. J., M. A. Stahmann, C. Lapresle, and P. Grabar: Immunochemistry of synthetic polypeptides and and polypeptidyl proteins. J. Immunol. 83, 552 (1959).

Gaddum, J. H., and P. S. Guth: A comparison of the kallikrein-kinin system in sheep and dogs. Brit. J. Pharmacol. 15, 181 (1960).

Goodfriend, T. H., L. Levine, and G. D. Fasman: Antibodies to bradykinin and angiotensin: A use of carbodiimides in immunology. Science 144, 1344 (1964).

Greenbaum, L. N., and T. Hosoda: Studies on the isolation of bradykininogen. Biochem. Pharmacol. 12, 325 (1963).

Nardi, G. L., (discussion after M. M. Forell): Therapy with kallikrein and protease inhibitors. Ann. N. Y. Acad. Sci. 104, 368 (1963).

Ouchterlony, O.: Antigen-antibody reaction in gels. Acta path. microbiol. scand. 26, 507 (1949).

Scheiddeggar, I. J.: Une micro-methode de l'immunoelectrophoresis. Int. Arch. Allergy 7, 103 (1955)

Webster, M. E., E. W. Emmart, W. A. Turner, H. Moriya, and J. V. Pierce: Immunological properties of the kallikreins. Biochem. Pharmacol. 12, 511 (1963).

KININOGENASES, KININASES, AND THEIR INHIBITORS

I. TRAUTSCHOLD, H. FRITZ, and E. WERLE

Klinisch-Chemisches Institut an der Chirurgischen Klinik der Universität München, Germany

During the last few years we have investigated the participation of the kinin system in acute pancreatitis and in shock, and the prevention of the liberation of kallikrein by naturally occurring proteinase inhibitors. We studied first some of the properties of kallikrein of the pancreas; the main purification was done according to the previously published methods (Werle and Trautschold, 1963). The last stage of the purification leading to pure preparations included fractionation on DEAE-cellulose, Sephadex columns, and finally column electrophoresis. During these purification steps there was no evidence of the existence of two kallikrein fractions, as demonstrated by Habermann (1962). We can confirm his statement that the appearance of several kallikrein fractions depends on the extent and effectiveness of the initial autolysis on pancreatic tissue.

By careful manipulations we tried to isolate the kallikrein in its native form. One kallikrein unit of this preparation corresponds to 0.8 μg protein. The molecular weight measured by the molecular sieve-gel filtration was between 34,000 and 35,000. This is in a good agreement with the calculate molecular weight of 34,500 according to the amino acid composition. We found 292 amino acids and 4.7% hexosamines and 5% free hexoses (Table 1). Alanine was the N-terminal amino acid. The isoelectric point was at pH 4.0. Habermann (1962) calculated the molecular weight from sedimentation constant of his preparation as 24,000; Moriya (1959) calculated his as 33,000.

This proves that the size of kallikrein molecules and some of its properties depend on the methods of isolation used. Our preparations of swine pancreatic kallikrein were quite stable at pH 5-9, but were destroyed in acid or alkaline solution. Pan-

TABLE 1
Amino acid composition of hog pancreas kallikrein

Amino acids	(%)	(moles)
Lys	4.4	12
His	3.9	10
Arg	1.7	4
Asp	11.6	35
Thr	6.55	22
Ser	5.25	20
Glu	11.0	30
Pro	7.5	26
Gly	5.2	28
Ala	4.4	20
Cys 1/2	3.6	12
Val	3.5	12
Met	1.1	3
Ileu	4.2	13
Leu	7.8	24
Tyr	4.0	9
Phe	4.9	12
Glucosamine	3.6	8
Galactosamine	0.9	2
Hexoses	5.0	10

Calculated molecular weight: 34,550

creas kallikrein is an enzyme with high substrate specificity and acts as a proteinase only with its physiological substrate the kininogen. Besides this, we found an infinitely low unspecific proteolytic activity corresponding to 4×10^{-10} g trypsin/kallikrein unit. This activity is partially inhibited by the kallikrein-trypsin inhibitor of beef organs and, to a lesser extent, by soybean trypsin inhibitor.

In order to replace the biological kallikrein unit with a chemical one we want to propose a new definition of the kallikrein unit in line with the international enzyme units, catalyzing the transformation of 1 μmole substrate per minute. We used BAEe for the estimation of the kallikrein activity (Trautschold and Werle, 1961) and expressed the amount of liberated ethyl alcohol in terms of reduced NAD (-the molar extinction coefficient at 366 mμ is 3.3×10^6 cm^2/mole). One international unit of pancreatic kallikrein is that amount which causes the change in light absorption at 25°C by 1.1 ΔE/min in a total test volume of 3.0 ml. With BAEe as substrate, the turnover number is 3.5×10^3 moles/min at 25°C, the Michaelis constant $= 2 \times 10^{-4}$ mole/liter and the maximal velocity = 0.067 ΔE/min/kallikrein unit. On the basis of kinin liberating activity, one international kininogenase milliunit liberates the amount of kinins per minute which produces a lowering of blood pressure or a contraction of smooth muscle equivalent to 1.06 μg synthetic bradykinin or 1.2 μg kallidin.

The four kallikreins from pancreas, submandibulary gland, urine, and serum are different in the following points: (1) the sources (tissues and body fluids); (2) behavior during purification procedure; (3) electrophoretic mobility; (4) stability; (5) proteolytic activity; (6) kinin-forming activity; (7) esterolytic activity; (8) biological activity and destruction in vivo and in vitro; (9) immunological properties; (10) inactivation by natural proteinase inhibitors.

Our second step in the investigation of the pathogenesis of pancreatitis and therapeutic effect of inhibitors in acute pancreatitis, was to clarify the behavior of inhibitors against kininogenases (Trautschold, 1965). We studied the properties of the kallikrein-trypsin inhibitor from beef organs, called Trasylol. In the course of our investigations we found three new naturally occurring trypsin inhibitors of the low-molecular type. As shown by Anderer (1965), the inhibitor from beef parotid gland and beef lung (Trasylol) is identical with the pancreatic trypsin inhibitor from Kunitz (Kassell et al., 1965; Chauvet et al., 1964), a fact which, after considering the identical properties of the inhibitors, we assumed a long time ago. We could detect this polypeptide consisting of 58 amino acids only in organs of beef. It is not secreted into the saliva or into the pancreatic juice. In the pancreas of various mammals, including the beef, we could demonstrate a specific trypsin inhibitor which is secreted into the pancreatic juice (Fritz et al., 1965). We isolated another trypsin inhibitor from submandibulary gland of dogs which also inhibits chymotrypsin but not kallikrein or plasmin (Trautschold et al., 1963 (see Table 3). This inhibitor seems to occur only in canines and felines. Finally we found a trypsin inhibitor which also inhibits plasmin but not chymotrypsin or kallikrein. This inhibitor occurs in accessory sex glands and the secretions of all the studied male mammals (Table 2). The highest inhibitor content was found

TABLE 2
Some Properties of New Porteinase Inhibitors

Isolated from:	Lung parotis	Pancreas	Submandib. gland	Accessory sex gland
Species	Beef	Mammals	Dog	Mammals (male)
Concentration in tissue ImE/g (trypsin, BAPA):	570	150	10,000	5,000
Inhibition of				
Trypsin	+	+	+	+
Chymotrypsin	+	–	+	–
Kallikrein	+	–	–	–
Plasmin	+	–	–	+
Resistant to pepsin	+	–	–	+
Molecular weight	6,500	6,900-7,200	14,000	6,800
Occurence in Secretion	–	+	+	+
Specific activity ImE/μg protein (trypsin, BAPA)	4	3	2.9	3.2
Isoelectric point	≈11	8-12	≈12	>8

in the vesicular gland (Table 3). Normal male mice treated with testosterone showed an increase in inhibitor content of vesicular gland from the normal of about 2,250 up to 9,000 inhibitor milliunits (ImU). The inhibitor content was nearly completely depressed in castrated animals (Haendle et al., 1965). After treatment with testosterone the inhibitor level rises again.

In contrast to the polyvalent trypsin-chymotrypsin-kallikrein-plasmin inhibitor of beef organs, the other inhibitors are more or less specific trypsin inhibitors. By means of ion-exchange chromatography and electrophoresis they were all isolated as pure preparations. The amino acid composition of only the submandibulary inhibitor, which shows two active fractions, was analyzed, this investigation resulted in 60

TABLE 3
Content of trypsin inhibitors of various tissues

Material	ImE per g i.e. ml	Author*
Submandibulary gland, dog	10,000	Trautschold et al. 1963
Seminal vesicle, guinea pig	5,000	Haendle et al. 1965
Bull sperm	3,000	Haendle et al. 1965
Soybean	2,200	Rackis and Anderson 1964
Colostrum	1,700	Laskowski et al. 1957
Ovomucoid	1,407	McDonnell et al. 1954
Serum	860-1,260	Bundy and Mehl 1958
Beef lung	570	Werle 1958
Mammalian pancreas	130-175	Fritz et al. 1965
Human sperm	125-150	Haendle et al. 1965
Urine	0-7.5	Faarvang 1959

*Trautschold, I., E. Werle, H. Haendle, and H. Sebening: Hoppe-Seylers Z. physiol. Chem. 332, 328 (1963).
Haendle, H. et al.: See references.
Rackis, J. J., and R. L. Anderson: Biochem. biophys. Res. Commun. 15, 230 (1964).
Laskowski, M., B. Kassell, and G. Hagerty: Biochim. biophys. Acta 24, 300 (1957).
McDonnell, L. R., E. D. Ducay, T. F. Sugihara, and R. E. Feeney: Biochim. biophys. Acta 13, 140 (1954).
Bundy, H. F., and J. W. Mehl: J.clin.Invest. 37, 947 (1958).
Werle, E.: in E. Werle and W. Appel:, Z.Naturforsch. 14b, 385 (1959).
Fritz, H., et al.: See references.
Faarvang, H. J.: Acta endocr. (Kbh.) 31, 117 (1959).

amino acids (Table 4). The calculated molecular weight of this inhibitor is about 6,500; however, determination by the gel filtration gave a value of 14,000 ± 300.

The calculated molecular weight of Trasylol based on its structure is 6,511, and the value obtained from its sedimentation constantly is 11,600. The value according to our gel filtra-

TABLE 4
Amino acid composition of dog's submandibulary trypsin inhibitor

	Fraction I*	Fraction II†
Lys	4	5
His	1	2
Arg	2	2
Asp	7	7
$CySO_3H$	2	
Thr	4	4
Ser	5	5
Glu	5	4
Pro	4	5
Gly	5	5
Ala	4	3
Cys 1/2	5	6
Val	2	1
Met		1
Ileu	3	3
Leu	3	3
Tyr	2	2
Phe	2	2
Number of amino acids	60	60
Calculated molecular weight	6,545	6,545

*Lower electrophoretic mobility.
†Higher electrophoretic mobility.

tion method is 6,700 ± 200. The pancreatic and sex gland inhibitors also show molecular weights between 6,500 and 7,000.

All four inhibitors are positively charged polypeptides, and very stable in acid and neutral pH.

On the basis of inhibition of trypsin, a property common to most of the proteinase inhibitors, we propose to replace the biological kallikrein-inhibitor unit (KIU) by a unit which is independent of the biological test conditions and of the degree of purity of the trypsin or inhibitor preparation used. One international inhibitor unit is the amount of inhibitor which inhibits the action of one international trypsin unit. With a new specific trypsin substrate, benzolyarginine-p-nitroanilide with a molar extinction coefficient of 9.95×10^6 cm^3/mole, one inhibitor unit is that amount which reduces the change of light absorption by 3.32 Δ E/min at 405 mμ in a total volume of 3 ml. The numerous other kinetic data we have obtained (Trautschold, 1965) cannot be discussed here.

The enzyme inhibitor complex dissociates more or less readily in acid solution. The Trasylol trypsin complex is the most stable one, with an extrapolated inhibitor constant at pH 7.8 of about 2×10^{-11} mole/liter, whereas the kallikrein complex is already completely dissociated at pH 4.0, with an inhibitor constant at pH 7.8 of 1.2×10^{-8}. The value for the chymotrypsin complex lies between these two values. According to these dissociation constants, the kallikrein may be displaced by trypsin in its binding with Trasylol. Trasylol, which was isolated from its complex with trypsin or kallikrein, is not different from the original inhibitor molecule. The trypsin-submandibulary inhibitor complex is quite unstable. At pH 7.8, its inhibitor constant is 3.4×10^{-7} mole and the complex completely dissociates at pH 5.0 (Table 5).

TABLE 5
Dissociation Constants of Trypsin and Kallikrein Complexes with the Inhibitors from Beef Organs and Dog's Submandibulary Gland at Various pH Values

	Trypsin-Trasylol complex		Kallikrein Trasylol complex		Trypsin-submandibulary inhibitor complex	
pH	K_i (mol/l)	pK_i	K_i (mol/l)	pK_i	K_i (mol/l)	pK_i
2.0	4.5×10^{-4}	3.4				
3.0	2.4×10^{-6}	5.6				
4.0	2.6×10^{-3}	8.6		0		
5.0			1.9×10^{-5}	4.7		0
6.0	1.3×10^{-10}	9.9	3×10^{-6}	5.5	2.2×10^{-6}	5.7
7.8	2×10^{-11} extrapolated	10.7	1.2×10^{-8}	7.9	3.4×10^{-7}	6.5

In serum of mammals we could demonstrate at least two specific high-molecular inhibitors of all kallikreins. The first causes an immediate inhibition of about 2-3 kallikrein units/ml serum. During further incubation of up to 4 hours, an inhibition of an additional 3-4 biological kallikrein units occurs.

Serum which was heated for 1 hour at 56°C shows no more time-dependent inhibition, but inhibits kallikrein spontaneously like untreated serum. In electrophoresis, both inhibitors migrate with the α-globulin region, the spontaneously acting inhibitor, however, a little faster than the time-dependent one.

The α_1-inter α- and α_2-serum trypsin inhibitors (pure preparations) do not inhibit kallikrein at all, not even serum kallikrein, which is closer to trypsin.

As a model for the events in acute pancreatitis we first used autolyzed homogenates of pancreatic tissue. At no time during autolysis could we detect active trypsin. We found a proteolytic activity, however, even in the freshly prepared homogenates, which could not be completely inhibited by Traysylol or by the other above-mentioned trypsin inhibitors. The increase of this "spontaneous proteolytic activity," as we call it, can be inhibited, however, by Trasylol. During autolysis, the amounts of trypsinogen and chymotrypsinogen were reduced; kallikreinogen was partially activated.

In agreement with other investigators, we obtained similar results with pancreas tissues of various animals at different stages of acute experimental pancreatitis.

The polyvalent trypsin and kallikrein inhibitor Trasylol has been used successfully for treatment of acute pancreatitis for several years. This positive therapeutic effect was also seen in experimental pancreatitis. The hypothesis for explaining the mode of action was based primarily on the inhibition of a premature intrapancreatic trypsinogen activation, which means inhibition of proteolytic autolysis. The reduction of kininogen level in blood, which we could show during experimental and human pancreatitis, and the additional finding that no activation of trypsin is demonstrable in pancreatic tissue, might be the proof that Trasylol acts by inhibiting the liberation of kinins by kallikrein. We distinguish two kinds of kinin liberation. The first is a local one which is catalyzed by specific tissue kininogenases, especially by kallikreins, or by a local extravasal activation of serum kallikrein system. The second type is a generalized kinin liberation by serum kallikrein, which is activated by unspecific mechanisms (Fig. 1).

The first general response of pancreatic tissue to any irritation is the formation of an intrapancreatic edema. The capillaries and connective tissue system are damaged in the same way as the edema progresses. The escape of fluid from the capillary system which also contains serum kallikreinogen

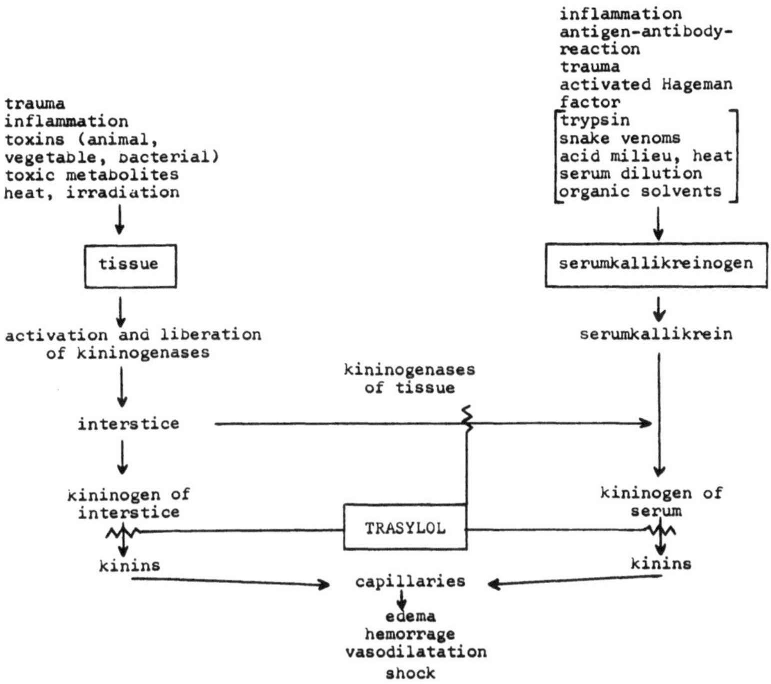

Fig. 1. Pathological kinin liberation and its inhibition by Trasylol.

enhances the edema. During these inflammatory reactions, serum kallikreinogen can be activated, which leads this and pancreatic kallikrein as well to a local kinin liberation (Fig. 2). The kinins in turn increase the edema by enhancing capillary permeability. The capillary-dilating effect will cause hemostasis in the inflamed area. By this time the edema reaches an irreversible state in which the maximally secreting acinus cells become relatively insufficient and may now be attacked by proteolytic enzymes. The activation of the still unknown autolytic enzyme system takes place during the course of events and causes micro- and macronecrosis. At this state of pancreatitis an escape of serum kallikreinogen-activating enzymes into the blood stream is possible and may lead to the generalized kinin liberation.

In these pathological conditions, Trasylol may have the following actions: it inhibits the local kinin liberation and

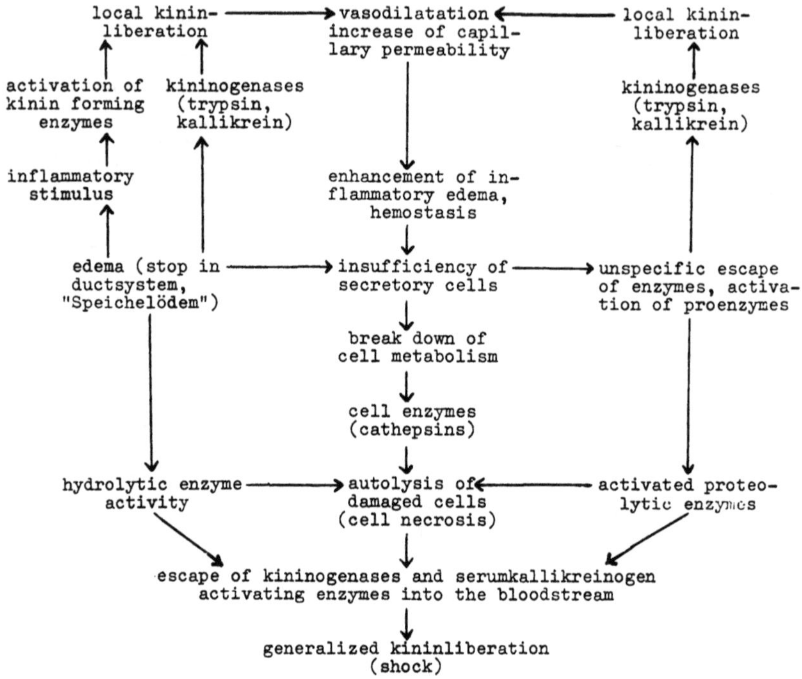

Fig. 2. Scheme of the pathophysiological reactions in the course of acute pancreatitis.

thereby stops the enhancement of edema. Furthermore, it inhibits the increase of the autolytic ("spontaneous proteolytic") activity. The third mode of action is an inhibition of generalized liberation on kinin and a prevention or reduction of shock. Another possible therapeutic action is a reduction of the secretion rate of excretory enzymes, as proposed by Schultis and Rick (1964). These mechanisms (Trautschold 1965) warrant an application of high Trasylol doses at the very beginning of acute pancreatitis.

Dogs and rats are not convenient experimental animals for studying the therapeutic action of Trasylol, since this substance as we found, does not inhibit the pancreatic kallikrein of these two species. The decrease of kininogen concentration in blood during shock is more pronounced than during acute pancreatitis. The normal range of 15-20 μg liberated bradykinin per ml, which we found with a modified Diniz-Carvalho (1963) method is reduced in shock patients to an average of 6.5 μg/bradykinin/ml (range 2-14 μg). In the experimental endotoxin shock in dogs which is induced by injection of celom fluid of *Ascaris lumbricoides* or by injection of Proteus endotoxin, the kininogen level is reduced to about 5 μg/ml. Trasylol given immediately after the induction of shock reduces the decrease of kininogen level and blood pressure and causes a faster return to normal kininogen blood level than in untreated animals. There is also a reduction of mortality rate of treated animals.

During pancreatitis and shock, we were not able to demonstrate free kinins in the blood stream, not even with a modification of existing methods which give a recovery of 80-90%. After the injection of 100 μg bradykinin into the femoral artery of a dog we found a maximal recovery of only 30% in the total venous outflow. If we injected 1 mg bradykinin into the aorta descendens by a catheter, the kinin content in the blood, withdrawn simultaneously from the femoral artery of one side, was 0.1 μg bradykinin/ml in the first minute; after 3 min it was only 0.04 μg/ml. The total recovery was less than 1% of the injected bradykinin. This is in no relation to the distribution of blood volume from the aorta descendens. The low recovery cannot be explained only by the inactivation of kinin by serum enzyme. We estimated the kinin-destroying capacity of different tissue extracts of dogs. The highest values with an inactivation of 220 μg bradykinin/g fresh tissue per min at 25°C were found in kidney, followed by liver, lung, spleen, brain, intestine, serum, and erythrocytes. In plasma the kinin destruction is lower than in serum. A rough differentiation of the enzymes which participate in kinin destruction is possible by studying the influence of some enzyme blocking agents like thiols or SH-blocking substances. Chelating agents and thiol compounds which inhibit mainly the kininases such as carboxypeptidases, according to Erdös and Wohler, and aminopeptidases are less effective in inhibiting kinin destruction by tissue kininases. Thioglycollic acid is not only a potent kininase inhibitor but also increases the yield of kinin liberation by transforming kininogen molecules into the SH-form, which then can be attacked by kininogenases. The fact that, for example, the kinin-destroying enzymes of liver are activated by cysteine and strongly inhibited by p-chloromercuribenzoic acid indicates the participation of a catheptic activity. We met similar conditions in splenic and intestinal tissues. Complex enzyme systems, probably containing peptidases and catheptic enzymes, exist in lung and kidney. Kininases of erythrocytes differ from those of serum, showing almost no inhibition by thiol compounds such as thioglycollic acid or mercaptoethanol, whereas ethylmaleinimide shows a pronounced inhibition.

REFERENCES

Anderer, F. A.: Zur Identität des Kallikrein-Inaktivators aus Rinderlunge und Rinderparotis. Z. Naturforsch. 20b, 499 (1965).

Chauvet, J., G. Nouvel, and R. Acher: Structure primaire d'un inhibiteur pancreatique de la trypsine. Biochim. biophys. Acta 92, 200 (1964).

Diniz, C. R., and I. F. Carvalho: A micromethod for determination of bradykininogen under several conditions. Ann. N. Y. Acad. Sci. 104, 77 (1963).

Erdös, E. G., and J. R. Wohler: Inhibition in vivo of the enzymatic inactivation of bradykinin and kallidin. Biochem. Pharmacol. 12, 1193 (1963).

Fritz, H., I. Trautschold, G. Hartwich, and E. Werle: Neue körpereigene Inhibitoren für tryptische Enzyme. Fed. European Biochemichal Soc., Second Meeting, Vienna (1965).

Habermann, E.: Trennung und Reinigung von Pankreaskallikreinen. Hoppe-Seyler's Z. physiol. Chem. 328, 15 (1962).

Haendle, H., I. Trautschold, H. Fritz, and E. Werle: Über einen neuen Trypsininhibitor. Herbstagung Ges. Physiol. Chem. Berlin (1965).

Kassell, B., M. Radicevic, M. J. Ansfield, and M. Laskowski: The basic trypsin inhibitor of bovine pancreas IV. Biochem. biophys. Res. Commun. 18, 255 (1965).

Moriya, H.: Chemical studies on callicrein VIII. Yakugaku Zasshi 79, 1451 (1959).

Schultis, K., and W. Rick: Zur Beeinflussung der Pankreasfunktion beim Menschen durch Proteinaseinhibitoren. Verh. dtsch. Ges. inn. Med. 70, 824 (1964).

Trautschold, I., and E. Werle: Spektrophotometrische Bestimmung des Kallikreins und seiner Inaktivatoren. Hoppe-Seyler's Z. physiol. Chem. 325, 48 (1961).

Trautschold, I.: Enzyme und Inhibitoren der Bildung und des Abbaus gefässaktiver Polypeptide (Kinine). Habilitationsschrift Med. Fak. Univ. München (1965).

Werle, E., and I. Trautschold: Kallikrein, kallidin, kallikrein inhibitors. Ann. N.Y. Acad. Sci. 104, 117 (1963).

DISCUSSION

Erdös: Dr. Massion and I have used Trasylol (2,000 KIU/kg) in dogs in endotoxin shock (CARI Symposium, Bacteremic Shock Mechanisms and Therapy; in press). Administration of Trasylol improved the rate of survival in these animals and reduced the hypotension.

Margolis: In human plasma, you found 15-20 μg bradykinin-Eq/ml. Our results show consistently less than 6 μg/ml, and Webster found even less. Diniz's latest values are lower than his previous estimates.

Trautschold: We are adding thioglycollic acid to the serum to be determined and add more of this substance after destroying the kininases. We found that thioglycollic acid not only inhibits the kininases which are still left in the serum and those from the added kininogenase but also increases kinin yield in transforming perhaps oxidized kininogen in the SH form which then can be attached in a specific way. Without these precautions our kininogen values would be below 10 μg bradykinin/ml.

Back: Are your serum kallikrein inhibitors different from the serum plasmin inhibitors? Norman has described a fast and slow natural inhibitor of plasmin in serum.

Trautschold: The isolated plasmin inhibitor of plasma is identical with the α-trypsin inhibitor which does not inhibit kallikrein. The two kallikrein-inhibiting principles do not only inhibit serum kallikrein but also glandular kallikrein or urinary kallikrein. There is no evidence that they can also inhibit plasmin.

Sicuteri: There are differences between the clinical and pharmacological effects of Trasylol in serum. The threshold dose (hypotension) of trypsin is not inhibited by 100,000 units of Trasylol. The threshold hypotensive dose of kallikrein (10-15 U of Padutin) is partially inhibited by Trasylol. In the clinical field encouraging results have been observed by using it in subarachnoid hemorrhage for reduction of headache and meningism. Others claim reduction of pain and improvement in shock in myocardial infarction.

Trautschold: There are differences between the inhibition of enzymes by Trasylol <u>in vitro</u> and <u>in vivo</u>. We also found only a partial inhibition of kallikrein <u>in vivo</u>. In the case of trypsin the administration of about 100,000 KIU of Trasylol will increase the trypsin-inhibiting capacity of blood by only a small percentage. In tissues the rise in inhibitor content due to injected Trasylol might be much higher. If there is trypsin activity left, it might be explained by Haverback's finding. According to him, trypsin is bound to a serum protein, and it is protected in this complex which is still active against inhibition by inhibitors.

INACTIVATION AND POTENTIATION OF THE EFFECTS OF BRADYKININ*

ERVIN G. ERDOS and HSIU Y. T. YANG

Department of Pharmacology,
University of Oklahoma School of Medicine,
Oklahoma City, Okla., U.S.A.

During the last few years we have been studying various aspects of the enzymatic metabolism of bradykinin and kallidin. Our aims were to find agents that potentiate the effect of kinins, to develop methods for blocking the activity of the peptides and releasing enzymes in vivo, and to characterize some enzymes that inactivate the peptides.

Potentiation in vivo

First of all we were searching for compounds that can enhance the cardiovascular effects of kinins, presumably by inhibiting the inactivating enzymes in vivo. Werle and Grunz showed in 1939 that cysteine potentiates the hypotensive effect of kallidin, although the mechanism of the enzymatic inactivation of the peptide was not known until much later. In 1961 we attributed the inactivation of bradykinin and kallidin in human blood plasma to a carboxypeptidase, named carboxypeptidase N (Erdös, 1961). Similar or identical enzymes were found in the plasmas of various animals (Erdös and Sloane 1962, Erdös et al. 1963a, 1964, 1965). This enzymatic hydrolysis of bradykinin and kallidin was inhibited by various com-

*This work was supported in part by Grants NB 05196 and HE 08764 from the National Institutes of Health, United States Public Health Service.

pounds; chelating agents were especially active in vitro. Ferreira and Rocha e Silva suggested (1962) the in-vivo use of BAL (2,3-dimercapto-1-propanol) to block the inactivation of bradykinin. Erdös and Wohler (1963a,b) found nine compounds that potentiate the cardiovascular effects of bradykinin in the guinea pig. Since many of these agents inhibit the inactivation in vitro of bradykinin in blood plasma, it is assumed that the potentiation of bradykinin and kallidin rests in the first place upon the inhibition in vivo of the enzymatic metabolism of the peptide.

This work was expanded, and other inhibitors were tested and found active in the various laboratory animals. Figure 1 shows the results of experiments done in the rat. Injection of

POTENTIATION OF THE CARDIOVASCULAR EFFECTS OF BRADYKININ AND KALLIDIN IN THE RAT BY 2-MERCAPTOETHANOL

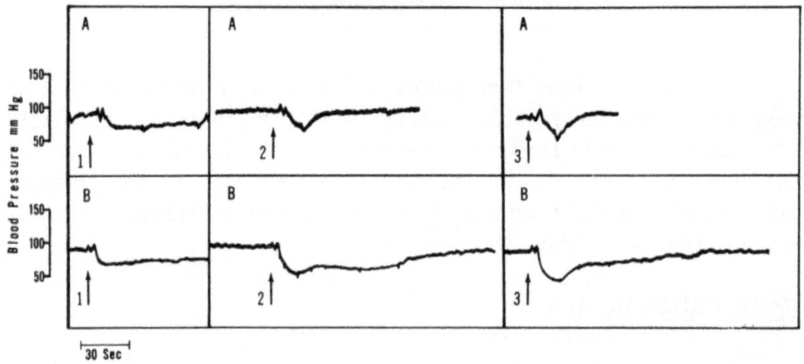

Fig. 1. Rat weighed 240 g. The blood pressure in the carotid artery was measured in urethan narcosis. Between A and B, 2-mercaptoethanol (13 mg/kg) was injected twice i.v. At 1, eledoisin, 0.4 μg/kg; at 2, bradykinin, 4 μg/kg; at 3, kallidin, 1 μg/kg.

2-mercaptoethanol potentiated the hypotensive effect of kallidin and bradykinin, without influencing the action of eledoisin. At the dose level of 10-15 mg/kg i.v., 2-mercaptoethanol is quite specific in that it sensitizes the animals to the kinins only. The effect of such an amount of 2-mercaptoethanol is short lasting. At the higher dose level, when 50-60 mg 2-mercaptoethanol/kg was injected i.v., the effect of eledoisin was also potentiated in some of the animals (experiments were done with the cooperation of Dr. G. Cano).

In the dog, various compounds were found which enhance the cardiovascular effects of bradykinin and kallidin (Erdös,

1966). This potentiation is particularly impressive when the peptides are injected into a tributary of the superior mesenteric vein of the dog. Bradykinin or kallidin has to be given in doses 2 to 5 times higher into the mesenteric vein than in the femoral vein for an equal hypotensive response in the systemic arterial blood pressure.

In the experiment shown in Fig. 2, 1 μg bradykinin/kg lowered the systemic arterial blood pressure about as much as 0.2 μg bradykinin/kg did when injected into the femoral vein of the same dog. This discrepancy in doses required probably could be explained by the increased inactivation of the peptide during the passage through the splanchnic circulation. When the animal was pretreated with two i.v. injections of BAL (12 resp. 23 mg/kg), the effect of bradykinin was greatly enhanced. This was particularly noticeable after bradykinin was administered via mesenteric vein. The drop in mean systemic arterial blood pressure increased and was longer lasting than before BAL was given.

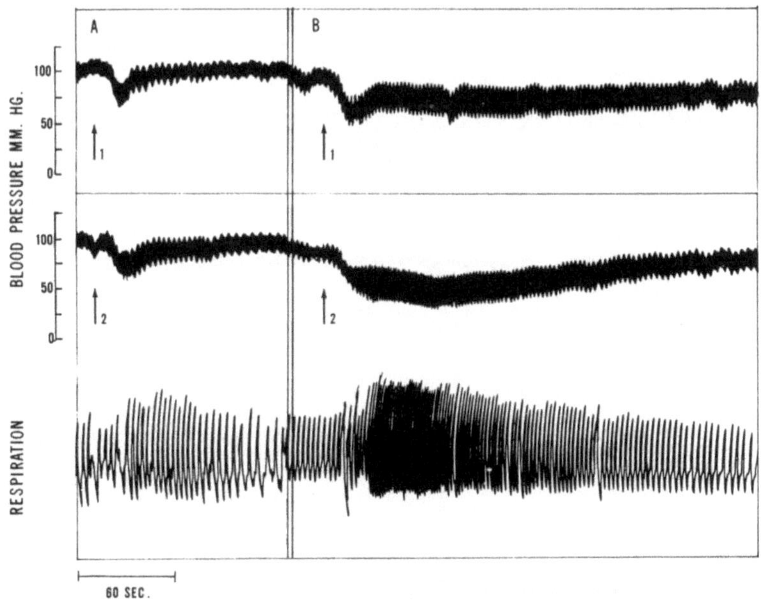

Fig. 2. Comparison of the hypotensive effect of bradykinin after injecting it into the femoral vein (at 1, 0.2 μg/kg) with the effect of the injection of the peptide in a tributary of the superior mesenteric vein (at 2, 1 μg/kg). Between A and B, BAL was given (11-kg male mongrel dog in pentobarbital narcosis). Blood pressure was measured in the carotid artery.

Blocking the effect of bradykinin

In contrast to such agents as histamine, serotonin, or acetylcholine, no specific antagonists of kinins have been yet found which could block their effects in a wide range of tests in various experimental animals. The observation of the very rapid hydrolysis in vitro of bradykinin by purified hog pancreatic carboxypeptidase B (Erdös, 1961) led to the application of this enzyme in vivo (Erdös et al., 1963a,b). Intravenous injection of carboxypeptidase B blocked the cardiovascular effects of bradykinin or kallidin in guinea pigs, cats, dogs, rabbits (Erdös et al. 1963a,b), and even in an alligator (the last experiment was done with the cooperation of Dr. P. S. Güth).

In the current studies, injection of 3-6 mg purified pancreatic carboxypeptidase B per kg abolished the hypotensive effects of bradykinin and kallidin in the rat (Fig. 3). Larger doses were needed, however, to block the activity of Met—Lys—bradykinin.

In the cat, after a single i.v. injection of carboxypeptidase B, the level of the enzyme decreases in the blood exponentially. The half-life of the enzyme is about 30 min. The decrease in

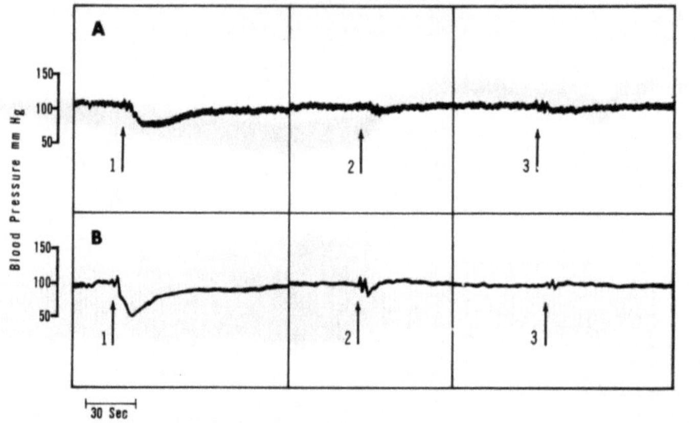

Fig. 3. Rat A weighed 270 g; rat B 300 g; urethan narcosis.
A: i.v. injection of 3.6 μg kallidin/kg at 1 and 3; 3.6 mg carboxypeptidase B/kg at 2.
B: i.v. injection of 3 μg bradykinin/kg at 1; 6.6 μg/kg at 3, and 6.7 mg carboxypeptidase B/kg at 2.

the level of the circulating enzyme and the return of the hypotensive effect of injected bradykinin are closely related (Erdös et al., 1963b). Using lines I and II of Fig. 4 we calculated a maintenance dose for carboxypeptidase B. Starting 20 min after the first injection, 15% of the original dose of the enzyme (2 mg/kg) was given in the femoral vein at 10-min intervals, and the high level of carboxypeptidase B was maintained in blood with the resulting complete block of the hypotensive effect of bradykinin. (The peptide was injected every 15 min into the contralateral femoral vein.)

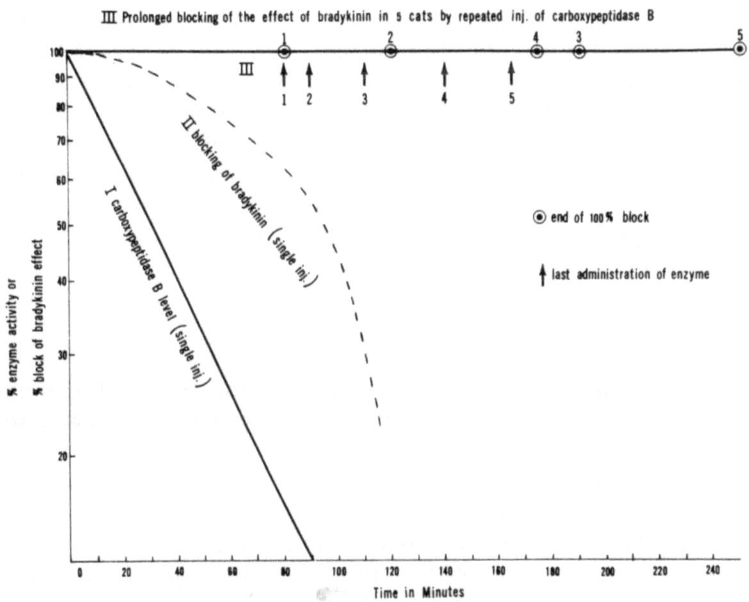

Fig. 4. Prolonged blocking of the effect of bradykinin on the mean arterial blood pressure in the cat. For details, see text. Lines I and II were taken from a publication of Erdös et al. (J. Pharmacol. exp. Ther. 142, 327, 1963).

Line III of Fig. 4 shows the time of injection of the last dose of the enzyme and the subsequent end of the 100% blocking of the effect of bradykinin in 5 cats. When the administration of carboxypeptidase B was stopped, the effect of the peptide started to return. Thus, repeated i.v. injections of carboxypeptidase B can block the effects of bradykinin in the cat for hours.

Fate of injected carboxypeptidase B

When exogenous carboxypeptidase disappears from the circulation of various animals, it is taken up by the tissues. In the dog hind limb it appears first in the lymph (Webster et al., 1965). In our current studies, injected carboxypeptidase B was recovered in active form from the kidney of dog, cat, guinea pig, and rat. Highest concentrations were found in the guinea pig and rat kidney. In nine guinea pigs, 3.4×10^{-2} μ-mole of hippuryl-L-arginine was hydrolyzed per minute per mg kidney (range $0.8 \times 9 \times 10^{-2}$). The concentration in the rat kidney was about the same as in the guinea pig. The rat liver showed a significant activity, but the amount of enzyme recovered from the liver of other animals was negligible. Cat and dog kidney also concentrated from the blood a sizable fraction of the injected carboxypeptidase B. Large amounts of the enzyme were excreted in the urine of guinea pig, rat, and cat. Spleen, brain, and lung showed negligible activity, indicating that carboxypeptidase B was not removed from the circulation by these organs.

Preparation of carboxypeptidase B

The swine pancreatic carboxypeptidase used in these experiments was purified in our laboratory from acetone-dried swine pancreatic powder (Folk et al., 1960). Usually, purified carboxypeptidase preparations have to be kept in frozen solution. The purified enzyme does not withstand repeated freezing and thawing, especially if it contains less than 10 mg protein/ml (Folk, personal communication). We found it convenient to prepare carboxypeptidase B in the form of a dry, stable powder. Carboxypeptidase B could be lyophilized in 0.3 M neutralized glycine with about 20% to 40% loss of the original activity. Of the three freeze-dried preparations tested, two were kept in a deep freeze and one at room temperature. No further loss of activity was observed in any of the preparations during several months of storage. (The activity of the enzyme was determined with hippuryl-L-arginine substrate.)

Esterase activity of plasma carboxypeptidase

Plasma carboxypeptidase in its native or 100-fold purified form has a very significant esterase activity in that it hydrolyzes the ester substrate hippuryl-L-argininic acid even faster than hippuryl-L-lysine. This ester is also cleaved by pancreatic carboxypeptidase. Cadmium strongly activated pancreatic carboxypeptidase B with the ester substrate, and cobalt inhibited it (Folk and Gladner, 1961). The esterase activity of the carboxypeptidase in swine or human plasma was inhibited by low concentration of cadmium sulfate (10^{-6} M). This finding is taken as additional evidence to show that the pancreatic and plasma carboxypeptidase are not identical.

Inactivation of bradykinin in the kidney

Bumpus et al. (1964) indicated that infused bradykinin which escapes inactivation in the circulation of the rat is taken up by the kidney. Various groups of investigators (Frey et al., 1950; Hamberg and Rocha e Silva, 1954; Fasciolo 1964; Carvalho and Diniz, 1964) have shown that homogenized kidney inactivates bradykinin and kallidin. We intended to characterize this enzymatic process, which is important since it obviously contributes to the homeostatic balance of the body (Erdös and Yang, 1965).

The first approach consisted of investigating the role of various subcellular fractions of the homogenized rat kidney. The inactivation of bradykinin by these extracts was assayed on the isolated rat uterus (Fig. 5). Most of the activity of the homogenized rat kidney was concentrated in a fraction that contained the microsomes. (These experiments were repeated with essentially similar results eleven times.) The bradykininase could be separated from the ribosomes in 0.026% (w/v) desoxycholate solution in the ultracentrifuge at 144,000 × *g* in 2 hours. In control experiments, the distribution of alkaline phosphatase in the subcellular fractions was very similar to the distribution of the enzyme or enzymes that inactivate bradykinin. Alkaline phosphatase in the rat kidney is localized mainly in the microsomal fraction (Shibko and Tappel, 1965). There was no correlation between the distribution of lysosomal acid phosphatase and bradykininase in the homogenized kidney. The relatively low activity of the lysosomal-light mitochondrial fraction was probably due to microsomal contamination. These findings were supported by preliminary studies with the electron microscope. (Dr. H. Helander of Göteborg cooperated in this phase of the work.) Hardly any bradykininase was found in the soluble fraction. This is in contrast to the high rate of

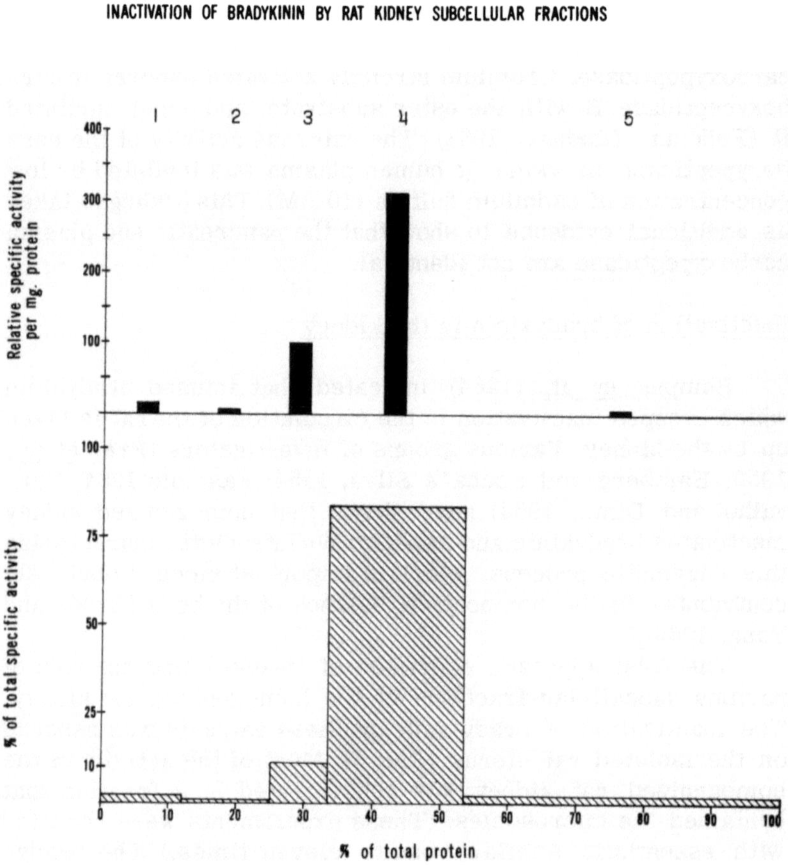

Fig. 5. Rat kidney was homogenized in 0.25 M sucrose. The inactivation of bradykinin was measured on the isolated rat uterus. Fraction 1: nuclei plus cell debris. Fraction 2· heavy mitochondria. Fraction 3: light mitochondria plus lysosomes. Fraction 4: microsomes. Fraction 5: final supernatant. Most of the activity was located in the microsomal fraction.

destruction of bradykinin activity by the soluble, final supernatant fraction of the homogenized rat liver as found by Amundsen and Nustad (1965) and confirmed by us. The enzymatic activity in the kidney was much higher than in blood plasma or in the liver. The enzymes in the microsomal and in the lysosomal fractions of rat kidney are active at a neutral but not acid pH. Practically all the enzymatic activity ceases at pH 5.

The microsomal enzyme or enzymes are inhibited completely by chelating agents such as 1,10-phenanthroline or EDTA and by $CdSO_4$. Urgocyton, a fluorescent material from liver which was previously used in another system to preserve the activity of bradykinin (Amundsen et al., 1964), inhibits the inactivation of bradykinin by the microsomal enzymes. The possibility exists that urgocyton acts as a chelating agent. When $CoCl_2$ is added to a kidney carboxypeptidase-urgocyton complex, the inhibitory effect disappears, and the activity of the enzyme is restored by $CoCl_2$. Another suggested mechanism for this inhibition would consist of binding of urgocyton by cobalt, and thereby preventing a direct inhibition of the enzyme by urgocyton.

In order to characterize the action of the enzymes in kidney we needed a fair amount of starting material. For large-scale extraction, swine kidney cortex was the source of enzymes prepared as a stable acetone-dried powder. This material was suspended in 2% NaCl and precipitated between 45% and 65% saturation of the solution with ammonium sulfate. The active material was adsorbed on a DEAE-Sephadex column and eluted stepwise with NaCl in a 0.05 M Tris buffer of pH 7.8. The activity peak appeared with 0.15 N NaCl. This fraction still contained at least two enzymes capable of hydrolyzing bradykinin, as shown by biological and chemical techniques.

The substrates used were bradykinin, Ac—Ser—Pro—Phe—Arg, a tetrapeptide representing the last four, and Ac—Phe—Arg dipeptide representing the last two amino acids in the nonapeptide chain of bradykinin. The N-terminal amino acids of the tetrapeptide and dipeptide were acetylated to prevent hydrolysis by aminopeptidases or dipeptidases. We used bradykinin and the other two peptide substrates of related structure in peptide mapping, which revealed that most of the activity is due to an endopeptidase. The other enzyme is an exopeptidase, a carboxypeptidase. Both are active at a neutral pH. They were inhibited by chelating agents such as EDTA, 1,10-phenanthroline, or BAL, but were not affected by *p*-chloromercuriphenyl sulfonic acid. Inhibitors of endopeptidases (Trasylol and DFP) were equally inactive. 2-Mercaptoethylamine, an activator of cathepsins, inhibited the endopeptidase. The properties of this endopeptidase are different from those of other previously described mammalian endopeptidases.

The results of peptide mapping are shown in Fig. 6. The endopeptidase in this kidney fraction splits the Pro^7—Phe^8 bond both in bradykinin and in its tetrapeptide derivative. (We have seen previously that the same bond is hydrolyzed by an enzyme in a microbial collagenase preparation from Clostridium histolyticum as well.) The Phe—Arg released by the kidney endopeptidase is broken down further by a dipeptidase contamination upon prolonged incubation. D-Pro[7] derivative of bradykinin is not hydrolyzed by the enzyme.

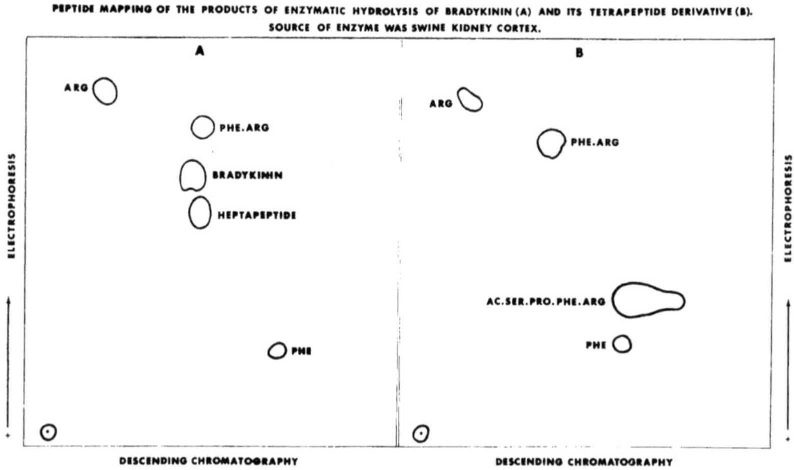

Fig. 6. Peptide mapping of the products of enzymatic hydrolysis of bradykinin (A) and its tetrapeptide derivative Ac—Ser—Pro—Phe—Arg (B). The incubation mixture for A contained: 0.1 M Tris buffer (pH 7.4), 500 μg bradykinin, 10 μg enzyme preparation (endopeptidase). The mixture was kept at 31°C for 1 hour. B was the same except that concentration of the tetrapeptide was 10^{-2} M. Descending chromatography was carried out on Whatman 3 MM paper in butanol:water:acetic acid (4:5:1). The chromatography was followed in the second dimension by a 90-min electrophoresis (60 v/cm) in a pyridine:acetic acid:water (1:10:289) buffer of pH 3.5.

Heat treatment of the extract of the acetone powder at 62°C for 10 min destroyed about 75% of the bradykininase activity. The endopeptidase in the extract became inactive, but the carboxypeptidase withstood the heat treatment. This heat-treated preparation was purified on a DEAE-cellulose column and was the source of kidney carboxypeptidase. The eluant in the chromatography column was 0.005 M Tris buffer, pH 7.8, with a stepwise-increased NaCl gradient. The active enzyme appeared with 0.06 N NaCl. The enzyme was purified 60- to 80-fold when the activity was assayed with Ac—Phe—Arg substrate and with a ninhydrin spectrophotometric technique. The enzyme hydrolyzed the bond of the C-terminal arginine in bradykinin and in various related peptides.

The hydrolysis of two substrates by the kidney carboxypeptidase is shown in Fig. 7. When Ac—Phe—Arg was the substrate, *p*-chloromercuriphenyl sulfonic acid was added to the incubation mixture in this experiment to inhibit an acylase contamination. In the current studies this acylase contamination was removed with ammonium sulfate fractionation. The enzyme released C-terminal arginine both from the di- and the tetrapeptide substrates. In addition, after prolonged incubation, a small amount of free phenylalanine appeared from the tetrapeptide substrate.

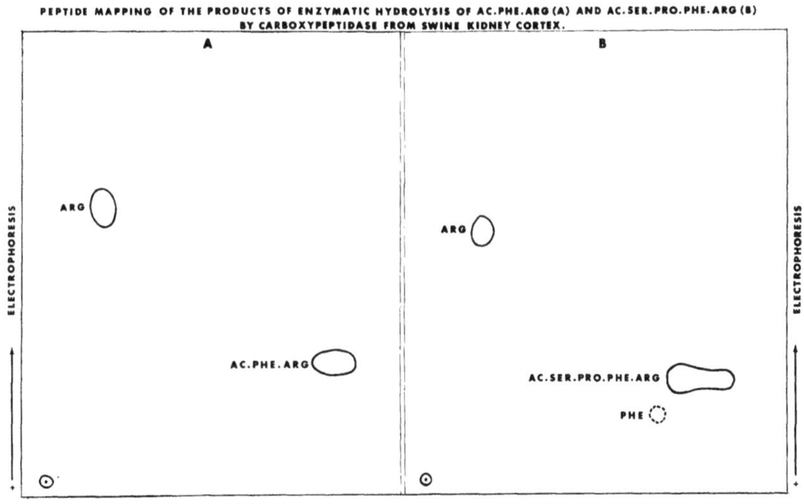

Fig. 7. Peptide mapping of the products of enzymatic hydrolysis of Ac—Phe—Arg (A) and Ac—Ser—Pro—Phe—Arg (B) by carboxypeptidase from swine kidney cortex. Source of enzyme was 30 μg of partially purified carboxypeptidase. The mixture was incubated for 4 hours. For details see Fig. 6.

The inhibition pattern of the kidney carboxypeptidase is shown in Table 1. The data indicate that the enzyme is different from chymotrypsin, cathepsins, catheptic collagenase (Schaub, 1964), and catheptic (Greenbaum and Sherman, 1962) or pancreatic carboxypeptidase (Folk et al., 1960). In Table 2, the activity of human plasma carboxypeptidase is compared with swine kid-

TABLE 1
Inhibition of the Hydrolysis of Bradykinin by Kidney Carboxypeptidase

Inhibitor	Method	
	Peptide mapping	Bioassay
EDTA	+	+
1, 10-Phenanthroline	+	+
DFP	0	0
Trasylol (100 U/ml)		0
p-Chloromercuriphenyl-sulfonic acid	0	
Iodoacetic acid		0
β-Mercaptoethylamine		+
2,2′-Dipyridyl		+

The concentration of the inhibitors fell within the range of 5.10^{-3} and 1.10^{-3} M; + = inhibition; 0 = no inhibition.

ney carboxypeptidase. The carboxypeptidase in human plasma releases the basic C-terminal amino acid when the amino acid preceding arginine is either phenylalanine or glycine, as shown by the high rate of hydrolysis of hippuryl-L-arginine, hippuryl-L-lysine, and acetyl-L-phenylalanyl-L-arginine. The kidney carboxypeptidase, however, cleaved the Ac — Phe — Arg bond much faster than Bz—Gly—Arg or Bz—Gly—Lys bonds. Its activity with hippuryl—L—arginine or lysine was much slower. Thus, it behaved differently from the carboxypeptidase of human and swine plasma or of swine pancreas. Swine kidney carboxypeptidase releases C-terminal arginine fastest at the concentrations of the substrates used (10^{-2} M), when it is linked to phenylalanine, that is, to an aromatic amino acid.

Swine kidney contains a third enzyme that can inactivate bradykinin. The enzyme is prolidase which cleaves the Arg^1—Pro^2 bond of the peptide. This has been proven with the sequential use of peptide mapping and a Spinco amino acid analyzer. The destruction of the biological effects was followed on the isolated rat uterus. The source of enzyme was a partially purified commercial prolidase preparation which was treated with DFP and then activated with $MnSO_4$. The preparation used was much less active than the other two kidney enzymes. The activity of prolidase decreased below neutrality.

TABLE 2
Hydrolysis of Bradykinin and Related Substrates by Carboxypeptidases of Plasma and Kidney

	Human plasma	Swine kidney
Bradykinin	+	+
Ac—Phe—Ser—Pro—Phe—Arg	+	
Ac—Ser—Pro—Phe—Arg	+	+
Ac—Phe—Arg	+	+
Hippuryl-L-lysine	+	slight*
Hippuryl-L-arginine	+	slight*
Hippuryl-L-argininic acid	+	
Hippuryl-L-ornithine	slow	
Hippuryl-L-phenylalanine	0	

The activity of plasma carboxypeptidase was determined in peptide mapping and with a spectrophotometric assay. The activity of swine kidney carboxypeptidase was determined in peptide mapping. The hydrolysis of Ac—Phe—Arg was also assayed with a spectrophotometric ninhydrin assay; + = hydrolysis; 0 = no hydrolysis.

*Compared with the hydrolysis of Ac—Phe—Arg

The enzyme was inhibited by chelating agents and by activators of cathepsins.

Figure 8 summarizes the results of our studies of kidney enzyme. The figure shows the bonds broken by the various enzymes in the peptide chain of bradykinin but does not attempt to evaluate the relative contribution of each enzyme to this inactivation process. The figure also indicates that, although the kidney can accumulate injected bradykinin from the circulation, it has a complex and effective enzyme system capable of inactivating the peptide.

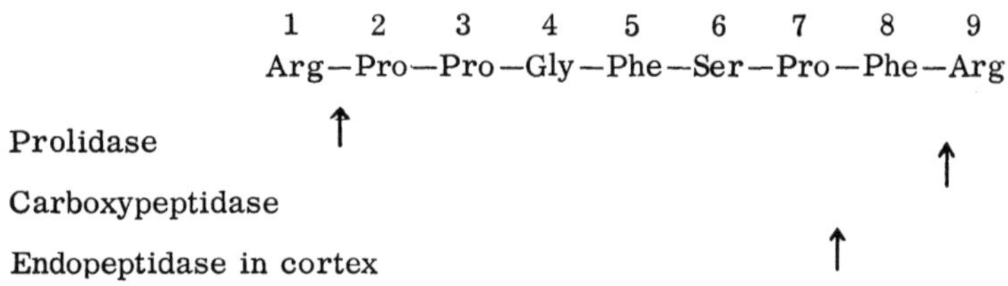

Fig. 8. Inactivation of bradykinin by enzymes in swine kidney

SUMMARY

Studies of the metabolism of kinins are presented. Various compounds were effective in potentiating the hypotension caused by bradykinin, presumably by inhibiting its enzymatic hydrolysis. 2-Mercaptoethanol, for example, enhanced the cardiovascular effect of bradykinin in the rat. In the dog, bradykinin was much less effective when injected into the splanchnic circulation instead of the femoral vein; BAL greatly potentiated the action of the peptide on the mean arterial blood pressure after either route of administration.

Purified pancreatic carboxypeptidase B blocked the effects in vivo of the kinins in rat and other animals, by inactivating the peptides very rapidly in the circulation. Repeated administration of the enzyme can block the hypotensive activity of the kinin in the cat for hours. The fate of the injected enzyme was also scrutinized in various animals. Significant amounts of carboxypeptidase B were recovered from the kidney.

The carboxypeptidase that occurs in blood plasma has an esterase activity as well.

Homogenized kidney hydrolyzed bradykinin very rapidly. Most of the enzymes involved are concentrated in the microsomal fraction of the rat kidney. Swine kidney contains at least three enzymes capable of inactivating bradykinin. One is a prolidase, another is a carboxypeptidase, and the third and most active enzyme is an endopeptidase. The last two have been purified partially and studied by means of various substrates and inhibitors.

ACKNOWLEDGEMENTS

The various peptides used were donated by the following individuals: R. Bircher, M. Bodanszky, J. E. Folk, N. Manning, R. B. Merrifield, E. Nicolaides, E. Schröder, J. M. Stewart, and K. Vogler. The assistance of W. J. Graham, E. M. Sloane, O. Sneed and L. L. Tague in some of the phases of this work is greatly appreciated.

REFERENCES

Amundsen, E., B. A. Waaler, J. Dedichen, B. Laland, S. Laland, and N. Thorsdalen: Kininase inhibition by a fluorescent substance prepared from liver. Nature (Lond.) 203, 1245 (1964).

Amundsen, E., and K. Nustad: Kinin-forming and destroying activities of cell homogenates. J. Physiol. (Lond.) 179, 479 (1965).

Bumpus, F. M., R. R. Smeby, I. H. Page, and P. A. Khairallah: Distribution and metabolic fate of angiotensin II and various derivatives. Canad. med. Ass. J. 90, 190 (1964).

De Carvalho, I. F., and C. R. Diniz: Kinin-forming enzyme (kininogenin) in rat kidney. Ann. N.Y. Acad. Sci. 116, 912 (1964).

Erdös, E. G.: Enzymes that inactivate peptides, in Metabolic Factors Controlling Duration of Drug Action, B. B. Brodie and E. G. Erdös, Eds. Proc. First Int. Pharmacol. Meeting (1961) Stockholm, Sweden, 6, 159 (1962).

Erdös, E. G., and E. M. Sloane: An enzyme in human blood plasma that inactivates bradykinin and kallidins. Biochem. Pharmacol. 11, 585 (1962).

Erdös, E. G., A. G. Renfrew, E. M. Sloane, and J. R. Wohler: Enzymatic studies on bradykinin and similar peptides. Ann. N.Y. Acad. Sci. 104, 222 (1963a).

Erdös, E. G., J. R. Wohler, and M. I. Levine: Blocking of the in vivo effects of bradykinin and kallidin with carboxypeptidase B. J. Pharmacol. exp. Ther. 142, 327 (1963b).

Erdös, E. G., and J. R. Wohler: Inhibitors of the in vivo enzymatic inactivation of bradykinin and kallidin. Life Sci. 2, 270 (1963a).

Erdös, E. G., and J. R. Wohler: Inhibition in vivo of the enzymatic inactivation of bradykinin and kallidin. Biochem. Pharmacol. 12, 1193 (1963b).

Erdös, E. G., E. M. Sloane, and I. M. Wohler: Carboxypeptidase in blood and other fluids. I. Properties, distribution and partial purification of the enzyme. Biochem. Pharmacol. 13, 893 (1964).

Erdös, E. G., and H. T. Yang: Studies of the metabolism of bradykinin. Biochem. Pharmacol. 14, 1391 (1965).

Erdös, E. G., I. M. Wohler, M. I. Levine, and M. P. Westerman: Carboxypeptidase in blood and other fluids. Values in human blood in normal and in pathological conditions. Clin. chim. Acta 11, 39 (1965).

Erdös, E. G.: Hypotensive peptides: Bradykinin, kallidin, and eledoisin, Advances in Pharmacology, S. Garattini and P. A. Shore, Eds. vol. IV. New York: Academic Press (1966).

Fasciolo, J. C.: Formation and destruction of plasma kinins. Acta physiol. lat.-amer. 14, 170 (1964).

Ferreira, S. M., and M. Rocha e Silva: Potentiation of bradykinin by dimercaptopropanol (BAL) and other inhibitors of its destroying enzyme in plasma. Biochem. Pharmacol. 11, 1123 (1962).

Folk, J. E., K. A. Piez, W. R. Carroll, and J. A. Gladner: Carboxypeptidase B. IV. Purification and characterization of the porcine enzyme. J. biol. Chem. 235, 2272 (1960).

Folk, J. E., and J. A. Gladner: Influence of cobalt and cadmium on the peptidase and esterase activities of carboxypeptidase B. Biochim. biophys. Acta 48, 139 (1961).

Frey, E. K., H. Kraut, and E. Werle: Kallikrein. Padutin. Stuttgart: Ferdinand Enke Verlag, (1950).

Greenbaum, L. M., and R. Sherman: Studies on catheptic carboxypeptidase. J. biol. Chem. 237, 1082 (1962).

Hamberg, U., and M. Rocha e Silva: Studies on the enzymatic inactivation of bradykinin. Acta physiol. scand. 30, 215 (1954).

Schaub, M. C.: Eigenschaften und intracelluläre Verteilung eines kollagenabbauenden Kathepsins. Helv. physiol. Pharmacol. Acta 22, 271 (1964).

Shibko, S., and A. L. Tappel: Rat-kidney lysosomes: Isolation and properties. Biochem. J. 95, 731 (1965).

Webster, M. E., N. S. Skinner, Jr., and W. J. Powell, Jr.: Role of kallidins in vasodilatation of skeletal muscle. Fed. Proc. 24, 488 (1965).

Werle, E., and M. Grunz: Zur Kenntnis der darmkontrahierenden, uteruserregenden und blutdrucksenkenden Substanz Dk. Biochem. Z. 301, 429 (1939).

DISCUSSION

Diniz: We are studying a kininogenase in kidney homogenate, and our method of assay depends very much on an efficient inhibition of kininase. We are using hydroxyquinoline as kininase inhibitor, as suggested by Ferreira and colleagues. This inhibitor protects bradykinin, only partially, from kininases. Have you any suggestions how to improve the yield of released bradykinin by kidney homogenate?

Erdös: We can prevent the enzymatic inactivation of bradykinin by the kidney homogenate with 1, 10-phenanthroline or 2.2′-dipyridil.

Collier: In searching for antagonists or inhibitors of kinins that can be used as tools in determining how far this group of peptides is involved in pathological reactions, specificity of inhibition is vital. Can you tell us how far your swine pancreatic carboxypeptidase B is specific toward kinins, compared with other biologically active peptides?

Erdös: Carboxypeptidase B will inactivate those peptides that have a free C-terminal basic amino acid.

THE ROLE OF CATHEPSINS IN THE INACTIVATION OF PLASMA KININS

LOWELL M. GREENBAUM,* and KEIKO YAMAFUJI

Department of Pharmacology, College of Physicians & Surgeons of Columbia University, New York, N. Y., U.S.A.

For several years our laboratory has been concerned with the properties and possible functions of the intracellular proteinases known as the cathepsins. These enzymes exist in the lysosomal particles of the cell (de Duve, 1959) and consist of a series of catalytic agents which, like pancreatic trypsin, chymotrypsin, and pancreatic carboxypeptidases, hydrolyze synthetic peptide substrates with somewhat exacting specificity. They are quite different when compared to their extracellular counterparts in terms of pH optima, requirements for cofactors, and their intracellular location. Despite our knowledge of the presence of cathepsins in almost all tissues, we lack understanding of them in two respects. First, a homogeneous cathepsin preparation has not been prepared, thus preventing comparison of their structure with that of other known proteases. Second, the actual function of these enzymes in the body remains to be elucidated. The cathepsins are of interest here since they seem to have properties necessary for participation in the formation and inactivation of kinins under conditions that may exist in inflammation or tissue injury, e.g. in an acidic environment. Such conditions favor catheptic action.

Figure 1 indicates the best-known pathway for the formation of bradykinin from the protein bradykininogen by catalysis through kallikrein enzymes. A pathway is also outlined for the destruction of bradykinin. One of the interesting things about the complement of cathepsins in a cell is that these are agents

*Career scientist of the Health Research Council of the City of New York.

which theoretically might be able to (a) convert bradykininogen to bradykinin; (b) inactivate bradykinin; and (c) convert zymogens of enzymes such as kallikreins to active enzymes. Such properties are indicated by the light lines drawn in Fig. 1. The experiments discussed below were designed to test some of these possibilities.

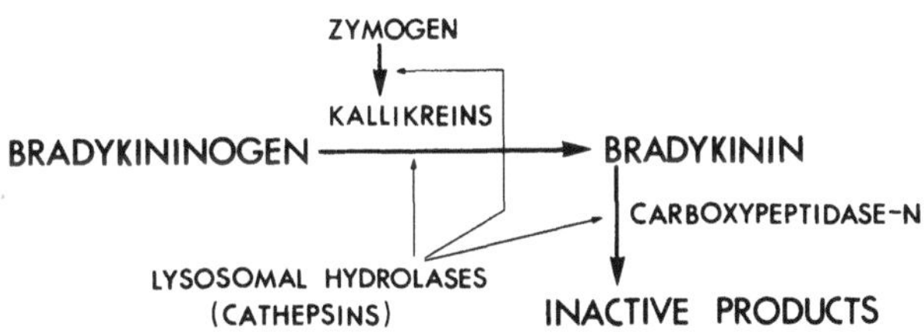

Fig. 1. In vivo formation and metabolism of bradykinin. The light arrows represent possible participation of catheptic enzymes in this scheme.

We first investigated the ability of a cathepsin preparation to degrade bradykinin and related compounds. As a source of enzyme we have used a highly purified preparation from beef spleen. The procedures for this preparation were worked out some years ago in collaboration with Professor Fruton (Greenbaum and Fruton, 1957). The fraction we have used is known as the "Hg-ethanol" fraction and contains the cathepsins listed in Fig. 2. A typical synthetic peptide substrate of each

Catheptic Enzymes in "Hg-ethanol" fraction	Typical Synthetic substrate
Catheptic carboxypeptidase A	Cbz-L-Glu-L-Tyr
Catheptic carboxypeptidase B	Bz-Gly-L-Arg
Cathepsin B	Bz-L-Arg-Am
Cathepsin C	Gly-L-Phe-Am

Fig. 2. The cathepsins present in the "Hg-ethanol" fraction of the spleen extract.

enzyme is also listed. Catheptic carboxypeptidases A and B are enzymes which resemble the pancreatic carboxypeptidases in terms of the synthetic substrates they hydrolyze (Greenbaum and Sherman, 1962). They differ from the extracellular enzymes in that the catheptic enzymes are maximally active at acid pH and require sulfhydryl activators. It should also be noted that catheptic carboxypeptidase B and the kininase of blood, carboxypeptidase N (Erdös et al., 1962), attack hippuryl-L-arginine, but differ considerably in their properties in terms of pH optima and requirements for cofactors. Cathepsin B is a trypsin-like enzyme and has been shown to convert the inactive zymogen of trypsin (trypsinogen) to trypsin (Greenbaum et al., 1959) at acid pH. Cathepsin C appears to be a dipeptidase which acts from the amino-terminal position of a polypeptide under specific conditions (Fruton, 1963).

If bradykinin is incubated with the catheptic preparation at pH 4 in the presence of an SH activator, two major products are

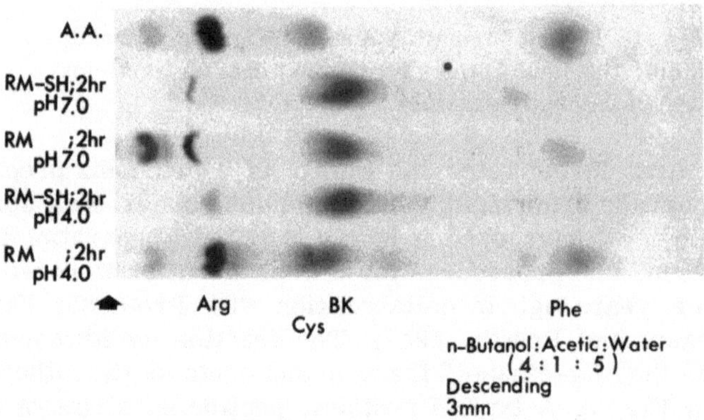

Fig. 3. Descending chromatography of the products of incubations of bradykinin and the cathepsin preparation. The complete reaction mixture consisted of 0.01 M bradykinin + 0.01 M cysteine + catheptic preparation in 0.05 M acetate (pH 4.0) or 0.05 M phosphate buffer (pH 7.0) after 2 hours at 35°C, 0.01 ml spotted on paper. Chromatogram was run for 18 hours and developed for ninhydrin-positive material as previously described (Greenbaum and Yamafuji, 1965). A.A. = Standards of cysteine (and cystine), arginine (darkest spot), and phenylalanine. RM = Complete reaction mixture. RM-SH = Reaction mixture without cysteine.

formed which have been identified by paper chromatography (Fig. 3) as free arginine and phenylalanine, the carboxyl-terminal amino acids of bradykinin. At neutral pH or in the absence of a thiol activator, much less degradation of bradykinin takes place. This property differentiates the cathepsin enzymes from carboxypeptidase N, since the latter enzyme is actually inhibited by sulfhydryl compounds. Similar findings were obtained when methionyl-lysyl-bradykinin was used as a substrate for the catheptic enzymes. Assay of pharmacological activity of bradykinin after incubation with the cathepsin preparation at acid pH and in the presence of sulfhydryl compounds has demonstrated that pharmacological activity is almost completely lost after the catheptic treatment (Greenbaum _et al._; unpublished observations). The conclusion that may be reached from these and other studies (Greenbaum and Yamafuji, 1965) is that the intracellular carboxypeptidases attack bradykinin sequentially at the carboxyl-terminal position of bradykinin (and other related kinins), inactivating the kinins and liberating the amino acids arginine and phenylalanine, as seen in Fig. 4.

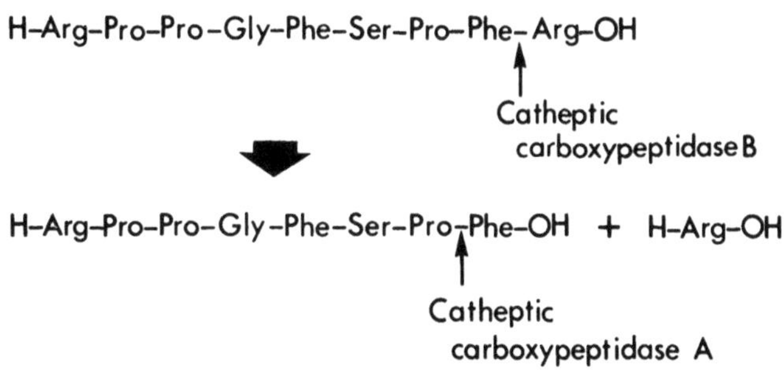

Fig. 4. Sequential degradation of bradykinin by catheptic carboxypeptidase B and catheptic carboxypeptidase A.

Our attention was then focused on the possibility that catheptic enzymes, or enzymes that act preferentially at acid pH, might produce kinins from the precursor protein, bradykininogen. For this study, partially purified bradykininogen prepared from bovine or human blood (Greenbaum and Hosoda, 1963) was incubated at varying pH with the cathepsin preparation and samples of the incubation mixtures tested on the guinea pig ileum for kinin activity. Thiol activators were omitted from these studies to prevent the action of the catheptic carboxypeptidases. As may be seen from Fig. 5, an agent is produced maximally at pH 5.0 which contracts the ileum. Other studies have shown that such an agent is destroyed when incubated with chymotrypsin or carboxypeptidase B, demonstrating its kinin nature. It should also be mentioned that no kininase activity was found at neutral pH in the cathepsin preparation, thus indicating that pH 5.0 was optimal for production of the kinin. It also indicates that the intracellular enzyme producing the kinin differs from the kallikreins of tissues in that little activity was obtained at neutral pH where kallikreins are most active (Webster and Pierce, 1961).

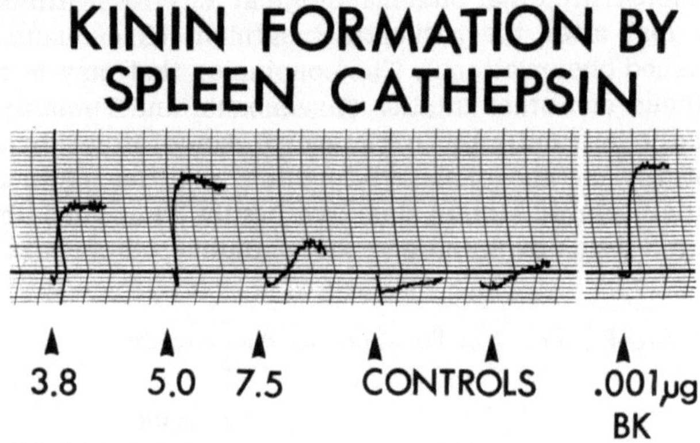

Fig. 5. Bradykininogen + spleen cathepsin preparation at different pH; 5 mg bradykininogen (human) dissolved in 1.0 ml of 0.05 M acetate or Tris buffer at varying pH; 0.05 ml of the cathepsin preparation added, and after 120 min a sample was removed, neutralized, and assayed for its ability to contract the guinea pig ileum. BK = Synthetic bradykinin. Controls, bradykininogen alone; cathepsin preparation alone.

In addition to investigating the activity of enzymes in tissues such as spleen, investigations were carried out with enzymes contained in polymorphonuclear leukocytes, cells which are known to invade inflamed areas. The object was to determine whether, at acid pH, an environment within the leukocyte cell that occurs during phagocytosis (Rogers, 1964), enzymes present in the cell could produce or inactivate kinins. The cells were collected from the peritoneal cavity of rabbits by the procedures of Cohn and Hirsch (1960) after the intraperitoneal injection of glycogen. One modification that was made in the procedure was to lyse the cells in acetate buffer at pH 3.6 rather than use citrate buffer which was found to interfere in the guinea pig ileum assay. The whole lysate was used as a source of enzyme. Human bradykininogen was incubated at acid pH with the lysate preparation and the resulting products tested on the ileum for kinin activity. The results as seen in Fig. 6 demonstrate the production of kinin (RM). Samples of control tubes containing the lysate itself or bradykininogen in the absence of the lysate showed no activity. When the active material produced by the reaction was incubated with pancreatic carboxypeptidase B or chymotrypsin for a short interval of time, all activity was abolished, demonstrating the polypeptide nature of the active material. Thus, strong evidence was obtained that a kinin, presumably bradykinin, may be formed by enzymes contained in leukocytes acting on brady-

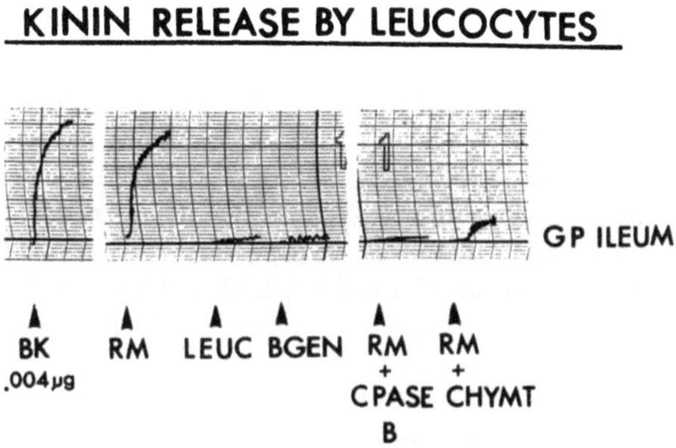

Fig. 6. Bradykininogen + lysate of polymorphonuclear leukocytes (PMN); 10 mg bradykininogen (human) was dissolved in 2.0 ml of 0.2 M acetate buffer at pH 3.5, 0.3 ml of a lysate of PMN added, and after 16 hours 0.01- to 0.05-ml samples were assayed for activity. Additional 0.2-ml samples of the reaction mixture after neutralizing were incubated with 20 μg chymotrypsin of carboxypeptidase B (pancreatic) for 15 min. BK = Synthetic bradykinin. RM = Complete reaction mixture. LEUC = Leukocyte preparation incubated alone. BGEN = Bradykininogen incubated alone. RM + CPASE B = Reaction mixture after carboxypeptidase B. RM + CHYMT = Reaction mixture after chymotrypsin.

kininogen at acid pH. While previous investigations have indicated that leukocytes contain kininases (Schwab, 1962), this is to our knowledge the first demonstration of kinin-producing enzymes that are active at acid pH being contained in polymorphonuclear leukocytes. Other studies have indicated that there is little or no kininase activity at acid pH, although kininase activity does exist at neutral pH in the PMN cells (Greenbaum, unpublished observations).

Since inflammation is known to be accompanied by an acidic environment, it is of interest to examine our current findings of enzymatic production of kinins at acid pH as possibly playing some role in this pathological condition. Kinins are known to mimic several of the cardinal signs of inflammation (Lewis, 1964). The current evidence that spleen extracts contain enzymes that produce kinins and inactivate kinins at acidic pH leads to the speculation that enzymes, normally present in the lysosomes within the cell, when liberated by injury may act upon the blood protein bradykininogen to yield bradykinin. Conditions under which activator (thiol) groups are in low concentration would favor the formation of kinin. On the other hand, the presence of compounds such as glutathione would favor degradation of kinins by catheptic carboxypeptidases. Although this was not studied in the experiments reported, it is also of interest to speculate whether catheptic enzymes liberated from tissues might activate the zymogen of kallikreins, kallikreininogen, which in turn might initiate kinin formation. Eisen and Keele (this volume) have also speculated about such a possibility, referring to cathepsin B as a possible activating enzyme. The eventual isolation of pure kallikreinogen will allow the testing of this hypothesis, which has support in that cathepsin B is known to activate trypsinogen (Greenbaum et al., 1959).

The finding of a protease in polymorphonuclear leukocytes that can form kinin from bradykininogen under acid conditions raises several possibilities, especially since the intracellular pH of these cells reaches rather low levels (pH 5.0) during phagocytosis and since these cells do migrate into inflamed areas. A possible relationship of PMN leukocytes, kinins, and inflammation are seen in Fig. 7. In this simplified scheme, kinins are being produced by the leukocyte protease, thus increasing capillary permeability, a known property of the kinins. This allows more leukocytes to migrate from the capillary blood stream into the inflamed area, increasing kinin

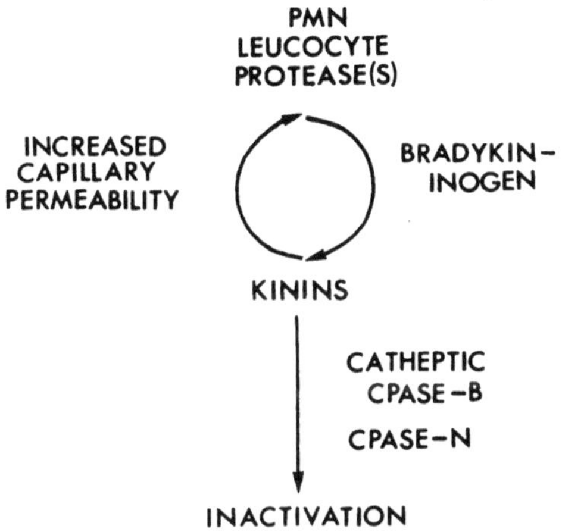

Fig. 7. Possible relationship of polymorphonuclear leukocytes, kinins, and inflammation.

formation. Kinin levels may be reduced by degradative reactions caused by carboxypeptidase N or catheptic carboxypeptidases, dependent on the conditions present.

Finally, it should be noted that on the basis of the studies reported, serious consideration must be given to several different proteases contained in the cell. These proteases may play a role in the regulation of localized kinin levels in conditions under which these enzymes might be liberated from intracellular stores or in inflammation where they may be present in polymorphonuclear leukocytes. The properties of these proteases are quite distinct from the known properties of kallikreins and carboxypeptidase N.

ACKNOWLEDGMENTS

Supported by Grant AM-09393 and a general research support grant, U. S. Public Health Service; Health Research Council of the City of New York; and Life Insurance Medical Research Fund.

We would like to thank Dr. R. B. Merrifield for his gift of solid bradykinin, and Dr. R. Bircher of Sandoz Corp. for a gift of ampules of bradykinin. We would also like to thank Mr. Richard Freer and Mr. Robert Ten Eick for their efforts in this investigation.

REFERENCES

Cohn, Z. A., and J. G. Hirsch: The isolation and properties of the specific cytoplasmic granules of rabbit polymorphonuclear leucocytes. J. exp. Med. 112, 983 (1960).

De Duve, C.: Lysosomes: A new group of cytoplasmic particles, in Subcellular Particles, T. Hayashi, Ed., p. 128. New York: Ronald (1959).

Erdös, E. G., E. M. Sloane, and I. M. Wohler: Carboxypeptidase in blood and other fluids. Biochem. Pharmacol. 13, 893 (1962).

Fruton, J. S.: In The Proteins, H. Neurath, Ed., Vol. 1, (1963).

Greenbaum, L. M., and J. S. Fruton: Purification and properties of beef spleen cathepsin B. J. biol. Chem. 226, 173 (1957).

Greenbaum, L. M., A. Hirshkowitz, and I. Shoichet: The activation of trypsinogen by cathepsin B. J. biol. Chem. 234, 2885 (1959).

Greenbaum, L. M., and R. Sherman: Studies on catheptic carboxypeptidase. J. biol. Chem. 237, 1082 (1962).

Greenbaum, L. M., and T. Hosoda: Studies on the isolation of bradykininogen. Biochem. Pharmacol. 12, 325 (1963).

Greenbaum, L. M., and K. Yamafuji: Catheptic degradation of bradykinin. Life Sci. 4, 657 (1965).

Lewis, G. P.: The role of peptides in the first stages of inflammation, in Injury, Inflammation, and Immunity, L. Thomas, J. Uhr, and L. Grant, Eds. Baltimore: Williams and Wilkins (1964).

Rogers, D. E.: Intracellular inflammation: Dynamics and metabolic changes in polymorphonuclear leucocytes participating in phagocytosis, in Injury, Inflammation, and Immunity, op. cit.

Schwab, J.: Kininases in leucocytes and other tissues. Nature (Lond.) 195, 345 (1962).

Webster, M. E., and J. V. Pierce: Action of the kallikreins on synthetic ester substrates. Proc. Soc. exp. Biol. (N.Y.) 107 186 (1961).

DISCUSSION

Diniz: We have found with Dr. I. F. Carvalho an enzyme in rat kidney homogenate which hydrolyzes BAMe and bradykininogen with the release of bradykinin. This enzyme is located in lysosome-like particles. (Ann. New York, Acad. Sci. 116, 912, 1964).

Werle: Many years ago we found an enzyme in kidney which releases kinin, but is inhibited by plasma. We called it kallikrein.

Greenbaum: The pH optimum of the enzyme activity I described is on the acid side. I believe this quite different from the activity you are referring to.

Oates: Does inflamed tissue reach a pH as low as the pH optimum of the kinin-forming enzymes in polymorphonuclear leukocytes?

Greenbaum: It is not necessary for the pH of inflamed tissues to reach the optimum pH of the enzymes. It is well known that leukocytes produce local concentrations of acid; consequently localized reactions might take place.

Collier: The processes you have described might provide a mechanism in injured tissue whereby kinins might first be released and later destroyed, if SH co-factors appear at the appropriate time. Have you any evidence that sulfhydryl compounds increase in concentration during the course of a local injury reaction and recovery?

Greenbaum: We do not have any direct evidence. However it is not unreasonable to assume that proteolysis caused by intracellular proteinases released from cells during injury might provide small molecular weight peptides containing SH groups which may act as activators.

Back: Have you studied the activity of the leukocyte proteases on any other substrates (casein, etc.) and the possibility that one may have fibrinolytic activity as well?

Greenbaum: We have not looked for fibrinolytic activity. These cells do have proteolytic enzyme activity toward denatured hemoglobin. They also have a carboxypeptidase B activity. We are continuing our studies with artificial substrates to characterize the proteolytic anzymes present.

Schachter: The cathepsin, which Dr. Greenbaum described, releases kinin from a special substrate in hours rather than seconds. It seems unlikely, therefore, that the tissue kininases could not effectively deal with small amounts of these substances released as slowly as this. Also, this cathepsin acts at pH 4 or thereabouts, and I imagine that by the time cells and their membranes are exposed to this pH one does not need to look for kinins or histamine as primary mediators of the inflammatory process. I think when we find these substances in fluids they are often, though not always, the consequence rather than the cause of the inflammatory reaction.

THE PHYSIOLOGICAL AND PATHOLOGICAL ROLE OF THE KALLIKREIN-KALLIDIN SYSTEM

MARION E. WEBSTER

Laboratory of Cardiovascular Physiology, National Heart Institute, Bethesda, U.S.A.

The biochemical characterization of the kallikrein-kallidin system has led to speculation that endogenous release of these potent vasodilator polypeptides may play a role in a variety of physiological and pathological conditions. The techniques necessary to support or refute proposals involving the activation of this enzyme system *in vivo*, however, are only now being evolved.

PHYSIOLOGICAL ROLE

Perhaps the most intriguing of the normal physiological roles proposed for the glandular kallikreins has been that they act as mediators of functional vasodilatation. Evidence for this view has been provided by Hilton (1962, 1963), who has studied the release of an endogenous kallikrein after stimulation of the salivary, sweat, and pancreatic glands, as well as from the glandular elements of the tongue. Evidence against this hypothesis has been presented by Schachter and his co-workers (Bhoola *et al.*, 1965; Schachter, 1966). There is, however, little question that the kallikreins are present in glandular tissue, and that their endogenous release could result in local formation of the kallidins.

That the kallikreins are potent vasodilators can readily be demonstrated. As shown in Fig. 1 (Webster and Gilmore, 1964) direct renal infusion of kallidin-10 1 μg/min produced

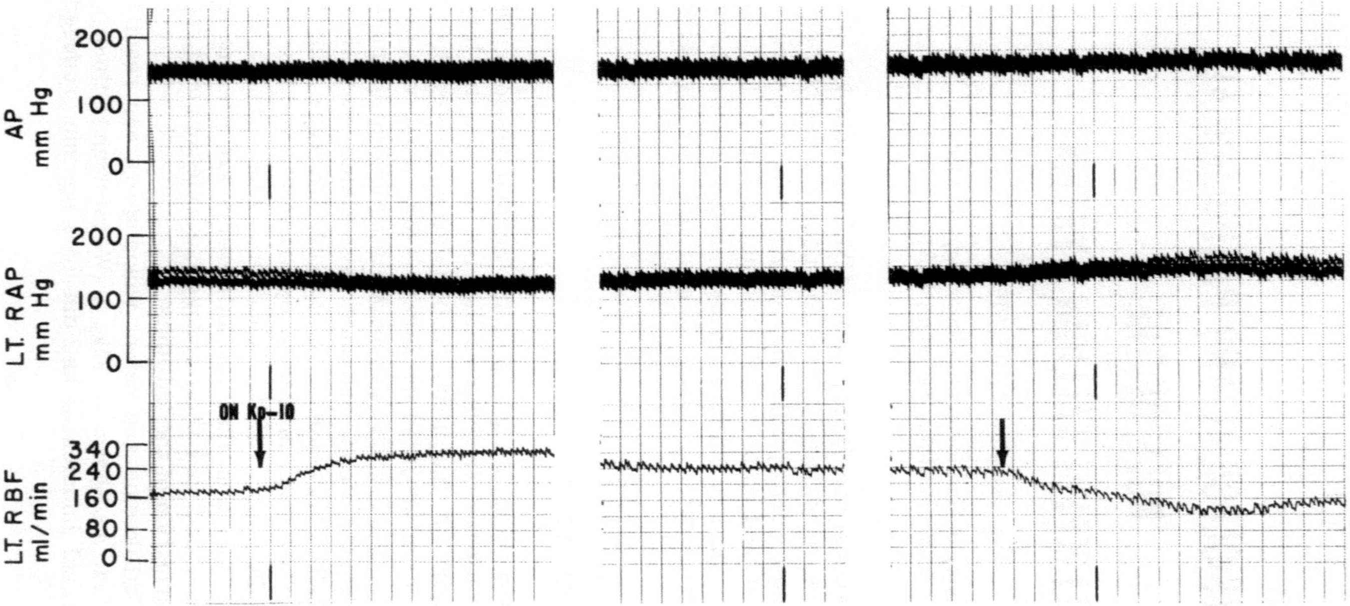

Fig. 1. Effect of infusion of kallidin-10 (KD-10) at 1 μg/min on arterial pressure (AP), left renal arterial pressure (LT RAP), and left renal blood flow (LT RBF) in a dog. The first panel shows the initial response to the infusion, the second panel shows the response after 15 min, and the third panel the response when the 30-min infusion was ended. First arrow indicates the onset of infusion, the second arrow the end of infusion. Paper speed = 0.25 mm/sec. Dog weight = 13 kg, left kidney weight = 45.5 g (Webster and Gilmore, Amer. J. Physiol. 206, 714, 1964).

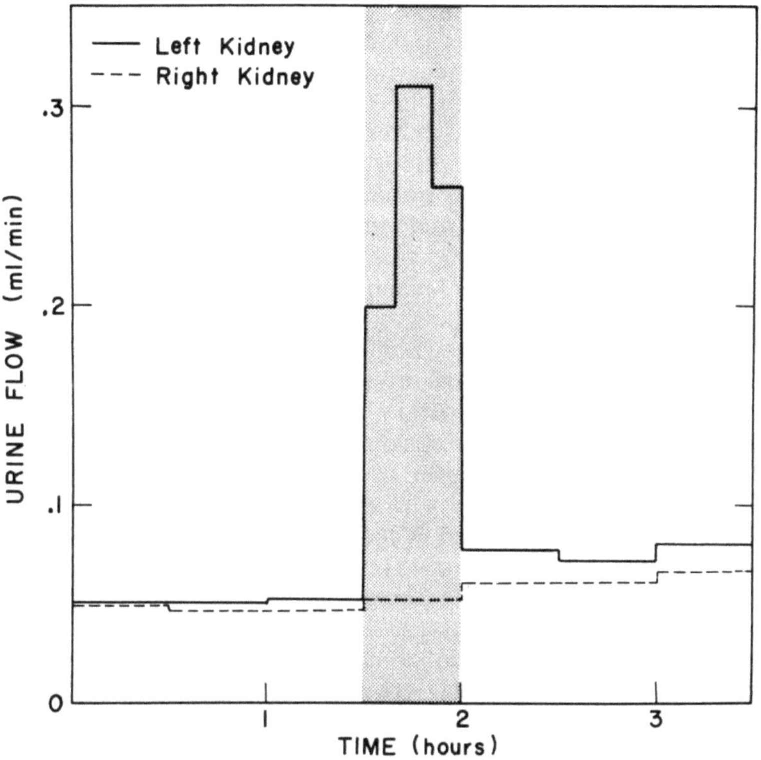

Fig. 2. Increase in urine flow during infusion of kallidin-10 at 1 μg/min in a dog receiving no fluid during the experiment. The shaded area indicates the duration of infusion. The solid line is a plot of the data obtained from the infused left kidney and the broken line the data obtained from the contralateral noninfused right kidney. Dog weight = 18 kg. (Webster and Gilmore, Amer. J. Physiol. 206, 714, 1964).

an increase in directly measured renal blood flow with little or no change in renal arterial pressure, and no alteration in systemic blood pressure. Concomitant with the increase in blood flow a diuresis was observed (Fig. 2), which was associated with substantial increases in sodium and chloride excretion but smaller increases in total solute and potassium excretion. Neither urinary pH nor free water clearance changed consistently. The pattern of water and electrolyte secretion suggested that the observed changes were secondary to an increase in glomerular filtration rate, even though an increase in endogenous creatinine or inulin clearance could not be detected. The contralateral kidney showed no changes during infusion of the decapeptide, indicating the rapidity with which kallidin-10 is destroyed in vivo. The kallidins, therefore, fulfill at least two of the criteria necessary for a local mediator of physiological functions. They are present in glandular tissue, as demonstrated by others, and are fully capable of producing vasodilatation and increased excretion.

For a substance to be a physiological mediator, however, it must be shown to be released under physiological conditions, and, when specific blocking agents are used, the vascular response should decrease. For these latter experiments, the vasodilatation induced by stimulation of the chorda tympani nerve in the cat submaxillary gland was chosen, since other investigators (Hilton and Lewis, 1956) had demonstrated the release of a kallikrein after stimulation of the nerve and during perfusion of the gland with Locke's solution. In the present experiments (Skinner and Webster, 1966), infusion of kallidin was compared with electrical stimulation of the nerve in the presence and absence of carboxypeptidase B. This enzyme has been shown (Erdös et al., 1963) to suppress the in vivo effects of the kallidins. Infusion of kallidin-9 at 2 μg/min directly into the gland consistently produced a pronounced vasodilatation which could be effectively blocked when the blood perfusing the gland contained carboxypeptidase B; however, the vasodilatation produced in response to electrical stimulation of the nerve was the same in the presence or absence of carboxypeptidase B. These data are in agreement with those of Bhoola et al. (1965), and suggest that kallidin is not responsible for functional vasodilatation in this gland.

The mediator of the atropine-resistant vasodilatation of this salivary gland has been unexplained since the original discovery of this phenomenon by Heidenhain (1872). For many years the view was held that the chorda tympani contained true vasodilator nerve fibers, while other authors have provided evidence that stable metabolic factor(s) may be responsible. No satisfactory mechanism, including that of activation of the kallikrein-kallidin system (Hilton and Lewis, 1956), has been proposed to explain the intense atropine-resistant vasodilatation produced by parasympathetic nerve stimulation. Although it is frequently stated that atropine abolishes the salivary secretion but does not alter the vasodilatation substantially, atropine (Hilton and Lewis, 1955; Skinner and Webster, 1966) was found to cause a significant decrease in the vascular response to chorda stimulation. Nevertheless, considerable residual vasodilatation remains. Recently, Bhoola et al. (1965) suggested the possibility that the sympathetic after-dilation may be due to stimulation of β-adrenergic receptors. The presence of these β-receptors in the cat submaxillary gland has been confirmed (Skinner and Webster, 1966), since injections of isoproterenol directly into the gland result in an intense vasodilatation. Also, the combination of adrenergic β-blocking agents (propranolol or dichloroisoproterenol) and atropine either partially or completely suppresses the vasodilatation produced by electrical stimulation of the chorda tympani nerve. Those glands, which continued to respond to stimulation after both propranolol and atropine blockade, also continued to respond to isoproterenol infusions. It appears likely that the atropine-resistant vasodilatation of this gland is the result of stimulation of adrenergic β-receptors.

Suggestions have also been made that the kallikrein-kallidin system may play a role in the decrease in vascular resistance seen during muscle contraction or after release of arterial occlusion. These suggestions have persisted despite the fact that Hilton and Lewis (1958) and Carretero et al. (1965) were unable to find evidence for the release of either a kallikrein or a kallidin after muscle contraction, and Horton (1964) was unable to detect kallidin in venous blood flow collected after reactive hyperemia. In the present experiments (Webster et al., 1966) infusion of kallidin-9 at 1-4 μg/min directly into the isolated, perfused gracilis muscle of the dog produced a pronounced decrease in resistance which could be effectively blocked by simultaneous perfusion of the muscle with blood containing 15-60 μg carboxypeptidase B per ml. The vasodilatation produced by muscle contraction, release of arterial occlusion, and lumbar cholinergic sympathetic nerve stimulation, however, was not altered by the presence of carboxypeptidase B. Also, the vasodilatation following occlusion was more prolonged when the occlusion was immediately preceded by muscle contraction than when preceded by kallidin infusion, suggesting that the polypeptide was not as stable as the "exercise" factor. Evidence that the failure of carboxypeptidase B to block functional vasodilatation was not the result of its inability to enter the perivascular space was obtained when this enzyme was found in hind-limb lymph after systemic infusion.

These data make it unlikely that activation of the kallikrein-kallidin system is responsible for functional vasodilatation either in glands or in skeletal muscle. This failure to substantiate a normal physiological role for the kallidins should stimulate investigation of a wider range of physiological conditions, such as increased capillary permeability or stimulation of smooth muscle. Teleological reasoning predicts that a normal physiological role is to be expected.

PATHOLOGICAL ROLE

Although a physiological role for these polypeptides is yet to be determined, activation of the kallikrein-kallidin system may be important in three pathological conditions. Thus patients with hereditary angioneurotic edema (Table 1) were shown to be deficient in an inhibitor of plasma kallikrein (Landerman et al., 1962). Initially the plasma inhibitor concentration was determined by incubating dilutions of plasma with human plasma kallikrein and determining residual kallikrein by its ability to produce vasodilatation in the dog. As is shown in Table 1, the mother and her elder son, both of whom were afflicted with the disease, had lower levels than those for the controls. However, it is apparent that substantial levels of

TABLE 1

Inhibition of Plasma Kallikrein by Plasma from Patients with Hereditary Angioneurotic Edema

Subject	Inhibitor levels of plasma kallikrein as measured by: Vasodilatation*	TAME**
Mother	75	.16
Older Son	75	.13
Younger Son	150	1.3
Father	200	2.0
Normal	150	1.0

*Reciprocal of plasma dilution necessary to cause 50% inhibition.
**Inhibitor units/ml plasma where 1 unit is approximately that found in 1.0 ml of plasma.

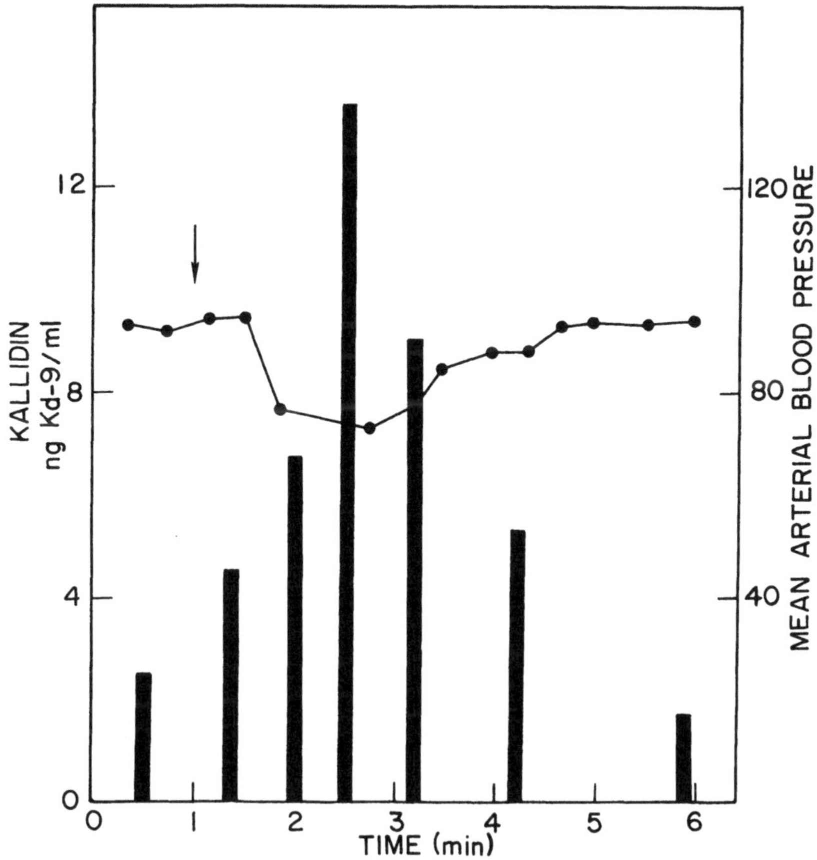

Fig. 3. Release of kallidin after injection of epinephrine in a patient with the carcinoid syndrome. Flushing was induced in the patient by injection of 2 μg epinephrine as indicated by the arrow. Solid line represents mean arterial blood pressure; bars represent levels of kallidin in peripheral arterial blood calculated in ng kallidin-9/ml.

inhibitor are still present. When measurements of the inhibitor levels were made, with p-toluenesulfonyl-L-arginine methyl ester (TAMe) as the substrate, it could readily be demonstrated that the patient and her older son had grossly lowered levels of inhibitor. These data suggest that two inhibitors of plasma kallikrein are present normally in plasma, both of which are measured when kallidinogen is used as the substrate (vasodilatation), whereas only one is measured when TAMe is used as the substrate. Subsequent investigations by Donaldson and

Evans (1963) and Kagen (1964) have suggested that the inhibitor to plasma kallikrein which is deficient in this disease is also an inhibitor to C′-esterase and to the permeability factor PF/dil, currently identified as prekallikreinase (Webster and Innerfield, 1965).

Another pathological condition in which dysfunction of the kallikrein-kallidin system has been implicated is as a mediator of carcinoid flushes (Oates et al., 1964). With the aid of a recently published method for the estimation of the kallidins (Webster and Gilmore, 1965), the release of a kallidin in a patient with the carcinoid syndrome after an epinephrine-induced flush has been confirmed. As shown in Fig. 3, after the injection of epinephrine, the levels of kallidin began to rise, reaching a peak in 1.5 min. Blood pressure fell concomitantly and returned to normal at the same time that the flush subsided, and the .kallidin levels returned to less than 3 ng/ml.

Also, a kallidin was found (Melmon et al., 1966) in synovial fluid taken from 13 patients suffering from arthritides of varying etiology. Synovial peptide ranged from 1.6 to 58 ng kallidin-9/ml, and in the initial samples in these studies correlated well with the degree of warmth, pain, and tenderness of the joint. However, subsequent knee effusions, particularly those obtained from patients with rheumatoid arthritis, failed to show such a correlation. In two patients suffering from gouty arthritis, an acute attack (spontaneous or induced by microcrystalline sodium urate) resulted in a dramatic increase in polypeptides 24 hours after the onset of symptoms. Intra-articular injection of synthetic kallidin-9 in the dog produced warmth, swelling, and pain. These observations suggest that kallidin is contributing to the inflammatory synovial reaction seen in arthritis.

SUMMARY

The following findings make it appear unlikely that activation of the kallikrein-kallidin system is responsible for functional vasodilatation either in glands or in skeletal muscles: the failure of carboxypeptidase B to alter the vasodilatation produced by stimulation of the chorda tympani nerve in the cat submaxillary gland; the identification of the factor responsible for the atropine-resistant vasodilatation of this gland as the result of stimulation of adrenergic β-receptors;

the failure of carboxypeptidase B to alter the decrease in resistance after muscle contraction, release of arterial occlusion and lumbar cholinergic sympathetic stimulation in the dog; the greater stability of the "exercise" factor as compared to kallidin; and the presence of carboxypeptidase B in perivascular fluid. Three pathological conditions (hereditary angioneurotic edema, carcinoid flushing, and arthritides of varying etiology) have been investigated and evidence found that the kallidins may be playing a pathological role. However, in each case the exact role has not been delineated by specific blocking agents. Further studies will be required to determine whether the proposed pathogenic role of the kallikrein-kallidin system will withstand such a critical evaluation.

REFERENCES

Bhoola, K. D., J. Morley, M. Schachter, and L. H. Smaje: Vasodilatation in the submaxillary gland of the cat. J. Physiol. (Lond.) 179, 172 (1965).

Carretero, O., A. Nasjletti, and J. C. Fasciolo: The kinin content of human blood at rest and during vasodilatation. Experientia (Basel) 21, 141 (1965).

Donaldson, V. H., and R. R. Evans: A biochemical abnormality in hereditary angioneurotic edema. Amer. J. Med. 35, 37 (1963).

Erdös, E. G., J. R. Wohler, and M. I. Levine: Blocking of the in vivo effects of bradykinin and kallidin with carboxypeptidase B. J. Pharmacol. exp. Ther. 142, 327 (1963).

Heidenhain, R.: Über die Wirkung eineger Gifte auf die Nerven der Glandula Submaxillaris. Pflüger's Arch. ges. Physiol. 5, 309 (1872).

Hilton, S. M.: Local mechanisms regulating peripheral blood flow. Physiol. Rev. Supp. 5, 42, 265 (1962).

Hilton, S. M.: A discussion of the evidence for kinins as the agents of vasodilator reactions. Ann. N.Y. Acad. Sci. 104, 275 (1963).

Hilton, S. M., and G. P. Lewis: The cause of the vasodilatation accompanying activity in the submandibular salivary gland. J. Physiol. (Lond.) 128, 235 (1955).

Hilton, S. M., and G. P. Lewis: The relationship between glandular activity, bradykinin formation and functional vasodilatation in the submandibular salivary gland. J. Physiol. (Lond.) 134, 471 (1956).

Hilton, S. M., and G. P. Lewis: Vasodilatation in the tongue and its relationship to plasma kinin formation. J. Physiol. (Lond.) 144, 532 (1958).
Horton, E. W.: An increase in butanol-extractable 5-hydroxytryptamine in venous blood during reactive hyperaemia. J. Physiol. (Lond.) 170, 101 (1964).
Kagen, L. J.: Some biochemical and physical properties of the human permeability globulins. Brit. J. exp. Path. 45, 604 (1964).
Landerman, N. S., M. E. Webster, E. L. Becker, and H. E. Ratcliffe: Hereditary angioneurotic edema. II. Deficiency of inhibitor for serum globulin permeability factor and/or plasma kallikrein. J. Allergy 33, 330 (1962).
Melmon, K. L., M. E. Webster, S. Goldfinger, and J. E. Seegmiller: The presence of a kallidin in inflammatory synovial effusions from arthritides of varying etiologies. To be published (1966).
Oates, J. A., K. Melmon, A. Sjoerdsma, L. Gillespie, and D. T. Mason: Release of a kinin peptide in the carcinoid syndrome. Lancet 1, 514 (1964).
Schachter, M.: Kallikrein in the submaxillary gland, in E. G. Erdös, N. Back, and F. Sicuteri, Eds. This volume.
Skinner, N. S., Jr., and M. E. Webster: Kallidin, beta-receptors and functional vasodilatation in the submaxillary gland of the cat. To be published (1966).
Webster, M. E., and J. P. Gilmore: Influence of kallidin-10 on renal function. Amer. J. Physiol. 206, 714 (1964).
Webster, M. E., and J. P. Gilmore: The estimation of the kallidins in blood and urine. Biochem. Pharmacol. 14, 1161 (1965).
Webster, M. E., and I. Innerfield: Interrelationship of human plasma kallikrein and plasmin in inflammation. Enzymol. biol. clin. 5, 129 (1965).
Webster, M. E., N. S. Skinner, Jr., and W. J. Powell, Jr.: The role of the kallidins in vasodilatation of skeletal muscle. To be published (1966).

DISCUSSION

Hilton: May I first comment on your experiments with carboxypeptidase. It is one thing to have a high enough concentration of carboxypeptidase in the blood to destroy systemically injected kinin and quite another to expect that a similar (or rather a smaller) concentration at the site of action will prevent the kinin from performing its function within the gland. Even with your intra-arterial injections, the kinin will probably have taken 20-30 sec to reach the gland, whereas in the gland it is being formed at the site of action.

We are very familiar with this problem in connection with a substance having a much shorter half-life in blood, i.e., acetylcholine. The concentrations of cholinesterase in blood, together with those at the end-plate region in skeletal muscle, make it quite a tricky matter to demonstrate the transmitter function of acetylcholine by systemic injection. Even with close arterial injection, abnormally large amounts of acetylcholine have to be given, very much more than one could expect to be released by the motorneurons. Similarly, after giving carboxypeptidase you could certainly have found a dose of kallidin that would work by intra-arterial injection to its gland.

Now, to turn to the atropine-resistant vasodilatation in the salivary gland. There is plenty of evidence that atropine does not block all the effects of the secretomotor nerve fibers on the gland cells; O_2 consumption still rises, as does CO_2 output. Glucose consumption increases also. The gland cannot push out fluid, but it is still made to work, and, as I will emphasize in my talk, it still releases kinin-forming enzyme. In explanation of this, one need look no further than into the question why atropine, which blocks the effects of injected acetylcholine, does not abolish the effects of secretomotor nerve stimulation.

Finally, I will be dealing in my talk with the evidence for release of kinin-forming enzymes and formation of kinin during activity in various glands, including the pancreas. Let me just say at this stage that, even before we looked into this, it was well known that pancreatic enzymes appear in the blood stream. They had to pass through the tissue fluid of the gland to get there, so how can you assert that kinin will not have been formed in the gland?

Webster: We quite agree with Dr. Hilton that it is much more difficult to block a substance when it is released endogenously than when it is infused. The concentration of carboxypeptidase B used in these experiments was sufficient to completely suppress the vascular response to kallidin infusion. Although these levels might not be sufficient to abolish the effect of its hypothetical in vivo release, we did expect to see some reduction in the vascular response to chorda stimulation. It was for this reason that we used multiple levels of stimulation of the nerve. If the kallikrein-kallidin system is importantly related to glandular bloodflow, one might expect some demonstrable reduction, particularly when only minimal vasodilatation was induced.

We certainly did not mean to leave the impression that we thought kallikrein was not released from the gland. Your own evidence of the release of a kallikrein in the venous perfusates of the gland during both sympathetic and parasympathetic stimulation as well as Dr. Schachter's observations of increased kallikrein concentrations in the saliva, would certainly indicate that a kallikrein had been released. The question we asked was relative to the vascular effect of the kallikrein-kallidin system. We cannot ignore the fact that a combination of cholinergic and beta-adrenergic blocking agents will essentially abolish the vasodilatation associated with chorda tympani stimulation with or without the presence of carboxypeptidase B. In view of the above findings, as well as those of Dr. Schachter, we find it difficult to accept the kallikrein-kallidin system as being of importance in controlling salivary gland blood flow.

Sicuteri: Non possiamo dire se le chinine hanno o non hanno azione fisiologica. E' importante osservare che il chininono arterioso e piu alto di quello venoso. Inoltre se noi eseguiamo una ischemia il chininogeno nel sangue venoso durante la iperemia reattiva contiene un po' piu di chinine ma soprattutto e diminuito il chininogeno. La carboxipeptidasi blocca la dilatazione de callidina; e stato fatto questo a flusso bloccato? Infatti ischemia significa abbassamento del pH e questo a sua volta significa una vita piu lunga delle chinine.

KALLIKREIN AND VASODILATATION IN THE SALIVARY GLAND

M. SCHACHTER

Department of Physiology, University of Alberta, Edmonton, Alberta, Canada

Although evidence for the existence of vasodilator nerve fibers was first presented by Ludwig in 1851, this basic fact is still in dispute. Ludwig, and later Claude Bernard in 1872, demonstrated the remarkably pronounced vasodilatation that occurs in the submaxillary gland when the chorda tympani nerve is stimulated. Both these early authors assumed that this vasodilatation was a "direct" action of the nerve fibers on the blood vessels.

The existence of dilator nerve fibers was first challenged by Barcroft and his colleagues (Barcroft and Piper, 1912; Barcroft, 1914) as a consequence of their experiments on the relations among metabolic activity, secretion, and blood flow in the submaxillary gland. From these experiments Barcroft (1914) concluded that the vasodilatation that is associated with sympathetic nerve stimulation (a vasodilatation usually follows a pronounced vasoconstriction) is entirely secondary to the secretory metabolism. He was somewhat more cautious in dismissing the presence of vasodilator fibers in the chorda tympani nerve, stating: "It is not impossible that under normal circumstances dilatation may be instituted by dilator fibers and maintained by metabolic products."

More recently, Barcroft's view has been extended by Hilton and Lewis (1955, 1956, 1958; Lewis 1959; Hilton, 1960). According to these authors all functional vasodilatation in the gland is due to a release of kallikrein, which, passing from secreting gland cells into the tissue spaces, splits off kallidin, the vasodilator peptide, from an α-2 globulin in plasma and lymph. These authors suggested further that similar systems of kinin-releasing enzymes are widespread and exist generally in glands, thus affording a mechanism whereby glandular organs obtain an increased blood supply when actively secreting. This generalization, however, has been questioned by Bhoola et al. (1962), who failed to find a kinin-releasing enzyme in an active form in many glandular organs of the guinea pig.

From further studies my colleagues and I further concluded that kallikrein is very unlikely to play a major role in the pronounced vasodilatation that accompanies stimulation of the chorda tympani nerve in the submaxillary gland. We have also found that sympathetic nerve stimulation is more effective than chorda-lingual nerve stimulation in releasing kallikrein from the cat's submaxillary gland. A brief summary of our results follows:

In the cat (Bhoola et al., 1965). (a) The close arterial injection of saliva into a gland with natural circulation does not duplicate the vasodilatation produced by stimulation of the chorda-lingual nerve. The increased blood flow produced by injected saliva was always slower in onset; it was never as great as that obtained by nerve stimulation (or by injected acetylcholine or bradykinin), and it was more protracted. Further, whereas marked duplication of the vasodilator response to nerve stimulation or to injections of acetylcholine occurred with repeated responses, a progressive reduction of the vasodilatation quickly occurred with repeated injections of saliva. (b) After desensitization of the blood vessels to a standard dose of bradykinin by close arterial infusion or by repeated injections of large amounts of this substance, the equivalent vasodilator response to chorda-lingual nerve stimulation or to injection of acetylcholine was reduced to a much lesser degree. (c) It was possible to produce a maximal, atropine-resistant vasodilatation by stimulation of the chorda-lingual nerve when the gland had been perfused for as long as 3 hours with horse serum from which cat saliva fails to release kinin. Under these conditions the close arterial injection of saliva failed to cause vasodilatation. (d) It was possible to produce a maximal, atropine-resistant vasodilatation by stimulation of the chorda-lingual nerve in a gland practically depleted of its kallikrein (Beilenson et al., 1964). The concentration of kallikrein in the submaxillary gland and in saliva was reduced to practically undetectable amounts after preliminary ligation of Wharton's duct for 3 days and subsequent nerve stimulation. Although the concentration of kallikrein in such a gland or in its saliva could be reduced more than 10,000-fold, the vasodilatation produced by chorda-lingual nerve stimulation was unaffected.

The above results in the cat are not in accord with the conclusion that parasympathetic vasodilatation is mediated exclusively by kallikrein, if, in fact, it plays a significant role at all in this response. The rapid dilator response on stimulation, and its usual abrupt cessation when stimulation is discontinued, suggest that a labile neurotransmitter substance, released in close relation to the receptor sites in vascular muscle, is concerned in its mediation. Our results provide no further clues in this direction. If it is assumed, however, that the relative resistance of chorda-lingual-induced vasodilatation to atropine is simply an extreme case of the well-known variation in the sensitivity of cholinergic receptors to this drug (cf. Ambache, 1955; Morley et al., 1963), then our results are consistent with the view that cholinergic vasodilator nerve fibers are present in this nerve.

In the rabbit (Morley et al., 1963). Contrary to the cat and dog, both the secretion and vasodilatation produced by stimulation of the chorda-lingual nerve in the rabbit are equally sensitive to inhibition by atropine, and they also recover in parallel after atropine blockade. When the vasodilatation produced by chorda-lingual nerve stimulation is blocked by atropine, the delayed vasodilatation produced by sympathetic nerve stimulation is unaffected. The parasympathetic and sympathetic vasodilatation, therefore, are not likely to be caused by the same chemical mediator.

The regulation of kallikrein secretion

The concentration and output of kallikrein in submaxillary saliva was compared after stimulation of the sympathetic and chorda-lingual nerves (Beilenson et al., 1964). Sympathetic nerve stimulation was more effective than chorda-lingual stimulation in releasing kallikrein from the gland. The kallikrein concentration in saliva may increase as much as 100-fold if sympathetic replaces chorda-lingual nerve stimulation.

These experiments provide new information about kallikrein secretion.

It is desirable that experiments are carried out with kallikrein in addition to those related to its vasodilator properties so that its physiological significance may be clearly established.

REFERENCES

Ambache, N.: Pharmacol. Rev. 7, 467 (1955).

Barcroft, J.: The Respiratory Functions of the Blood. London: Cambridge University Press, (1914).

Barcroft, J., and H. Piper: J. Physiol. (Lond.) 44, 359 (1912).

Beilenson, S., M. Schachter, and L. H. Smaje: J. Physiol. (Lond.) 177, 61P (1964).

Bhoola, K. D., R. May May Yi, J. Morley, and M. Schachter: J. Physiol. (Lond.) 163, 269 (1962).

Bhoola, K. D., M. Schachter, J. Morley, and L. H. Smaje: J. Physiol. (Lond.) 179, 172 (1965).

Hilton, S. M.: in Polypeptides Which Affect Smooth Muscles and Blood Vessels, ed. M. Schachter. pp. 258-262. London: Pergamon (1960).

Hilton, S. M., and G. P. Lewis: J. Physiol. 129, 253 (1955).

Hilton, S. M., and G. P. Lewis: J. Physiol. (Lond.) 134, 471 (1956).

Hilton, S. M., and G. P. Lewis: J. Physiol. (Lond.) 144, 532 (1958).

Lewis, G. P.: J. Physiol. (Lond.) 147, 458 (1959).

Morley, J., M. Schachter, and L. H. Smaje: J. Physiol. (Lond.) 167, 29P (1963).

DISCUSSION

Hilton: I hope to be able to show how I would deal with most of the points raised by Dr. Schachter. I think I can show that there is plenty of life in the young horse yet. I would therefore only want to discuss one of Dr. Schachter's main lines of experiment at this stage. This concerns the experiments in glands perfused with horse serum, in which chorda stimulation still produced some vasodilatation. First, the turnover of plasma proteins in tissue fluid is so slow that one would not expect the cat's own proteins to have been eliminated in 3 hours of perfusion. Second, as I remember from your published paper, the experiment you showed in your slide (as in most of such experiments) you had to have the preparation under the influence of eserine. So there will have been plenty of acetylcholine accumulating in the vicinity of the blood vessels of the gland. So much that the dose of atropine you gave might have been too small to hinder its effect. I would not like to say that acetylcholine, released from secretomotor nerve endings, cannot play a part in the vasodilatation under normal conditions, but in these experiments of yours its effects will certainly have been exaggerated. I do not think your results should affect our main conclusion that, in the normal gland, plasma kinins play the major role.

Back: With a dog salivary gland perfusion preparation, which provided for cannulation of the artery leading directly into the gland, thus allowing for immediate recording of vascular changes and administration of test agents directly into the gland, the following observations were made: (1) Under constant chorda tympani stimulation atropine was able to block the vasodilation response. (2) The vasodilatation response of chorda stimulation was not blocked by Trasylol or carboxypeptidase B.

Schachter: Vasodilation still occurred after chorda stimulation in glands depleted of kininogen substrate.

Collier: Two speculative questions: (1) Does any other anticholinergic drugs than atropine, such as decamethonium or hexamethonium, block the chorda vasodilatation? (2) Is the function of kallikrein in the salivary gland perhaps a digestive rather than a vasodilator one? Is it possible, for example, that kallikrein liberates kinin in the stomach or stomach wall and this facilitates absorption? Alternatively, does kallikrein carry out some special proteolytic action?

Schachter: The interpretation of the effects of blocking drugs is difficult, since we stimulate the chorda tympani fibers preganglionically. Hexamethonium blocks the effect, but of course it does so by ganglionic blockade. We have tried curare, large doses of atropine, etc. injected close-arterially, but the results are difficult to interpret because of possible ganglionic block by these substances. Kallikrein has much in common with other exocrine enzymes. A possible unknown function for salivary and pancreatic kallikrein is, I think, a definite possibility.

Werle: It is difficult for kallidin to be a physiological regulator of the blood supply of the salivary glands because only the submaxillary gland has larger amounts of kallikrein. In some species the kallikrein content of the submaxillary gland is extremely high; in others, extremely low. In the saliva the kallikrein content is in general low, so that it is not likely that kallikrein has an important function in the upper digestive tract.

FURTHER EXPERIMENTS ON THE ROLE OF PLASMA KININS AS MEDIATORS OF FUNCTIONAL VASODILATATION IN GLANDULAR TISSUES

S. M. HILTON

Department of Physiology, University of Birmingnam, England

It is a common cliché that modern life is becoming impossibly fast, but so far man has not become sufficiently ingenious to extend this benefit to the field of publishing. When asked many months ago for a summary of my paper, I had very little idea what I was going to say. I knew no more than that the time was surely coming for a spirited defense of the theory relating plasma kinins with functional vasodilatation in glandular organs and tissues. This premonition has been borne out by what we have already heard. It is my excuse for my vague title; but now I must show how it can be given some substance.

As you now know, Schachter (1966) has endeavored to accumulate evidence that might tell against the theory (Bhoola et al., 1965). But much of what he has presented so far is of questionable value. For instance, we need hardly take too seriously the fact that a close arterial injection of saliva into the gland fails to duplicate the vasodilatation produced by stimulation of the chordo-lingual nerve. Likewise, it should not surprise anyone that close arterial infusion of a large dose of bradykinin will desensitize the gland more to a subsequent injection of a smaller dose of bradykinin than it will to stimulation of the chordo-lingual nerve. Findings like these have long been adduced against the theory of humoral transmission anywhere and, if we accepted them at their face value, we would not believe that acetylcholine (ACh) was a transmitter even at the neuromuscular junction. At this time, therefore, it may be more appropriate to mention some additional findings, or in other words to reconsider briefly those pieces of positive evidence on which the theory is based.

POSITIVE EVIDENCE FOR THE THEORY

The first I would recall, though indirect, shows that some substance more stable than acetylcholine is produced within the submandibular salivary gland during secretory activity. In some of our earliest experiments, Lewis and I (1955) showed that, if the chorda is stimulated for the first 15 sec of a 1-min period of circulatory arrest, by occlusion of the artery and vein, the hyperemia seen on re-establishing the circulation is at least as great as that seen after chorda stimulation alone. By contrast, the reactive hyperemia following a 1-min period of vascular occlusion alone is quite small. Moreover, if the gland is activated by a close arterial injection of 1 μg ACh — the material believed to be liberated by the secretomotor nerve endings — the same kind of evidence can be obtained for the production within the gland of the stable vasodilator substance.

It can be demonstrated, as a control, that the direct vasodilator action of acetylcholine would not outlast such a period of circulatory arrest. Lewis and I showed this in skeletal muscle, in which acetylcholine exerts a direct vasodilator action in doses too small to cause muscular contraction. In experiments in which the circulation was arrested immediately after an arterial injection of 1 μg ACh into the gastrocnemius muscle of the cat, removal of the occlusion clamp revealed a small residual effect — no more than the tail end of the vasodilatation which is seen at the corresponding time after the injection alone. Therefore, even if one concedes (as one must) that ACh released from secretomotor nerve endings may make a contribution to the vasodilatation within the active gland, there can be little doubt of the formation of some other vasodilator substance which is much more stable. I would also like to recall that this substance is apparently formed not in the blood but in the interstitial fluid of the gland and that it is carried away not by the venous outflow but by the lymph. If we stimulated the chorda during a period of venous occlusion, subsequent removal of the venous clamp was not followed by any hyperemia. This can be explained by reference to the fact that venous occlusion would, if anything, increase the flow of lymph from gland. It was only after we had abolished lymphatic drainage from the gland, by carefully separating it from the surrounding tissue, that we could retain the stable vasodilator material within the gland by occlusion of the venous outflow alone.

I would now like to turn to some of our later direct evidence, obtained in experiments in which we were perfusing salivary glands with Locke's solution and assaying the perfusate for kinin-forming activity (Hilton and Lewis, 1956). We showed that an increase in output of kinin-forming enzyme could be initiated as readily by an intra-arterial injection of ACh as by stimulation of the chordo-lingual nerve. But what I would particularly like to emphasize is that in these perfused glands after an injection of atropine, the output of kinin-forming enzyme is no longer increased by intra-arterial ACh. It is still increased by chorda stimulation, though the latter increase is smaller than normal. This supplements other evidence that, in the cat, atropine does not abolish the metabolic effects of chorda stimulation in the salivary gland and, even though no saliva is being formed, kinin-forming enzyme is being released into the interstitial fluid where, under conditions of normal circulation, plasma kinin may be formed.

This, it seems to me, deals with one argument which should be considered in relation to our experiments. One could pose the question this way: Is it possible that, under artificial conditions of perfusion with Locke's solution, substances being secreted by the gland may somehow find their way into the perfusate? Now, as we have just seen, in the atropinized gland, in which no secretion is formed on chorda stimulation, increased amounts of kinin-forming enzyme are nevertheless released into the perfusate.

The skeptic's question just posed can be examined again but from the opposite point of view, after considering experiments which Jones and I (1963, 1966) later carried out on the perfused pancreas. When the pancreas was activated by ACh or by pancreozymin, large amounts of kinin-forming enzyme were released into the perfusate coming from the gland. But what happened when we looked at the effect of injected secretin? This sample of secretin was very pure. It had no vasodilator action when injected into the gland with its normal arterial blood supply. It caused a secretion of pancreatic juice with high kinin-forming activity. Yet, in experiments with the perfused gland, injection of secretin gave rise to no significant release of kinin-forming activity into the perfusate. Thus, in this case, when secretion had been stimulated, the kinin-forming content of the juice was not reflected in the perfusate, and the result was in keeping with the lack of vasodilator activity of the secretin which we had previously found.

There can hardly be any doubt that in the specified conditions of activity in the salivary gland and pancreas of the cat under which vasodilatation occurs, plasma kinin is formed in the interstitial fluid of these glands and will contribute to the vascular response. This is exactly what Fox and I (1958) found in the human forearm skin — kinin formation in the subcutaneous tissue space when the sweat glands were activated. These are all positive findings which any critic of the theory relating plasma kinins with functional vasodilatation in glandular organs and tissues will have to take into account.

KININASE ACTIVITY IN SECRETIONS

Some years ago, before Jones and I had started our work on the pancreas, Schachter (1960) had already placed on record his doubt whether plasma kinins were involved in functional vasodilatation in that organ, on the grounds that pancreatic juice contains a kallikrein which is largely in an inactive form. This was based on a conclusion that Werle (1937) had reached after some of his early work on the vasodepressor activity of pancreatic juice injected intravenously. He had found that this juice possessed some vasodepressor activity but that this activity was enhanced if the secretion was previously incubated with fluid from the duodenum. Of course, Schachter's argument was not strictly relevant to the question whether or not plasma kinin was involved in functional vasodilatation in the pancreas, but, since we were working on this organ, it was an opportune occasion to look into the matter again. We mixed diluted pancreatic juice with the plasma substrate and tested the incubated mixture on the isolated rat uterus in the usual way. We found that the juice formed appreciable amounts of kinin even when diluted 1:100 and sometimes in a dilution of 1:1,000.

However, juice diluted 1:10 did not usually form as much kinin as might have been expected from the results with the 1:100 dilution. In such experiments, the undiluted juice formed even less. Now when undiluted juice was incubated with bradykinin for 1 min, it was found to destroy appreciable amounts.

Some small kininase activity was also present in juice diluted 1:10. It seems, therefore, that the apparent low kinin-forming activity usually found in undiluted pancreatic juice is due to the presence of this kininase. In some of our experiments, undiluted pancreatic juice had high kinin-forming activity, and in these cases the juice possessed no detectable kininase.

These findings differ from those of Amundsen and Nustad (1964) who investigated the kininase activity found in human saliva. They detected no such activity in saliva obtained by cannulation of the parotid or submandibular ducts, and they found that the enzyme originated from the squamous epithelial cells of the oral mucous membrane. Naturally, in our experiments the juice was collected from the cannulated pancreatic duct, but our results should not be surprising in view of the known kinin-destroying activity of carboxypeptidase B in pancreatic juice (Erdös et al., 1963).

BRADYKININ AND THE SALIVARY GLAND WITH LIGATED DUCTS

Now I would like to turn to the interesting question of what may be happening in the salivary gland with ligated ducts. As Schachter (1966) mentioned, Werle and Trautschold (1963) reported a few years ago that if they tied the ducts of the submandibular salivary gland, the concentration of several intracellular enzymes fell sharply within a few days. Among these was the kinin-forming enzyme itself. Schachter and his colleagues then followed this up by examining the effect of chordo-lingual stimulation on the blood flow through a gland whose duct had been tied for three days (Beilenson et al., 1965). They reported that a normal vasodilatation was obtained. This was really intriguing, and I became even more interested recently when I repeated these experiments, for it seemed to me that the vasodilatation was even enhanced in glands which had been submitted to this form of treatment. I then tried to quantitate the results in the following way: using fixed parameters of stimulation of the chordo-lingual nerve, in both normal and operated glands, only the duration of the stimulus (i.e. the number of pulses applied to the nerve) was varied. The venous effluent was measured as the number of drops of blood issuing from the gland during the first minute following the onset of stimulation. On a semilogarithmic plot of the results prepared much as one would plot a log dose-response curve, it is then seen that the operated glands are behaving quite differently from the normal glands. The effects for a given stimulus are greater and the slope of the curve is steeper. It is as if the operated glands were much more sensitive to the substance causing the vasodilatation. The next step was to test the responses to close arterial injections of bradykinin. When the dose-response curves were plotted, they showed the same dramatic difference between the operated and normal glands. These are, of course, the first experiments on this new effect, but they have all given the same results. They seem to show that when degenerative processes are occurring within the gland, its blood vessels become much more sensitive to bradykinin which (dare I say?) is the normal substance causing vasodilatation. As these experiments are continued, it will be interesting to look further and see what has happened to the response to ACh in these operated glands, for, if this too is enhanced, the amplification factor will be very great indeed.

THE INNERVATION OF THE BLOOD VESSELS OF THE SALIVARY GLAND

Finally, I would mention a problem which, although not strictly to do with peptides, has very much to do with the matter in hand. I refer to whether or not there are cholinergic nerve fibers innervating the blood vessels of the salivary gland. It has been very tempting on several occasions to speculate what would be the position now if no one had ever suggested that there were vasodilator fibers in the chorda tympani. Would anyone have cared de novo to argue the case vigorously today? As always, the final task will be not simply to continue the argument but to look and see. Dr. Hall, of the Physiology Department in Birmingham, has started examining salivary glands histologically, putting them through the procedure of histochemical staining for cholinergic nerve endings. I hope we will soon be able to state whether there are any such endings on the blood vessels.

In conclusion then, it is perhaps not surprising that we cannot provide the final answer today. We seem to have come a little closer to it, but perhaps not so close as to prevent us from enjoying the argument for several conferences still to come.

REFERENCES

Amundsen, E., and K. Nustad: Kinin-forming and destroying activities of saliva. Brit. J. Pharmacol. 23, 440 (1964).

Beilenson, S., M. Schachter, and L. H. Smaje: Kallikrein in the submaxillary gland of the cat. J. Physiol. (Lond.) 177, 61P (1965).

Bhoola, K. D., J. Morley, M. Schachter, and L. H. Smaje: Vasodilatation in the submaxillary gland of the cat. J. Physiol. (Lond.) 179, 172 (1965).

Erdös, E. G., A. G. Renfrew, E. M. Sloane, and J. R. Wohler: Enzymatic studies on bradykinin and similar peptides. Ann. N.Y. Acad. Sci. 104, 222 (1963).

Fox, R. H., and S. M. Hilton: Bradykinin formation in human skin as a factor in heat vasodilatation. J. Physiol. (Lond.) 142, 219 (1958).

Hilton, S. M., and M. Jones: Plasma kinin and functional vasodilatation in the pancreas. J. Physiol. (Lond.) 165, 35P (1963).

Hilton, S. M., and M. Jones: The role of plasma kinin in functional vasodilatation in the pancreas. Submitted for publication (1966).

Hilton, S. M., and G. P. Lewis: The cause of the vasodilatation accompanying activity in the submandibular salivary gland. J. Physiol. (Lond.) 128, 235 (1955).

Hilton, S. M., and G. P. Lewis: The relationship between glandular activity, bradykinin formation and functional vasodilatation in the submandibular salivary gland. J. Physiol. (Lond.) 134, 471 (1956).

Schachter, M.: Some properties of kallidin, bradykinin and wasp venom kinin, in Polypeptides Which Affect Smooth Muscles and Blood Vessels, M. Schachter, Ed. London: Pergamon Press (1960).

Schachter, M.: This volume (1966).

Werle, E.: Über den Aktivitätszustand des Kallikreins der Bauchspeicheldrüse und ihres äusseren Sekretes beim Hund. Biochem. Z. 290, 129 (1937).

Werle, E., and I. Trautschold: Kallikrein, kallidin, kallikrein inhibitors. Ann. N.Y. Acad. Sci. 104, 117 (1963).

DISCUSSION

Procacci: I am interested in the effects of local heating of a small area of the forearm. This kind of stimuli causes a vasodilatation that already forms in the heated region next to the areas of the stimulus. Do you think that kinin formation could be responsible for this phenomenon?

Hilton: This is a question I have long wondered about. Local heating or deformation of skin can activate sweat glands, so one could expect kinin formation in response to these stimuli, even if moderate, in skin areas where sweat glands abound. The only direct evidence for kinin formation in heated skin, however, comes from experiments on the skin of rats. This investigation was by Rocha e Silva and Antonio (1960) who found they had to heat to rather high temperatures (42-45°C), i.e. enough to cause a mild burn, in order to provoke a large release of kinins.

EFFECT OF BRADYKININ ON SUBMANDIBULAR SALIVARY GLAND PERMEABILITY

D. C. KROEGER and W. KRIVOY

Departments of Pharmacology, University of Texas Dental Branch, Houston, and Baylor University College of Medicine, Houston, Texas, U.S.A.

Previous reports from our laboratories have shown that electrical stimulation of the trigeminal nerve supply to the dog's mandibular canine tooth causes a release of a bradykinin-like polypeptide within the tooth pulp (Weatherred et al., 1963a; Weatherred and Kroeger, 1965). This kinin release is associated with a vasodilator response within the tooth pulp (Weatherred et al., 1963b). This information coupled with the fact that similar nerve stimulation of the rat's incisor caused a hyperpolarization of the membrane potential of these odontoblasts (Kroeger, 1962) suggested the possibility that a kinin-like polypeptide might be involved in altering the excitability and permeability of these cells within the tooth.

Because of a variety of unmastered technical difficulties associated with a study of permeability changes of the odontoblast cells within the tooth, we chose the submaxillary salivary gland of the dog as a suitable model for the study of glandular permeability. Studies by Martin and Burgen (1962) and Martin (1964) had indicated that stimulation of the cervical sympathetic ganglia or intravascular injections of epinephrine caused a marked increase in the permeability of the submaxillary salivary gland, of cats and dogs, to certain nonelectrolytes. They used a series of carbohydrates with known molecular radii and were able to estimate the increase in size of the "intercellular gaps" in the gland necessary to allow passage of these molecules from blood into saliva. Glucose with a molecular radius (M.R.)

of 3.4Å penetrated the gland after the injection of epinephrine, whereas inulin (M.R. 14.7 Å) did not. Their data indicated that the maximal opening of the intercellular gap was approximately 25 Å, allowing for the passage of molecules with a radius of 12 Å or less.

This is a report of a study concerning the effects of the plasma kinins, bradykinin and kallidin, in altering the permeability of the submaxillary gland of the dog to the carbohydrates, glucose and sucrose.

METHODS

Dogs, weighing 12 to 20 kg, were anesthetized with either a mixture of chloralose and urethane (Martin and Burgen, 1962) or a combination of morphine, i.m., and pentobarbital, i.v., 10 and 15 mg/kg respectively. In most experiments, blood pressure was monitored from the femoral artery, and drugs were administered into the femoral vein except for the intra-arterial injections given via the cannulated thyroid artery. Blood samples for carbohydrate analysis were obtained from the femoral vein contralateral to that vein used for drug administration.

The submaxillary salivary gland ducts (Wharton's) were cannulated intraorally with fine polyethylene tubing. Electrical stimulation of the chorda tympani proved to be the most satisfactory method for obtaining saliva samples. The intensity of the electrical stimulus was adjusted (3-5 V, 0.5 msec duration at 20 cps, Grass S-4 stimulator) to elicit a saliva flow of from 40 to 60 drops/min. In some experiments pilocarpine, 0.1 mg/kg i.v. was used to elicit saliva flow.

Sucrose (M.R. 4.4 Å), the usual marker substance, was administered i.v. as a 10% solution at the rate of 0.5 g/kg every 90 min to maintain adequate blood levels. Sucrose was quantitatively analyzed in both plasma and saliva by the method of Deane (1952). In certain experiments glucose permeability was studied by the enzymatic method of glucose oxidase (Glucostat, Worthington) for the analysis of both fluids.

Both a crystalline bradykinin triacetate (Cal Biochem.) and a solution of bradykinin (Sandoz) were used with no differences noted between them. Kallidin solution (Sandoz) and an epinephrine HCl solution (Adrenalin or Eastman) were also employed in the study.

RESULTS

Unlike the results published by Martin and Burgen (1962) the control samples of saliva, prior to any drug, showed higher saliva/plasma ratios (S/P) for both glucose and sucrose. In seven of twelve dogs, the salivary concentration of the sugar was greater than 5% of the plasma concentration, ($\overline{X}$ S/P = 0.090, range 0.060-0.136). Five other dogs showed concentrations less than 5% ($\overline{X}$ S/P = 0.031, range 0.009-0.050). No apparent difference appeared between the methods of inducing salivation, by chorda tympani stimulation or with pilocarpine. Changing the anesthetic mixture did not alter the situation, nor did changing the pre-anesthetic handling of the animal, or allowing several hours for the anesthetized animal to equilibrate. It was noted, however, that the dogs generally were highly infested with both "heart" and intestinal hookworms.

The intravenous injection of epinephrine, bradykinin, or kallidin, in doses ranging from 1 to 5 μg/kg, caused marked in-

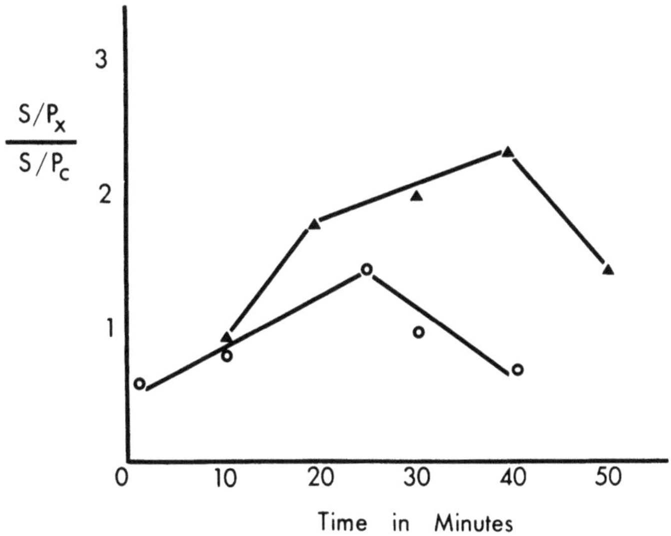

Fig. 1. The effects of epinephrine and bradykinin on the permeability of the salivary gland to sucrose. Post-drug S/P ratio: pre-drug S/P ratio is plotted vs. time after the injection. Epinephrine effects are represented by (▲) and bradykinin by (●).

creases in the S/P ratios. Figure 1 illustrates the changes of permeability resulting from the i.v. injection of bradykinin, 1 μg/kg, and epinephrine, 5 μg/kg. In this diagram, the saliva/plasma ratio at various time intervals after drug administration is compared to the predrug ratio S/P_x:S/P_c The effect of either drug reached a maximum within 20 to 30 min, then decreased toward preinjection levels by the end of a 90-min period. In the doses employed, no apparent dose-response effect was observed except possibly a reversed trend for epinephrine, 1 μg often being more effective initially than 5 μg/kg. Possibly the degree of vasoconstriction might have had an adverse effect on salivary formation, since flow rates were not noticeably affected.

In experiments in which both epinephrine and one of the peptides were given to the same animal, the order in which they were given appeared to be critical. If epinephrine was given 30 to 60 min before the peptide, the S/P_x : S/P_c values for the peptides at the 1 μg dose level had a mean value of 3.45 (range in 7 determinations, 1.50-8.00). If, however, the peptide was administered prior to epinephrine, a mean value for this ratio of 1.44 (range in 5 determinations, 1.25-1.75) was obtained.

These results strongly suggested that the polypeptide effect on salivary gland permeability was due at least in part to the release of epinephrine. To clarify this question, several procedures were employed, including the administration of adrenergic blocking agents as well as attempting to deplete the kinin-labile catecholamine stores by the repeated injections of bradykinin.

Figure 2 illustrates the effect of epinephrine and kallidin on blood pressure before and after phenoxybenzamine (5 mg/kg i.v.). It should be noted that, whereas the vascular responses were completed within 5 min, the permeability effect persisted for many minutes or even hours.

Table 1 shows the changes of salivary gland permeability that accompanied the changes in the blood pressure illustrated in Figure 2. It is observed that there was only a partial antagonism of the epinephrine effect, not a complete block as reported by Martin (1964). We did, however, observe a potentiation of the effect of bradykinin on the permeability to sucrose.

The inability of phenoxybenzamine to block the effects of epinephrine on permeability completely suggested that a β – receptor response might also be involved. Several experiments were

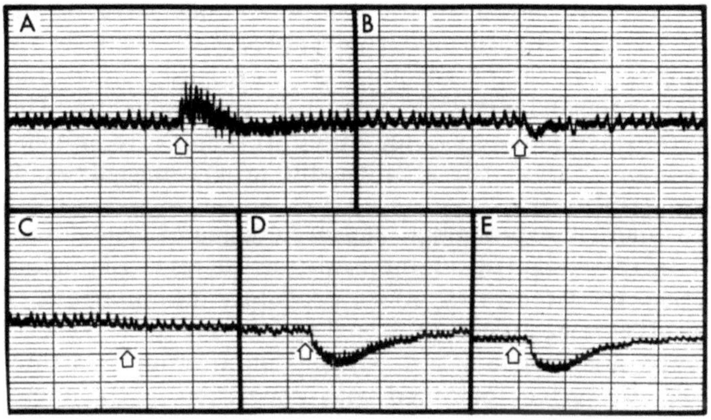

Fig. 2. Effects of epinephrine and kallidin on the blood pressure of the dog before and after phenoxybenzamine. Corresponding salivary gland permeability changes to these drugs are shown in Table 1. Dog, 19 kg. Sucrose was used as the marker. All drugs administered i.v. A and D: epinephrine, 3 μg/kg; B and E: kallidin, 1 μg/kg; C: phenoxybenzamine, 5 mg/kg.

TABLE 1
The effects of epinephrine and kallidin on the permeability of the submaxillary salivary gland to sucrose before and after phenoxybenzamine

	Saliva/plasma ratio					
	Control period			After phenoxybenzamine (5mg/kg i.v.)		
	Time after injection (min)			Time after injection (min)		
Treatment	0	15	30	0	15	30
Control	0.025			0.034		
Kallidin (1 μg/kg i.v.)		0.050	0.064		0.272	0.149
Epinephrine (3 μg/kg i.v.)		0.092	0.041		0.081	0.045

The saliva/plasma ratio was determined for sucrose, 0.5 g/kg, administered 15 min before the start of the experiment. Sixty min. was allowed between each injection of the test drugs.

conducted that used a dual block of both phenoxybenzamine, 5 mg/kg i.v., and dichloroisoproterenol (DCI), 6 mg/kg i.v. Accordingly, the values for the S/P_x: S/P_c ratios were 0.99 and 0.93 for epinephrine and 1.15 and 1.40 for bradykinin. Thus the dually-blocked salivary gland failed to respond to epinephrine, whereas bradykinin was still capable of showing some increase in permeability.

In another experiment, repeated doses of bradykinin were given intra-arterially at intervals of 20 min. Beginning with 1 μg/kg, its effect, S/P_x:S/P_c, gradually diminished for three consecutive doses from 3.0 to 2.7 to 1.9. The next dose of 3 μg produced a response of 5.4. These results would indicate that, although the effect obtained with repetition of the same dose may diminish, its value did not fall to 1.0 or below and when a higher dose was given, an increase in the permeability could still be elicited.

DISCUSSION AND CONCLUSIONS

The data presented here are in agreement with observations previously reported from our laboratories, that bradykinin produces a change in the permeability of cellular membranes. Studies of the influence of this and other polypeptides on neural (Krivoy et al., 1963) and cardiac function (Vick et al., 1965) indicate that their action is that of altering the pre-existing or steady state of the membrane. Thus it appears that the plasma kinins modulate tissue excitability, an action which is best explained by a selective alteration in the permeability of the cell membrane.

The injection of both bradykinin and kallidin into dogs causes an increase in the permeability of the submaxillary gland to both sucrose and glucose. The site at which permeability was altered could not be determined by these studies. Martin and Burgen (1962) suggested that the intercellular space between the ductal epithelium was at least one of the sites of action. The short latencies of the appearance of the marker into the saliva formed the basis of their supposition. Our data would support this concept, but neither these experiments nor those of the previous authors permit any assessment of possible acinar permeability changes.

Our data indicate that some of the effects of these polypeptides on permeability may be accounted for by the release of catecholamines. Increasingly, evidence is being presented that some if not many of the actions of the plasma kinins are the result of releasing catecholamines from certain storage sites in the body. The work of Feldberg and Lewis (1965) and Staszewska-Barczak and Vane (1965) specifically points to the adrenal medulla, although other chromaffin tissue might be affected. The ability of the β-receptor blocking agents to block the increase in permeability of the salivary gland in addition would support the idea that the principal catecholamine involved is epinephrine.

We were, however, able to show a change in permeability which was due to the kinins that persisted after the actions of epinephrine had been prevented by use of the two adrenergic blocking agents. Our data would also indicate that chorda tympani stimulation, necessary to cause salivation, would activate a plasma kinin (Hilton and Lewis, 1955) that would alter the "resting" permeability levels of the salivary gland. This would be true irrespective of whether the polypeptides act directly or indirectly through the release of epinephrine.

Phenoxybenzamine was only partially capable of preventing an increase in permeability in our series of dogs. Dichloroisoproterenol was noted to increase the permeability of the salivary gland seemingly by itself. It is known that this blocking agent does possess some ability to stimulate the β-receptors prior to blocking them (Moran and Perkins, 1958). Further studies of this problem with some of the newer β- adrenergic blocking agents which do not initially stimulate the receptors would help to clarify this problem.

In conclusion, (1) The polypeptides, bradykinin and kallidin, have been shown to alter the permeability of the submaxillary gland to certain carbohydrate molecules. (2) Some of the action of these polypeptides on permeability appear to be due to the release of epinephrine from certain storage sites in the body. (3) Bradykinin, however, can still alter permeability of the submaxillary salivary gland to sucrose or glucose when effects of epinephrine on permeability have been blocked by a combination of an α- and a β-adrenergic receptor blocking agent.

ACKNOWLEDGMENTS

Supported by U.S. Public Health Service Grants DE 01390 and MH 3477, and the General Research Fund of the University of Texas Dental Branch. We are indebted to Dr. R. P. Bircher of the Sandoz Pharmaceutical Co., Hanover, N.J., for a generous supply of synthetic bradykinin and kallidin; and to Dr. E. J. Fellows of Smith, Kline & French Laboratories, Philadelphia, Pa., for the supply of phenoxybenzamine used in these studies.

REFERENCES

Deane, N.: II. Methods of study of body water compartments (space techniques). Meth. med. Res. 5, 164 (1952).

Feldberg, W., and G. P. Lewis: Effects of peptides in the adrenal medulla of cats. J. Physiol. (Lond.) 178, 239 (1965).

Hilton, S. M., and G. P. Lewis: The cause of the vasodilation accompanying activity in the submandibular salivary gland. J. Physiol. (Lond.) 128, 235 (1955).

Krivoy, W. A., M. Lane, and D. C. Kroeger: The actions of certain polypeptides on synaptic transmission. Ann. N.Y. Acad. Sci. 104, 312 (1963).

Kroeger, D. C.: Effect of certain drugs on the odontoblast membrane potential. Biochem. Pharmacol. 9, 221 (1962).

Martin, K.: Observations on the increase in permeability induced by adrenaline in the submaxillary gland. J. Physiol. (Lond.) 172, 50 (1964).

Martin, K., and A. S. V. Burgen. Changes in the permeability of the salivary gland caused by sympathetic stimulation and by catecholamines. J. gen. Physiol. 46, 225 (1962-63).

Moran, N. C., and M. E. Perkins: Adrenergic blockade of the mammalian heart by a dichloro analogue of isoproterenol. J. Pharmacol. exp. Ther. 124, 223 (1958).

Staszewska-Barczak, J., and J. R. Vane: The release of catecholamines from the adrenal medulla by peptides. J. Physiol. (Lond.) 177, 57P (1965).

Vick, R. L., W. A. Krivoy, and D. C. Kroeger: The antagonism of chloroform-epinephrine induced cardiac arrhythmia by bradykinin and 6-glycine bradykinin. Arch. int. Pharmacodyn. 156, 100 (1965).

Weatherred, J. G., and D. C. Kroeger: Neurogenic activation of a plasma kinin in the dog tooth pulp. Amer. J. Physiol. In press.

Weatherred, J. G., D. C. Kroeger, and E. L. Smith: Non-adrenergic response of the dental pulp to inferior alveolar nerve stimulation. J. dent. Res. 37, 16 (1963a).

Weatherred, J. G., D. C. Kroeger, and E. L. Smith: Pressure response in the dental pulp chamber to inferior alveolar nerve stimulation. Fed. Proc. 22, 2 (1963b).

A RADIO-IMMUNOASSAY FOR BRADYKININ

JOCELYN SPRAGG, EDGAR HABER, and K. FRANK AUSTEN

Department of Medicine, Harvard Medical School; and the Medical Services, Massachusetts General Hospital, Boston, Mass., U.S.A.

INTRODUCTION

The current bioassay methods for such vasoactive peptides as bradykinin and angiotensin all present problems of specificity, quantitation, reproducibility, and convenience. Haber et al. (1965) have recently reported a new method for the quantitation of angiotensin II, which depends upon the binding of the polypeptide by specific antiangiotensin antibody. This paper describes a similar method for the assay of bradykinin.

METHODS AND RESULTS

In the present study, the nonantigenic polymer, poly-L-lysine, was chosen as the carrier molecule. The amino-terminal amino groups of bradykinin were coupled to the ϵ-amino groups of poly-L-lysine by a modification of the method described by Schick and Singer (1961), in which toluene-2, 4-diisocyanate is used as the coupling agent. The density of bradykinin attachment to the carrier protein, determined by the method of Spackman et al. (1958) for amino acid analysis, was 1 bradykinin molecule /5 lysine residues. This antigen was prepared as an emulsion with complete Freund's adjuvant and injected into the footpads of albino rabbits. Three and six months later, subcutaneous boosting with emulsified antigen was performed. Four and seven months after the initial immunization, all animals were boosted intravenously on alternate days during a 1-week period. Each animal received a total of 5 mg antigen in emulsion and 0.4 mg antigen intravenously. Each animal was bled from the central artery of the ear 1 week after the termination of intravenous boosting.

The presence of antibody against bradykinin was determined by the microcomplement fixation method described by Wasserman and Levine (1961). After overnight cold incubation of the antiserum and varying concentrations of the antigen with complement, sensitized erythrocytes were added. After a 37°C incubation period, the samples were centrifuged, and the optical density of the supernatants was measured at 412 mμ.

The percentage of complement fixed by the bradykinin-antibradykinin system was plotted as a function of the antigen concentration (Fig. 1). Inhibition of complement fixation in the region of antigen excess indicates that fixation is not a non-

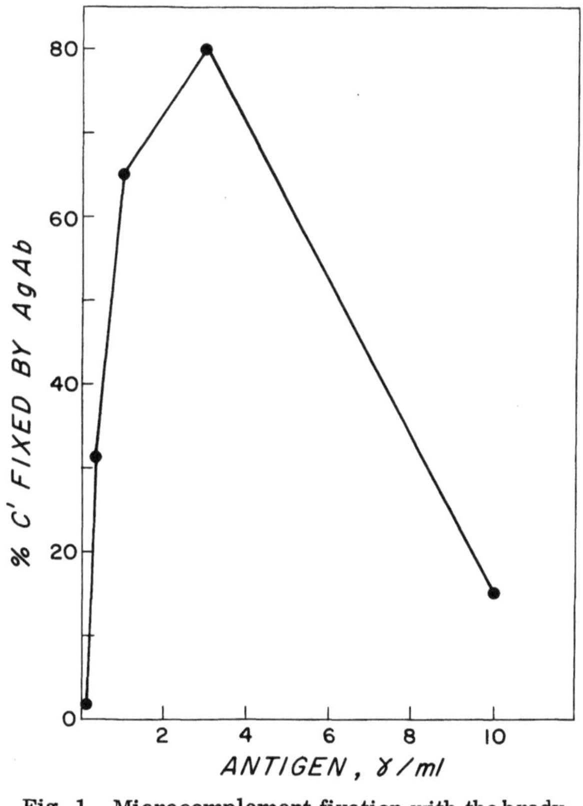

Fig. 1. Microcomplement fixation with the bradykinin-antibradykinin system. The antigen concentrations are shown on the abscissa. Antibody was used at a dilution of 1:50 and guinea pig complement at a dilution of 1:150.

specific function of antigen concentration. When bradykinin coupled to rabbit serum albumin was used as the antigen in this system, an almost superimposable curve was obtained. Unsubstituted polylysine or rabbit serum albumin did not fix complement. These results are consistent with the view that the antibody is specific for bradykinin and is not directed against the carrier molecule.

The presence of antibody against bradykinin was also ascertained by its ability to bind isotopically labeled bradykinin. Dinitrophenol end group analyses (Sanger, 1945) have shown that the amino-terminal amino group of bradykinin is readily acetylated by acetic anhydride. [H^3]-acetyl-bradykinin was prepared according to an acetylation procedure described by Vratsanos (1960). After incubation of the labeled material with antiserum, the mixture was applied to a Sephadex G-25 column, and the column was developed with 0.1 M tris acetate buffer, pH 7.5, in the cold, at a flow rate of 1.5 ml/min. Radioactivity in the effluent fractions was determined in a liquid scintillation counter, with Bray's solution (1960). Radioactivity in the first peak, or void volume, was taken to be that [H^3]-acetyl-bradykinin which had become bound to antibody. This peak was in the same location as the protein peak determined by measuring at 280 mμ the optical density of the effluent volumes. Radioactivity in the second peak, or included volume, was that [H^3]-acetyl-bradykinin which remained unbound. When normal rabbit serum was used in place of immune serum, there was no demonstrable radioactivity in the void volume.

In order to show antibody specificity in this system, an experiment was performed in which unlabeled bradykinin was allowed to compete with [H^3]-acetyl-bradykinin for binding sites on the antibody. The addition of biologically active unlabeled bradykinin to an incubation mixture of antiserum and labeled bradykinin, and subsequent separation of the mixture on Sephadex G-25, resulted in a decrease in radioactivity bound in the first effluent peak and a concomitant increase in unbound radioactivity in the second effluent peak (Fig. 2). Neither vasopressin nor angiotensin II was significantly able to displace radioactivity from the first effluent peak.

The radio-immunoassay is based upon the principle of competition between [H^3]-acetyl-bradykinin and unlabeled bradykinin for antibody sites. With the use of increments of bradykinin, a standard inhibition curve was developed, plotting the decrease in the amount of radioactivity bound by the anti-

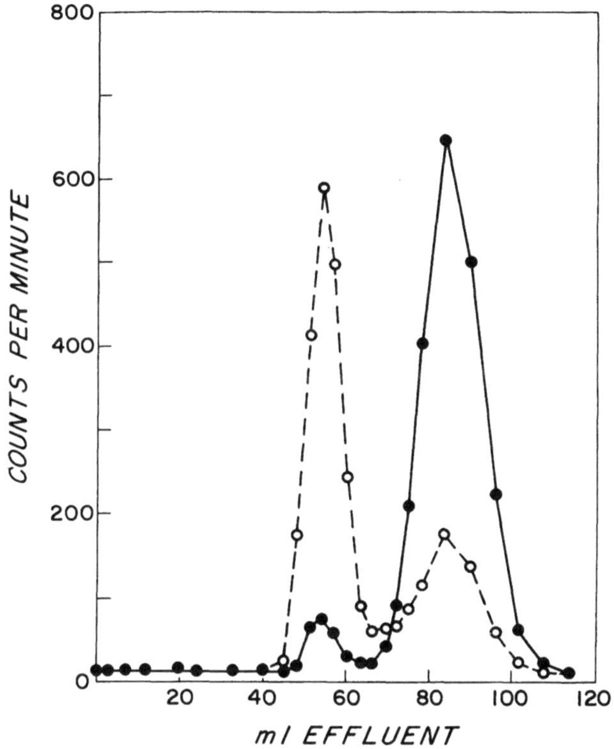

Fig. 2. Binding of $[H^3]$-acetyl-bradykinin by antibradykinin antibody. The broken line represents radioactivity in the column effluent after an incubation of labeled bradykinin with antiserum. The solid line represents radioactivity in the column effluent after the addition of 10 μg bradykinin to the incubation mixture. A portion of the radioactivity bound by the antibody has been displaced by the addition of unlabeled bradykinin. These plots are from early studies with the antibradykinin antibody. Since then, the sensitivity of the assay has been improved by a factor of 10^4.

body (per cent inhibition) as a function of the amount of bradykinin added to the incubation mixture. At the present time, this curve lies within the range of 1-25 ng of bradykinin. In routine analytical work, the amount of bradykinin contained in properly prepared biological fluids or in buffer can be measured by determining the extent to which it inhibits binding of $[H^3]$-acetyl-bradykinin by the antibody.

CONCLUDING COMMENTS

Since bradykinin is a low-molecular weight, nonantigenic polypeptide, it must first be rendered antigenic by coupling it to a high-molecular weight carrier molecule. Successful production of a rabbit antibradykinin antibody was first reported by Goodfriend et al. (1964), who used as an antigen a conjugate of bradykinin and rabbit serum albumin prepared by carbodiimide condensation. Using their antibody in the microcomplement fixation procedure these investigators reported bradykinin detection in the microgram range. However, since the carrier molecule and by-products of the condensation reaction were antigenic, and since the polypeptide was coupled at either end to the carrier, the antibody population studied had multiple specificity.

In the present assay, bradykinin was attached to a nonantigenic backbone in a specific fashion. Data from the microcomplement fixation study and from the competitive binding study are consistent with the view that an antibody population with a high degree of specificity is being employed in the radio-immunoassay. In addition to being specific, the radioimmunoassay is quantitative, reproducible, and convenient to use. At present, it equals the sensitivity of the most effective bioassay methods.

ACKNOWLEDGEMENTS

Supported by Grants AI-04967, AI-04536, and K3-AI-9702 from the National Institutes of Health, United States Public Health Service.

The effort and thought contributed to this project by Mr. Herbert C. Morse III are gratefully acknowledged. The authors also wish to acknowledge the generous gifts of synthetic bradykinin from Dr. R. Bircher of Sandoz Pharmaceuticals and from Dr. R. Bruce Merrifield of the Rockefeller University.

REFERENCES

Bray, G. A.: A simple efficient liquid scintillator for counting aqueous solutions in a liquid scintillation counter. Analyt. Biochem. 1, 279 (1960).

Goodfriend, T. L., L. Levine, and G. D. Fasman: Antibodies to bradykinin and angiotensin: A use of carbodiimides in immunology. Science 144, 1344 (1964).
Haber, E., L. B. Page, and G. A. Jacoby: Synthesis of antigenic branch-chain copolymers of angiotensin and poly-L-lysine. Biochemistry 4, 693 (1965).
Sanger, F.: The free amino groups of insulin. Biochem. J. 39, 507 (1945).
Schick, A. F., and S. J. Singer: On the formation of covalent linkages between two protein molecules. J. biol. Chem 236, 2477 (1961).
Spackman, D. H., W. H. Stein, and S. Moore: Automatic recording apparatus for use in the chromatography of amino acids. Analyt. Chem. 30, 1190 (1958).
Vratsanos, S. M.: On the mechanism of enzyme action. LXXI. Acylations of trypsin in organic solvents. Arch. Biochem. 90, 132 (1960).
Wasserman, E., and L. Levine: Quantitative micro-complement fixation and its use in the study of antigenic structure by specific antigen-antibody inhibition. J. Immunol. 87, 290 (1961).

DISCUSSION

Back: Have you studied your bradykinin antibody with gel diffusion or immunoelectrophoresis techniques?

Spragg: The high positive charge of the antigenic bradykinin-polylysine molecule has made the use of any techniques in which gel diffusion is involved virtually impossible.

Back: What do you think the eventual sensitivity will be in relationship to the bioassay method?

Spragg: From the angiotensin assay of Haber and co-workers, the effective range of this bradykinin assay should eventually fall within the micromicrogram range. The production of labeled bradykinin with a higher specific activity will contribute greatly to the increase in sensitivity of the assay.

Collier: When comparing bradykinin and kallidin on guinea pig lung, we found that they were cross-tachyphylactic and probably acted on the same receptor. Is the specificity of your bradykinin antibody great enough to exclude kallidin or other kinin?

Spragg: We are preparing an antibody directed against the NH_2-terminal end of the bradykinin molecule. It should be able to distinguish between bradykinin and kallidin.

Stewart: Is it actually known whether bradykinin itself actually occurs in the rabbit, or is a kinin of different structure present?

Spragg: According to Brocklehurst, a biologically indistinguishable peptide is present in the rabbit. Structurally, it has not yet been identified.

Stewart: In the second step of the antigen synthesis, do you find any significant degree of reaction of isocyanate with guanidine groups of arginine at the higher pH used?

Spragg: No. The excess diisocyanate is removed before the pH is increased. Also, the pK of the arginine guanidine group is considerably above the pH of this reaction step.

Stürmer: It would be of great interest to have an antibody against kininogens. Then perhaps it would be possible to deplete the kininogen of an animal to study the significance of kinin liberation in anaphylactic shock and in other conditions in which we believe the kinins are involved.

Back: With regard to Dr. Stürmer's comment on the therapeutic possibility of antisera against bradykininogen as opposed to bradykinin, our group has prepared an antiserum against human bradykininogen obtained from Dr. Greenbaum. Our antiserum was unable to inhibit the activation of bradykininogen with trypsin after incubation of the antiserum with the substrate.

Austen: Antibodies to kininogen would not necessarily protect against kallikrein action. Antibodies to bradykinin must be used in vivo in species of origin; otherwise they would yield antibodies to anti-bradykinin antibodies and result in the development of serum sickness. Specificity of antibradykinin assures its usefulness in radioimmunoassay in biologic fluids.

SELF-ANTAGONISM OF BRONCHOCONSTRICTION INDUCED BY BRADYKININ AND ANGIOTENSIN

H. O. J. COLLIER

Department of Pharmacological Research, Parke-Davis & Co., Hounslow, Middlesex, England

Bradykinin might affect respiration by acting on four different effector systems, either directly or through the nerve supply. These systems are (1) the ciliated and the secretory epithelial cells of the respiratory tract; (2) the muscles of ribs and diaphragm; (3) the pulmonary blood vessels; and (4) the pulmonary smooth muscle.

So far as I know, no studies of the action of bradykinin on ciliary action or mucus secretion in the respiratory tract have been published. Gjuris et al. (1964a,b) have described effects of bradykinin on nerves controlling respiratory movements. They claim that bradykinin induces tachypnoea by stimulation of sensory vagal endings; and apnoea by direct action on the respiratory center. Since no other group of workers appears to have reported on this topic, these findings are as yet unconfirmed.

The effects of bradykinin on the pulmonary blood vessels are discussed in other papers in this volume (Konzett, and Hauge et al. 1965). The present paper concerns the bronchoconstrictor action of bradykinin and, to a lesser extent, of angiotensin, in the guinea pig. These actions have been studied mainly by measuring the resistance of the lungs to inflation in vivo. Various factors might contribute to such resistance, but several reasons suggest that it is largely due to contraction of pulmonary smooth muscle.

Characteristics of bradykinin and angiotensin bronchoconstriction

Compared with the bronchoconstriction induced by histamine or 5-hydroxytryptamine, that induced by bradykinin shows several unusual features, some of which angiotensin shares. First, the intensity of the effect depends on the route or site of administration. In the guinea pig, although bradykinin and histamine are about equiactive when injected intravenously, bradykinin is ineffective and histamine effective when each is administered by aerosol (Herxheimer and Stresemann, 1961). The response of human asthmatics to inhaling bradykinin aerosol contrasts with that of guinea pigs. Inhaled bradykinin effectively reduces the vital capacity of chronic asthmatics, and in some subjects it is even more active after prior ingestion of alcohol (Herxheimer and Stresemann, 1963).

When both substances are dropped onto the pleural surface of guinea pig lung, bradykinin is active at a lower dose than histamine, although the dose-response line of histamine is steeper (Bhoola et al., 1962). On the contrary, the isolated trachea of the guinea pig is less sensitive to bradykinin than to histamine.

In another group of observations recorded in the literature, bradykinin shows mixed bronchoconstrictor and bronchodilator or incomplete bronchoconstrictor actions. When the bronchoconstrictor action of bradykinin is suppressed by an antipyretic drug, a bronchodilator action is sometimes unmasked (Collier and Shorley, 1960). Again, when the bronchoconstrictor response to bradykinin is recorded on a fast drum, a kink in the record of increasing resistance to inflation is sometimes seen with bradykinin, but not with histamine or with slow-reacting substance in anaphylaxis (SRS-A) (Berry and Collier, 1964).

By bronchography, Jänkälä and Virtama (1961, 1963) provided another example of the incompleteness of bradykinin bronchoconstriction in the guinea pig. Whereas 5-hydroxytryptamine completely shuts down all but the main trunk of the bronchial tree, bradykinin closes off only the smaller twigs.

Consistent with this observation are those on the dose-response curves of bradykinin in vitro and in vivo. In the guinea pig isolated trachea, the curve of bradykinin flattens off at less than half the maximum reached by histamine (Bhoola et al., 1962). Likewise, in vivo, the curve of bradykinin flattens at about the 50% level, whereas those of 5-hydroxytryptamine and eledoisin proceed to nearly 100% reduction of inflation volume (Gjuris and Westermann, 1965).

Recently, Mrs. Piper and I have found that the dose-response curve of angiotensin bronchoconstriction in vivo is even flatter than that of bradykinin, whereas the curves of histamine and SRS-A resemble that of 5-hydroxytryptamine. Thus endogenous bronchoconstrictor substances appear to fall into two groups in respect of their dose-response curves: those able to reduce and those not able to reduce the inflation volume to nil, when the inflation pressure is of 10 cm of water.

Another characteristic of bradykinin bronchoconstriction is the intense tachyphylaxis that sometimes develops. This may be seen in vivo (Collier et al., 1960) and in vitro (Berry and Collier, 1964).

A fourth feature of bradykinin bronchoconstriction in the guinea pig is its ready inhibition by antipyretic drugs, even after severing the external nervous connections of respiration. Bradykinin shares this response to antipyretics with SRS-A, but not with acetylcholine, angiotensin, histamine, 5-hydroxytryptamine, or substance P. Although antipyretics selectively inhibit the bronchoconstriction, they do not inhibit the increase in capillary permeability due to bradykinin. Therefore the bronchoconstrictor effect is unlikely to be due to pulmonary edema.

To summarize the characteristics of bradykinin bronchoconstriction that are unusual compared with those of histamine and 5-hydroxytryptamine, bradykinin shows (1) gross differences in potency according to the route of administration; (2) incompleteness of the response, with dose-response curves that flatten out at or below the 50% mark and with occasional unmasking of a bronchodilator effect; (3) tachyphylaxis in vivo and in vitro; and (4) inhibition by antipyretic drugs. Angiotensin bronchoconstriction is characterized by incompleteness of response and tachyphylaxis, as the present studies show.

Release of catecholamines by bradykinin and angiotensin has been observed in several animal species other than the guinea pig (Lecomte et al., 1961; Feldberg and Lewis, 1964; Staszewska-Barczak and Vane, 1965). Does it also occur in the guinea pig? If so, how far does it explain the unusual features of bradykinin and angiotension bronchoconstriction in that species? The results below represent an attempt to answer these questions (Collier et al., 1965).

Effects of adrenalectomy and β-adrenergic blockade

Guinea pigs were anesthetized by intraperitoneal injection of urethane (1.25-2.5 g/kg) or pentobarbital (50-80 mg/kg) as the sodium salt, and artificially ventilated through an intratracheal cannula with a Starling pump. The volume of air overflowing through a water valve connected with the sidearm of the cannula was recorded by the method of Konzett and Rössler (1940) and the percentage depression of the inflation volume derived from the record. In this apparatus a water valve opening at a pressure of about 10 cm of water, controls the pressure at which air is pumped into the lungs during recording.

Pronethalol and propranolol, which block β-adrenergic bronchodilator receptors (Nagasaka et al., 1964), increase the intensity of bronchoconstriction induced by a maximally effective dose of bradykinin and lessen the dose needed for a given submaximal effect. That the catecholamines thus blocked come mainly from the adrenals is shown by the fact that bilateral adrenalectomy has a comparable effect to pronethalol. Injecting adrenaline into the inferior vena cava of an adrenalectomized guinea pig counteracts these effects of removing the adrenals. After pithing the spinal cord and cutting the vagus nerves in the neck, pronethalol is still effective, showing that these effects are due to a direct action of bradykinin on the adrenal glands or its nerve supply.

Pronethalol or bilateral adrenalectomy intensifies the bronchoconstrictor response to angiotensin in much the same way as each does that to bradykinin. Both before and after pronethalol, angiotensin is tachyphylactic. As with bradykinin, pithing the spinal cord and vagotomy do not prevent potentiation by pronethalol.

When dose-response curves of bradykinin and angiotensin are plotted, adrenalectomy and pronethalol are found to shift the curves to the left and to raise the maximal degree of bronchoconstriction. After adrenalectomy or pronethalol, the maxima of the bradykinin curves exceed 90%, but the angiotensin curves do not quite reach this level, however high the dose used. With histamine and 5-hydroxytryptamine, adrenalectomy or pronethalol also shifts the dose-response curves, but this is purely a lateral shift, with no vertical component. As far as we have been able to determine with the limited supplies available, the dose-response curve of SRS-A behaves like those of histamine and 5-hydroxytryptamine, both before and after adrenalectomy or pronethalol.

In a series of experiments to test the significance of shifts in the position of the curves obtained with bradykinin and 5-hydroxytryptamine, sham instead of real adrenalectomy was performed, or saline instead of pronethalol was injected. Real adrenalectomy shifted the bradykinin dose-response curves very significantly more than did sham adrenalectomy ($P < 0.001$), and pronethalol was likewise significantly more effective than saline ($P < 0.05$). As far as we have determined so far, the shifts of the 5-hydroxytryptamine curves, however, after adrenalectomy or pronethalol did not differ significantly from those in control animals.

In another group of experiments the same dose of bradykinin was given repeatedly at 5-min intervals until tachyphylaxis developed. The animal was then submitted to bilateral adrenalectomy or treated with pronethalol and, after a short recovery period, the same dose of bradykinin was again given repeatedly. The first response to bradykinin after adrenalectomy or pronethalol was always much higher than the preceding one; but thereafter tachyphylaxis sometimes developed to succeeding doses of bradykinin, although a more complete adrenergic blockade might still abolish tachyphylaxis. A slow rise in the base line, accompanied by difficulty in reinflating the lungs, usually developed also when bradykinin was given repeatedly after adrenalectomy or pronethalol.

The question might be asked whether aspirin inhibits bradykinin bronchoconstriction by facilitating release of catecholamines. The failure of aspirin to antagonize angiotensin makes this most unlikely, and the improbability is much increased by the finding that aspirin is an effective antagonist of bradykinin after pronethalol. After adrenalectomy also, aspirin remains effective against bradykinin.

SUMMARY AND CONCLUSIONS

In the guinea pig, adrenalectomy and pronethalol shift the curves relating dose to bronchoconstrictor response obtained with bradykinin, angiotensin, histamine, and 5-hydroxytryptamine. For histamine and 5-hydroxytryptamine the shift is small and horizontal, whereas for bradykinin and angiotensin it is larger and has a vertical component. After vagotomy and pithing the spinal cord, pronethalol remains effective. These findings suggest that bradkykinin and angiotensin directly liberate catecholamines from the adrenal glands.

In contrast to histamine and 5-hydroxytryptamine, bradykinin and angiotensin do not normally, even at high doses, reduce the inflation volume to nil when the inflation pressure is 10 cm of water. After adrenalectomy or pronethalol, however, both bradykinin and angiotensin come close to preventing any air at all from entering the lungs at this pressure. Thus the incompleteness of bradykinin and angiotensin bronchoconstriction may, to a great extent, be explained by release of catecholamines.

Although the dose-response curves of bradykinin and angiotensin are raised vertically after adrenalectomy or pronethalol, their shapes remain much the same as before, and tachyphylaxis persists. These findings suggest that some other factor than release of catecholamines from the adrenals is involved for these characteristics of the bronchoconstrictor responses to bradykinin and angiotensin.

One may speculate on the physiological meaning of the release of catecholamines by bradykinin and other bronchoconstrictor substances. The findings, that pronethalol intensifies anaphylactic bronchoconstriction in the guinea pig (Collier and James, 1965), and that in adrenalectomized animals bradykinin produces a raised base line with persistent bronchoconstriction, suggest that release of catecholamines by bradykinin may be a protective mechanism against systemic effects of local injury reactions. Evidence that bradykinin mediates some injury reactions is now accumulating; the fact that angiotensin is also a potent liberator of catecholamines suggests that it too may act as a local hormone.

REFERENCES

Berry, P. A., and H. O. J. Collier: Bronchoconstrictor action and antagonism of a slow reacting substance from anaphylaxis of guinea pig isolated lung. Brit. J. Pharmacol. 23, 201 (1964).

Bhoola, K. D., H. O. J. Collier, M . Schachter, and P. G. Shorley: Action of some peptides on bronchial muscle. Brit. J. Pharmacol. 19, 190 (1962).

Collier, H. O. J., J. A. Holgate, M. Schachter, and P. G. Shorley: The bronchoconstrictor action of bradykinin in the guinea pig. Brit. J. Pharmacol. 15, 290 (1960).

Collier, H. O. J., and G. W. L. James: Analysis by drugs of anaphylactic bronchoconstriction in the guinea pig. Brit. J. Pharmacol. 25, 283 (1965).

Collier, H. O. J., G. W. L. James, and P. J. Piper: Intensification by adrenalectomy or by β-adrenergic blockade of the bronchoconstrictor action of bradykinin in the guinea pig. J. Physiol. (Lond.) 180, 13P (1965).

Collier, H. O. J., and P. G. Shorley: Analgesic antipyretic drugs as antagonists of bradykinin. Brit. J. Pharmacol. 15, 601 (1960).

Feldberg, W., and G. P. Lewis: The action of peptides on the adrenal medulla. Release of adrenaline by bradykinin and angiotensin. J. Physiol. (Lond.) 171, 98 (1964).

Gjuris, V., B. Heicke, and E. Westermann: Uber die Stimulierung der Atmung durch Bradykinin und Kallidin. Naunyn-Schmiedebergs Arch. exp. Path. Pharmak. 247, 429 (1964a).

Gjuris, V., B. Heicke, and E. Westermann: Apnoe nach Bradykinin und Kallidin. Naunyn-Schmiedebergs Arch. exp. Path. Pharmak. 248, 540 (1964b).

Gjuris, V., and E. Westermann: Bronchoconstrictor action of bradykinin, kallidin and eledoisin. Nature (Lond.), 205, 402 (1965).

Hauge, A., P. K. M. Lunde, and B. A. Waaler: Bradykinin and pulmonary vascular permeability in isolated blood-perfused rabbit lungs. Biochem. Pharmacol. 14, 1387 (1965).

Herxheimer, H., and E. Stresemann: The effect of bradykinin aerosol in guinea pigs and in man. J. Physiol. (Lond.) 158, 38P (1961).

Herxheimer, H., and E. Stresemann: Bradykinin and ethanol in bronchial asthma. Arch. int. Pharmacodyn. 144, 315 (1963).

Jänkälä, E. O., and P. Virtama: On the bronchoconstrictor effect of serotonin. Bronchographic studies on rabbits and guinea pigs. J. Physiol. (Lond.) 159, 381 (1961).

Jänkälä, E. O., and P. Virtama: Bronchographic demonstration of the bronchoconstrictor effect of bradykinin in the guinea pig. Ann. Med. exp. Fenn. 41, 436 (1963).

Konzett, H.: The action of hypotensive polypeptides on the pulmonary arterial pressure. Biochem. Pharmacol. 14, 1387 (1965).

Konzett, H., and R. Rössler: Versuchsanordnung zu Untersuchungen der Bronchialmuskulatur. Naunyn-Schmiedebergs Arch. exp. Path. Pharmak. 195, 71 (1940).

Lecomte, J., J. Troquet, and A. Dresse: Stimulation medullo-surrenalienne par la bradykinine. Arch. int. Physiol. 69, 89 (1961).

Nagasaka, M., A. F. De Schaepdryver, and C. Heymans: The anti-adrenergic property of nethalide on the isolated guinea pig lung. Arch. int. Pharmacodyn. 149, 232 (1964).

Staszewska-Barczak, J., and J. R. Vane: The release of catecholamines from the adrenal medulla by peptides. J. Physiol. (Lond.) 177, 57P (1965).

DISCUSSION

Zetler: How do your older results with substance P fit into the frame of your present results? Is its action bradykinin-like or eledoisin-like? I think it would be valuable to include substance P, for, according to Feldberg and Lewis, this polypeptide does not release catecholamines.

Collier: We have done no work on the potentiation of substance P by adrenalectomy or pronethalol in the guinea pig. We would like to work on this when time and supplies of substance P allow. I agree it is interesting, in view of the findings of Feldberg and Lewis. It would also be interesting to know the shape of the bronchoconstrictor dose-response curve for substance P.

Konzett: Does a β-blocking agent influence the action of bradykinin on isolated preparations (e.g. tracheal chains)? Have you any suggestions about the differences in bronchoconstrictor action of histamine and bradykinin, when applied to the pleural surface?

Collier: We were unable to obtain any potentiating effect with pronethalol on the action of bradykinin on isolated trachea, but very few experiments have been done. I think our original suggestion still holds that bradykinin and histamine act with different intensities on different varieties of smooth muscle. Presumably, at the pleural surface of the lung, the main smooth muscle is interstitial and terminal bronchiolar muscle. These are presumably very sensitive to bradykinin.

Greenbaum: Have you tried the experiments with reserpinized guinea pigs?

Collier: No.

Back: Have you found histamine to be tachyphylactic in your preparations?

Collier: Histamine shows some tachyphylaxis, but it is much less obvious than the tachyphylaxis of bronchoconstriction to bradykinin or angiotensin.

Back: How do you explain the low maximum activity demonstrated by the in-vitro data of bradykin on the trachea?

Collier: This is difficult to explain. We have no evidence that there is release of catecholamines in isolated trachea. The best hypothesis seems to be that in this situation not all smooth muscle fibers respond to bradykinin. Possibly there are relatively few bradykinin receptors in tracheobronchial muscle at this end of the bronchial tree.H

THE EFFECT OF BRADYKININ AND ANGIOTENSIN ON GANGLIONIC TRANSMISSION

W. HAEFELY, A. HÜRLIMANN, and H. THOENEN

Department of Experimental Medicine, F. Hoffmann-La Roche & Co. Ltd., Basle, Switzerland

It has been shown recently that bradykinin (Lecomte et al., 1961; Feldberg and Lewis, 1964) and angiotensin (Feldberg and Lewis, 1964) release catecholamines from the adrenal medulla. The mode of action, however, is not yet clear. It seems to be different from that of the classic ganglionic stimulants and of indirectly acting sympathomimetic amines.

We became interested in the ganglionic effects of these peptides through reports suggesting a possible action of angiotensin on the peripheral sympathetic nervous system. Benelli et al. (1964) observed a potentiation of vas deferens contractions induced by electrical stimulation of the hypogastric nerve in the guinea pig and postulated a facilitated norepinephrine release from sympathetic nerve endings in the presence of the peptide. In our experiments with the isolated perfused spleen of the cat, angiotensin failed to increase the output of norepinephrine when the splenic nerves were stimulated, although the peptide augmented the effects of nerve stimulation on the splenic smooth muscles (Thoenen et al, 1965). The hypogastric nerve-vas deferens preparation of the guinea pig however, differs from that of the isolated cat spleen. In the former, electrical stimulation of the hypogastric nerve involves to a large extent preganglionic fibers. It is therefore possible that the effect of angiotensin observed on the vas deferens is the result of an action on ganglionic cells rather than on the sympathetic nerve endings.

Both the chromaffin cells of the adrenal medulla and the

ganglionic cells of sympathetic ganglia originate from the neural crest. Despite their morphological and functional differentiation in fetal life they resemble each other in their responsiveness to many drugs. The results of Feldberg and Lewis (1964) show that angiotensin and bradykinin could also be expected to stimulate ganglionic cells. If, however, the stimulant action of the peptides on the adrenal medulla was a selective one on chromaffin cells, angiotensin and kinins would possibly act as well on extramedullary chromaffin cells also present in sympathetic ganglia and depress ganglionic transmission by the local release of catecholamines. Very recently Lewis and Reit (1965) found that angiotensin and bradykinin had indeed a ganglionic stimulant effect which was resistant to the action of hexamethonium and atropine.

The present paper reports the action of bradykinin and angiotensin on the resting and stimulated superior cervical ganglion of the cat studied by recording simultaneously the electrical activity of postganglionic nerves and the response of the nictitating membrane.

METHODS

Experiments were performed on 40 cats either spinalized under ether or anesthetized with urethane-chloralose intraperitoneally. The movements of both nictitating membranes were recorded either isotonically under a tension of 4 g with an ink-writing frontal lever giving a 7-fold magnification or (when postganglionic action potentials were recorded simultaneously) isometrically with a resting tension of 4 g, with strain gauges and a dual-channel ink-writing recorder. Drug injections were made either toward the ganglion or toward the nictitating membrane through two fine polyethylene tubes (0.3 mm internal diameter) introduced into the lingual artery. One of them was advanced into the external and common carotid arteries until its tip lay 1 cm proximal to the carotid bifurcation. Injections through this tube were made during occlusion of the external carotid artery distal to the lingual artery by a mechanically driven clamp firmly fixed on the artery. They will be called injections "to the ganglion." Injections "to the nictitating membrane" were done through the second tube with its tip some millimeters distal to the external carotid artery and without clamping the latter. The dead volume of both tubes was about

0.05 ml. All substances were diluted with Tyrode's solution and injected in a total volume of 0.15 ml.

The cervical sympathetic was cut low in the neck (proximal to the thyroid artery which has a branch accompanying the vago-

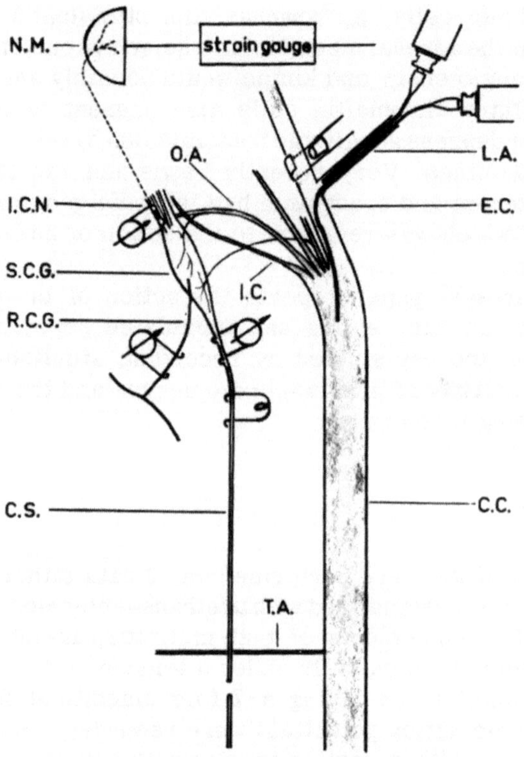

Fig. 1. Schematic diagram of the experimental setup used in this study. The localizations of stimulating and recording electrodes and the tubes used for drug injections are shown. Blood supply according to Chungcharoen et al. (1952). Abbreviations used: C.S. cervical sympathetic, S.C.G. superior cervical ganglion, I.C.N. internal carotid nerves, R.C.G. ramus communicans griseus to upper cervical nerve, N.M. nictitating membrane, C.C. common carotid artery, T.A. thyroid artery, E.C. external carotid artery, L.A. lingual artery, I.C. internal carotid artery, O.A. occipital artery.

sympathetic trunk up to the superior cervical ganglion). Stimulation was bipolar with platinum electrodes and a Grass S4 stimulator with stimulus isolation unit. Monophasic pulses usually of 0.5 msec duration were applied. Voltage, stimulus frequency, and duration of trains were varied according to the special needs.

Evoked compound action potentials or drug-induced asynchronous firing of pre- and postganglionic nerves (usually a small strand of the internal carotid nerve and/or a ramus communicans to the upper cervical nerves) were picked up with platinum electrodes, amplified with capacity-coupled preamplifiers and displayed on a dual-beam oscilloscope (Tektronix 502A) for visualization and photographic recording. The arrangement of tubes and electrodes is schematically depicted in Fig. 1.

The substances used were angiotensin amide (Hypertensin, CIBA), bradykinin (ampoules of synthetic bradykinin, Sandoz, Basle), all-D-bradykinin and all-D-angiotensin (synthetized in the Chemical Research Department of F. Hoffmann-La Roche,

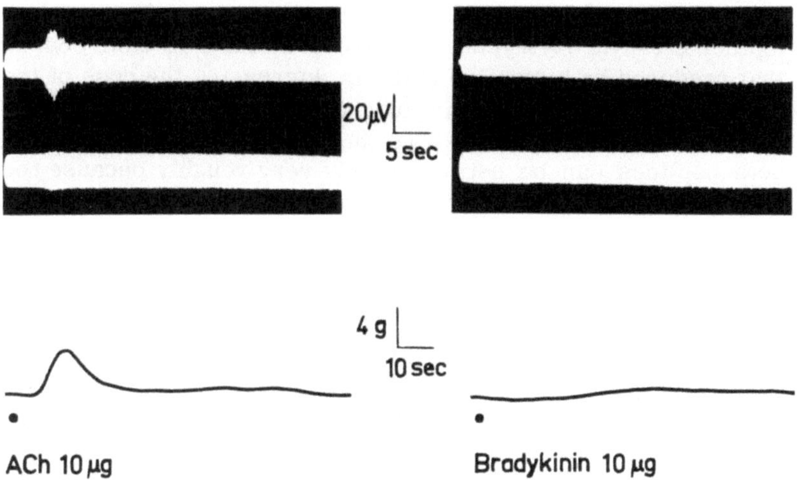

Fig. 2. Pre- and postganglionic discharges induced by acetylcholine (ACh) 10 µg and bradykinin 10 µg. Cat, 4 kg, anesthetized with 12 g urethane and 30 mg chloralose/kg i.p. The delayed onset of the increased activity after bradykinin contrasts with the rapid onset of discharge after acetylcholine. Preganglionic discharges are much weaker after both substances. Upper beam: ramus communicans griseus. Lower beam: cervical sympathetic.

Basle, by Drs. Vogler and Studer), epinephrine, serotonin (5-hydroxytryptamine creatinine sulfate, F. Hoffmann-La Roche, Basle), histamine dihydrochloride (F. Hoffmann-La Roche, Basle), DMPP (1,1-dimethyl-4-phenylpiperazinium iodide, Parke, Davis & Co., Ann Arbor), acetylcholine chloride (F. Hoffmann-La Roche, Basle).

RESULTS

Resting ganglia

The injection of bradykinin and angiotensin to the superior cervical ganglion induces asynchronous firing in postganglionic nerves (Figs. 2 and 3). These discharges appear after a latency of about 8 sec and are followed closely by a contraction of the nictitating membrane (Fig. 2). The increased asynchronous activity lasts from 1 to 2 min after either peptide, whereas for acetylcholine the usual duration is 1 to 3 sec. The firing produced by both peptides is also of considerably longer duration than those produced by DMPP (dimethylphenylpiperazinium iodide) or 5-hydroxytryptamine and resembles in its time course that produced by histamine (Fig. 3). Increasing the dose of the peptides increases the amplitude and shortens the latency of the evoked nerve activity (Fig. 3). The relative potencies of both peptides can be estimated only very roughly because the sensitivity to bradykinin varies independently from that to angiotensin in different ganglia. Usually angiotensin is more potent than bradykinin by a factor of 1:100. The smallest dose of bradykinin necessary for evoking postganglionic discharge was 10 μg, but usually 30 or 100 μg had to be injected. The threshold dose for angiotensin was not determined. If, however, ganglia responded to the peptide at all, they did so with 1 or 3 μg. We found only about half of all ganglia studied to be responsive to bradykinin and even less to angiotensin. Unresponsiveness to one or both of the peptides was unrelated to the sensitivity of the ganglia to acetylcholine (Figs. 4 and 5), which produced excitation in all ganglia studied in doses between 3 and 30 μg, or to histamine or 5-hydroxytryptamine. Some ganglia were completely unresponsive to 5-hydroxytryptamine. The largest contraction of the nictitating membrane observed with either peptide was less than half the maximal contraction obtainable with optimal stimulation of the cervical sympathetic or with optimal doses of acetylcholine or DMPP.

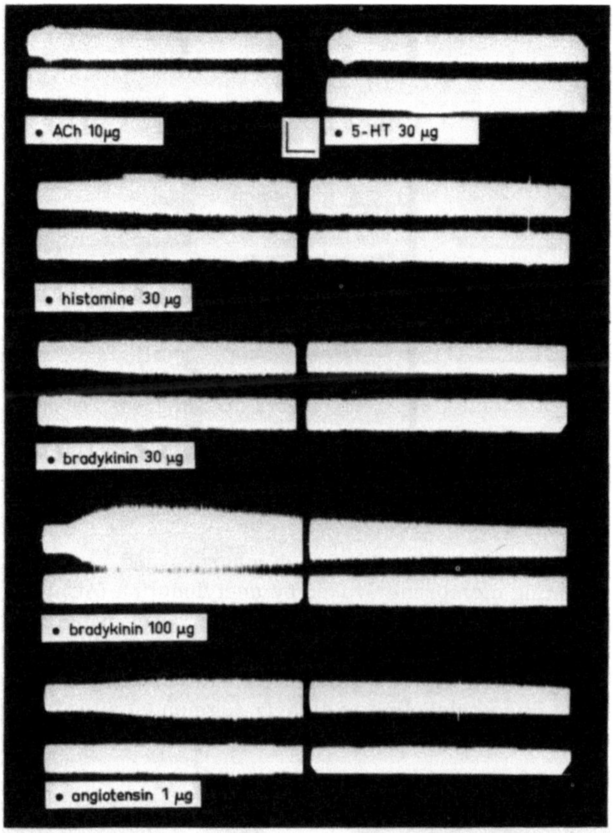

Fig. 3. Pre- and postganglionic discharges produced by injection to the ganglion of acetylcholine (ACh) 10 µg, 5-hydroxytryptamine (5-HT) 30 µg, histamine 30 µg, bradykinin 30 and 100 µg, and angiotensin 1 µg. Cat, 3.6 kg; treated with 1 mg reserpine/kg i.p. 24 hours before the experiment, anestnetized with 0.8 g urethane and 20 mg chloralose/kg i.p. Upper beam: ramus communicans griseus. Lower beam: cervical sympathetic.

Evoked discharges in the preganglionic trunk occurred with the peptides only in those preparations in which other stimulants also produced increased preganglionic activity. When there was a response in the preganglionic nerve, it was always much weaker than in postganglionic fibers. There was

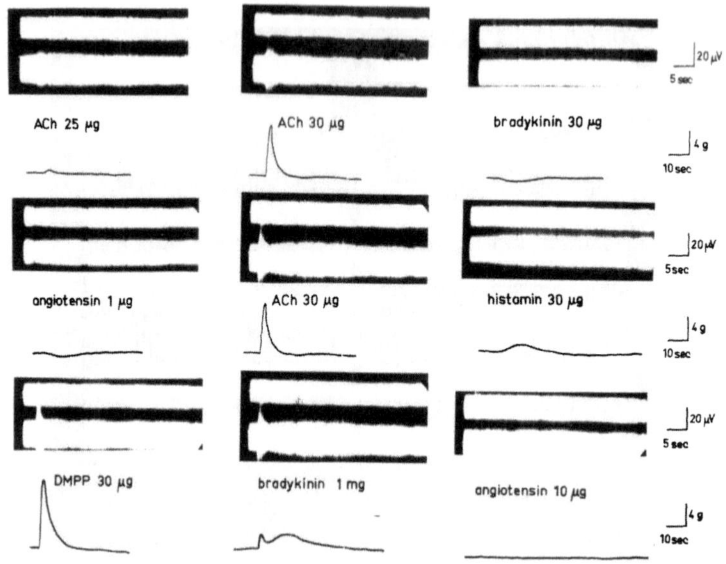

Fig. 4. Pre- and postganglionic discharges and contractions of the nictitating membrane evoked by acetylcholine (ACh) 25 µg and 30 µg, histamine 30 µg, bradykinin 30 µg, and angiotensin 1 µg. Cat, 3.5 kg, anesthetized with 1.2 g urethane/kg i.p. Upper beam: cervical sympathetic. Lower beam: ramus communicans griseus. The effect of acetylcholine is clearly augmented after 1 µg angiotensin which produced no discharges by itself.

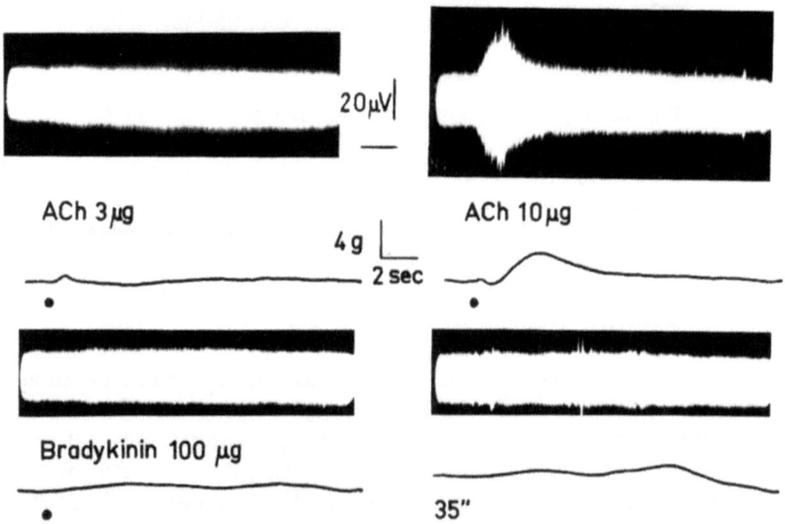

Fig. 5. Virtual absence of postganglionic discharge and nictitating membrane contraction after a high dose of bradykinin in a ganglion fairly sensitive to acetylcholine (ACh). Cat, 3 kg, anesthetized with 1.2 g urethane/kg i.p. Recording electrodes on the ramus communicans griseus.

no difference in the response of the ramus communicans griseus and the internal carotid nerve, provided the strand of the latter used was small. Therefore recording from the ramus communicans — which is easier for technical reasons — can be taken to be representative for the whole postganglionic output.

The ganglionic stimulant action of bradykinin and angiotensin was observed also in two cats pretreated with reserpine (Fig. 3). In such a cat the antipodes of bradykinin and angiotensin, all-D-bradykinin and all-D-angiotensin, were tested and found to be completely devoid of excitatory activity (Fig. 6) and to have no influence on discharges evoked by the L-isomer.

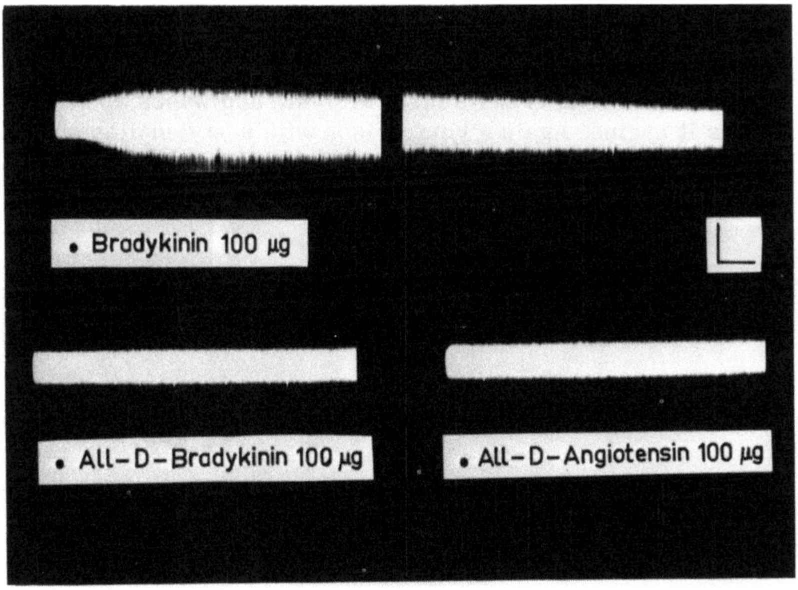

Fig. 6. Comparison of the effects of bradykinin, all-D-bradykinin, and all-D-angiotensin on postganglionic activity. Cat, 3.6 kg, treated with 1 mg reserpine/kg one day before the experiment. Anesthetized with 0.8 g urethane and 20 mg chloralose/kg i.p. Mean arterial blood pressure about 90 mm Hg. Recording electrodes on the ramus communicans griseus. The postganglionic discharge evoked by bradykinin in this preparation was the largest ever observed in our experiments. There was no response of the nictitating membrane.

Stimulated ganglia

The effect of bradykinin and angiotensin was studied on ganglia stimulated repetitively in doses which produced excitation of the resting ganglion. Since experiments mainly with sympathomimetic amines had shown that the nictitating membrane response may diverge in an unpredictable manner from the electrical activity of postganglionic nerves, their effect was tested on the evoked compound action potential of postganglionic nerves. The cervical sympathetic was stimulated repetitively at a frequency of 5 shocks/sec at submaximal strength as judged by the postganglionic spike height.

The typical effect of bradykinin is shown in Fig. 7. After a latency of about 8 sec the postganglionic action potential begins to increase, and a concomitant augmentation of the nictitating membrane response occurs. This facilitation of ganglionic transmission usually lasts for 1 to 3 min and wanes as gradually as it begins. Again a comparison with acetylcholine shows that the effect is less pronounced in intensity but longer in dura-

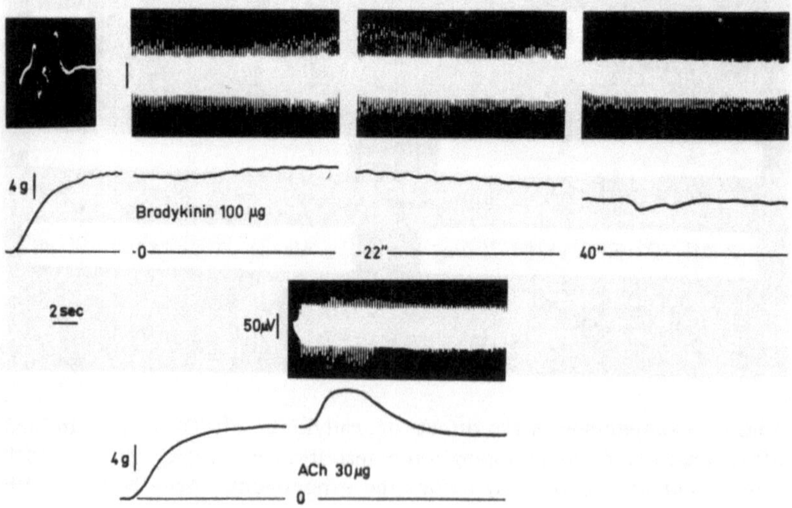

Fig. 7. Facilitation of ganglionic transmission by 100 μg bradykinin and 30 μg acetylcholine (ACh). Cat, 3 kg, anesthetized with 1.2 g urethane/kg. Stimulation of the cervical sympathetic at about 50% of the maximal strength for spike. Recording electrodes on the ramus communicans. Figures at base line of nictitating membrane indicate time in seconds after injection.

tion. Facilitation by bradykinin seems to be of longer duration than the evoked postganglionic discharge.

Angiotension had the same enhancing effect on ganglionic transmission and again was more potent than bradykinin (Figs.

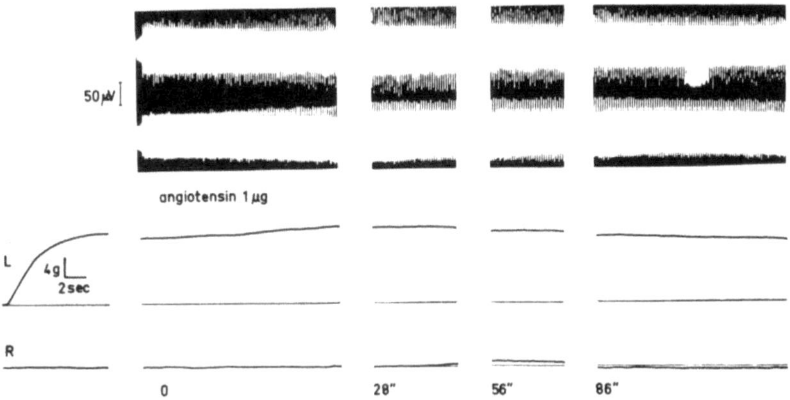

Fig. 8. Facilitation of ganglionic transmission by angiotensin 1 µg. Cat, 3 kg, anesthetized with 1 g urethane and 20 mg chloralose/kg i.p. Stimulation of the cervical sympathetic with 5 shocks/sec at half maximal strength for second spike. From top to bottom: ramus communicans, strand of internal carotid nerve, left and right nictitating membranes. The figures at the base line of the left nictitating membrane indicate time after injections in seconds.

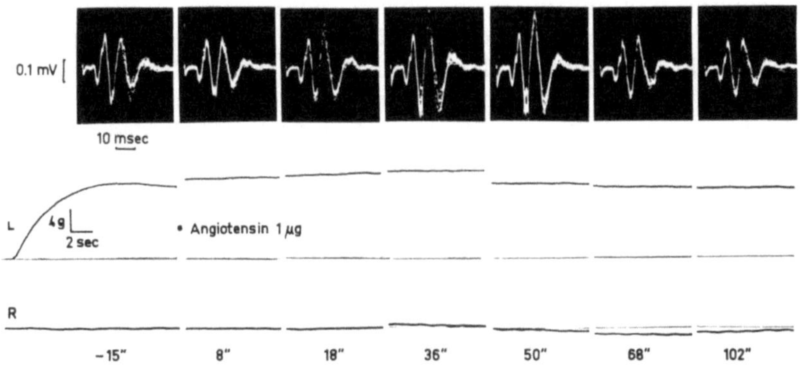

Fig. 9. Facilitation of ganglionic transmission by angiotensin 1 µg. Same cat as in Fig. 8. About five superimposed evoked compound potentials taken at different intervals before and after injection of the peptide are shown on the photographs.

8 and 9). With both peptides, facilitation was present in virtually all ganglia, i.e. also in cats with ganglia which did not respond with postganglionic discharges in the resting state to doses up to 10 times higher than necessary for facilitation.

Both bradykinin and angiotensin, even in the absence of an excitatory action on resting ganglia, increased the effects of acetylcholine injected within several minutes after the peptide (Fig. 4).

DISCUSSION

By recording the postganglionic nerve activity, the excitatory effect of bradykinin and angiotensin on the superior cervical ganglion of the cat has been demonstrated. This confirms the recent findings of Lewis and Reit (1965), who used the contractions of the nictitating membrane as a parameter for the ganglionic effect. Both peptides initiate on close-arterial injection to the ganglion an asynchronous firing of postganglioonic nerves. These action potentials which appear with a latency of about 8 sec and increase slowly in amplitude are seen for 1 to 2 min. Simultaneously with the postganglionic discharge, the tone of the nictitating membrane increases. This excitatory action often occurs without detectable impulses in the preganglionic trunk, which indicates a postsynaptic site of action, as has already been concluded by Lewis and Reit (1965) from the persistence of nictitating membrane contractions after preganglionic denervation. Occasionally, repetitive activity of low amplitude was seen in the preganglionic trunk. This occurred in those cats in which other ganglionic stimulants also produced preganglionic discharges. It may be explained by the varying presence of recurrent postganglionic fibers in the cervical sympathetic (Douglas et al., 1960), or by the presence of aberrant ganglion cells proximal to the superior cervical ganglion (we found aberrant ganglia in the cervical sympathetic visible by naked eye in about 5% of our cats).

The doses of the peptides necessary for inducing postganglionic discharges are of the same order as those found by Lewis and Reit (1965) for causing contraction of the nictitating membrane. On a molar basis angiotension seems to be about 100 times more potent than acetylcholine, whereas bradykinin is equal to or less potent than acetylcholine. However, postganglionic discharges were set up by acetylcholine in all experiments (the threshold dose varying between 1 and 30 μg), whereas angiotensin produced postganglionic discharges and contractions of the nictitating membrane in less than half the preparations studied, and responses to bradykinin occurred only a little more often. The responsiveness to the peptides of the ganglia is independent of their reactivity to other ganglionic stimulants. Ganglia sensitive to bradykinin may not respond to angiotensin. Furthermore, the efficacy of the peptides is very weak when their effect on the nictitating membrane is compared with that obtained with nicotinic stimulants or sympathetic nerve stimulation. This is explained by the more asynchronous discharge which never reaches amplitudes as high as after acetylcholine. The all-D-isomers of bradykinin and angiotensin produce no discharge and do not influence the activity evoked by their natural isomers.

In orthodromically stimulated ganglia (repetitive stimulation of the cervical sympathetic at submaximal strength) both peptides cause a marked facilitation of synaptic transmission. This effect lasts somewhat longer than the postganglionic discharges evoked by the same doses. Most strikingly, facilitation of ganglionic transmission was a very regularly observed effect of both peptides and occurred even in those ganglia which were not excited by them in the resting state. Facilitation was not inhibited by atropine and by pretreating cats with reserpine. Catecholamines seem therefore not to be involved in the ganglionic actions of the peptides.

The most likely explanation of the ganglionic effects of bradykinin and angiotensin seems a depolarizing action on ganglion cells. Experiments to prove this depolarizing effect are under way. We would conclude from the results obtained that depolarization by the peptides is of longer duration than that produced by acetylcholine. The degree of depolarization, however, seems to be too low to induce discharges of the resting ganglia in most cats but sufficiently high to increase the effect of orthodromic stimuli by recruitment.

Whether the ganglionic effects of bradykinin and angiotensin have any physiological meaning can not be concluded from the present investigation. The relatively high doses of the peptides necessary for an action in the superior cervical ganglion are not an argument against it, as has already been discussed by Lewis and Reit (1965). Postganglionic cells elsewhere in the body may be more easily accessible to drugs than in the fairly large superior cervical ganglion. Thus, threshold concentrations for increased contractions of the vas deferens of guinea pig at the frequency of 3/sec are about 3 ng/ml for angiotensin and about 300 ng/ml for bradykinin (our unpublished results).

SUMMARY

Bradykinin and angiotensin induced postganglionic discharges and contractions of the nictitating membrane in about half the cats studied when they were injected into the blood supply of the superior cervical ganglion of these animals. All-D-bradykinin and all-D-angiotensin are without stimulant effect and do not influence the discharges evoked by their L-isomers. A regular effect of bradykinin and angiotensin is a facilitation of ganglionic transmission and an augmentation of the ganglionic stimulant effect of acetylcholine. The actions described seem not to involve the release of catecholamines.

ACKNOWLEDGMENT

The valuable technical assistance of Mr. M. Fischer, Miss U. Hotz, and Miss S. Vögtlin is gratefully acknowledged.

REFERENCES

Benelli G., D. Della Bella, and A. Gandini: Angiotensin and peripheral sympathetic nerve activity. Brit. J. Pharmacol. 22, 211 (1964).

Chungcharoen D., M. de Burgh Daly, and A. Schweitzer: The blood supply of the superior cervical sympathetic and the nodose ganglia in cats, dogs and rabbits. J. Physiol. (Lond.) 118, 528 (1952).

Douglas W. W., D. W. Lywood, and R. W. Straub: On the excitant effect of acetylcholine on structures in the preganglionic trunk of the cervical sympathetic: With a note on the anatomical complexities of the region. J. Physiol. (Lond.) 153, 250 (1960).

Feldberg W., and G. P. Lewis: The action of peptides on the adrenal medulla. Release of adrenaline by bradykinin and angiotensin. J. Physiol. (Lond.) 171, 98 (1964).

Lecomte J., J. Troquet, and A. Dresse: Stimulation médullo-surrénalienne par la bradykinine. Arch. int. Physiol. 69, 89 (1961).

Lewis, G. P., and E. Reit: The action of angiotensin and bradykinin on the superior cervical ganglion of the cat. J. Physiol. (Lond.) 179, 538 (1965).

Thoenen H., A. Hürlimann, and W. Haefely: The effect of angiotensin on the response to postganglionic sympathetic stimulation of the cat's spleen; Lack of facilitation of norepinephrine liberation. Med. Pharmacol. exp. 13, 379 (1965).

DISCUSSION

Vick: Were there changes in arterial pressure which might have influenced your results, particularly with the rather large doses of vasodilators?

Haefely: I do not think so. The arterial pressure was about 150 mm Hg in the cats used and did not fall below 100 mm Hg. Furthermore, the cervical ganglion was cut and no reflex activity could occur at the postganglionic fibers.

Lembeck: Burn and Bülbring showed some years ago a facilitation of the effect of acetylcholine on the ganglion for about 30 minutes after a close intra-arterial injection of epinephrine to the ganglion. Could you observe such a potentiation of preganglionic stimulation or acetylcholine injection during a bradykinin infusion into the ganglion?

Haefely: The discharges produced by acetylcholine were markedly increased within several minutes after the injection of the peptides.

Konzett: Is anything known about the kininase content of the cervical superior ganglion of the cat? A ganglion can be sensitized to stimulants by denervation; perhaps by this procedure the action of the peptides could also be potentiated. Have you done such an experiment?

Haefely: I have no knowledge of the kininase content of the superior cervical ganglion. We have done no experiments with the peptides on denervated ganglion.

Habermann: You are using relatively high doses of bradykinin. Could such doses produce the effects observed by changing the ganglionic circulation or permeability?

Haefely: Bradykinin and angiotensin certainly have an effect on the circulation within the ganglion. It seems very unlikely, however, that the qualitatively similar actions on postgangliar neurons could be the consequence of the different circulatory effects of bradykinin and angiotensin.

POLYPEPTIDE RECEPTOR MECHANISMS: INFLUENCE OF pH

EDWARD J. WALASZEK and DONALD C. DYER

Department of Pharmacology, Kansas University Medical Center, Kansas City, Kansas, U.S.A.

This study was undertaken to ascertain the influence of pH on the effects of various pharmacologically active peptides. We believed that such a study would shed light upon polypeptide receptor mechanisms and hoped it would provide evidence regarding peptide-receptor interactions. In addition, information would be found that may indicate whether the peptide bearing an anionic or cationic charge would be more apt to combine with cationic or anionic receptors respectively. It is well known that conformational changes in proteins can be induced by changes in pH (Bernal, 1962). It may be that the conformation of the peptide and/or receptor would be optimal at certain pH values.

Obviously, changes in pH would affect the distribution of charge in the individual polypeptides. The ionization of receptors is another matter. If they are intracellular we may expect a great deal of buffering; however, if they are on the cell surface (as is more likely) we may expect that the receptors may change their degree of ionization as the pH is changed.

In this study we attempted to correlate the biological activity of four polypeptides on the isolated guinea pig ileum as influenced by changes in pH, with the isoelectric points of these polypeptides.

METHODS

Preparations of guinea pig ileum or rabbit duodenum were suspended in a modified Tyrode's solution at 35°C in a 10-ml bath. The contact time of the tissue with the polypeptides was 60 sec. The time interval between applications was 5 min. There were 58 experiments with the isolated guinea pig ileum and 6 experiments with the isolated rabbit duodenum.

The solution used was Tyrode's without $NaHCO_3$. The pH was adjusted by titration with 0.1 N HCl or 0.1 N NaOH. The pH values varied from 5 to 10. The pH was also varied by oxygenating the bath fluid with a 95% O_2-5% CO_2 mixture rather than with 100% O_2.

Two different procedures were utilized. In the first, the actions of the peptides were studied at each pH value for 45 min, starting with pH 5, increasing by one pH unit to pH 10, after which recovery was obtained at pH 7. In the other design the effects of the peptides were studied at pH 7, 8, 9, and 10, after which the Tyrode's solution was returned to pH 7, then decreased to pH 6 and 5.

Three different experimental designs were used. Three peptides were studied at one equiactive concentration that was obtained at pH 5. Otherwise two peptides were studied at two different dose levels, or one peptide was studied at three to five dose levels at each pH value, as required by the experiment.

The peptides were bradykinin (Sandoz), eledoisin (Sandoz), substance P (a preparation from our laboratory), and angiotensin (angiotensinamide, CIBA). In a few experiments angiotensin (free acid) was also used.

The peptides were dissolved in distilled water, and the injection volume was usually 0.05 to 0.2 ml. On very rare occasions it was 0.4 and 0.8 ml.

TABLE 1
Isoelectric Points of Pharmacologically Active Polypeptides

Bradykinin*	10.0-10.5
Angiotensin†	6.8 (6.3)
Eledoisin**	7 approx.
Substance P‡	8.6

*Rocha e Silva (1960).
†Bumpus et al. (1954).
**Personal communication (Sandoz).
‡Vogler et al. (1963).

RESULTS

Table 1 reports the isoelectric points of the four pharmacologically active polypeptides. The isoelectric point of angiotensin was reported in 1954 and probably represents primarily the decapeptide. Our own determination gave values of 6 to 6.3 for the angiotensin octapeptide.

Figures 1 and 2 represent portions of experiments illustrating the effects of angiotensin, bradykinin, and eledoisin at various pH values. In general, an increase in pH increases the effectiveness of angiotensin and eledoisin and decreases the effectiveness of bradykinin. One may easily distinguish bradykinin from eledoisin by their responses at pH 5 and pH 10.

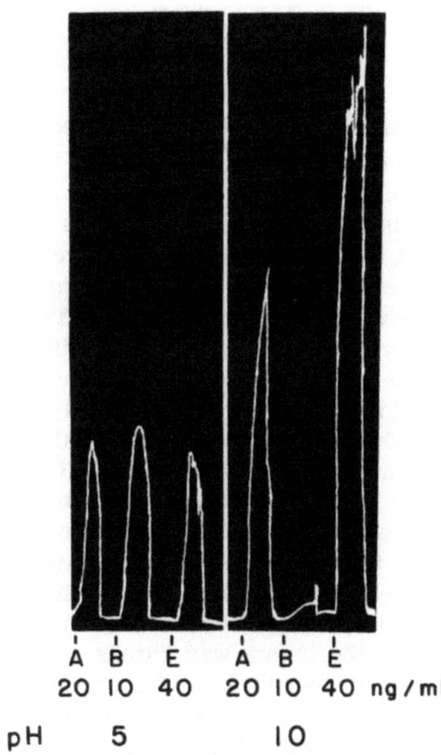

Fig. 1. Effects of angiotensin (A), bradykinin (B), and eledoisin (E) on the isolated guinea pig ileum at pH 5 and pH 10.

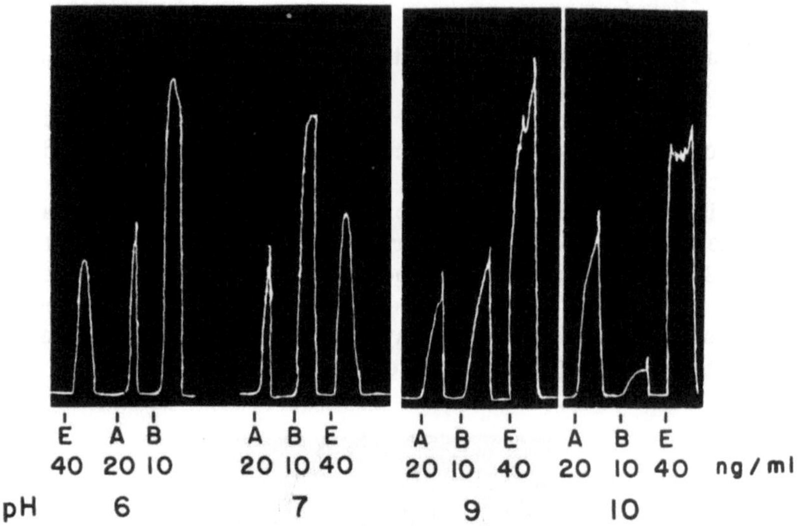

Fig. 2. Effects of angiotensin, bradykinin, and eledoisin on the isolated guinea pig ileum at pH 6, 7, 9, and 10.

Figure 3 illustrates the ineffectiveness of angiotensin and eledoisin at pH 5.8, whereas bradykinin is very effective. When the solution is returned to pH 9, eledoisin and angiotensin become effective while bradykinin loses its efficacy.

Figure 4 represents the isobole plot of the concentration of bradykinin vs. the pH. The quantitatively identical effects were obtained from dose-response curves and were arbitrarily chosen to be a 50-mm contraction. Maximal activity was seen at pH 6; the minimal activity was seen at pH 10.

Table 2 reports values obtained from four separate isolated guinea pig ileum experiments with angiotensin. As can be seen, the effects over the range pH 5-8 are essentially the same. Figure 5 represents the isobole plot of the concentration of angiotensin vs. the pH. Maximal activity was found at pH 10. The remainder of the curve was rather stable, indicating no point of true maximal activity. However, in experiments in which the pH was lowered from 7 to 5, angiotensin was least active at pH 5. Isobole plots representing experiments on the rabbit duodenum showed essentially the same characteristics, with a sharper increase in activity beginning at pH 9. Experiments comparing the angiotensin acid to the angiotensinamide demonstrated no differences in their behavior at various pH values.

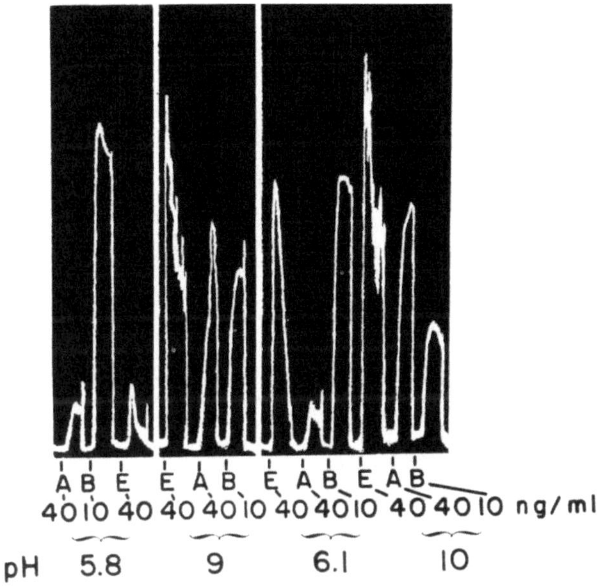

Fig. 3. Effects of angiotensin, bradykinin, and eledoisin on the isolated guinea pig ileum. The pH of 5.8 was obtained by oxygenating the modified Tyrode's solution at pH 9 with 95% O_2 and 5% CO_2. The pH 9 was obtained by oxygenating the modified Tyrode's solution at pH 9 with 100% O_2. The pH of 6.1 was similarly obtained by oxygenating the modified Tyrode's solution at pH 10 with 95% O_2 and 5% CO_2.

Figure 6 represents isobole plots of the concentration of eledoisin vs. the pH on the guinea pig ileum and rabbit duodenum. The point of maximal activity was seen at pH 9 in both preparations, while the point of minimal activity was seen at pH 5.

The results with substance P parallel those obtained with eledoisin. Maximal effects occurred at pH 9-10. However, two points of maximal activity were seen many times, one at pH 7 and one at pH 9.

DISCUSSION

In general, these results would indicate that the pH value at which the greatest biological activity is observed is also furthest away from the isoelectric point of the peptide. Thus

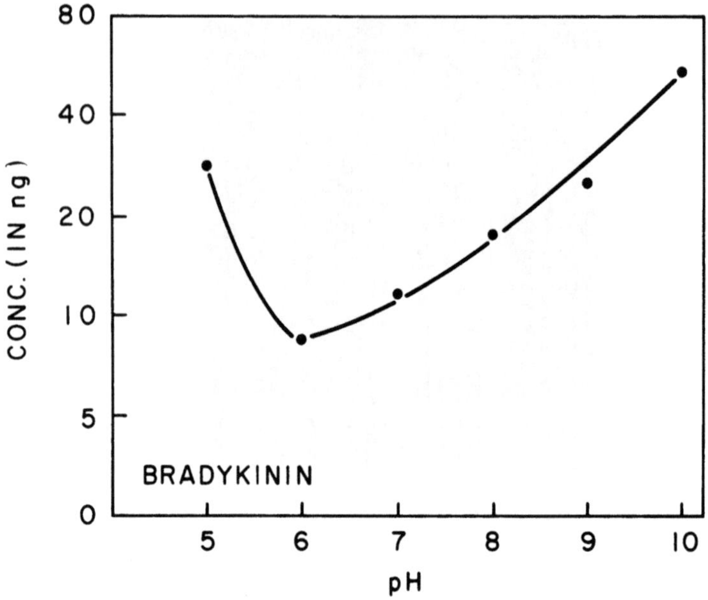

Fig. 4. Isobole representing the relation between pH and the concentration of bradykinin for constant effects. The constant effect was a 50-mm contraction as determined from dose-response curves.

TABLE 2
Response of Isolated Guinea Pig Ileum to 20 ng Angiotensin at Various pH Values

	Response in mm in four different isolated preparations*			
pH	1	2	3	4
5	48	30	48	30
6	52	18	47	28
7	48	21	47	55
8	45	20	48	60
9	58	23	52	(170)†
10	130	50	150	(170)†

*Mean of three responses at each pH.
†Maximal response.

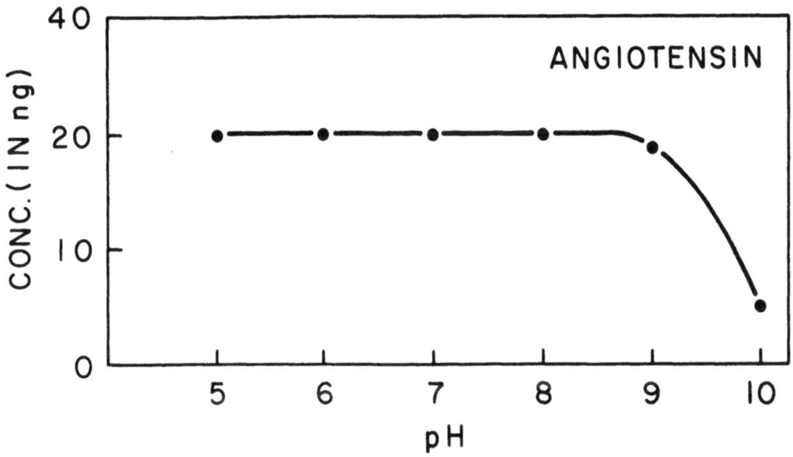

Fig. 5. Isobole representing the relation between pH and the concentration of angiotensin for constant effects. The constant effect was a 50-mm contraction as determined from dose-response curves.

bradykinin with an isoelectric point of 10 has least activity at pH 10, but maximal activity at pH 6. At pH 6 one would expect bradykinin to be a rather strongly positively charged particle that would react with a receptor site rich with carboxylic acid or other anionic groups. This type of receptor would have a negative charge. At pH 10, its point of least activity, bradykinin would be neutral. However, we must not lose sight of the fact that in experiments such as these we also may be changing the degree of ionization of the receptor.

The results with angiotensin are difficult to interpret since its effect is rather constant over a large pH range (5-9) and then drastically changes. Thus there is no point of minimal activity, although as mentioned above, pH 5 under certain conditions may be the point of minimal activity. The point of maximal activity is pH 10. One explanation of these results is the general agreement that angiotensin has two different actions on the guinea pig ileum. One is cholinergic and is probably mediated via acetylcholine, and the other is a direct effect of the peptide (Khairallah and Page, 1961). In small doses the primary effect is the one dealing with a release of acetylcholine. Acetylcholine (a highly charged molecule) would have the same effect over a wide pH range (Ariens, 1964). Thus the constant effect of angiotensin observed over the pH range 5-9 is probably due

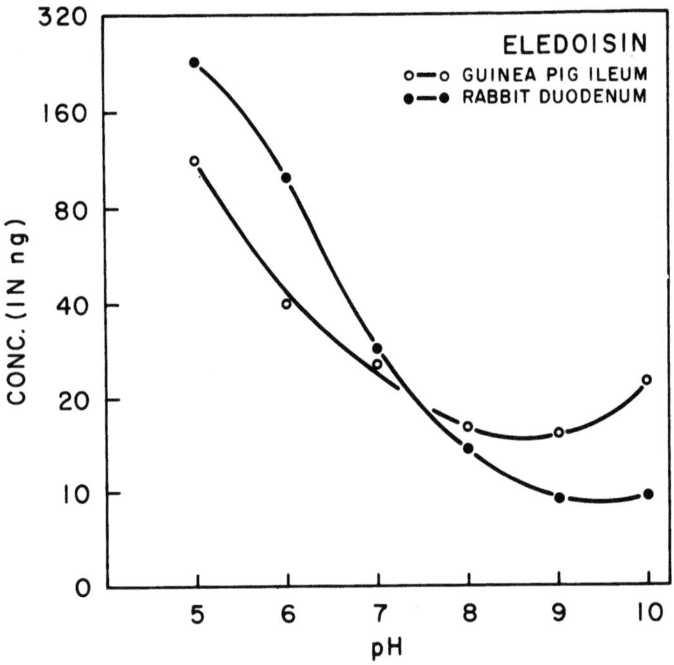

Fig. 6. Isoboles representing the relation between pH and the concentration of eledoisin for constant effects. The constant effect was a 50-mm contraction as determined from dose-response curves.

to the cholinergic component. An assumption must be made that the release of acetylcholine is not influenced by changes in pH. This point will be studied further by means of an atropinized guinea pig ileum. The maximal effect seen at pH 10 may be due to the direct effect of the peptide. Huidobro and Paladini (1963) have reported previously that angiotensin was more effective at pH 9.3, but this enhanced effect was not seen at pH values lower than 9. They explained this effect as a change in the conformation of the peptide at higher pH values. The work of Paiva et al. (1963), who used optical rotatory dispersion and other physiocochemical determinations, indicated no conformational transitions between pH 2.5 and 8.5. Thus it would seem likely to us that if conformational changes do occur they would affect the receptor substance rather than the peptide. Unfolding of proteins may unmask spare angiotensin receptors or make the available ones more accessible. One point in favor of a change in the receptor was that potentiation of angiotensin at pH 10 was greatest when the isolated guinea pig ileum was most labile (difficult to relax after washing and a slightly elevated base line); when it was stable the potentiation was minimal and sometimes absent. The increase in activity may also be due to the interaction of a highly negatively charged peptide with a positively charged receptor. The positivity of the receptor becomes apparent only at pH 10. It is of interest that the half-ionization pH of the tyrosine group in the angiotensin octapeptide is 10.2 (Smeby et al., 1962).

The results with eledoisin indicate a maximal activity at pH 9 and a minimal activity at pH 5. Since the isoelectric point of eledoisin is 7 it would be negatively charged, while the receptor would most probably be cationic with perhaps a lysine group. The results indicated a similar effect on the rabbit duodenum.

The results with substance P are not conclusive. Our results would indicate that the maximal effect is seen at the isoelectric point of substance P (8.6). This may be true and if so it would be the only peptide that behaves in this fashion. Since the substance P preparation used was not pure, doubt may be cast upon these results. Furthermore, on some occasions a double peak was seen. That is, greatest activity was seen at pH 7 and at pH 9 or 10. This may indicate the presence of at least two peptides with two different isoelectric points. If it is true that the maximal activity is seen at its isoelectric point, one may postulate a receptor accepting both a negative charge and a positive charge at a point of least water solubility of the peptide. Such a site could, for example, be a phospholipid containing a negative phosphatidic acid group and a positively charged choline moiety.

SUMMARY

The influence of pH on the pharmacological effects of bradykinin, angiotensin, eledoisin, and substance P was studied on the isolated guinea pig ileum. Angiotensin and bradykinin demonstrated their maximal activity at pH 10 and 6 respectively which was essentially opposite to their respective isoelectric points. Eledoisin with an isoelectric point of 7 demonstrated its maximal effect at pH 9. Substance P resembled eledoisin in these studies.

REFERENCES

Ariens, E. J.: Molecular Pharmacology, p. 381. New York, Academic Press (1964).

Bernal, J. D.: The structure of Molecules, in Comprehensive Biochemistry, Vol. 1, p. 113. Amsterdam, Elsevier Press (1962).

Bumpus, F. M., A. A. Green, and I. H. Page: Purification of angiotensin. J. biol. Chem. 210, 287 (1954).

Huidobro, H. V., and A. C. Paladini: Potentiation of angiotensin action on smooth muscle by alkaline pH. Experientia (Basel) 19, 572 (1963).

Khairallah, P. A., and I. H. Page: Mechanism of action of angiotensin and bradykinin on smooth muscle in situ. Amer. J. Physiol. 200, 51 (1961).

Paiva, T. B., A. C. M. Paiva, and H. A. Scheraga: The conformation of angiotensin II in aqueous solution. Biochemistry 2, 1327 (1963).

M. Rocha e Silva: Biochemistry and Pharmacology of Bradykinin, in Polypeptides Which Affect Smooth Muscles and Blood Vessels, p. 210, M. Schachter, Ed. New York, Pergamon Press (1960).

Smeby, R. R., K. Arakawa, F. M. Bumpus, and M. M. Marsh: A proposed conformation of isoleucyl 5-angiotensin II. Biochem. biophys. Acta 58, 550 (1962).

Vogler, K., W. Haefely, A. Hürlimann, R. O. Studer, W. Lergier, R. Strässle, and K. H. Berneis: A new purification procedure and biological properties of substance P. Ann. N.Y. Acad. Sci. 104, 378 (1963).

DISCUSSION

Lembeck: Adjusting the pH by adding NaOH changes the Na/K ratio in the Tyrode's solution. Do you know the effect of a change in this ratio on the action of polypeptides?

Walaszek: The change of Na/K balance is very small indeed even at pH10. Changing the Na/K ratio by this amount in normal Tyrode's at pH7 did not seem to affect the action of the polypeptides.

Pratesi: Esistono differenze di influenza del pH sulla azione del bradichinina tra un tessuto e un altro? Particolarmente vorremmo sapere se a questo riguardo esistono differenze tra i diversi territori vascolari?

Walaszek: Certainly an acidosis exists in shock. Hence bradykinin should be more effective at the acid pH. We also have an acid pH in inflammation. We intend to study the circulatory effects of bradykinin in animals in acidosis and alkalosis.

Greenbaum: Changing the pH of the environment would introduce additional variables in muscle contractions aside from the charge on the substances to be tested. How would you resolve this problem?

Walaszek: In experiments such as those we believe that receptors would also ionize. Thus it would be difficult to separate a certain effect. However, our experiments indicate that the action of the peptides can be detected over a wide pH range. I believe that we are fortunate that the guinea pig ileum is so hardy a preparation.

Oates: Would not determination of maximal response at various pH's provide a better way of separating the action of pH on the receptor-muscle system from those on the peptide?

Walaszek: No. Contractions are always more quantitative at about 20-40% of maximal.

Meinardi: 1) The isoelectric point of our substance P is 10.6. Did you check your own preparation? (2) Does the crude substance P preparation used have a possible buffering action?

3) Could substance P be broken down during the course of the experiment at the high pH?

4) Is the guinea pig ileum preparation stable at low pH?

Walaszek: 1) No. We used the data of Vogler et al.

2) Yes. The impurities in the substance P preparation may add some buffering action.

3) No. We do not believe that there is a breakdown.

4) The guinea pig ileum is not stable at pH5 for very long periods. In our experiments we kept it at pH5 for only 35-45 min. Under these conditions the ileum continues to be stable.

SENSITIZATION OF SMOOTH MUSCLE TO THE ACTION OF PLASMA KININS BY CHYMOTRYPSIN

H. EDERY
Israel Institute for Biological Research, Ness Ziona, Israel

Various proteins, including enzymes, have been examined for their influence on the sensitivity of smooth muscle preparations to stimulating substances (Edery, 1963, 1964, 1965).

Chymotrypsin (100-500 μg), even after a contact as short as 5 sec with the guinea pig ileum suspended in 5 ml Tyrode's solution, strongly sensitized the organ to plasma kinins. These include not only those already synthetized, such as bradykinin, kallidin (Fig. 1), and methionyl-lysyl bradykinin, but also some others liberated from dog plasma pseudoglobulin by plasma kinin forming enzymes of diverse origin, not yet chemically defined.

The sensitization caused by chymotrypsin was remarkably specific, since it did not occur toward acetylcholine, histamine, serotonin, eledoisin, substance P, ATP, or inorganic contracting substances. This specificity has permitted the identification of plasma kinin traces in unknown mixtures. In addition, it was possible to distinguish between plasma kinins and *Vespa orientalis* kinin (Edery and Ishay, 1965), because chymotrypsin did not sensitize the rat isolated uterus to the latter. No sensitization was observed with heated or DFP-treated chymotrypsin, or with carboxypeptidase, proteose-peptone, plasmin, bovine thrombin, and lipases, or with some nonenzymic proteins. In contrast, trypsin and chymotrypsinogen produced such an effect.

More recent work (Grunfeld and Edery, 1965) showed that the sensitivity of rat isolated fundus and jird isolated ileum toward plasma kinins could be greatly and specifically increased by chymotrypsin.

The bronchoconstriction elicited by bradykinin in Konzett-Rössler's guinea pig preparation was strongly enhanced by previous intravenous injection of 4 to 8 mg chymotrypsin/kg. However, this effect was not as constant as in isolated organs. No obvious reason for the variability was found. In several of

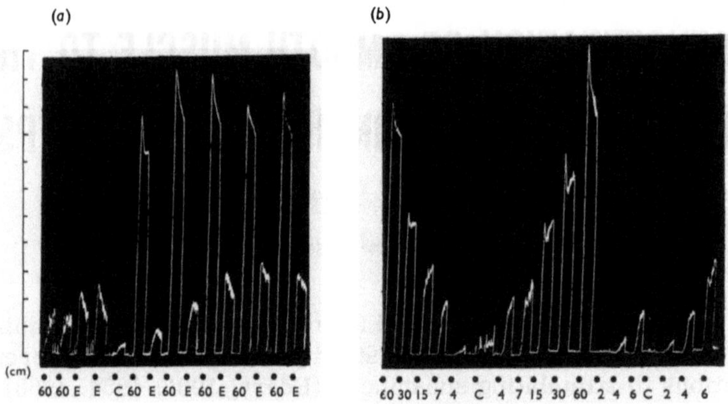

Fig. 1. Guinea pig isolated ileum suspended in 5 ml of atropinized Tyrode's solution. Time of contact, 1 min; time cycle, 5 min. The preparation was washed twice after each application of drug; (a) and (b) are successive tracings from the same experiment. Responses to synthetic kallidin (at marks, doses in ng) and to eledoisin (E, 5 ng). Chymotrypsin (C, 500 μg) sensitized the preparation to kallidin so much that 2 ng of the peptide produced a contraction. Responses to eledoisin returned to their control height after a temporary reduction (from Edery, 1965).

these experiments the response to bradykinin after repeated administration of chymotrypsin became biphasic; i.e. the peptide first produced bronchodilatation which was followed by bronchoconstriction. In such instances bradykinin reversed the bronchoconstrictive effect of histamine. These findings are illustrated in Fig. 2.

The studies reported suggest that chymotrypsin may act by selectively uncovering the specific receptors of plasma kinins in smooth muscle. Advantage may therefore be taken of this phenomenon as a simple new test to decide whether or not a contracting substance belongs to the class of plasma kinins. Only their responses on smooth muscle preparations will be enhanced after chymotrypsin.

REFERENCES

Edery, H.: Israel J. exp. Med. 11, 204 (1963).
Edery, H.: Brit. J. Pharmacol. 22, 371 (1964).

Edery, H.: Brit. J. Pharmacol. 24, 485 (1965).
Edery, H., and I. Ishay: Proc. Israel physiol. pharmacol. Soc. 1, 17 (1965).
Grunfeld, Y., and H. Edery: Proc. Israel physiol. pharmacol. Soc. 1, 64 (1965).

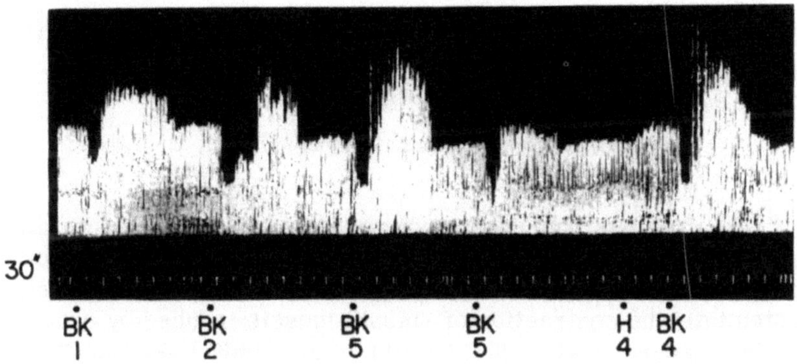

Fig. 2. Guinea pig anesthetized with urethane. Record of lungs' resistance to inflation (Konzett-Rössler's method). The animal had been given intravenously three separate doses of chymotrypsin totaling 6 mg/kg. Note the biphasic response elicited by bradykinin (BK, doses in μg) and the antagonism of bronchoconstriction caused by histamine (H).

POTENTIATION OF THE EFFECT OF BRADYKININ

J. A. GLADNER

National Institutes of Health, Bethesda, Md., U.S.A

In previous reports we have demonstrated that certain naturally occurring polypeptides released during the thrombin catalyzed conversion of fibrinogen to fibrin could markedly potentiate the contraction of smooth muscle induced by bradykinin (Gladner et al., 1963; Osbahr et al., 1963; Laki and Gladner, 1964). This critical stage of blood coagulation is illustrated in Fig. 1, of which only the first line need be of interest to this discussion. In every species where this reaction has been studied to date (Laki and Gladner, 1964) the action of thrombin has always resulted in the release of two major acidic peptides, A and B (sometimes referred to as fibrinopeptides or cofibrin). This paper proposes to extend and expand some of the previous work as well as to include some of our more recent findings.

In the initial studies, bovine peptides A and B (Fig. 2) were examined for their possible potentiating action upon bradykinin. It should be stressed that in this, and examples to follow, these compounds have been fully characterized by enzymatic, chemical, and physical chemical procedures. It is obvious from their structure that these peptides are acidic in nature, as will be shown to be the case for all peptides A and B of all species examined thus far. Our first experiments were carried out at the National Heart Institute with Dr. E. Lenard (Gladner et al., 1963). On an isolated rabbit carotid artery strip we observed that bovine peptide B at a concentration of 250 mg/ml caused a slow rise in the tension exerted by the muscle fibers. However, incubation of the muscle with one tenth this concentration of the peptide developed tension in the fibers which was markedly increased after each electrical stimulation. Extensive washings over prolonged periods could return the muscle strip to a normal electrical stimulation response.

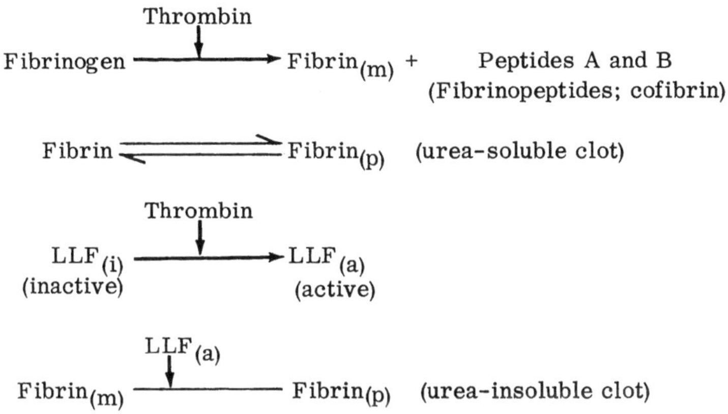

Fig. 1. The overall series of biochemical reactions of the thrombin-catalyzed conversion of fibrinogen to fibrin.

For more definitive and less complicated assays, now employed routinely, we are utilizing the apparatus described by Erdös, which measures the contraction of rat uterus in the estrus cycle (Erdös, et al., 1963). Rat uterus in estrus, induced by an injection of diethylstilbesterol 18 hours prior to sacrifice, is suspended in a 5-ml volume of continuously oxygenated deJalon's solution kept thermostated at 29°C. Construction of the vessel is such that the muscle strip can be very rapidly washed with fresh deJalon's at the same temperature. Figure 3 illustrates a typical result. Our working solutions of bradykinin, made up fresh daily, contained 1.65×10^{-5} μg/liter or 2×10^{-5} μg/liter. At the indicated times, the large figures give the number of total microliters of this solution added. From right to left one may note the response given by the smooth muscle by additions ranging from 10 to 50 μliters; 10 μliters appears to be the subthreshold dosage. The interruption of the graph indicates the incubation of the isolated uterus with 4×10^{-3} mg of pure bovine peptide-B/

glu.asp.gly.ser.asp.pro.pro.ser.gly.asp.phe.leu.thr.glu.gly.gly.gly.val.arg.

(*a*)

N-Acetyl.thr.glu.phe.pro.asp.tyr(SO_4).asp.glu.gly.glu.asp.asp.arg.pro.lys.val.gly.leu.gly.ala.arg.

(*b*)

Fig. 2. The amino acid sequence of bovine peptide-A and -B.

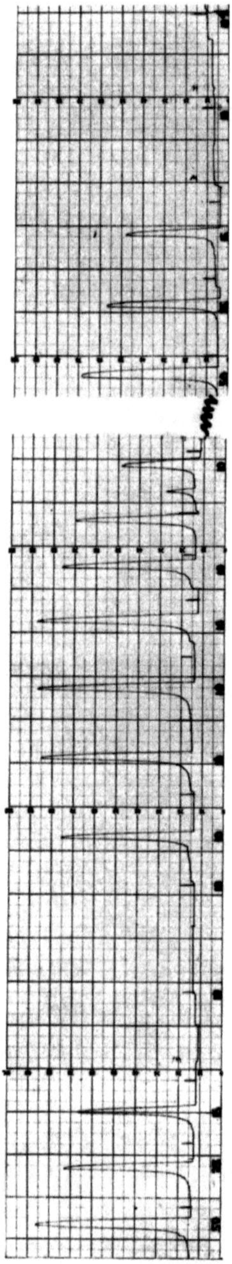

Fig. 3. The effect of bovine peptide-B on the bradykinin-induced contraction of rat uterus. See text for details.

ml for a period of 1 hour. At the cessation of this time the addition of 10 μliters (1.65×10^{-5} μg/ liter) now produces a total contraction, in sharp contrast to its prior ineffectiveness. Repeated additions and washings eventually return the dose response to normal levels. At present, our techniques are refined to the levels in which incubation times can be reduced to 10 min and concentration of bovine peptide-B lowered to 4×10^{-4} mg/ml to produce a potentiating effect. On the other hand, bovine peptide-A is completely without effect even at higher concentrations and longer incubation times.

Our interest, naturally, was directed to investigating peptides released by thrombin from other species of fibrinogen to ascertain: (1) if this potentiating effect was distributed throughout the species; (2) if one could find similarities in structures that might be responsible; and (3) to attempt to correlate structure and function by chemical and enzymatic modifications of some of the active peptides. It should also be of interest to attempt to correlate these results to evaluation.

Studies of other peptide species are in progress, and we can report some of our later findings. The structures for human peptide-A and -B are given below (Blombäck et al., 1962, 1963)

H—Ala—Asp—Ser—Gly—Glu—Gly—Asp—Phe—Leu—Ala—Glu—Gly—
—Gly—Gly—Val—Arg—OH

Human peptide-A

NH_2 (on Asp⁴) NH_2 (on Asp⁶)

Pyr—Gly—Val—Asp—Asp—Asp—Glu—Gly—Phy—Phe—Ser—
—Ala—Arg—OH

Human peptide-B

Actually two human peptide A's were found, differing only in that one contained a phosphorylated serine in position 3 from the amino terminal. Subsequently, the peptide containing the phosphorylated serine was designated human-α, the nonphosphorylated, human-β (Osbahr et al., 1963). Surprisingly, the human-β peptide exhibited a potentiating effect on bradykinin smooth-muscle contraction superior to that shown by the bovine peptide-B (Osbahr et al., 1964a). This effect can be shown in Fig. 4; conditions of bradykinin concentration and incubation are identical with those in Fig. 3. A very recent assay of human peptide-B is without any effect. Evidently, nature had made a switch somewhere in the millions of years separating man and ox, yet the similarities between the bovine and human peptide-A are striking.

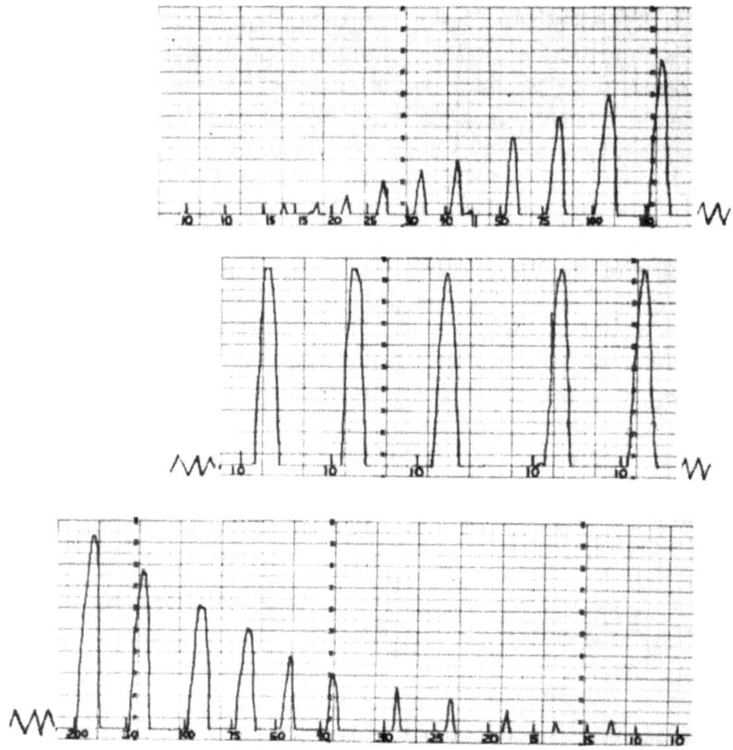

Fig. 4. The effect of human-B peptide on the bradykinin-induced contraction of rat uterus. See text for details.

Enzymatic and chemical modifications were performed on the bovine-B peptide and the human-β peptide; these were then assayed as described above. The following modifications were performed on bovine peptide-B:

1. The Lys-Val bond was completely hydrolyzed by trypsin, the peptides containing 6 and 15 amino acid residues were separated by appropriate ion-exchange chromotography.
2. The carboxyl terminal arginine was completely removed by the use of carboxypeptidase B and the split products separated.

3. The sulfate was chemically removed from both the intact 21-membered peptide and 15-membered one by standing in 1 N HCl for 20 hours at 25°C (see 1).
4. The peptide was hydrolyzed by pepsin for 20 hours. Peptide mapping disclosed the presence of a large number of products; no attempt was made to separate this mixture.
5. The peptide was hydrolyzed in 6 N HCl for 22 hours at 105°C.

After these modifications, all preparations were taken to dryness, redissolved in water to their original concentrations, and assayed for possible potentiation on an equivalent molar basis. In modification 1, the 15-membered peptide remains fully active, whereas the one containing 6 amino acid residues is unreactive. The peptides retain full activity after modifications 2 and 3. The peptic hydrolysate, 4, still contains a moderate potentiating ability. Complete acid hydrolysis, 5, abolishes all activity. These experiments extend our previous findings on bovine peptide-B (Gladner et al., 1963).

Human-β peptide was altered first by removal of the C-terminal arginine; this produced no difference in the potentiating effect. The peptide could be cleaved at the Phe-Leu and Leu-Ala bonds by incubation with chymotrypsin, resulting in four easily separable peptides, two of which contained C-terminal arginine. Both arginine-containing peptides were devoid of any potentiating activity; in sharp contrast to this result, the other two peptides containing the N-terminal portions showed no loss in potentiating activity (Osbahr et al., 1964a).

This mode of investigation is being extended to peptides of other species as they become available. Recently the α and β peptides released from canine fibrinogen have been isolated and characterized (Osbahr et al., 1964b). Like their human counterparts, the only difference between the α and B, respectively, is the phosphorylation of the position-3 serine in the β. The structure of the B-canine peptide is given below:

$$\begin{array}{l} \qquad\qquad\qquad\qquad\qquad\qquad\qquad\quad NH_2 \\ \qquad\qquad\qquad\qquad\qquad\qquad\qquad\quad | \\ H{-}Thr{-}Asp{-}Ser{-}Glu{-}Gly{-}Lys{-}Glu{-}Phe{-}Ileu{-}Ala{-}Gly{-}Gly \\ \qquad\qquad\qquad\qquad\qquad\qquad\qquad\qquad\qquad\qquad\quad {-}Gly{-}Val{-}Arg{-}OH \end{array}$$

The potentiating action of these peptides appears similar to the human. Modification studies on the order described above appear to be yielding the same results. The peptides released from lamprey fibrinogen have been reported (Doolittle, 1963). We have recently isolated the peptides from the fibrinogen of frog. Their amino acid analyses are given in Table 1. Very preliminary assays of the unpurified mixtures of these lamprey and frog peptides indicate the presence of possible potentiating activity. It is interesting to note that the B-peptides of both species are the largest reported to date.

TABLE 1
Amino Acid Analysis of Frog and Lamprey Fibrinopeptides

	Lamprey		Frog	
Amino acid*	A	B	A	B
Aspartic acid	3	13	4	11
Threonine	0	3	1-2	4
Serine	2	2-3	0	8
Glutamic acid	0	3	1-2	4
Proline	0	2	0	2
Glycine	0	2	3	4
Alanine	0	4	4	2
Valine	0	3	0	5
Isoleucine	2	1	0	0
Leucine		3-4	0	0
Tyrosine	0	1†	0	1†
Phenylalanine	0	0	1	0
Lysine	0	1	0	1
Arginine	1	1	1	1
NH_3	‡	‡	‡	(3-4)
Total residues	8	39-41	15-17	43

*Indicated as number of residues found in each peptide.
†Present as tyrosine-O-sulfate.
‡Not reported.

Figures 5 and 6 show some representative sequences of peptides A and B of various fibrinogens (Laki and Gladner, 1964). It can be easily discerned that between the respective series of peptides of each species there are many similarities in structure.

At the present time we are investigating the potentiating effect of some of the peptides of the following species: bovine, human, dog, rabbit, monkey, horse, chicken, lamprey, and frog. These studies are far from complete, and findings will be reported in future publications.

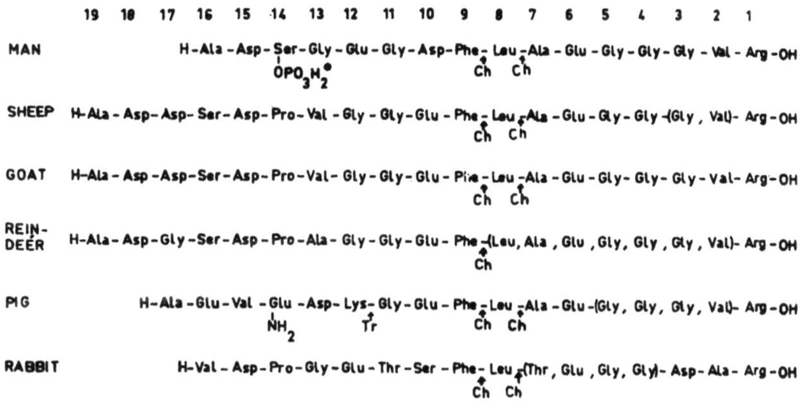

Fig. 5. Sequences of peptide-A released from various fibrinogens. (Blombäck and Doolittle, 1963; Laki and Gladner, 1964).

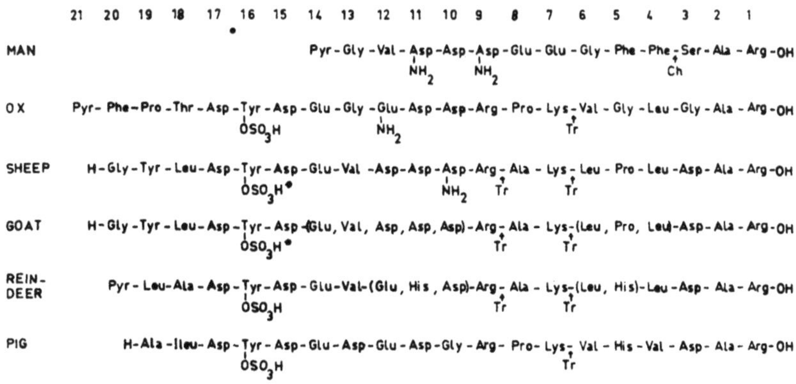

Fig. 6. Sequences of peptide-B released from various firbinogens. (Blombäck and Doolittle, 1963; Laki and Gladner, 1964).

In addition, we were interested to see whether any synthetic bradykinin analogs could be potentiated in a manner analagous to the natural product. Table 2 presents only four of the many existing analogs. Each synthetic peptide, when assayed at the appropriate levels, could be potentiated as if it were bradykinin in our assay system.

TABLE 2
Bradykinin Derivatives* Examined for Possible Potentiation by Bovine Fibrinopeptide-B

Bradykinin	Threshold dose on rat uterus (g/ml Tyrode's solution)
Gly^{6}	1.10^{-9}
Tyr^{5}	5.10^{-7}
Gly^{9}	10^{-6}-10^{-7}
Lys^{9}	5.10^{-7}

*We wish to express our appreciation to Dr. E. Schröder for a generous supply of these compounds.

DISCUSSION

Two questions may be asked at this point: (1) Is the action of certain of the peptides A and B related in any way to a true physiological phenomenon? (2) Is there within each of these active peptides a specific number of residues or "sites" responsible for the potentiating activity?

It should be emphasized that these are naturally occurring peptides which are released from a naturally occurring substrate, fibrinogen, by a specific enzyme, thrombin. These peptides thus may take their place among the other naturally occurring biologically active peptides. Because it is believed that the fibrinogen-to-fibrin conversion goes on constantly to a small extent in the blood stream, these peptides could always be present.

Many investigators have been especially interested in microcirculation of the blood. It has been postulated (Zweifach, 1957, 1959) that the smooth muscle cell in the capillary walls "is regulated not only by substances that directly cause them to contract and relax, but also by other substances that modify the capacity of the cells to react to stimuli and do work." It is thus possible that the blood-clotting reaction may have some relation to control of smooth muscle in the capillary wall. One might suggest from our studies that certain peptides released during blood clotting may play a vital role in the microcirculation of the blood.

As for the second question, modification studies carried out on the aforementioned bovine, human, and canine peptides indicate that the locus of the activity of these peptides might reside in a smaller sequence or "active site." This concept is intriguing, especially since the potentiating activity lies within the N-terminal portion of the active human-β peptide chymotryptic fragments and contains the sequence -Ala-Asp-Ser-Gly-Glu-. This sequence has been found to be the "active site" of the proteolytic enzymes chymotrypsin (Schaffer et al., 1957), trypsin (Schaffer at al, 1958), and thrombin (Gladner and Laki, 1958). Of all the serine residues in these three enzymes, only one, the serine of the "active sites," can react stoichimoetrically with DFP, resulting in its phosphorylation with concomitant complete inactivation of the enzymes. Possibly the inactivity of the α-peptides, which contain a phosphorylated serine, may be inactive, owing to a similar biosynthetic mechanism. On the other hand, the dephosphorylation of the serine may act in the role of a regulator, its removal by certain enzymes being necessary when activity is required by the organism.

From further studies now in progress it is hoped the answers to our questions will be realized.

REFERENCES

Blombäck, B., and R. F. Doolittle: Amino acid sequence studies of fibrinopeptides from several species. Acta chem. scand. 17, 1819 (1963).

Blombäck, B., M. Blombäck, P. Edman, and P. Hessell: Amino-acid sequence and the occurrence of phosphorus in human fibrinopeptides. Nature (Lond.) 193, 883 (1962).

Blombäck, B., M. Blombäck, and P. Edman: On the structure of human fibrinopeptides. Acta chem. scand. 17, 1184 (1963).

Doolittle, R. F.: Characterization of lamprey fibrinopeptides. Biochem. J. 94, 742 (1963).

Erdös, E. G., A. G. Renfrew, E. M. Sloane, and J. R. Wohler: Enzymatic studies on bradykinin and similar peptides. Ann. N.Y. Acad. Sci. 104, 222 (1963).

Gladner, J. A., and K. Laki: The active site of thrombin. J. Amer. chem. Soc. 80, 1263 (1958a).

Gladner, J. A., P. M. Murtaugh, J. E. Folk, and K. Laki: Nature of peptides released by thrombin. Ann. N.Y. Acad. Sci. 104, 47 (1963).

Laki, K., and J. A. Gladner: Chemistry and physiology of the fibrinogen-fibrin transformation. Physiol. Rev. 44, 127 (1964).

Osbahr, A. J., J. A. Gladner, and K. Laki: The action of human thrombin on human fibrinogen. Biochem. biophys. Res. Commun. 13, 462 (1963).

Osbahr, A. J., J. A. Gladner, and K. Laki: Studies on the physiological activity of the peptide released during the fibrinogen-fibrin conversion. Biochim. biophys. Acta 86, 535 (1964a).

Osbahr, A. J., R. W. Colman, K. Laki, and J. A. Gladner: The nature of the peptides released by canine fibrinogen. Biochem. biophys. Res. Commun. 14, 555 (1964b).

Schaffer, N. K., L. Simet, S. Harshman, R. R. Engle, and R. W. Drisko: Phosphopeptides from the acid hydrolyzed P_{32}-labeled diisopropylphosphoryl chymotrypsin. J. biol. Chem. 225, 197 (1957).

Schaffer, N. K., R. P. Lang, L. Simet and R. W. Drisko: Phosphopeptides from the acid hydrolyzed P -labeled isopropyl methylphosphon—fluroid-inactivated trypsin. J. Biol. Chem. 230, 185 (1958).

Zweifach, B. W.: The microcirculation. Amer. J. Med. 23, 684 (1957).

Zweifach, B. W.: The microcirculation of the blood. Sci. Amer. 200, 54 (1959).

DISCUSSION

Erdös: The fibrinopeptides used originate from various animals. They sensitize a mammalian tissue to bradykinin. Some of the animals which were the source of fibrinopeptides may not even react to bradykinin. Do fibrinopeptides potentiate the effect of bradykinin in mammals only?

Gladner: This can certainly be true. For example, we encounter great difficulties with the preparation of our peptides released from chicken fibrinogen. Species cross-reaction should be studied. We are investigating the fish system currently.

Back: Have you studied the effect of other enzymes, particularly plasmin, to form by virtue of its fibrinogenolytic action peptides that potentiate bradykinin action?

Gladner: This has not yet been studied, but I would say it is very possible that peptides containing the right sequence and size released by plasmin would give the same result. This would tend to agree with our results if found.

Prado: I believe it would be necessary to extend the study of the potentiation to other tissues and to experiments in vivo. The phenomenon seems closely related to the potentiation by *Bothrops jaracaca* venom.

Gladner: I am in complete agreement with Dr. Prado's comments. Some in-vivo studies are in progress. We have also assayed *bothrops* venom factor and find a similar potentiation effect. We are certain *bothrops* factor is a polypeptide, but I must stress that our fibrinopeptides are naturally occurring in all species studied.

Ferreira: I would like to suggest that you use guinea pig ileum instead of rat uterus. The guinea pig ileum is a much more sensitive preparation for testing potentiating agents. I would like to know if the fibrinopeptide was able to potentiate in vivo the pharmacological action of bradykinin.

Gladner: I agree with your first comment and in the future will certainly attempt to use ileum, but our facilities, space, and techniques are limited. Thus we use uterus, since Dr. Erdös assisted us in our initial experiments. In-vivo experiments are just now under consideration with various collaborators.

BRADYKININ-POTENTIATING FACTOR

S. H. FERREIRA

Department of Pharmacology, Faculty of Medicine, University of Sao Paulo, Ribeirão Preto, São Paulo, Brazil

The bradykinin-potentiating factor (BPF) is a component of *Bothrops jararaca* venom which potentiates in vitro as well as in vivo some pharmacological effects of bradykinin (Ferreira, 1964; 1965). This communication deals with its purification, biochemical characteristics, pharmacological actions,and some considerations of the mechanism of bradykinin potentiation by BPF.

Partial purification. The factor was extracted with ethanol (90%) from a heated solution of *B. jararaca* venom. To this alcoholic solution four volumes of ethyl ether were added, the precipitate dissolved in distilled water, and lyophilized. The dry powder is referred to in this communication as "bradykinin-potentiating factor," BPF.

This procedure brings about an 8- to 10-fold increase in the bradykinin-potentiating activity as compared to the crude venom. The factor so obtained was no longer contaminated with pharmacologically active amines, did not release kinin when incubated with plasma, and was unable to clot heparinized plasma as the original venom does.

Biological estimation. Presence of BPF was detected in eluates obtained in electrophoresis and chromatography by adding an aliquot of the test solution to the bath of the isolated guinea pig ileum before a standardized dose of bradykinin was given. Quantitative determinations of BPF was performed with arbitrary "potentiating units", one unit corresponding to the dose of BPF which increases the effect of a single dose of synthetic bradykinin (BRS 640, Sandoz) to equal that of a double dose. Although the absolute values of the "potentiating unit" may vary from one gut to the other, they remain constant in the same preparation. A regression line between concentration of BPF in a solution and the biological estimation was found. The potentiating unit in our experiments ranged from 0.3 to 0.6 μg/ml, final concentration in the bath of the isolated preparation (Ferreira, 1964).

Occurrence of bradykinin potentiating activity in some snake venoms. In order to compare the bradykinin-potentiating activity of some venoms the specific-bradykinin potentiating activity of *B. jararaca* venom was taken as 1, and the ratio (R) of activity of any preparation was determined in relation to this value. The bradykinin-potentiating activity was also found in the venoms of *B. atrox* (R = 0.42), *B. neuwiedii* (R = 0.30), and *Crotalus, viridi viridi* (1.7) in which the highest ratio was observed. The ratios for other venoms were lower than 0.10 (*B. cotiara*, *B. alternata; C. terrificus terrificus;* white and yellow venoms, *C. atrox*, *C. adamanteus*, *C. viridis helleri*). These results are preliminary since only one sample of each venom was available.

Physicochemical characteristics. BPF is thermostable; dialyzable; soluble in water, in 5% tricholoracetic acid, and in absolute ethanol; and insoluble in absolute acetone, in chloroform, and in ether. BPF activity was destroyed by acid hydrolysis and long incubation with trypsin. This property would point to the polypeptide nature of the factor(s).

The results of chromatography (DEAE Sephadex A-50) and paper electrophoresis of BPF showed that the potentiating activity could be detected in two different fractions. All these observations led us to suggest that BPF activity could be due to the presence of at least two components (Ferreira, 1964, 1965).

Experiments with isolated preparations. BPF potentiated the action of bradykinin upon the guinea pig ileum, rat uterus, and rabbit duodenum, without any apparent effect on the contrations elicited by histamine, acetylcholine and eledoisin. Figure 1 shows the data recorded in two experiments with the guinea pig ileum (Ferreira and Rocha e Silva, 1965). BPF, in concentration of 0.5 μg/ml in the bath fluid, produced a definite potentiation of bradykinin and kallidin, without action on eledoisin

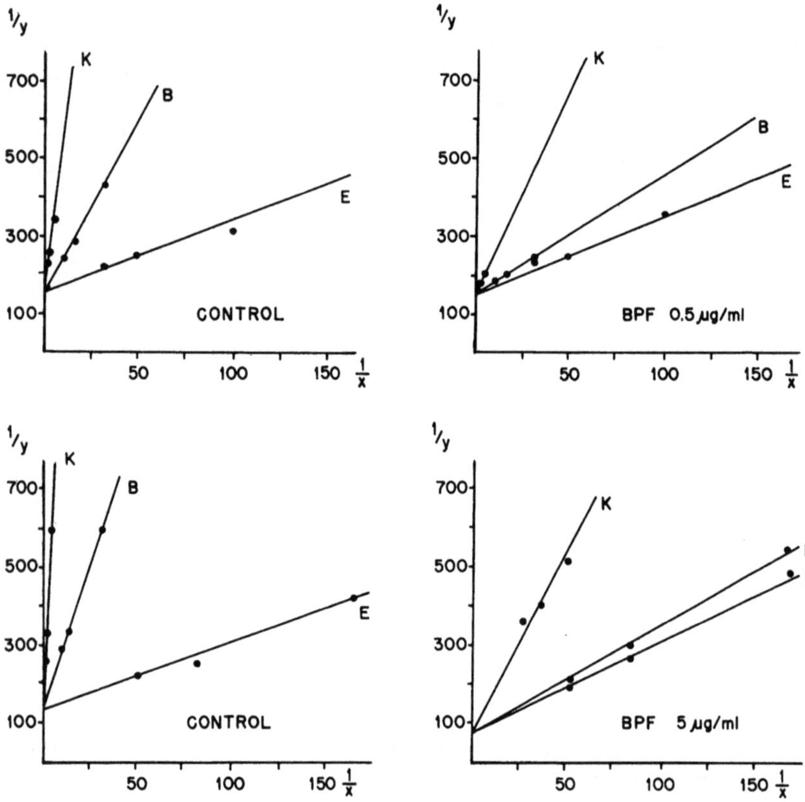

Fig. 1. Two experiments represented by double reciprocal plots. In ordinates, the reciprocal of the responses (1/y), and in abcissae, the reciprocal of the final concentration of the agonists (1/x). Note that kallidin (K) and bradykinin (B) lines tend to displace toward the eledoisin (E) line during the treatment of the guinea pig ileum with BPF. Tyrode's containing atropine (1.10^{-4} mg/liter and diphenhydramine (5.10^{-5} mg/liter) (Ferreira and Rocha e Silva, Experienta 21,347, 1965).

responses. It must be noted that when the concentration of BPF reached 5 μg/ml, the line of bradykinin responses matched that of eledoisin, indicating that the sensitivity of the isolated preparation became almost the same for the two polypeptides.

The bronchoconstrictor as well as the coronary dilating effects of bradykinin were also potentiated by BPF. The potentiation of bradykinin coronary vasodilatation by BPF can be observed in Fig. 2. The upper record shows the action of BPF upon bradykinin effect. The lower record shows potentiation of bradykinin when given after vasopressin (Pitressin)-induced

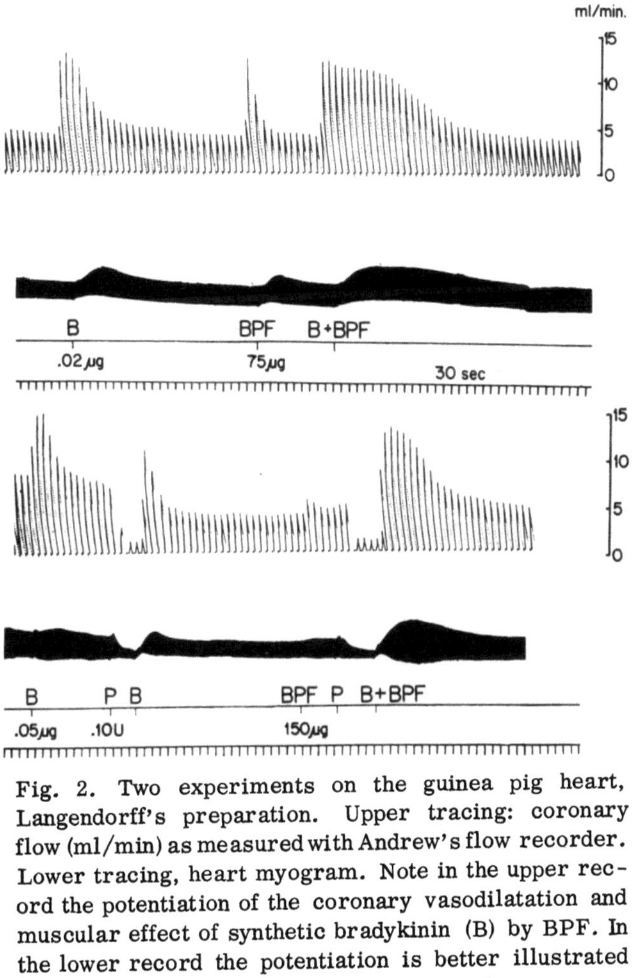

Fig. 2. Two experiments on the guinea pig heart, Langendorff's preparation. Upper tracing: coronary flow (ml/min) as measured with Andrew's flow recorder. Lower tracing, heart myogram. Note in the upper record the potentiation of the coronary vasodilatation and muscular effect of synthetic bradykinin (B) by BPF. In the lower record the potentiation is better illustrated after the coronary vasoconstriction induced by vasopressin (P) (A. Antonio, to be published).

vasoconstriction. (This corrective effect of bradykinin upon coronary spasm produced by vasopressin was first demonstrated in this laboratory by Dr. Antonio.)

Experiments in vivo. BPF was found to potentiate arterial hypotension, cerebral vasodilatation, capillary vascular permeability, hemoconcentration, and the effects on the central nervous system induced by bradykinin.

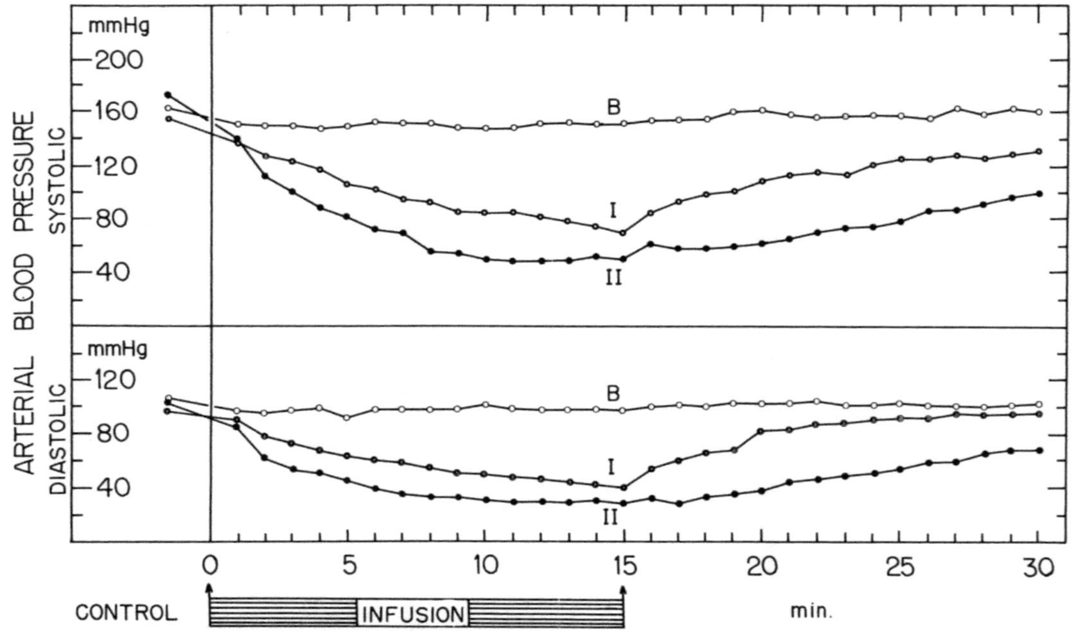

Fig. 3. Action of BPF upon the effect induced by a slow continuous infusion of synthetic bradykinin on the femoral arterial blood pressure in anesthetized dogs. Comparative responses show a striking difference between bradykinin (B = 1 μg/kg/min, 12 animals) and combined bradykinin plus BPF. The factor was administered at two different dose levels: I, 0.0375 mg/kg/min; II, 0.075 mg/kg/min (6 animals). BPF per se elicits no appreciable effect on vascular pressures (Amorin et al., 1965).

The factor, per se, when given in rapid injection (intravenous, intracarotid, or into the right atrium) caused a slight fall in systemic blood pressure which returned rapidly to control levels. This hypotensive effect could not be observed when BPF was administered in a slow continuous infusion.

Differences in responses to bradykinin infusion given alone or potentiated by BPF are shown in Figure 3. Administration of bradykinin in the concentration indicated caused but a slight and brief decrease in systemic blood pressure. Both the intensity and duration of hypotension were much more pronounced, however, when the peptide was combined with BPF. Under these circumstances a complete recovery was not observed until 60 min after completion of the infusion (Amorim et al., 1965).

Under normal conditions eledoisin is 8 times more active than bradykinin on cat arterial blood pressure and 20-50 times more active in the dog (Olmstedt and Page, 1962: Stürmer and Berde, 1963). If the animals were treated with equipotent doses of both peptides before and after administration of BPF, bradykinin became more potent in the cat and half as potent as eledoisin in the dog. In these experiments doses of bradykinin (0.03 to 0.06 μg/kg) below the threshhold to produce any measurable effect induced, after the treatment of the dog with BPF, a hypotension equivalent to 0.015-0.03 μg eledoisin/kg. It must be pointed out that the hypotensive effects of eledoisin, histamine, and acetylcholine were not potentiated by BPF (Ferreira and Rocha e Silva, 1965).

The vasodilating effect of bradykinin upon cerebral vessels was also potentiated by the factor (Graeff et al., 1965).

In an attempt to correlate the central effects of the intraventricular injection of bradykinin (Feldberg and Sherwood technique) with changes in electrocorticogram, a series of experiments in unanesthetized cats is being carried out in our department by Drs. Corrado and Graeff. Immediately after administration of 20 μg synthetic bradykinin (diluted in 0.1 ml saline) the cats vocalized and became restless, and the EEG presented a desynchronized pattern. Within 2-3 min the cats returned to normal. Saline and BPF (150 μg) evoked similar, but less pronounced responses lasting for about 20-30 sec. If bradykinin was injected 5 min after the BPF treatment it promoted initially a similar sequence of events, but the cats became much more restless, miaouing very frequently and presenting deambulatory movements; EEG also showed a higher

TABLE 1

Action of BPF Upon the Bradykininolytic Activity of Guinea Pig Plasma

The data show the mean recovered activity expressed as per cent of the initial activity present in the following incubation mixtures: (a) BRS, 2 μg; (b) guinea pig plasma 1/10, 0.2 ml; (c) BPF in the indicated final concentration; (d) Tris buffer, pH 7.5, 0.2 ml; (e) the 2-ml samples were made up with distilled water. The number of assays performed is indicated in parentheses. Temperature of incubation: 37° C (Ferreira, 1964).

Time of incubation (min)	Per cent recovery						
	Experiment I			Experiment II			
		With BPF			With BPF		
	Control	(1 μg/ml)	10 μg/ml	Control	(40 μg/ml)	(200 μg/ml)	(750 μg/ml)
	(3)	(2)	(2)	(4)	(3)	(3)	(2)
1	75.0	89.5	84.0	60.0	72.3	73.6	70.0
4	4.0	47.0	45.0	37.0	62.7	62.6	64.5
7	0	12.0	23.0	0			
10					49.5	47.6	55.5
16					38.5	41.1	40.0

degree of desynchronization. This excitation phase was replaced after 4-6 min by a motor sedation: the animals became prostrated, with signs of catatonia, relaxation of the nictitating membrane, and narrowing of the pupil slot. The electrocorticogram showed a light sleep pattern. The behavior of the cats gradulally became normal and after 2 hours no sign of the bradykinin action could be observed.

Inhibition of bradykininolytic activity. The results presented in Table 1 show that BPF concentrations, ranging from 1 to 750 μg/ml, partially blocked the inactivation of bradykinin by guinea pig plasma. No increase of the inhibition could be observed by increasing the concentrations of BPF from 40 to 750 μg/ml (experiment II). It has been indicated (Ferreira, 1964) that BPF is able to partially block the bradykinin inactivation by a solution of hemolyzed red cells (guinea pig). This partial inhibition of the kininase activity by BPF was also observed in the plasma of the rat, the cat, and the dog. It must be pointed out that no loss of BPF-potentiating activity could be observed in the incubating mixtures at the end of the assays.

In order to correlate these findings in vitro with the bradykinin potentiation observed in vivo, dogs were treated with the factor, and the bradykininolytic plasma activity was determined at several time intervals. Figure 4 shows that the intensity of hypotension induced by bradykinin may be correlated with the degree of inhibition of the kininolytic activity of the plasma after BPF administration.

COMMENTS

Our results show that the venom of *B. jararaca* contains a factor (BPF) that potentiates some pharmacological actions of bradykinin in vitro as well as in vivo. It has been suggested that the potentiating activity of the factor might be due to the presence of some polypeptide components (Ferreira, 1964; 1965). Similar bradykinin-potentiating activity has been found in other snake venoms.

Acetylcholine, histamine, and eledoisin effects were not enhanced by BPF on isolated preparations or upon arterial blood pressure. These findings indicated that the observed potentiation differed from a "general sensitization" of smooth-muscle preparations and could not depend on interference with

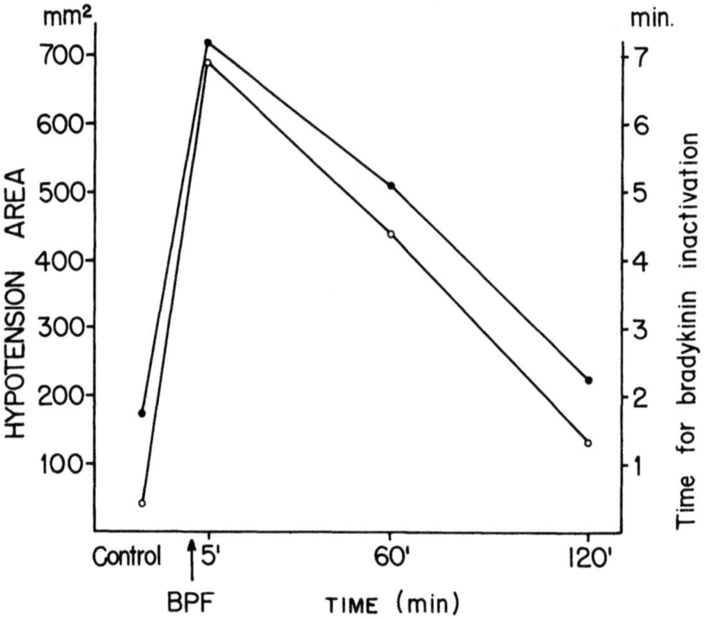

Fig. 4. Comparison between hypotension induced by bradykinin and bradykininolytic activity of dog's plasma before and after BPF (2 mg/kg) treatment. Hypotension area (o—o) was measured by calculating the area below the original base line and the tracing after bradykinin administration (1 μg/kg). The bradykininolytic activity of plasma (o—o) was estimated by determining the time of 50% inactivation of 1 μg of bradykinin by 0.5 ml of plasma, at 37° C. (Final volume of 1 ml made up with 0.2 ml of Tris buffer, pH 7.5, and saline.) The samples of plasma were collected just after bradykinin administration. For each plasma sample two assays were performed on the guinea pig ileum. The points represent the results observed on 3 animals. Note that potentiation of bradykinin effects runs parallel with the inhibition of bradykininolytic activity of plasma (Ferreira, unpublished results).

the mechanism of restoration of arterial blood pressure.

Eledoisin was found to be more active than bradykinin when assayed on guinea pig ileum and arterial blood pressure of the dog. According to Sicuteri et al. (1963) and Stürmer and Berde (1963), the higher resistance of eledoisin to kininases could be one of the reasons for the difference of activity between the two polypeptides. A first explanation of the mechanism of the potentiation of bradykinin by BPF was based on the observation that BPF reduced the activity ratio between eledoisin and bradykinin to almost 1. Thus, after BPF, the isolated guinea pig ileum became equally sensitive to both polypeptides, and the ratio of activity eledoisin/bradykinin in the arterial blood pressure, which varied from 15 to 41 in the control, was reduced to 1 to 2 after treatment of the animal with the factor (Ferreira and Rocha e Silva, 1965).

Further support to this suggestion was given by the observation that BPF partially inhibited the inactivation of bradykinin by plasma and hemolyzed red cells. As shown here, the bradykinolytic activity of plasma was reduced by small concentrations of BPF (1 and 10 × μg/ml), but the degree was not increased after raising the concentrations (from 40 to 750 μg/ml). This fact may be explained by the presence of two or more kininases, of which only one would be inhibited by BPF. In fact, Erdös and Sloane (1962) observed that, besides the fraction IV-1 of human plasma which contains an enzyme showing some similarity to carboxypeptidase B, two other fractions had kininase activity.

A general correlation between potentiation of the hypotensive effects of bradykinin and inhibition of kininase activity of the plasma was found after administration of BPF. But, in considering the partial inhibition of plasma kininolytic activity caused by the factor, it seems difficult to explain the strong potentiation in vivo by BPF as entirely due to inhibition of bradykinin destruction by plasma. If the potentiation is due to the blockade of peptide inactivation, one should admit that under physiological conditions bradykinin is also inactivated in tissues other than the blood by a kininase highly sensitive to the action of BPF, at sites close to the pharmacological receptors.

We have described earlier how some thiocompounds, such as dimercaprol and thioglycollic acid, potentiated the hypotensive actions of bradykinin in the rabbit and the dog (Ferreira and Rocha e Silva, 1962; Ferreira et al., 1962; Corrado, 1963; Rocha e Silva, 1963). Erdös and Wohler (1963), found dimercaprol one of the most potent bradykinin potentiators in the series of thiocompounds tested upon the arterial blood pressure of the guinea pig. Since BPF is more effective in vivo than dimercaprol, it could be considered the most effective bradykinin potentiator so far described (Ferreira, 1964).

Doses of bradykinin (0.03 to 0.06 μg/kg), normally below the threshold to produce hypotension, were definitely hypotensive after the treatment of the dog with BPF. A liberation of bradykinin into the blood stream of the order of 5 μg/kg probably could not produce a detectable variation of the level of plasma bradykininogen. Thus the animal treated with BPF should provide a highly sensitive experimental subject in the study of the participation of bradykinin or bradykinin-like substances in some physiological or pathological conditions.

ACKNOWLEDGMENT

The author is indebted to Professor M. Rocha e Silva for advice and criticism. He also wishes to thank Dr. S. Schemberg for samples of snake venoms. This work was aided by grants from the Fundacão de Amparo á Pesquisa and from the United States Public Health Service, National Institutes of Health (NB 02953-03).

REFERENCES

Amorim, D. S., S. H. Ferreira, J. C. Manco, A. Tanaka, A. Sader, and S. S. Cardoso: Potenciacão dos efeitos circulatórios da infusão contínua de bradicinina. XXI Congresso Brasileiro de Cardiologia. Arquivos bras. Cardiol. in press (1965).

Corrado, A. P.: Recent investigations of the pharmacology of bradykinin and their physio-pathological implications. Settim. Med. Suppl., 1, 51 (1963).

Erdös, E. G., and E. M. Sloane: An enzyme in human blood plasma that inactivates bradykinin and kallidins. Biochem. Pharmacol. 11, 585 (1962).

Erdös, E. G., and J. R. Wohler: Inhibition in vivo of the enzymatic inactivations of bradykinin and kallidin. Biochem. Pharmacol. 12, 1193 (1963).

Ferreira, S. H.: Potenciacão da bradicinina por fator presente no veneno da bothrops jararaca. Thesis, Fac. Med. Ribeirão Prêto (1964).

Ferreira, S. H.: A bradykinin potentiating factor (BPF) present in the venom of *Bothrops jararaca.* Brit. J. Pharmacol. 24, 163 (1965).

Ferreira, S. H., and M. Rocha e Silva: Potentiation of bradykinin by dimercaptopropanol (BAL) and other inhibitors of its destroying enzyme in plasma. Biochem. Pharmacol. 11, 1123 (1962).

Ferreira, S. H., and M. Rocha e Silva: Potentiation of bradykinin and eledoisin by BPF (bradykinin factor) from *Bothrops jararaca* venom. Experientia (Basel) 121, 347 (1965).

Ferreira, S. H., A. P. Corrado, and M. Rocha e Silva: Potenciacão do efeito hipotensor da bradicinina por inibicão do enzima bradicininolítico plasmatico. Ciência e Cultura 14, 238 (1962).

Graeff, F. G., S. H. Ferreira, A. P. Corrado, and M. Rocha e Silva: Potentiation of the cerebral vascular action of bradykinin by the "Bradykinin Potentiating Factor" (BPF) in the dog. Experientia (Basel) 21, 607 (1965).

Olmstedt, F., and I. H. Page: Hemodynamic effects of eledoisin kallidin II and bradykinin in unanesthetized dogs. Amer. J. Physiol. 203, 951 (1962).

Rocha e Silva, M.: The physiological significance of bradykinin. Ann. N.Y. Acad. Sci. 104, 190 (1963).

Sicuteri, F., P. Fanciullacci, G. Franchi, and S. Michelacci: The endecapetide eledoisin as powerful vasodilating and hypotensive agent in man. Experientia (Basel) 19, 44 (1963).

Stürmer, E., and Berde, A.: A comparative pharmacological study of synthetic eledoisin and synthetic bradykinin. J. Pharmacol. exp. Ther. 140, 349 (1963).

DISCUSSION

Erdös: In various animals various compounds potentiate the cardiovascular effects of bradykinin. In the dog, for example, 8-hydroxy-5-quinoline sulfonic and diethyldithiocarbamic acid is effective. In the rat, low doses of mercaptoethanol (10-15 mg/kg) and in the cat intravenous injection of glutathione achieve the same effect.

Edery: Dr. Grunfeld and I found that glutathione potentiates specifically the action of bradykinin on smooth muscle. Apparently the SH group in glutathione is critical because methyl-glutathione is inactive. It is possible that in tissue injury when intracellular glutathione is liberated some interaction with bradykinin may occur. Do you have data on the chemical structure of the BPF? Have you done the amino acid analysis?

Ferreira: I also observed that glutathione potentiated bradykinin effects on guinea pig ileum a little. The problem is whether enough glutathione is released from the injured cells to potentiate bradykinin effects. At the present stage of purification, I can only say that BPF activity might be due to the presence of two polypeptides. No amino acid analysis was performed.

THE INFLUENCE OF INTRA-ARTERIAL INFUSIONS OF SYNTHETIC BRADYKININ ON FLOW AND COMPOSITION OF LYMPH IN DOGS

E. STÜRMER

Biological and Medical Research Division,
Sandoz Ltd., Basle, Switzerland

The plasma kinins, bradykinin (kinin-9) and kallidin (kinin-10), not only induce hypotension and stimulate extravascular smooth muscle but also exert a powerful effect on capillary permeability. As we have shown (Konzett and Stürmer, 1960) pure synthetic bradykinin causes a leakage of Evan's blue dye from the blood of guinea pigs into the tissues even in doses as low as 1 ng/0.1 ml given intracutaneously. Using radioactive (Au^{198}) Frimmer showed in 1961 that bradykinin rendered the capillary walls of rabbit skin readily permeable to particles with a mean diameter of 50 to 100 Å. Moreover, Witte et al. (1961), using direct u.v. microscopy on the rat mesentery, noted the spread of fluorochromated plasma proteins into the interstitial space after local application of 50 ng bradykinin, without any sign of capillary damage. The edema of the paw we found (Stürmer and Cerletti, 1961) in rats after a subplantar injection of 1-100 μg bradykinin, may be regarded as indirect evidence for increased capillary permeability.

Increased passage of plasma proteins into the interstitial space resulting from the action of plasma kinins ought to increase drainage into the lymphatics. Experiments to test this hypothesis have yielded conflicting results. Edery and Lewis (1963) failed to detect an elevated flow of lymph in anesthetised dogs after injections of bradykinin in doses of 0.1-1 mg into the femoral artery. However, Vogel and Ströcker (1965) administered kallidin intravenously in a dose of 50 μg/kg to rats and found a considerable increase in lymph flow in the thoracic duct associated with an enhanced passage of polyvinylpyrrolidone across the plasma-lymph barrier.

Owing to the short half-life of bradykinin (Saameli and Eskes, 1962), it is possible that a single intravascular injection is capable of only partially simulating the effect of locally released plasma kinins. For this reason we were prompted to reinvestigate the action of kinins on lymph secretion, administering the kinins by intra-arterial infusion.

METHOD

The experiments were performed on 12 dogs, body weight ranging between 11.4 and 19.1 kg, mean weight 14.3 kg, anesthetised with butallylonal (65 mg/kg administered s.c. by the method of Rein and Schneider, 1934). The main lymphatic of the right hind leg which runs between the femoral artery and the femoral vein was cannulated with a polyethylene tube of 1.2 mm external diameter; the remaining lymphatics, which pass through the inguinal fossa, had previously been occluded as far as possible by two ligatures, care being taken not to damage the femoral vein, artery, and nerve. The infusions were given into the short muscular branch of the femoral artery, and the rate of infusion was 1 ml/min in all experiments. In order to obtain a steady flow of lymph, the calcaneum was pierced with a wire which was clamped in a Kirschner caliper (1927). The latter was fixed excentrically by means of a bearing to a disk which was rotated at 28 rpm and imparted a "cycling" movement to the leg. Sagittal elastic fixation of the toes ensured rhythmic movement of the foot. No artificial respiration was used. The blood pressure was recorded from the carotid artery by a mercury manometer.

The lymph issuing over a period of 15 min was collected in a separate polysterene vessel. The following parameters were measured. (1) Lymph flow per 15-min period was determined gravimetrically on a direct reading balance. (2) The lymphocyte count was performed in a Schilling chamber and expressed as the total number of lymphocytes for the 15-min sampling time. (3) The optical refraction was determined with a Zeiss refractometer. (4) Total protein was determined by the Kingsley-Weichselbaum biuret method (1939; 1946) by the ultramicrotechnique of M. C. Sanz. This method requires a sample of only 5 μ liters lymph.

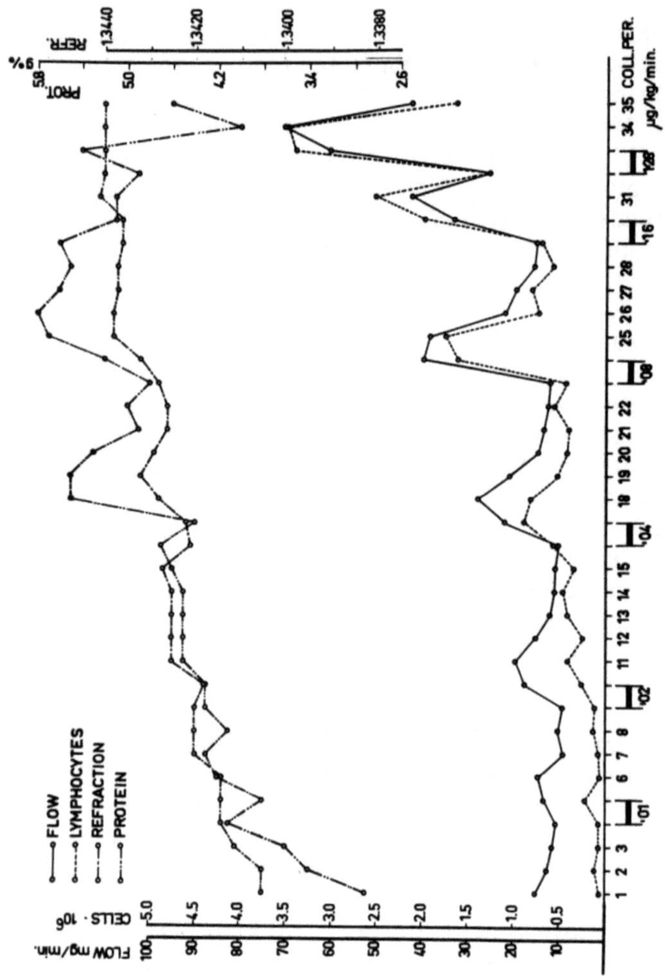

Fig. 1. Effect of i.a. infusions of bradykinin on lymph flow and composition in a dog (representative experiment). The 33 15-min collection periods (COLL.PER.) are plotted on the abscissa. The unbroken curve represents lymph flow, the ordinate indicating the actual flow values in milligrams per minute. Intra-arterial bradykinin infusions are shown as black bars beneath the abscissa; doses are given in μg/kg/min at a rate of 1 ml/min. Total number of lymphocytes is denoted by the dotted line; the ordinate indicates the cell count in millions per 15-min collection period. The refractometer values are represented by the — · — line, and the total protein values in grams per cent by the — ·· — line.

RESULTS

Intra-arterial infusion of bradykinin in doses of 0.005 to 1.28 μg/kg/min over 15 min raises the flow of lymph in a reproducible and dose-dependent manner. Simultaneously and almost parallel with the increase in flow, there is a rise in the total number of lymphocytes. The refractometer values and the total protein values show a certain degree of correlation, and the fact that they rise during and after infusion of bradykinin indicates that the lymph is not only flowing more rapidly but also has an increased protein content. Figure 1 illustrates the results of one such representative experiment.

Table 1 contains the results of the same experiment. The changes brought about by bradykinin are expressed in per cent increase of the preinfusion values. On plotting the percentage increases in lymph flow against the logarithmic infusion doses, an approximately linear dose-effect relation is obtained, at least for a certain dose range, as is shown in Figure 2.

A comparison between bradykinin and kallidin — four-point assay design — reveals that, in this experimental setup, kallidin is roughly as active as bradykinin.

It might be added that in preliminary experiments, doses of epinephrine up to 1 μg/kg/min, 5-hydroxytryptamine up to 10 μg/kg/min, and histamine up to 1 μg/kg/min elicited no definite increase in lymph flow.

SUMMARY

The plasma kinins, bradykinin and kallidin, administered in dogs by intra-arterial infusion increase the lymph flow. At the same time, the total lymphocyte count and the protein content of the lymph are raised. These findings show that plasma kinins may influence the flow and the composition of lymph and are compatible with the view that they affect the permeability of membranes. A pharmacological analysis of these observations, including the use of antagonists of endogenous transmitter substances and tissue hormones such as histamine, serotonin, epinephrine, etc., will be the next step in our studies.

TABLE 1
Effect of Bradykinin Infusions in a Dog
Changes in % of the Initial Values

Dose µg/kg/min	Lymph-flow	Lymphocyte count	Total protein	Blood pressure
0.01	+ 25	+190	+ 5	±0
0.02	+110	+298	+ 9	±0
0.04	+169	+ 56	+17	±0
0.08	+228	+300	+21	+19 (short)
0.16	+178	+256	?	±0
1.28	+320	+173	+10	−36 (short)

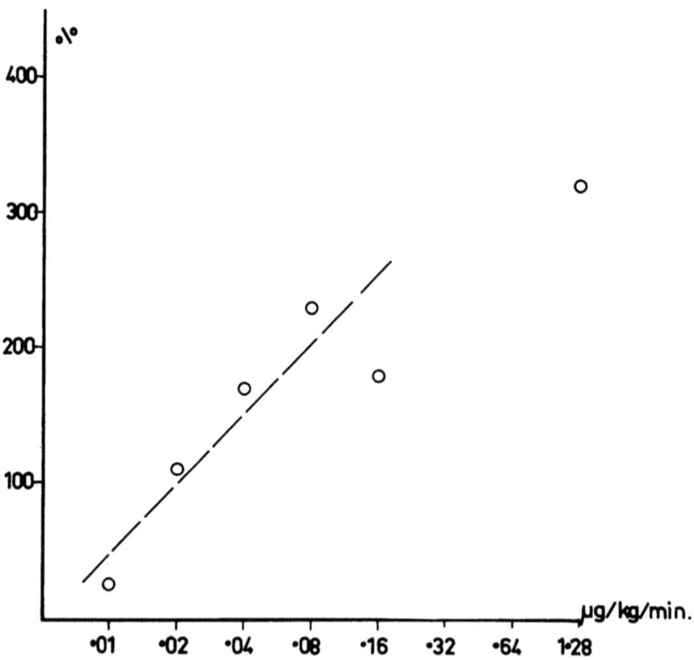

Fig. 2. Lymph flow of a dog, increase after i.a. infusions of bradykinin. Ordinate: per cent increase in lymph flow calculated from the preinfusion values. Abscissa: doses of i.a. bradykinin infusions in µg/kg/min, logarithmic scale.

REFERENCES

Edery, H., and G. P. Lewis: Kinin-forming activity and histamine in lymph after tissue injury. J. Physiol. (Lond.) 169, 568 (1963).

Frimmer, M.: Beeinflussung der Durchlässigkeit von Blut-Capillaren der Kaninchenhaut für Makromoleküle durch einige biogene, gefässaktive Substanzen. Naunyn-Schmiedeberg's Arch. exp. Path. Pharmak. 242, 390 (1961).

Kingsley, G. R.: The determination of serum total protein albumin and globulin by the biuret reaction. J. biol. Chem. 131, 197 (1939).

Kirschner, M.: Verbesserungen der Drahtextension. Langenbecks Arch. klin. Chir. 148, 651 (1927).

Konzett, H., and E. Stürmer: Biological activity of synthetic polypeptides with bradykinin-like properties. Brit. J. Pharmacol. 15, 544 (1960).

Rein, H., and D. Schneider: Erfahrungen über Narkosen zu wissenschaftlichen Versuchszwecken. Klin. Wschr. 13, 870 (1934).

Saameli, K., and T. K. A. B. Eskes: Bradykinin and cardiovascular system: Estimation of half-life. Amer. J. Physiol. 203, 261 (1962).

Stürmer, E., and A. Cerletti: Das Bradykinin-Oedem der Rattenpfote. Helv. physiol. Acta 19, C32 (1961).

Vogel, G., and H. Ströcker: Gesteigerte Capillarpermeabilität im tierexperimentellen Modell und ihre Beeinflussung durch Pharmaka. Naunyn-Schmiedeberg's Arch. exp. Path. Pharmak. 251, 180 (1965).

Weichselbaum, T. E.: An accurate and rapid method for determination of proteins in small amounts of blood serum and plasma. Amer. J. clin. Path. 10, 40 (1946).

Witte, S., K. T. Schricker, and E. Schmid: Zur Wirkung von synthetischem Bradykinin auf die Gefässpermeabilität. Arzneimittel-Forsch. 11, 619 (1961).

DISCUSSION

Guth: You and others used the term "capillary permeability" when speaking of bradykinin effects. D. A. Rowley has demonstrated by direct visualization that bradykinin permeability-increasing effect is primarily on the venule. Have you or others evidence that bradykinin acts on capillaries?

Stürmer: From our experimental data we are unable to localize the bradykinin effect on the so-called microcirculation. Therefore we concluded that "our findings are compatible with the view that bradykinin effects the permeability of membranes," thus avoiding the term capillary permeability.

Edery: In the experiments Lewis and I carried out bradykinin was injected in a single intra-arterial injection and in comparatively large doses. The difference in results might be explained by the difference in doses administered, since you have been giving quite large amounts in continuous slow infusion.

Stürmer: I agree with you that the difference in results is probably due to the different methods of kinin administration; i.e. that a single intra-arterial injection of bradykinin will not produce the same effects as an infusion. Concerning the doses, we administered bradykinin in doses per minute 100 to 2,000 times smaller than your smallest single injection dose and 1,000 to 20,000 times smaller than your largest single injection dose.

Waaler: Did you see any signs of edema formation during these infusions? Others found irreversible edema in some cat hind limb preparations during bradykinin infusions.

Stürmer: In our experiments we did not measure the actual volume of the dog's hind leg and this would be rather difficult to perform during the "cycling" movement. There was no sign of edema after an experiment of about 10 hours' duration.

Trautschold: Your findings on stimulation of lymph flow could explain the rapid kinin elimination from blood after injection, which is much faster than expected from the kininase activity. Could you detect an escape of injected kinins into the lymph?

Stürmer: We have not yet studied this.

THE ACTION OF HYPOTENSIVE POLYPEPTIDES ON THE PULMONARY ARTERIAL PRESSURE

H. KONZETT and G. BAUER

Pharmakologisches Institut der Universität Innsbruck, Austria

It is well established that the kinins bradykinin and kallidin produce, as a rule, a fall of systemic arterial blood pressure in various species (Elliott et al., 1960; Konzett and Stürmer, 1960; Gersmeyer and Spitzbarth, 1961; Stürmer and Berde, 1962). Eledoisin also lowers the arterial blood pressure of many animals (Stürmer and Berde, 1963; Gersmeyer et al., 1965). With regard to the depressor action, bradykinin, kallidin, and eledoisin differ quantitatively; eledoisin has the strongest hypotensive action. Under certain experimental conditions, however, a rise in systemic blood pressure may occur after administration of these polypeptides; e.g. the chicken blood pressure, after an initial fall, is increased by eledoisin (Stürmer and Berde, 1963).

The action of these hypotensive polypeptides on the pulmonary circulation is not as well understood as on the systemic circulation. In isolated perfused lung preparations of different species, bradykinin elicits different effects; it contracts the lung vessels of the rabbit (Lecomte and Troquet, 1960), of the guinea pig, and of the rat (Greef and Moog, 1964), but it usually causes a small or moderate pulmonary vasodilatation in the dog (Waaler, 1961). Isolated perfused cat lungs were insensitive toward bradykinin (Greef and Moog, 1964). In isolated preparations of guinea pig lungs containing mainly lung veins, bradykinin and eledoisin produced a greater reduction in flow than in preparations containing mainly lung arteries (Moog and Fischer, 1964). Lung veins, so far, have been more sensitive toward bradykinin and eledoisin than lung arteries (Moog and

Fischer, 1964). In anesthetized dogs, bradykinin decreases pulmonary arterial pressure (Maxwell et al., 1962), whereas eledoisin increases it (Nakano, 1964). In man, bradykinin may produce a rise or a fall of pulmonary arterial pressure (Gersmeyer and Spitzbarth, 1961; Bishop et al., 1963).

In the present experiments the effects of bradykinin, kallidin and eledoisin on systemic and pulmonary arterial pressure were studied in anesthetized guinea pigs, rabbits, and cats. In addition, the action of these polypeptides was investigated on isolated arteries from the same species.

METHODS

Guinea pigs were anesthetized with 1.2 g urethan/kg i. p., rabbits with 1 g urethan/kg i.v., and cats with a mixture of 0.06 g chloralose/kg and 0.3 g urethan/kg i.v. The animals were ventilated by a Starling respiration pump. After opening the thorax and administering heparin, the mean systemic, pulmonary arterial, and sometimes left atrial pressure were measured continuously with Statham pressure transducers connected to cannulas placed in a carotid artery, in the main stem of the pulmonary artery next to the obliterated *ductus arteriosus Botalli*, and in the left atrium.

The synthetic (Sandoz) polypeptides bradykinin, kallidin, and eledoisin were injected intravenously. Other drugs used were: acetylcholine chloride, 1-epinephrine bitartrate, 5-hydroxytryptamine creatinine sulfate, phenylbutazone, mefenamic acid, and flufenamic acid.

Isolated pulmonary arteries of guinea pigs, rabbits, and cats were prepared according to a procedure described by Engelhardt and Hahn (1957). The effects of substances added to the bath (oxygenated Starling-Ringer solution) were recorded isotonically.

RESULTS

In guinea pigs, bradykinin produced, besides a fall in systemic blood pressure, an increase in pulmonary arterial pressure when given in doses from 0.5-8 μg/kg (Fig. 1). In most experiments there was a fairly good dose-response relationship. Kallidin had a qualitatively similar effect. Eledoisin in

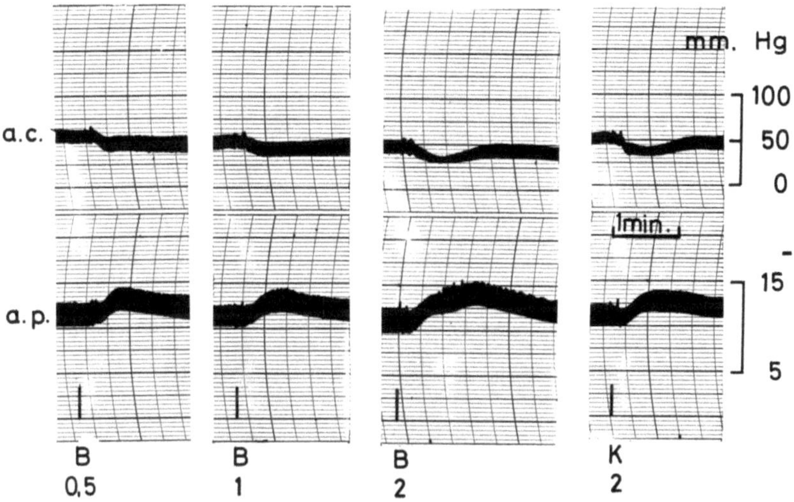

Fig. 1. Effect of bradykinin (B) and kallidin (K) on the systemic arterial blood pressure in the carotid artery (a.c.) and on the pulmonary arterial blood pressure (a.p.) of an anesthetized guinea pig (0.61 kg). The doses in μg/kg are indicated below B and K.

doses from 0.1-1.0 μg/kg did not increase, but sometimes even slightly decreased, the pulmonary arterial pressure of guinea pigs; higher doses of eledoisin were not given, in order to avoid a deep fall of systemic blood pressure.

In guinea pigs, the rise in pulmonary arterial pressure after bradykinin was independent from changes in the left atrial pressure which either was not influenced at all, was slightly decreased, or was increased. In the last instances, these changes began later than the rise in pulmonary arterial pressure.

The antipyretic and anti-inflammatory agents phenylbutazone, mefenamic acid, and flufenamic acid (in doses from 1-10 mg/kg) diminished or inhibited completely the effect of bradykinin on the pulmonary arterial pressure for 30-60 min. After pretreatment with these analgesic substances, the dose of bradykinin had to be increased up to sixfold to produce a response of the same height as before the antagonist was given.

In rabbits, bradykinin in doses from 1-10 μg/kg elicited only a small rise in pulmonary arterial pressure without a meaningful dose-response relationship. Eledoisin in doses from 0.05 to 1.6 μg/kg, however, increased pulmonary arterial pressure in the rabbit regularly (Fig. 2). The effect was de-

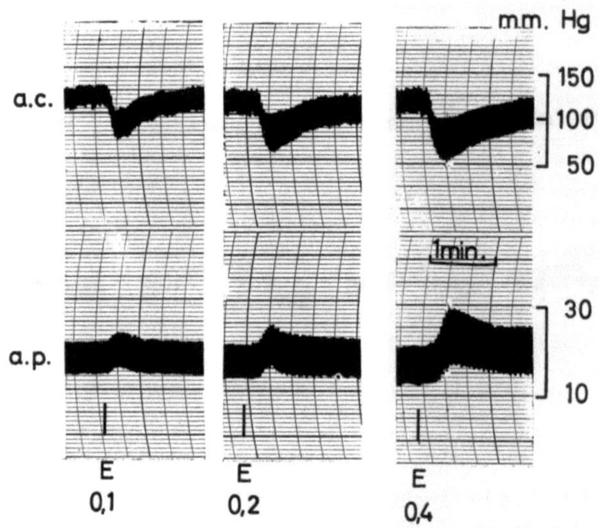

Fig. 2. Effect of different doses (μg/kg) of eledoisin (E) on the systemic arterial blood pressure in the carotid artery (a.c.) and on the pulmonary arterial blood pressure (a.p.) of an anesthetized rabbit (3.6 kg).

pendent on the dose and easily reproducible.

Phenylbutazone, mefenamic acid, and flufenamic acid (in doses from 1-8 mg/kg) diminished or abolished completely the effect of these small doses of eledoisin on the pulmonary arterial pressure (Fig. 3); after pretreatment with these substances, up to 10 times higher doses were needed to restore the response on the pulmonary arterial pressure. The response to 5-hydroxytryptamine was not altered by these compounds.

In cats, bradykinin in doses from 0.3-10 μg/kg produced mostly a slight rise in pulmonary arterial pressure which was antagonized by phenylbutazone only occasionally. Eledoisin in doses from 0.2-10 μg/kg did not influence or decrease the pulmonary arterial pressure of cats.

Bradykinin in high doses (10 μg/ml) contracted strips from pulmonary arteries of guinea pigs (Fig. 4). Previous addition of phenylbutazone (0.4 mg/ml) to the bath abolished this effect, whereas the vasoconstriction caused by epinephrine was not affected. Bradykinin (up to 20 μg/ml) on strips from pulmonary arteries of rabbits and cats was without effect.

Eledoisin was ineffective on isolated pulmonary arteries of guinea pigs and cats. On strips from pulmonary arteries of

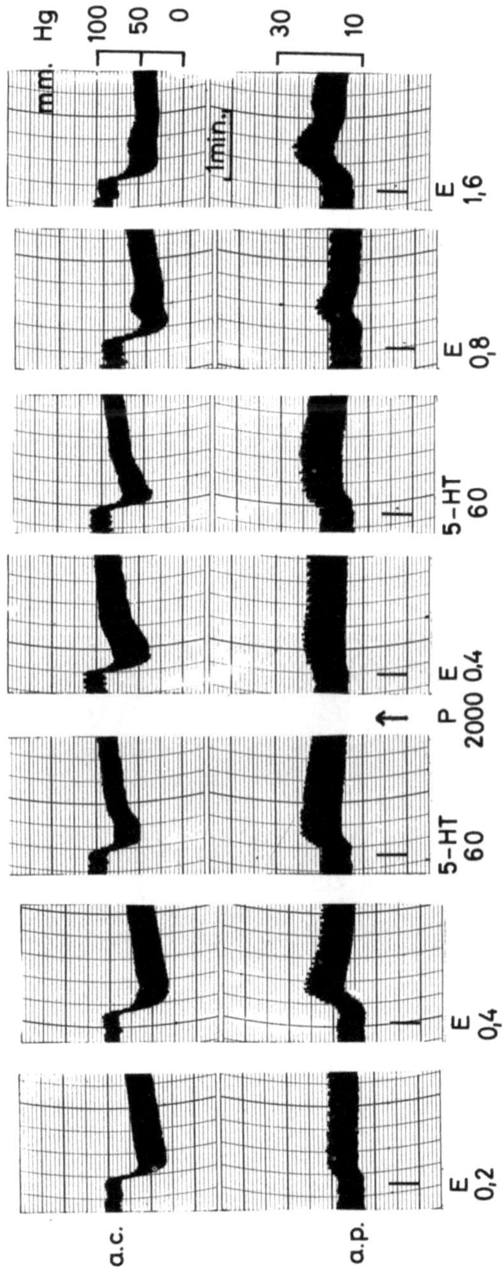

Fig. 3. Effect of eledoisin (E) and 5-hydroxytryptamine (5-HT) on the systemic arterial blood pressure in the carotid artery (a.c.) and on the pulmonary arterial pressure (a.p.) of an anesthetized rabbit (3.9 kg.) before and after administration of phenylbutazone (P). The figures below E, 5-HT, and P indicate the doses in μg/kg. Time intervals: from P to E, 0.4:4 min; to 5-HT, 60:11 min, to E, 0.8:22 min, to E, 1.6:30 min.

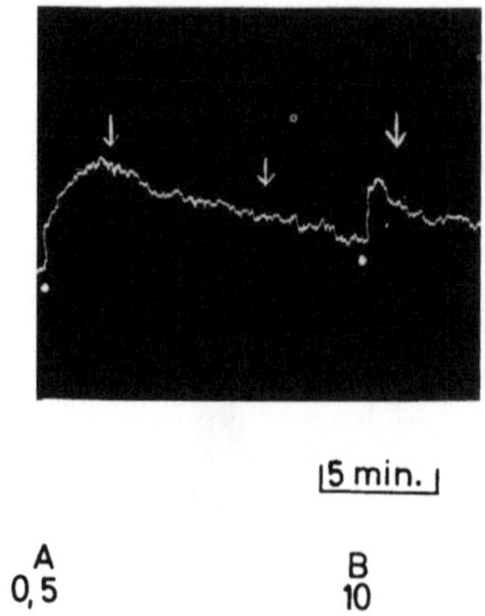

Fig. 4. Effect of epinephrine (0.5 μg/ml at A) and bradykinin (10 μg/ml at B) on an isolated pulmonary artery of a guinea pig. Arrows indicate washing.

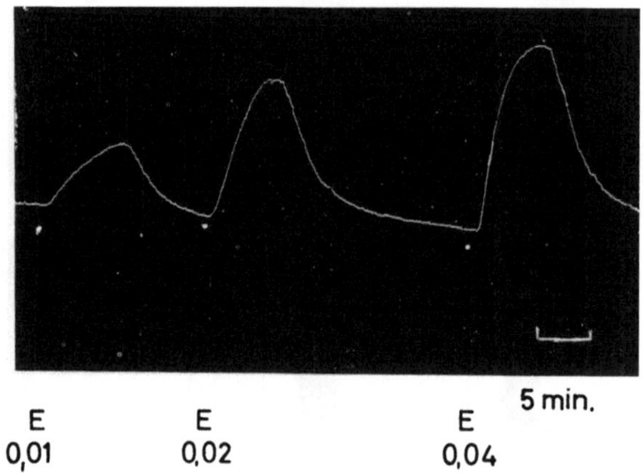

Fig. 5. Effect of different doses of eledoisin (0.01, 0.02, and 0.04 μg/ml) on an isolated pulmonary artery of a rabbit. The preparation was washed at the peak of response.

rabbits, however, eledoisin in doses from 0.0025 to 0.1 μg/ml produced contractions which showed a good dose-response relationship (Fig. 5). Phenylbutazone (0.4 mg/ml), mefenamic acid, and flufenamic acid (0.1 mg/ml) antagonized the effect of eledoisin on the strips from pulmonary arteries of rabbits. The antagonistic property of these three compounds toward eledoisin is probably a specific one since the contractions due to epinephrine or acetylcholine were not antagonized.

DISCUSSION

Bradykinin increases the pulmonary arterial pressure of the guinea pig, rabbit, and cat, whereas the systemic arterial pressure is lowered. In the heart-lung preparation of the guinea pig, bradykinin caused also an increase in pulmonary arterial pressure, accompanied only by slight changes in right auricular pressure and in cardiac output (Klupp and Konzett, 1965). A direct action of this polypeptide on the pulmonary artery was therefore assumed. During the rise of the pulmonary arterial pressure after bradykinin, the left atrial pressure remained practically unchanged; this again is in favor of a direct action of the polypeptide. It may be added that bradykinin contracts the isolated preparations of pulmonary arteries of guinea pigs and cattle (Klupp and Konzett, 1965) as well.

The effect of bradykinin on the pulmonary arterial pressure in the guinea pig is accompanied by a reduction in tidal air, e.g. a bronchoconstriction (Collier et al., 1960). Eledoisin, which is an even stronger bronchoconstrictor agent on the guinea pig lungs than are bradykinin and kallidin (Stürmer and Berde, 1963), does not raise pulmonary arterial pressure in this species. Therefore, a rise in pulmonary arterial pressure is not necessarily connected with a bronchoconstriction.

The antipyretic and anti-inflammatory substances phenylbutazone, mefenamic acid, and flufenamic acid, which have been found to antagonize the bronchoconstrictor effect of bradykinin in the guinea pig (Collier and Shorley, 1960; 1963) have also an antagonistic action against its effect on pulmonary arterial pressure. The rise of the pulmonary arterial pressure in the rabbit due to eledoisin was also antagonized by phenylbutazone, mefenamic acid, and flufenamic acid as was the contraction of the isolated pulmonary artery of the rabbit, produced by eledoisin. The antagonism of phenylbutazone, mefenamic acid, and flufenamic acid toward bradykinin and eledoisin is probably a specific one, since larger doses of both agonists restored the response, and effective doses of the antagonist did not reduce the responses to other hypertensive agents

SUMMARY

Bradykinin and kallidin raise pulmonary arterial pressure in anesthetized guinea pigs, rabbits, and cats. Eledoisin increases the pulmonary arterial pressure in anesthetized rabbits. The effect of bradykinin on the pulmonary arterial pressure in the guinea pig is probably a direct one.

Antipyretic and anti-inflammatory substances, e.g. phenylbutazone, mefenamic acid, and flufenamic acid, which have been found to antagonize the bronchoconstrictor action of bradykinin in guinea pig lungs, antagonize also the effect of bradykinin on the pulmonary arterial pressure in guinea pigs and the effect of eledoisin on the pulmonary arterial pressure in rabbits.

REFERENCES

Bishop, J. M., P. Harris, and N. Segel: The effect of synthetic bradykinin on the pulmonary and systemic circulation in man. J. Physiol. (Lond.) 165, 37P (1963).

Collier, H. O. J., and P. G. Shorley: Analgesic antipyretic drugs as antagonists of bradykinin. Brit. J. Pharmacol. 15, 601 (1960).

Collier, H. O. J., and P. G. Shorley: Antagonism by mefenamic and flufenamic acids of the bronchoconstrictor action of kinins in the guinea-pig. Brit. J. Pharmacol. 20, 345 (1963).

Collier, H. O. J., J. A. Holgate, M. Schachter, and P. G. Shorley: The bronchoconstrictor action of bradykinin in the guinea pig. Brit. J. Pharmacol. 15, 290 (1960).

Elliott, F. D., E. W. Horton, and G. P. Lewis: Action of pure bradykinin. J. Physiol. (Lond.) 153, 473 (1960).

Engelhardt, G., and G. Hahn: Pharmakologische Beeinflussung anaphylaktischer Reaktionen an Gefässpraparaten. Naunyn-Schmiedeberg's Arch. exp. Path. Pharmak. 231, 507 (1957).

Gersmeyer, E. F., and H. Spitzbarth: Über Kreislaufwirkungen von synthetischem Bradykinin beim Menschen und beim wachen Hund. Klin. Wschr. 39, 122 (1961).
Gersmeyer, E. F., A. Castenholz, and M. Nicolay: Kreislaufuntersuchungen über Eledoisin mit Therapieresultaten. Klin. Wschr. 43, 309 (1965).
Greef, K., and E. Moog: Vergleichende Untersuchungen über die bronchoconstrictorische und gefässconstrictorische Wirkung des Bradykinins, Histamins und Serotonins an isolierten Lungenpräparaten. Naunyn-Schmiedeberg's Arch. exp. Path. Pharmak. 248, 204 (1964).
Klupp, H., and H. Konzett: Der Einfluss von Bradykinin und Kallidin auf den Druck in der Arteria pulmonalis. Naunyn-Schmiedeberg's Arch. exp. Path Pharmak. 249, 479 (1965).
Konzett, H., and E. Stürmer: Biological activity of synthetic polypeptides with bradykinin-like properties. Brit. J. Pharmacol. 15, 544 (1960).
Lecomte, J., and J. Troquet: Antagonisme entre bradykinin et phenylbutazone chez le lapin. C. R. Soc. Biol. (Paris) 154, 1115 (1960).
Maxwell, G. M., R. B. Elliott, and G. M. Kneebone: Effects of bradykinin on the systemic and coronary vascular bed of the intact dog. Circulat. Res. 10, 359 (1962).
Moog, E., and J. Fischer: Die Wirkung des Histamins, Serotonins, Bradykinins, Eledoisins und Hypertensins auf isolierte durchströmte Arterien-, Venen- und Bronchienpräparate der Meerschweinchenlunge. Naunyn-Schmiedeberg's Arch. exp. Path. Pharmak. 249, 384 (1964).
Nakano, J.: Studies on the cardiovascular effects of synthetic eledoisin. J. Pharmacol. exp. Ther. 145, 71 (1964).
Stürmer, E., and B. Berde: Kallidin und Bradykinin, vergleichende pharmakologische Untersuchungen. Naunyn-Schmiedeberg's Arch. exp. Path. Pharmak. 243, 355 (1962).
Stümer, E., and B. Berde: A comparative pharmacological study on synthetic eledoisin and synthetic bradykinin. J. Pharmacol. exp. Ther. 140, 349 (1963).
Waaler, B. A.: The effect of bradykinin in an isolated perfused dog lung preparation. J. Physiol. (Lond.) 157, 475 (1961).

DISCUSSION

Waaler: In our laboratory we found similar effects with isolaed, perfused rabbit lungs. Bradykinin, kallidin, and eledoisin cause vasoconstriction. Tachyphylaxis was observed with all three substances, but no cross-tachyphylaxis was seen between bradykinin and kallidin on the one hand, eledoisin on the other. Eledoisin was very active in a concentration of 10^{-9} g or less.

Shimamoto: In an in-vitro perfusion experiment (Locke-Ringer solution) we found two types of venoconstriction caused by bradykinin. Cutaneous veins (saphenous or marginal ear veins) of rabbit exhibited a spastic constriction, whereas coronary veins and liver veins of dog showed a tonic constriction to bradykinin. In the latter preparations bradykinin reduced the flow by roughly 5%. Increasing the doses of bradykinin from 1 ng to 10,000 ng did not decrease the flow further, but prolonged the effect. Aspirin, cyproheptadine, nialamide, and pyridinolcarbamate antagonized this venoconstriction. They were noncompetitive antagonists, except pyridinolcarbamate, which antagonized competitively and reversibly.

Melmon: Are you certain that autonomic reflexes were ruled out in your experiments with peptides? The rise in pulmonary arterial pressure followed the fall in systemic pressure in the guinea pig and rabbit and correlated with a decrease in cardiac output. A carotid or aortic baroreceptor may still be active here. Isolated muscle preparations from the pulmonary artery and veins of the calf contract only with very large doses of bradykinin. Furthermore, other investigators find that bradykinin in man produces profound decrease in pulmonary arterial pressure. Did you use a potent adrenergic blocker to determine the role of the autonomic nervous system?

Konzett: The administration of atropine, dihydroergotamine KO 592 (a β-blocking agent of C. H. Boehringer Sohn, Ingelheim), or hexamethonium did not diminish the rise in pulmonary arterial pressure caused by bradykinin in the guinea pig. In our experiments on isolated bovine pulmonary artery, bradykinin produced occasionally a contraction in doses from 70-100 μ g/ml.

BRADYKININ AND PULMONARY VASCULAR PERMEABILITY IN ISOLATED BLOOD-PERFUSED RABBIT LUNGS

A. HAUGE, P. K. M. LUNDE, and B. A. WAALER

Institute of Physiology, University of Oslo, Norway

Bradykinin causes a marked vasodilatation in most areas of the systemic circulation. In the pulmonary vascular bed of various species, however, bradykinin seems to cause vasoconstriction, as is shown by several investigators (Lecomte and Troquet, 1960; Gersmeyer and Spitzbarth, 1961; Klupp and Konzett, 1963; Hauge et al., 1964). Another important effect of bradykinin in systemic vascular areas is the increase in capillary permeability, which has been demonstrated by various techniques. Kjellmer and Odelram (1965) found that a marked and irreversible edema might develop in skeletal muscle on intravascular administration of bradykinin. We have been interested in the possible effect of intravascularly administered bradykinin on permeability of the small vessels in the lung, since vasoconstriction is the response of some larger pulmonary vessels to this substance. The problem related to such a permeability effect seemed to be of considerable practical, as well as theoretical, interest.

We have used an isolated perfused rabbit lung preparation; the details of the method are given elsewhere (Hauge et al., 1966) and will be described here only briefly. Both lungs, together with the heart and the large thoracic vessels, were removed from an anesthetized animal, which was kept alive up to the point of lung removal. The perfusion inflow cannula was tied into the pulmonary artery. Other large vessels were ligated near the heart. The outflow cannula was placed in the left atrium. The whole preparation was placed inside a plastic casing, where it hung freely on a piece of tape tied around the ventricles. The upper end of the tape was connected to the base of a Mettler balance with suitable movement damping. The weight of nonpulmonary structures, including pieces of glass and rubber tubing, was nearly constant throughout an experiment. The experimental arrangement therefore allowed weight changes of the lungs to be followed continuously, whereas absolute weight values for the lungs could not be recorded.

Perfusion of the pulmonary vascular bed was carried out with heparinized whole blood from donor rabbits and at constant volume pulsatile inflow, by means of a Dale and Schuster perfusion pump. The blood from the left atrium was drained back into the reservoir. Thermostatically controlled water baths surrounded the various pieces of equipment, and the blood temperature was kept at about 37-38°C.

Positive pressure ventilation with 4% CO_2 in air was carried out with a Starling "Ideal" pump and the ventilation overflow arrangement of Konzett and Rössler (1940). In addition to the weight changes, pulmonary arterial pressure and tidal volume were followed continuously. Pulmonary arterial pressure was recorded with a water manometer system on a smoked kymograph drum. Tidal volume was obtained indirectly by recording the ventilation overflow on the kymograph. Left atrial pressure was usually kept constant at about 3-5 cm of water, but could be changed at will.

Injections of bradykinin into the pulmonary arterial tubing in doses from 0.5 to 5 μg caused a constrictor response in the preparation. There was a relatively marked tachyphylaxis on repeated injections (Hauge et al., 1964). Injections of as much as 50 μg bradykinin caused only small and reversible weight changes in the preparation, which we judged were due to capacity alterations in some parts of the pulmonary vascular bed.

Figure 1 illustrates that marked edema might develop in this preparation, and that the edema development might be followed continuously by observing changes in weight and tidal volume. Edema was here provoked by addition of sodium edetate (Na_2 EDTA) to the 250 ml of blood perfusate in three doses of 150 mg each. The addition of the first two doses of this chelating agent caused a moderate fall in vascular resistance, a moderate increase in tidal volume, and no significant change in the weight. Subsequent to the addition of the third dose, however, a marked edema formation started, as indicated in the figure by the increase in weight and by the reduction in tidal

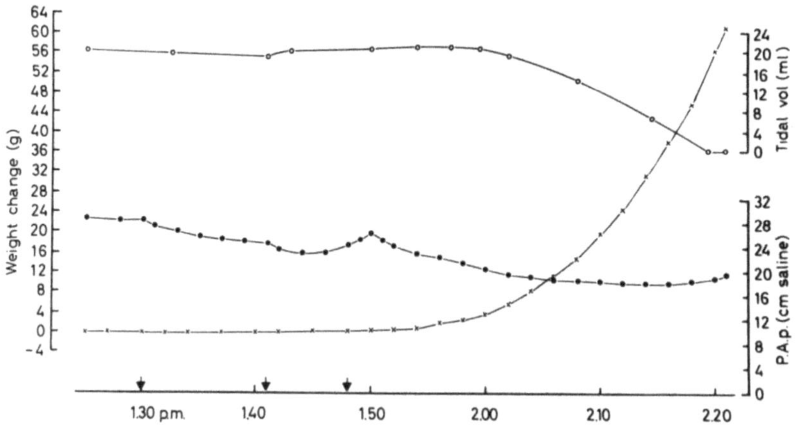

Fig. 1. Formation of pulmonary edema in isolated lung preparation after addition of disodium-edetate (Na_2 EDTA) to the perfusate.

Isolated perfused lungs (IPL) from female rabbit, 3.3 kg. At each arrow addition of 150 mg NA_2 EDTA to blood reservoir. Perfusion (constant volume inflow) started at 11:07 a.m. Blood flow (BF) through preparation: 117 ml/min. Symbols: o——o = tidal volume (TV); •——• = pulmonary arterial pressure (P.A.p.); x——x = weight change.

volume. The weight of a pair of rabbit lungs at the start of an experiment is about 15-20 g, and the increase in weight can be as much as 64 g as is shown in Fig. 1. Tidal volume diminished from about 20 ml to zero. It is worth noting that weight increase of about 2-3 g could be recorded before any measurable reduction in tidal volume took place.

Our next experimental step was to infuse bradykinin into the pulmonary arterial tubing over periods of 4 to 10 min. Twenty-two such intusions were carried out in 13 experiments. The doses injected varied from 6 to 200 μg per min. The blood infusion levels. We believe that these weight changes are due min. The effects of such bradykinin infusions are illustrated in Figs. 2 and 3.

Figure 2 shows the effects of two infusions of 15 μg bradykinin/min for 8 min in a preparation with a flow of 240 ml/min. The weight was gradually but slowly increasing in this preparation, which had been perfused for several hours at this stage. The bradykinin infusion caused a definite vasoconstrictor response, whereas hardly any weight changes could be seen. After each infusion of bradykinin, the left atrial pressure (and thereby the capillary pressure) was increased for a limited period of time. The elevation of left atrial pressure naturally caused some increase in pulmonary arterial pressure, and we aimed at a left atrial pressure elevation which gave a pulmonary arterial pressure similar to that reached during the preceding infusion of bradykinin. Such increases in

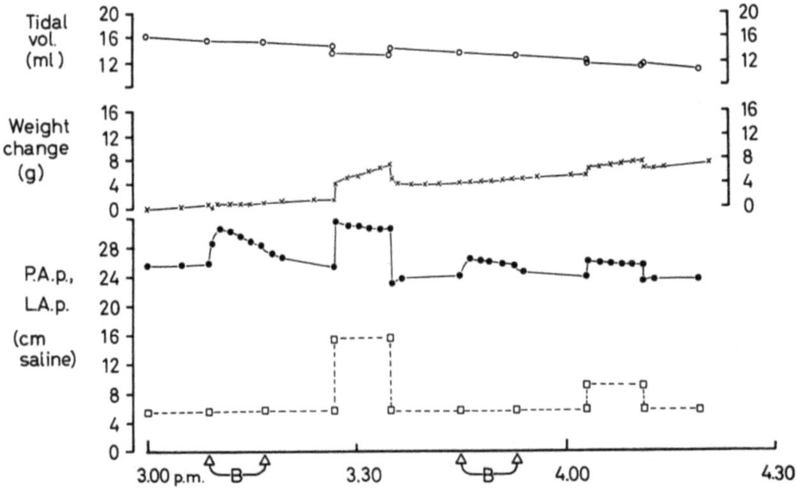

Fig. 2. Effects of bradykinin infusions and of subsequent left atrial pressure elevations in isolated lung preparation.

IPL from male rabbit, 2.9 kg. Between arrows, infusion of bradykinin, 15 μg/min (B), into the pulmonary arterial tubing. Perfusion (constant volume inflow) started at 10.58 a.m. BF: 240 ml/min. Symbols: o——o = TV; x——x = weight change; •——• = P.A.p.; □------□ = left atrial pressure (L.A.p.). Abbreviations as in Fig. 1.

left atrial pressure caused, however, changes in the weight curve which differed from that seen during bradykinin infusions. Initially a rapid change in the weight occurred, and the subsequent slope of the curve was definitely steeper than before the left atrial pressure elevation. This is interpreted as an indication of an initial capacity increase in pulmonary vessels and in the left atrium on the pressure elevation. Thereafter, presumably, an increased net outflux of fluid from the capillaries must have taken place. When the pressure was again reduced, the capacity of the vessels and the atrium

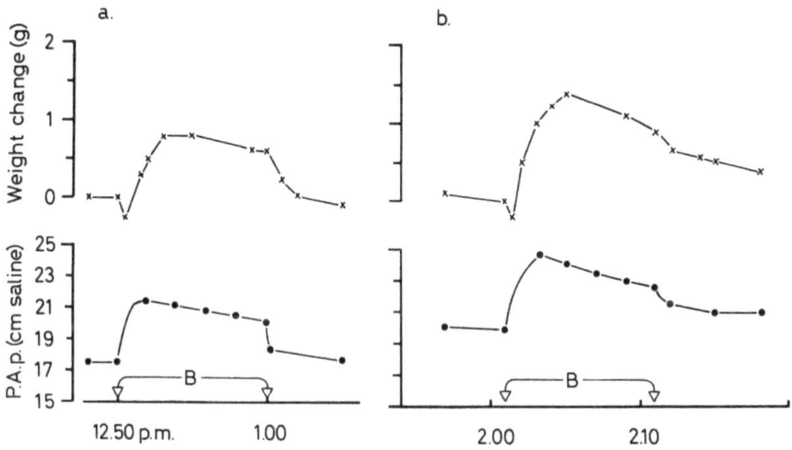

Fig. 3. The effects of bradykinin infusions on weight and on pulmonary arterial pressure in isolated lung preparation.

IPL from male rabbit, 3.2 kg. At *a* 15 μg bradykinin/min (B) was infused into the pulmonary arterial tubing, between arrows. BF: 150 ml/min. At *b*, 30 μg bradykinin/min (B) was infused into the pulmonary arterial tubing, between arrows. BF: 230 ml/min. Perfusion (constant volume inflow) started at 12.30 p.m. (Between *a* and *b* the setting of the perfusion pump was increased.) Symbols: x——x = weight change, •——• = P.A.p. Abbreviations as in Fig. 1.

diminished, and the former slope of the weight curve was again obtained.

The findings here suggest, first, that the bradykinin infusions cause no increased net outward movement of fluid from the small pulmonary vessels. There is no measurable increase in vascular permeability. Second, the vasoconstriction caused by bradykinin apparently must occur at precapillary sites. Had postcapillary constriction occurred, one would have expected weight changes similar to those obtained during elevation of the left atrial pressure.

Figure 3 shows the effect of two 10-min infusions of bradykinin in another experiment. The weight change scale is more extended than in Fig. 2, making the details on the corresponding curve more pronounced. In this case the constrictor response to bradykinin, which is indicated by the rise in pulmonary arterial pressure, was accompanied by relatively marked changes in the weight. Initially there was a small drop in the weight, followed by an increase which closely paralleled the change in pulmonary arterial pressure. When the infusion stopped, the weight and pressure came back toward the preinfusion levels. We believe that these weight changes are due to capacity changes of some pulmonary vessels. First a small capacity reduction takes place as vessels constrict. Then an increase in capacity occurs, corresponding to about 1 or 1.5 g, presumably caused by blood accumulation proximal to the constricted vessels, at a precapillary site. Capacity changes of this size in the larger pulmonary arterial vessels of our preparation are feasible. Increased capillary permeability with increased outflux of water can hardly have taken place during these infusions in Fig. 3.

After having shown that plain arterial infusions of bradykinin in the doses mentioned caused no detectable progressive weight changes, indicating increased permeability of the small vessels, we tried to infuse bradykinin in situations where a marked net outflux of water was already going on. Such situations were created in two ways: (1) we elevated the left atrial pressure to a certain level, thereby elevating the capillary pressure; (2) we perfused the lungs in the reverse direction. Having thus induced, in one of these ways, increased capillary pressure with a continuous net outflux of fluid from the small vessels, we superimposed infusions of bradykinin, as shown in Fig. 4. Blood flow was 130 ml/min. Infusion of 30 μg bradykinin/min caused a moderate vasoconstrictor response, but it hardly changed the slope of the weight curve. If anything, the slope was moderately reduced. Such infusions were carried out ten times in seven different experiments. The doses infused varied from 25 to 200 μg/min, and the blood flow during the infusions from 130 to 350 ml/min. In none of the experiments did we get any evidence of an increased outward flux of fluid during infusions. Such bradykinin infusions thus did not seem to increase capillary permeability in the preparation, as evaluated in this way.

In our last type of experiments the blood flow through the preparation was reversed, with inflow through the left atrium and the pulmonary veins. In such a situation a continuous weight increase of the preparation can be recorded, indicating increased pressure in the capillaries and net outward filtration of fluid from vessels. Nine infusions of 6 to 50 μg bradykinin/min in five experiments of this type have been carried out. The blood flow values during infusions were between 120 and 275 ml/min. Again, no indication could be obtained from any of these infusions of an increased vascular permeability caused by bradykinin.

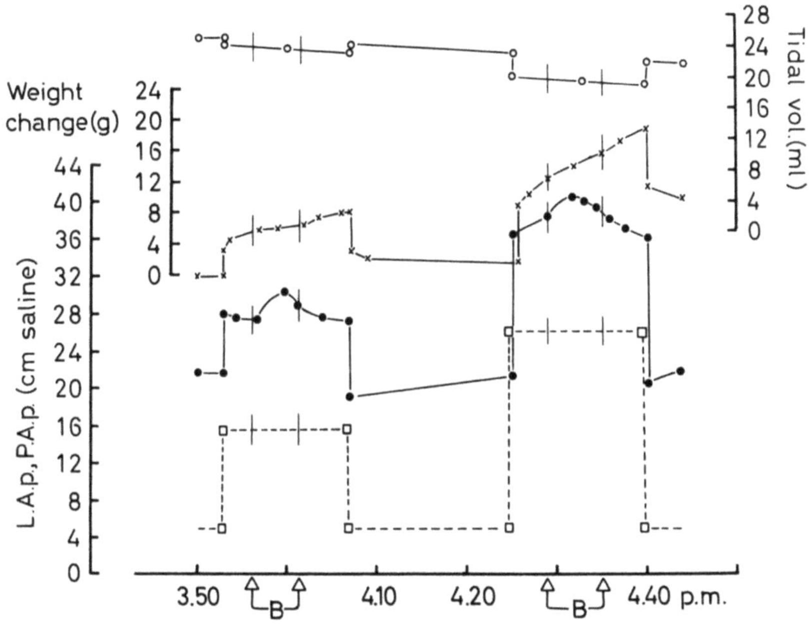

Fig. 4. Effects of bradykinin infusions in isolated lung preparation during periods of increased left atrial pressure.

IPL from female rabbit, 3.0 kg. Between arrows, infusion of 30 μg bradykinin/min (B) into the pulmonary arterial tubing. Perfusion (constant volume inflow) started at 11.59 a.m. BF: 130 ml/min. Symbols: o——o = TV; x——x = weight change; •——• = P.A.p.; □------□ = L.A.p. Abbreviations as in Figs. 1 and 2.

One example of such an experiment is shown in Fig. 5. The flow here was 160 ml/min. In episode a, a reversal of the perfusion took place during R. An initial rapid weight increase, presumably due to capacity changes of vessels and the left atrium can be seen on reversing the perfusion. A subsequent continuous increase in the weight occurred then during the reversed perfusion. When, as in episode b, bradykinin infusion (25 µg/min) was carried out during such a period of reversed perfusion, no apparent additional changes in the increase of the weight could be seen. In this experiment bradykinin had a diphasic vasomotor effect, as seen sometimes during reversed perfusion. A small pressor response was followed by a longer-lasting depressor response. Whether

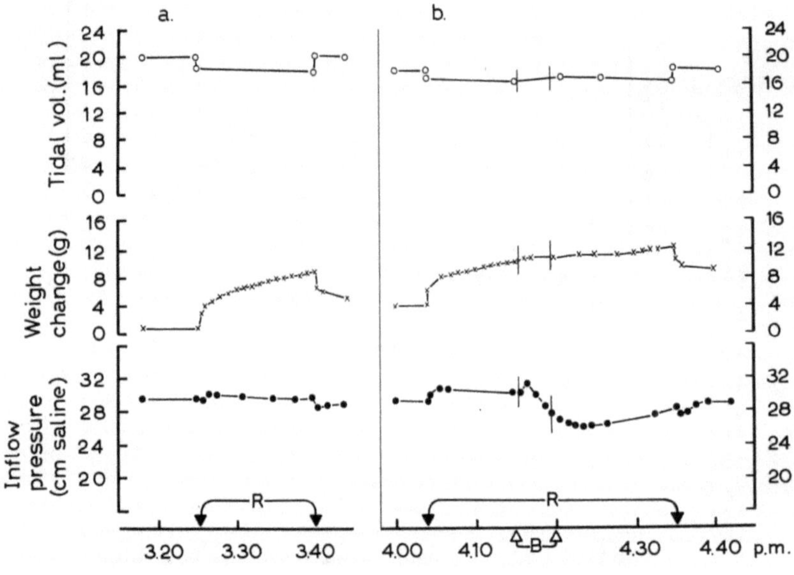

Fig. 5. Effects of reversing perfusion in isolated lung preparation and of superimposing bradykinin infusion during such reversal.

IPL from male rabbit, 3.2 kg. At *a.* and *b.* perfusion was reversed between black arrows (R). At *b*, infusion of 25 µg bradykinin/min (B) into the inflow tubing, between white arrows. Perfusion (constant volume inflow) started at 11.53 a.m.. BF 160 ml/min. Symbols: o——o = TV; x——x = weight change; •——• = pulmonary inflow pressure. Abbreviations as in Fig. 1.

only constriction or such a diphasic response occurred, there were no additional changes in the weight development during bradykinin infusion.

Di Mattei (1962) has found that injections in vivo of 25-50

μg bradykinin/kg to rabbits subsequent to an injection of epinephrine (50 μg/kg) provoked acute pulmonary edema. In an isolated lung preparation from a 3-kg rabbit we injected into the pulmonary arterial tubing 100 μg bradykinin 2 min after having injected 200 μg epinephrine. No edema formation could be observed on such a combined injection.

We must conclude from our experiments that infusions of bradykinin in doses large enough to cause marked vasoconstrictor responses caused no observable increase in vascular permeability in isolated perfused lungs of the rabbit. Edema formation could not be provoked in the preparation by the arterial injection of a large dose of bradykinin 2 min after the administration of epinephrine. The vasoconstriction caused by bradykinin in the pulmonary vessels of the rabbit lung must occur mainly at precapillary sites.

REFERENCES

Di Mattei, P.: Occurrence of kinins in epinephrine-induced pulmonary oedema in rabbits. Arch. int. Pharmacodyn. 140, 368 (1962).

Gerzmeyer, E. F., and H. Spitzbarth: Über Kreislaufswirkungen von synthetischem Bradykinin beim Menschen und beim wachen Hund. Klin. Wschr. 39, 1227 (1961).

Hauge, A., P. K. M. Lunde, and B. A. Waaler: The effect of bradykinin, kallidin, and eledoisin upon the pulmonary vascular bed of isolated blood-perfused rabbit lungs. J. Physiol. (Lond.) 173, 33P (1964).

Hauge, A., P. K. M. Lunde, and B. A. Waaler: Vasoconstriction in isolated blood-perfused rabbit lungs and its inhibition by cresols. Acta physiol. scand. 66, 226 (1966).

Kjellimer, I., and H. Odelram: The effect of some physiological vasodilators on the vascular bed of skeletal muscle. Acta physiol. scand. 63, 94 (1965).

Klupp, H., and H. Konzett: Zur Wirkung von Bradykinin und Kallidin auf den Druck in der Arteria pulmonalis. Naunyn-Schmiedeberg's Arch. exp. Path. Pharmak. 246, 19 (1963).

Konzett, H., and R. Rössler: Versuchsanordnung zu Untersuchungen an der Bronchialmuskulatur. Naunyn-Schmiedeberg's Arch. exp. Path. Pharmak. 195, 71 (1940).

Lecomte, J., and J. Troquet: Antagonisme entre bradykinine et phenylbutazone chez le lapin. C. R. Soc. Biol. (Paris) 154, 1115 (1960).

DISCUSSION

Guth: Rowley has demonstrated that bradykinin induces increases in permeability only when the veins are constricted and the blood backs up into the arteries. In the pulmonary bed, where the arteries are constricted and nobody knows what happens to the veins, one does not have the conditions necessary for producing increased passage of substrates out of the vessel walls.

Waaler: When the left atrial pressure was elevated in our experiments (Fig. 4), this must have caused blood accumulation in the veins and venules, and no outward flux of fluid from the vascular bed was going on. Still we found no evidence of bradykinin causing increased permeability of the small pulmonary vessels.

Hilton: I think there is little doubt that with his elegant techniques, Prof. Waaler has shown the constriction to be restricted to the precapillary vessels. Incidentally, Dr. Guth, it cannot be said that bradykinin causes venoconstriction in all vascular beds. In a recent analysis of the vascular effects in skeletal muscle it was found that bradykinin caused increased capillary permeability and filtration without any evidence of venoconstriction. My question is about the experiments with retrograde perfusion. Here the arterioles are the "postcapillary" vessels. So, if the perfusion pressure was high enough one would expect extravasation of fluid after giving bradykinin.

Waaler: Retrograde perfusion per se caused outward flux of fluid, as shown in the last figure. This is presumably due to the capillary pressure being high during this perfusion. When bradykinin also was infused, no additional increase in weight was seen. Vasoconstrictor responses to bradykinin during reversed perfusion were small, however. Sometimes diphasic vasomotor responses to bradykinin were seen during reversed perfusion, as demonstrated in Fig. 5.

VENOCONSTRICTION INDUCED BY BRADYKININ IN THE RABBIT EAR

P. S. GUTH, R. BOBBIN, G. CANO, and J. AMARO

Department of Pharmacology, Tulane University School of Medicine, New Orleans, U.S.A.

Bradykinin is widely regarded as a potent vasodilator capable of increasing blood flow, increasing vascular permeability, and lowering systemic blood pressure through arteriolar and capillary dilation. Minority reports have appeared in the literature which indicate that some vessels may be constricted by bradykinin. These reports have not received sufficient attention because they conflict with the established view. They will be briefly reviewed here.

Vasoconstriction in the isolated rabbit lung has been demonstrated (Lecomte and Troquet, 1960), in the absence of bronchomotor effects (Hauge et al., 1964). Similar observations have been made in the intact human (Gersmeyer and Spitzbarth, 1961) and in cat and guinea pig (Klupp and Konzett, 1963, 1965). Klupp and Konzett (1965) adduced evidence which may be taken to mean a direct action by bradykinin on pulmonary vessels, namely that it causes isolated pulmonary arterial strips to contract. Furthermore, in the guinea pig heart-lung preparation, bradykinin causes an increase in pulmonary arterial pressure without an increase in cardiac output.

Most of the constant-volume perfusion experiments have been concerned with pressure changes on the arterial side in response to bradykinin. Hashimoto et al. (1964), using a donor dog to perfuse the dog hind limb, demonstrated a biphasic pressor response in the perfusion circuit, but when the skin of the

*Supported by a grant from the Louisiana Heart Association.

recipient limb was removed, only the pressor response remained. Chou et al. (1965), employing constant-volume perfusion in dogs, found that bradykinin caused a rise in mesenteric venous pressure.

Burch and DePasquale (1962), employing the technique called rheoplethysmography in the human finger, suggested that bradykinin caused a constriction of arteriovenous anastomoses. This closure of shunt circulation causes passive dilation of capillaries. Venoconstriction could induce an analogous increase in resistance to flow, causing passive capillary dilation. Rowley (1964) in a particularly impressive paper, demonstrated in the rat that 5-hydroxytryptamine, histamine, and bradykinin all produce venoconstriction and increased permeability. He found the venules intensely dilated and concluded that this dilation was passive and secondary to venoconstriction. He produced further evidence that these agents do not cause an increase in permeability directly. When perfusing the veins in retrograde direction, 5-hydroxytryptamine, histamine, and bradykinin did not cause an increase in permeability. Therefore, he concluded that these agents do not cause permeability increases directly but that an apparent permeability increase occurs as a consequence of active venoconstriction and the resulting passive dilation of the venules. Haddy (1960) also found that the intravascular infusion of histamine and 5-hydroxytryptamine caused an increase in venous tone. Thus it may be a general phenomenon that what is termed "capillary permeability" is really venule and perhaps capillary dilation secondary to venoconstriction, and all three tissue hormones produce their permeability effects in this way.

The experiments previously reported from this laboratory (Guth et al., 1963) in which bradykinin caused a decrease in outflow from rabbit's ear, cat and dog hind limb, and rat hindquarter, were all performed in isolated animal appendages perfused with saline solutions. It is known (Burn and Dale, 1926) that in such preparations anomalous reactions to drugs do occur. Therefore it was decided to investigate the actions of bradykinin on the vasculature of an intact animal. To this end, it was injected into the marginal ear veins of rabbits, and the ipsilateral or contralateral ear veins were visualized by means of low-power microscopy (Guth et al., 1963). In the intact rabbit, it was found that bradykinin likewise was able to cause a constriction of the veins.

These experiments in the intact animal had the disadvantage

that the venoconstriction following bradykinin may have had a variety of causes other than direct action of bradykinin, such as stimulation of baroreceptor reflexes, catecholamine release, or stimulation of pain receptors.

Under certain circumstances, bradykinin induces a rise in blood pressure. As a matter of fact, Parratt (1964) reported that its repeated administration caused a qualitative change in the character of the response from depressor to pressor. He concluded that bradykinin raised the blood pressure whenever the animal was made hypotensive. Furthermore, the pressor response was abolished by adrenalectomy, reserpine, or phentolamine pretreatment, leading to the conclusion that it was caused by catecholamines released by bradykinin. Several other investigators (Lecomte et al., 1964; Croxatto and Belmar, 1961, 1962; Rosas et al., 1965; Feldberg and Lewis, 1963, 1964, 1965) have similarly demonstrated a bradykinin pressor effect. Feldberg and Lewis (1963, 1964, 1965) have shown that ganglion blocking agents do not inhibit epinephrine release by bradykinin.

Therefore it seemed quite likely that bradykinin induces its venoconstriction by the local release of catecholamines. In order to test this possibility, an adrenergic blocking agent, phentolamine, and denervation, including sympathectomy, have been employed. Rabbit's ears were deprived of sensory fibers as described below to remove the possibility that bradykinin might induce its venoconstriction by initiating pain reflexes. Finally by sympathectomy and by the use of very small doses of bradykinin, the possibility of venoconstriction as a result of baroreceptor-stimulation was obviated.

METHODS

Portions of the following nerves were removed from male albino rabbits: the anterior auricular nerve, the posterior auricular nerve, the auricular-temporal branch of the trigeminal nerve, and the caudal auricular branches of the facial nerve. The superior cervical ganglion was also excised. At least one week was allowed for nerve degeneration before the animal was used in an experiment.

The ears were tested for lack of sympathetic innervation by demonstrating that the ear vessels no longer reacted to changes in body temperature. A faradic current was applied to test for the severance of the sensory nerves. The methods were essentially those of Grant (1935), Armin and Grant (1953), and Grant and Thompson (1963).

For the experimental procedure, the animals were anesthetized with allobarbital plus urethane (0.6 ml/kg, i.p.), with ad lib. supplementation. The ear was glued to a plastic stage (Atlantic Surgical Co.) such that the marginal vein was situated over an opening in the plastic.

A small branch of the marginal vein was cannulated with a 30-gauge needle and the cannula advanced to the point of junction with the marginal vein. The area to be observed and photographed was about 1 cm downstream along the vein. A heat filter was placed between the microscope light and the ear.

Ear and rectal temperatures were monitored, and body temperature was regulated by means of a heating pad controlled by the rectal thermometer.

Injection volumes never exceeded 0.04 ml.

RESULTS

As demonstrated in Fig. 1, bradykinin produces a roughly sigmoid log dose-response curve. The responses plotted in Figs. 1 and 2 are the maximum noted constrictions. The curve exhibits a plateau at about 45-50% constriction. A similar

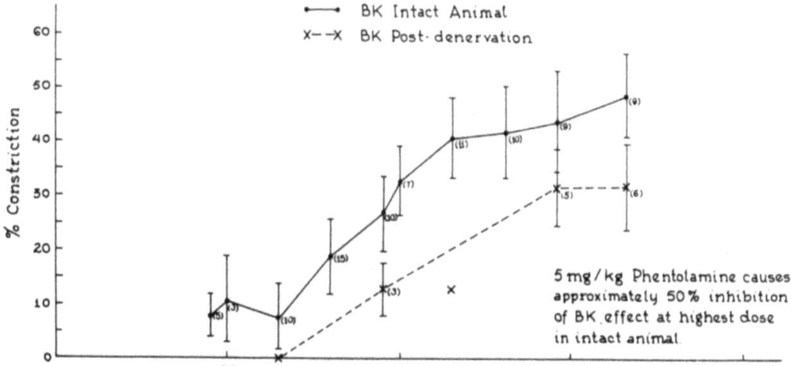

Fig. 1. Dose-response curve of bradykinin in the intact and denervated rabbit's ear. I represents the standard deviation. The number of observations is given in parentheses. The intact ear is represented by o——o and the denervated ear by x——x.

maximum effect occurs with norepinephrine. Doses as low as 10 ng (10^{-9} g) have on occasion produced a response of this magnitude. Injections of 5 mg phentolamine/kg reduce the response to various bradykinin doses in the intact animal. Parenthetically, in order to produce a fall of 10 to 15 mm Hg in systemic blood pressure in the intact rabbit, upward of 200 ng bradykinin is required.

After denervation, the bradykinin dose-response curve is shifted to the right and downward, indicating a decrease in effectiveness and a decrease in the maximal elicitable effect to about 35% to 40%. Bradykinin is roughly ten times as potent a venoconstrictor on a weight basis or seventy times as potent on a molar basis as norepinephrine.

The dose of phentolamine employed throughout these experiments, 5 mg/kg, was sufficient to block all doses of norepinephrine used in both the intact and denervated animals, completely, but blocked bradykinin only partially.

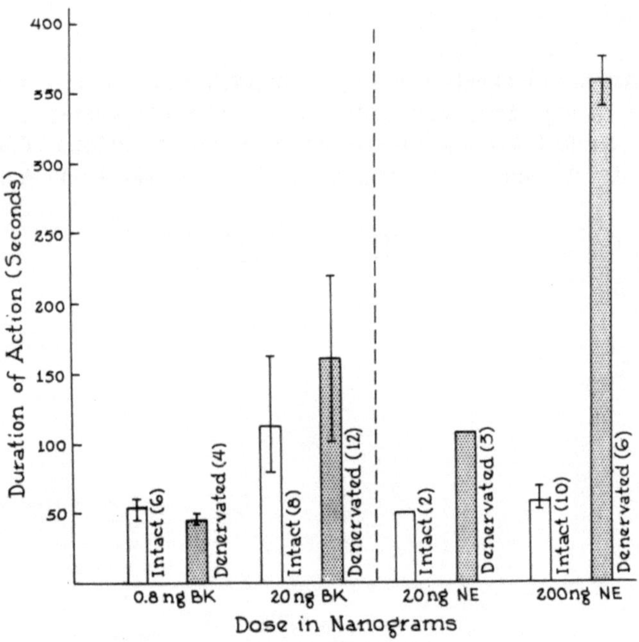

Fig. 2. The duration of the constrictor responses to bradykinin (BK) and norepinephrine (N.E.). I represents the standard deviation. The number of observations is given in parentheses.

Figure 2 is the histogram of the duration of action of two doses each of bradykinin and norepinephrine in the intact and totally denervated rabbit's ear. The doses chosen were from the rising phases and the plateaus of the bradykinin and norepinephrine dose-response curves. It may be seen that norepinephrine, particularly in the high dose employed, produces a much longer effect in the denervated than in the intact animal. On the other hand, no such denervation supersensitivity may be seen with bradykinin.

Control injections of saline solution and undiluted bradykinin vehicle (kindly supplied by Sandoz Pharmaceuticals, Hanover, N. J.) produced no venoconstriction.

DISCUSSION OF RESULTS

The bradykinin-induced venoconstriction reported here and elsewhere finds at least five possible explanations. (1) Bradykinin may be releasing catecholamines, histamine, and/or serotonin locally. (2) It may be inducing a baroreceptor-mediated reflex venoconstriction. (3) It may be causing sensory stimulation and thus reflex venoconstriction. (4) It may act directly to cause venoconstriction. (5) Some combination of the above. The results will be discussed in relation to these five explanations so that perhaps some may be eliminated.

It has been demonstrated that norepinephrine is completely antagonized by phentolamine in both the denervated and intact animals. On the other hand, bradykinin is only partially antagonized in the intact animal. Therefore, at best, the release of norepinephrine by bradykinin may be only partially responsible for the venoconstriction. Furthermore, in the denervated animal, bradykinin is still active but less so. Denervation removes a store of catecholamines. Therefore, on the basis of partial antagonism by phentolamine and reduction in effect following denervation, it may be tentatively concluded that bradykinin acts partially by means of catecholamines released from sympathetic nerve endings. The possibility also exists that phentolamine antagonizes bradykinin directly. It may be argued that the remainder of the bradykinin effect is accounted for by the release of catecholamines from chromaffin cells in the ear. The evidence against this argument is based on the partial nature of the phentolamine blockade and on drug reaction caused by denervation supersensitivity. The catecholamines released from chromaffin cells should be antagonized by phentolamine as well as those released from nerve endings. Supersensitivity to catecholamines in the denervated ear is shown by large increases in the duration of action of injected norepinephrine. If bradykinin was releasing chromaffin catecholamines in the dernervated ear, prolongation of venoconstriction would be expected also. Bradykinin, however, shows no prolongation of action in the denervated ear. Thus it may be concluded that the venoconstriction induced in the denervated ear by bradykinin does not depend on the release of catecholamine from chromaffin tissue. In fact, should phentolamine be found to be a weak bradykinin antagonist as are dibenamine and dibenzyline (Rocha e Silva and Garcia Leme, 1963), then catecholamine release need not be invoked at all. Reflex venoconstriction, whether induced by baroreceptor or sensory mechanisms, is not seriously considered as an explanation for bradykinin venoconstriction. In the doses employed in these experiments, bradykinin has no effect on systemic blood pressure. Moreover, reflex venomotor changes in response to sensory stimulation are characterized by a constriction followed by a dilation. Finally, in this regard, if complete denervation is assumed, neither reflex mechanism could operate.

The explanations remaining are that bradykinin causes a direct venoconstriction and/or that it releases catecholamines, histamine, and 5-hydroxytryptamine. Experiments employing animals depleted of these amines are planned.

SUMMARY

Bradykinin causes a constriction of the marginal vein of the intact and denervated rabbit ear in doses of 1-20 ng. Local release of catecholamines, probably from nerve endings, is partially responsible for the venoconstriction. It is suggested that venoconstriction is a primary vascular event. Subsequently venular dilation as well as an increase in vascular permeability occur.

REFERENCES

1. Armin, J., and R. T. Grant: The artery of the denervated rabbit ear as a sensitive pharmacological test object. J. Physiol. (Lond.) 121, 593 (1953).

2. Burch, G. E., and N. P. dePasquale: Bradykinin, digital blood flow and the arterio-venous anastomoses. Circulat. Res. 10, 105 (1962).
3. Burn, J. H., and H. H. Dale: The vasodilator action of histamine and its physiological significance. J. Physiol. (Lond.) 61, 185 (1926).
4. Chou, C. C., E. D. Frohlich, and E. C. Texter: A comparative study of the effects of bradykinin, kallidin II, and eledoisin on segmental superior mesenteric resistance. J. Physiol. (Lond.) 176, 1 (1965).
5. Croxatto, H., and J. Belmar: Hypertensive effects of bradykinin in rats. Nature (Lond.) 192, 879 (1961).
6. Croxatto, H., and J. Belmar: Hypertensive effects of bradykinin in rats. Acta physiol. lat. Amer. 12, 122 (1962).
7. Feldberg, W., and G. P. Lewis: Release of adrenalin from cat's suprarenals by bradykinin and angiotensin. J. Physiol. (Lond.) 167, 46P (1963).
8. Feldberg, W., and G. P. Lewis: The action of peptides on the adrenal medulla. Release of adrenalin by bradykinin and angiotensin. J. Physiol. (Lond.) 171, 98 (1964).
9. Feldberg, W., and G. P. Lewis: Further studies on the effect of peptides on the suprarenal medulla of cats. J. Physiol. (Lond.) 178, 239 (1965).
10. Gersmeyer, E. F., and H. Spitzbarth: Uber Kreislaufwirkungen synthetischem Bradykinin beim Menschen und beim wachen Hund. Klin. Wschr. 39, 1227 (1961).
11. Grant, R. T.: Further observations on the vessels and nerves of the rabbit's ear with special reference to the effects of denervation. Clin. Sci. 2, 1 (1935).
12. Grant, R. T., and R. H. S. Thompson: Cholinesterase and the nerve supply to blood vessel in the rabbit's external ear. J. Anat. (Lond.) 97, 7 (1963).
13. Guth, P. S., G. Cano, and J. Jaramillo: The effect of bradykinin on vascular smooth muscle. Ann. N.Y. Acad. Sci. 104, 69 (1963).
14. Haddy, F. J.: Effect of histamine on small and large vessel pressures in the dog foreleg. Amer. J. Physiol. 198, 161 (1960).
15. Hashimoto, K., S. Kumakura, and N. Taira: Vascular reflex responses induced by an intraarterial injection of aza azepino-phenothiazine, andromedotoxin, veratridine, bradykinin and kallikrein, and blocking action of sodium salicylate. Jap. J. Physiol. 14, 299 (1964).

16. Hauge, A., P. K. M. Lunde, and B. A. Waaler: The effect of bradykinin, kallidin and eledoisin upon the pulmonary vascular bed of isolated blood-perfused rat lungs. J. Physiol. (Lond.) 173, 33P (1964).
17. Klupp, H., and H. Konzett: Zur Wirkung von Bradykinin und Kallidin auf den Druck in der Arteria pulmonalis. Naunyn-Schmiedeberg's Arch. exp. Path. Pharmak. 246, 19 (1963).
18. Klupp, H., and H. Konzett: Der Einfluss von Bradykinin und Kallidin auf den Druck in der Arteria pulmonalis. Naunyn-Schmiedeberg's Arch. exp. Path. Pharmak. 249, 479 (1965).
19. Lecomte, J., and J. Troquet: Antagonisme entre bradykinine et phenylbutazone chez le lapin. C. R. Soc. Biol. (Paris) 154, 1115 (1960).
20. Lecomte, J., J. Troquet, and A. Cession-Fossion: Sur la natur de l'hypertension arterielle provoque par la bradykinine et la kallidine. Arch. int. Pharmacodyn. 147, 518 (1964).
21. Parratt, J. R.: Vasopressor responses to kallidin, bradykinin and eledoisin in hypotensive rats. J. Pharm. Pharmacol. 16, 132 (1964).
22. Rocha e Silva, M., and J. Garcia Leme: Antagonists of bradykinin. Med. exp. (Basel) 8, 287 (1963).
23. Rosas, R., D. Montague, M. Gross, and D. F. Bohr: Cardiac action of vasoactive peptides, 1. Bradykinin. Circulat. Res. 16, 150 (1965).
24. Rowley, D. A.: Venous constriction as the cause of increased vascular permeability produced by 5-hydroxytryptamine, histamine, bradykinin and 48/80 in the rat. Brit. J. exp. Path. 45, 56 (1964).

DISCUSSION

Sicuteri: Phentolamine blocks 5-hydroxytryptamine, epinephrine, and norepinephrine. It would be important to observe the effect of a "pure" 5-hydroxytryptamine antagonist such as LSD. Bradykinin in doses that affect the general circulation does not cause local venoconstriction in man. Could these be correlated with a difference in species?

Guth: According to Burch and de Pasquale, such a constriction occurs in man. Perhaps the inconsistencies in results arise from differences in the venous circulatory beds.

Vogt: Would you think that the formation of edema by bradykinin is mainly the result of venoconstriction which showed so impressively and not a consequence of an effect on the capillary wall?

Guth: My answer would have to be based on Rowley's work demonstrating that bradykinin causes increased vascular permeability only when veins are constricted and the venules dilated. According to his report, application of bradykinin by retrograde perfusion, does not allow for passive venule dilation does not increase vascular permeability. It is possible to conclude that bradykinin does not act directly on the capillaries to increase their permeability.

Back: In studies at present in progess in our laboratories we have observed that bradykinin causes vasoconstriction and consequent decrease in blood flow in the isolated perfused salivary gland preparation in the dog. The bradykinin was administered under constant perfusion technique.

Would you care to speculate on the physiologic implications of this peculiar but perhaps more common action of bradykinin?

Guth: I'm not certain that bradykinin has any physiologic role in the sense of a function in healthy animals. However, as mentioned by Drs. Greenbaum and Eisen, subsequent to trauma, where tissue pH falls and lysosomal enzymes are activated, bradykinin release followed by venoconstriction could certainly lead to edema formation as well as pain, itch, etc.

Nakano: Can you apply your observations made on rabbit ears to the remainder of the venous system?

Guth: We have not studied the veins of other circulatory beds in the rabbit and are therefore not able to generalize from our own work. However, examples of venoconstriction in other species and other beds have been mentioned at this symposium and are appearing in the literature.

Shimamoto: I have used your in-vivo technique and found an antagonistic effect of pyridinolcarbamate. However, the effect of this compound in patients may be a very complicated puzzle that must be analyzed carefully. I would like to point out the importance of further research on the significance of the increase of venous tone by bradykinin; either tonic or spastic, which might possibly be involved in the nutritional uptake by tissue, or in the exudative or hemorrhagic states.

Concerning the problem of venous return during hypotension caused by bradykinin, we found that the hypotensive effect of bradykinin was partially antagonized by pyridinolcarbamate in rabbits. However, the antagonistic effect of this compound in the veins might possibly be involved in this effect. Elevated venous pressure in patients suffering from congestive heart failure was lowered by pyridinolcarbamate.

Sicuteri: Nelle vene del dorso della mano dell'uomo non sono documentabili effetti venocostrittori. Si tratta di diversita di specie animali oppure di diversita di sede, di distretto vascolare?

Piovella: Indubbiamente le venule della rana e del ratto da noi osservate sono poco confrontabili con le grosse vene del braccio nell' uomo. Inoltre i reperti da noi osservati sulle venule del mesentere possono differire anche per distretto vasale, come e stato dimostrato da Beretta e Coll. anche con queste interssanti sostanze.

CARDIOVASCULAR ACTIONS OF PHYSALAEMIN

A. H. GLASSER and G. B. FREGNAN

Farmitalia, Istituto Ricerche, Milan, Italy

Physalaemin and eledoisin and perhaps substance P belong to a group of peptides having similar pharmacological actions (Bertaccini et al., 1965a,b) both in vitro and in vivo. In vivo, these peptides affect the systemic arterial blood pressure in a similar way, but physalaemin is two to four times more active than eledoisin in decreasing blood pressure in dogs and increasing it in chickens. As previously reported, both the increase of blood flow and the ypotension provoked by eledoisin in several animal species, including man, could be explained by the peripheral vasodilating property of this peptide. Therefore, since physalaemin seemed to be closely related to eledoisin, we decided to investigate the peripheral vascular effects of physalaemin as well. In addition, since we had no data for proving whether the dilator action of eledoisin on vascular beds was a direct or an indirect action via the peripheral nervous system, we performed experiments on denervated vascular beds and made comparisons with other vasodilator drugs.

METHODS

Mongrel dogs of either sex were anesthetized with thiopental sodium (30 mg/kg i.v.) followed by α-chloralose (80 mg/kg i.v.) and kept under artificial respiration by a Starling pump. Procedures to record systemic arterial blood pressure, total femoral blood flow, and skin and muscle blood flows in normal, innervated hind limbs were described previously (Bergamaschi and Glässer, 1964). Chronic denervation of the leg was performed by cutting the femoral and the sciatic nerves and by stripping gently both femoral artery and vein at the femoral triangle 8 to 10 days before blood flow studies. The isolation of the chronically denervated leg was performed by dissecting completely all the tissues and by sawing the bone so that the leg was connected to the rest of the body only through the femoral artery and vein.

In other experiments the peripheral vascular activity of eledoisin, bradykinin, nitroglycerin, carbaminoylcholine, and papaverine was studied on the innervated and denervated beds of a muscle preparation in situ (Stainsby and Renkin, 1961).

In all experiments mean blood flow changes (ml/min) were recorded with a Shipley and Wilson rotameter (1951), and drugs as well as control solvent solutions (0.1 ml) were injected intra-arterially in few seconds. The minimal changes of flow brought by the solvents were subtracted from the values for drugs. Only in a few experiments were the drugs injected intravenously to study their action on systemic arterial blood pressure.

RESULTS

a) Systemic arterial blood pressure. In accordance with previous results, rapid intravenous injections of physalaemin and eledoisin caused a dose-related reduction of the mean blood pressure, threshold doses being respectively 10^{-10} to 5×10^{-10} g/kg (0.1 to 0.5 ng/kg) for physalaemin and 5×10^{-10} to 10^{-9} g/kg (0.5 to 1 ng/kg) for eledoisin. The hpotension was short lasting for both peptides. However, at larger doses the hypotension caused by physalaemin was of shorter duration than that caused by eledoisin, while at small and medium doses, physalaemin was two to five times more active than eledoisin.

b) Blood flow in normal and chronic denervated hind limbs. Control: In a group of normal animals average peripheral vascular resistance values were 8 PRU for the muscle bed and 6 PRU for the total-femoral bed; corresponding basal mean flow rates were 23 and 26 ml/min. In other normal animals, on the contrary, these figures were respectively 3.6 and 3 PRU with basal flow rates of 52 and 55 ml/min. Skin beds had a high peripheral resistance of 9.3 PRU with a flow of 18 ml/min. After chronic denervation and acute isolation it was found that total femoral blood flow averaged 30 ml/min, with a vascular resistance of about 5 PRU. Vascular reactivity as indicated by re-

TABLE 1.
Per Cent Increases of Blood Flow in Different Beds After Physalaemin and Eledoisin

	Physalaemin (g/kg i.a.)							Eledoisin (g/kg i.a.)				
	10^{-15}	10^{-14}	10^{-13}	10^{-12}	10^{-11}	10^{-10}	10^{-9}	10^{-13}	10^{-12}	10^{-11}	10^{-10}	10^{-9}
Total femoral flow												
a) High PRU	0	+22	+ 86	+114	+135			0	+12	+ 53	+164	
b) Low PRU			+ 4	+ 16	+ 60	+104	+168		+ 4	+ 8	+ 60	+150
c) Chronic denervated, isolated		+ 5	+ 10	+ 25	+ 59	+110	+215	0	+13	+ 28	+ 87	+155
Muscle flow												
a) High PRU	0	+25	+106	+128	+150			0	+31	+108	+174	
b) Low PRU		+ 8	+ 12	+ 33	+ 62	+ 98	+160	0	+ 8	+ 20	+ 61	+150
Skin flow			+ 6	+ 16	+ 40	+ 56	+ 76			+ 4	+ 28	+ 70

Drugs were injected intra-arterially and blood flow changes recorded by rotameter.

$$PRU = \frac{\text{mean arterial pressure (mm Hg)}}{\text{mean blood flow (ml/min)}}.$$

active hyperemia following 30-sec arterial occlusion was present in total femoral beds of normal and isolated legs and in muscle districts. As expected, this type of hyperemia was absent in skin beds. In addition, the peak hyperemia flow rates were particularly pronounced when basal vascular resistance was high.

Intra-arterial injection: In normal and denervated animals physalaemin and eledoisin increased blood flow in all tested beds. The changes of blood flow started a few seconds after injections with a rapid increase; blood flow then returned to the control values within a maximum of 10-20 min. The total response was greater as the dose was increased except when systemic arterial blood pressure was strongly affected by recirculating peptides. Results obtained with physalaemin and eledoisin are summarized in Table 1.

As can be seen from Table 1 and from Fig. 1, intra-arterial injection of physalaemin and eledoisin regularly increased total femoral and muscle blood flows, even at doses as low as 10^{-14} g/kg (0.01 pg/kg) for physalaemin and 10^{-12} g/kg (1 pg/kg) for eledoisin. Only larger doses of peptides affected skin flow as well. Interestingly enough, the differences in vascular reactivity were noted. Muscle and femoral beds had stronger reaction than skin beds to both peptides. In addition, the reactivity of muscle and femoral beds to both peptides became quite impressive when the basal vascular tone was high. Hence, on muscle and femoral beds with a high vascular resistance to flow (high PRU), small and medium doses of physalaemin increased blood flow even in doses about 100 to 200 times lower than those which are equally active on partially relaxed beds (low PRU). However, when higher doses of physalaemin were used, beds with a high tone reacted only 50 times more strongly than those having a low tone. Eledoisin was also stronger on tonic vessels, but differences were less marked, since the activity ratio between equally active doses was always less than 50.

Because of the convergence of the dose-response lines of physalaemin and eledoisin, the activity ratio between them was wider at low doses and smaller at the highest doses in our experiments. Therefore physalaemin was only five to ten times more active than eledoisin on femoral and muscle flow with high PRU. On the same beds but with low PRU the ratio varied only from 20 to 5. The sensitivity of skin beds was not so remarkable, and dose-response curves of both peptides were

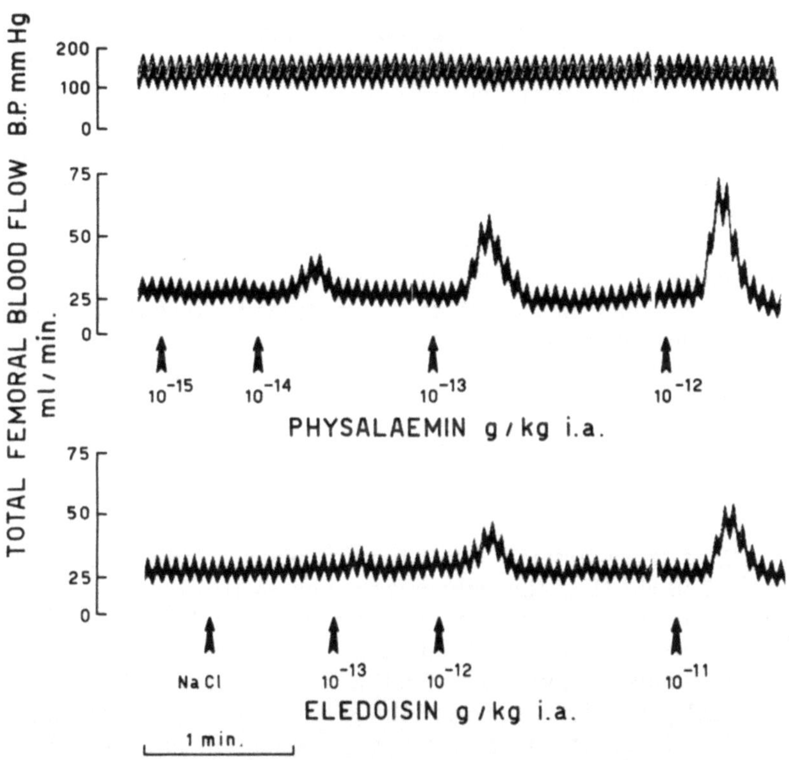

Fig. 1. Effect of increasing intra-arterial doses of physalaemin (above) and eledoisin (below) on femoral blood flow as recorded by rotameter.

flatter. Nevertheless, physalaemin was stronger than eledoisin at least in the lower dose range.

Physalaemin and eledoisin were active on legs chronically denervated and acutely isolated, as on the innervated femoral vessels.

c) Blood flow in intact and acutely denervated muscle preparations in situ. Control blood flow through the innervated muscle was 27 ml/min/100 g muscle, with a mean arterial perfusion pressure of 132 mm Hg. Two hours after denervation, blood flow was 26 ml/min/100 g muscle, with a perfusion pressure of 114 mm Hg. Reactive hyperemia was present to a high degree either just before or after denervation. Graded intra-arterial doses of eledoisin (10^{-12} to 10^{-8} g/100/g muscle, bradykinin (10^{-12} to 10^{-8} g/100 g muscle), nitroglycerin (10^{-9} to 10^{-7} g/100 g muscle), carbaminoylcholine (10^{-9} to 10^{-7} g/100 g muscle), and papaverine (10^{-6} to 10^{-5} g/100 g muscle), increased the local blood flow without significantly affecting the perfusion pressure. All drugs were less active than eledoisin when the doses increasing flow up to 100% over control were compared. Bradykinin and carbaminoylcholine, however, became nearly equally or even more active when larger doses were used.

After denervation, all drugs increased blood flow more than in the intact preparation, but their activity ratios were almost the same in the denervated as in the innervated preparation.

DISCUSSION

Since intra-arterial injections of physalaemin increased arterial blood flow through muscle and skin beds as well as through the coronary artery (report to be published elsewhere) without affecting the systemic arterial blood pressure, this peptide seems to have a vasodilating action like that described previously for eledoisin and nitroglycerin (Bergamaschi and Glässer, 1963, 1964). Moreover, physalaemin and eledoisin were still active on isolated chronically denervated vascular beds. Hence a direct smooth muscle-relaxing action on resistance vessels seems to be the most satisfactory explanation for their local vasodilating activity and for the arterial hypotension produced by intravenous doses of both peptides. Likewise, the vasodilating actions of eledoisin, bradykinin, papaverine, carbaminoylcholine, and nitroglycerin were not reduced by acute nerve section. Perhaps these agents, also, act through a mechanism similar to that of physalaemin and eledoisin. The increased responsiveness of vessels possessing higher vascular resistance values may be explained by a greater room for dilatation in partially constricted beds. However, skin vessels, in spite of their high basal resistance values, are poorly sensitive to both peptides. It is still to be explained why the activity ratio between physalaemin and eledoisin varies considerably with the state and type of the vessels and with the route of administration, and why the dose-response curves for physalaemin, eledoisin, bradykinin, and carbaminoylcholine have different shapes.

At any rate, at present physalaemin is by far the most active vasodilator, particularly at the minimal and medium active doses.

REFERENCES

Bergamaschi, M., and A. H. Glässer: Effect of the endecapeptide eledoisin on the coronary blood flow: Comparison with nitroglycerin, bradykinin, and epinephrine in the dog. Circulat. Res. 13, 329 (1963).

Bergamaschi, M., and A. H. Glässer: Peripheral vasodilator action of eledoisin, bradykinin, and nitroglycerin in anesthetized dogs. Circulat. Res. 15, 371 (1964).

Bertaccini, G., J. M. Cei, and V. Erspamer: Occurrence of physalaemin in extracts of the skin of Physalaemus fuscumaculatus and its pharmacological action on extravascular smooth muscle. Brit. J. Pharmacol. 25, 363 (1965a).

Bertaccini, G., J. M. Cei, and V. Erspamer: The action of physalaemin on the systemic arterial pressure of some experimental animals. Brit. J. Pharmacol. 25, 380 (1965b).

Shipley, R. E., and C. Wilson: An improved rotameter. Proc. Soc. exp. Biol. (N.Y.) 78, 724 (1951).

Stainsby, W. N., and E. M. Renkin: Autoregulation of blood flow in resting skeletal muscle. Amer. J. Physiol. 201, 117 (1961).

HEMODYNAMIC EFFECTS OF PHYSALAEMIN AND OF SOME SYNTHETIC ELEDOISIN AND PHYSALAEMIN-LIKE POLYPEPTIDES

G. BERTACCINI, A. BORGHETTI, G. DE CARO, and G. LA GRECA

Istituto di Patologia Speciale Medica, Istituto di Famacologia, dell'Università di Parma, Italy

The study of some biologically vasoactive substances has suggested the possibility of additional research work and clinical applications.

In previous publications (1-4) we described the cardiovascular effects of eledoisin in man, taking into consideration its pharmacologically characteristic actions on the systemic and regional blood flow as shown by Erspamer et al. (5-9). Later we studied the effects of a new polypeptide isolated by the same authors from amphibian skin, physalaemin (10) and other eledoisin and physalaemin-like substances. The activity of these new polypeptides has been examined in parallel assays with eledoisin.

The peptides were screened for hypotensive effect by administering them to man and anesthetized dog (Table 1). The table shows that the hypotensive activity of physalaemin is three to four times stronger than that of eledoisin. Eledoisin or physalaemin-like substances have a similar action in man and dog, but on the whole man is not so sensitive as dog.

In these cardiovascular studies we have used not only eledoisin and physalaemin but also eledoisin-nonapeptide (El IX) and physalaemin-octapeptide (Ph VIII) in man, in order to examine both the hypotensive action and the systemic and regional blood flow in various clinical conditions.

MATERIALS AND METHODS

Hypotensive effect (eledoisin, physalaemin, El IX, Ph VIII)

Six normotensive healthy subjects (age 25-40) and five hypertensive subjects (age 40-60) were studied. The systolic and diastolic blood pressures were measured by sphygmomanometer every minute during the infusion period (10 min) and for 10 min after the infusion. Heart rate was determined by electrocardiogram.

Cardiac output

Man (eledoisin and physalaemin). In five normotensive subjects, cardiac output was determined every minute during infusion (10 min) and for 10 min afterward, with the sphygmographic method (11). Simultaneous recording of carotid and femoral sphygmogram, phono- and electrocardiogram, were done with a Hellige multichannel recorder.

Dog (eledoisin, physalaemin, and polypeptides marked with an asterisk in the table). Ten anesthetized dogs (pentobarbital i.v. 30 mg/kg) were used. Kymographic recording of blood pressure was done with a mercury gauge directly connected to femoral arteries. Measurement of O_2 consumption and A-V difference (venous mixed pulmonary blood, arterial femoral blood) were taken into consideration for determining the cardiac output by the Fick method.

Polygraphic studies (eledoisin, physalaemin, El IX, Ph VIII)

In five normotensive and in five hypertensive subjects various parameters were registered with a Hellige multichannel recorder.

Electrocardiographic findings (eledoisin)

Subjects were divided as follows: 14 with normal electrocardiogram; 14 showing secondary T-wave alterations; 12 with either primary (8) or postinfarctual (4) ischemia; 5 with electrocardiographic signs of injury. In one subject changes induced by physalaemin, El IX, and Ph VIII were also studied.

The substances were given in i.v. infusion and the EKG changes were continuously recorded.

TABLE 1

Hypotensive effect of eledoisin, physalaemin, and related peptides in dog and man †

Nº	STRUCTURE (positions 1 2 3 4 5 6 7 8 9 10 11)		% HYPOTENSION DOG	% HYPOTENSION HUMAN
1	H – Pyr – Pro – Ser – Lys – Asp(OH)-Ala – Phe – Ile – Gly – Leu – Met – NH_2 * (*Eledoisin*)		100	100
2	H – Pyr – Ala – Asp – Pro – Asp(NH_2)-Lys – Phe – Tyr – Gly – Leu – Met – NH_2 * (*Physalaemin*)		350	450
3	H —— Pro – Ser – Lys – Asp(OH)-Ala – Phe – Ile – Gly – Leu – Met – NH_2 *	Eledoisin-like	150	35
4	H —— Ser – Lys – Asp(OH)-Ala – Phe – Ile – Gly – Leu – Met – NH_2 *	Eledoisin-like	170	155
5	H —— Lys – Asp(OH)-Ala – Phe – Ile – Gly – Leu – Met – NH_2 *	Eledoisin-like	90	30
6	H —— Asp(OH)-Ala – Phe – Ile – Gly – Leu – Met – NH_2	Eledoisin-like	30	1,5
7	H —— Ala – Phe – Ile – Gly – Leu – Met – NH_2	Eledoisin-like	15	<2
8	H —— Phe – Ile – Gly – Leu – Met – NH_2	Eledoisin-like	1	<2
9	H —— *(D) Aps(NH_2)*-Ala – Phe – Ile – Gly – Leu – Met – NH_2 *	Eledoisin-like	75	
10	H —— *Aps(NH_2)* —— Phe – Ile – Gly – Leu – Met – NH_2	Eledoisin-like	10	≃4
11	H —— *Lys* —— Ala – Phe – Ile – Gly – Leu – Met – NH_2	Eledoisin-like	35	20
12	H —— *Lys* —— Phe – Ile – Gly – Leu – Met – NH_2	Eledoisin-like	50	20
13	H —— *Lys* —— Phe – *Phe* – Gly – Leu – Met – NH_2 *	Eledoisin-like	210	110
14	H —— *(D) Lys* —— *(D) Phe* – *(D)Phe* – *(D)Gly* – *(D)Leu* – *(D)Met* – NH_2	Eledoisin-like	<0,05	<2
15	H —— *Lys* —— Ile – Gly – Leu – Met – NH_2	Eledoisin-like	<0,2	<1
16	H —— *Lys* —— Phe – Ile – Gly – Leu – *Eti* – NH_2 *	Eledoisin-like	180	110
17	H —— Ala – *(D)Phe* – Ile – Gly – Leu – Met – NH_2	Eledoisin-like	<0,2	<2
18	H —— Ala – Phe – Ile – Gly – Leu – *(D)Met* – NH_2	Eledoisin-like	<0,1	<1
19	H —— Ala – Phe – Ile – Gly – *Ileu* – Met – NH_2	Eledoisin-like	<0,1	<2
20	H —— Ala – Phe – Ile – Gly – Leu – *Eti* – NH_2	Eledoisin-like	90	8
21	H —— Ala – *Ala* – Ile – Gly – Leu – Met – NH_2	Eledoisin-like	<0,1	<1
22	H —— Ala – *Tyr* – Ile – Gly – Leu – Met – NH_2	Eledoisin-like	~1	<1
23	H —— Ala – Phe – Ile – Gly – Leu – *Hcys* (*S-Pro*) – NH_2	Eledoisin-like	60	40
24	H —— Ala – Asp – Pro – Asp(NH_2)-Lys – Phe – Tyr – Gly – Leu – Met – NH_2 *	Physalaemin-like	240	150
25	H —— Asp – Pro – Asp(NH_2)-Lys – Phe – Tyr – Gly – Leu – Met – NH_2 *	Physalaemin-like	130	200-400
26	H —— Pro – Asp(NH_2)-Lys – Phe – Tyr – Gly – Leu – Met – NH_2 *	Physalaemin-like	300	400
27	H —— Asp(NH_2)-Lys – Phe – Tyr – Gly – Leu – Met – NH_2 *	Physalaemin-like	50	25
28	H —— Lys – Phe – Tyr – Gly – Leu – Met – NH_2 *	Physalaemin-like	300	
29	H —— Phe – Tyr – Gly – Leu – Met – NH_2	Physalaemin-like	0,5	<2
30	H —— Ala – Asp – Pro – Asp(NH_2)-Lys – Phe – *Ile* – Gly – Leu – Met – NH_2 *	Physalaemin-like	200	80
31	H – Pyr – Ala – Asp – Pro – Asp(NH_2)-Lys – Phe – Tyr – Gly – Leu – *Eti* – NH_2 *	Physalaemin-like	280	130
32	H —— Lys – Phe – *(m)Tyr* – Gly – Leu – Met – NH_2 *	Physalaemin-like	150	90
33	H —— Lys – Phe – Tyr – Gly – Leu – *Eti* – NH_2 *	Physalaemin-like	350	50

†Activity of eledoisin = 100. Peptides marked with asterisk were tested for cardiac output in the dog.

Regional circulatory changes (eledoisin, physalaemin)

Plethysmography by venous occlusion in regions of predominant cutaneous (hallux) and muscular (leg) blood flow was studied in five normotensive subjects without arteriopathy and

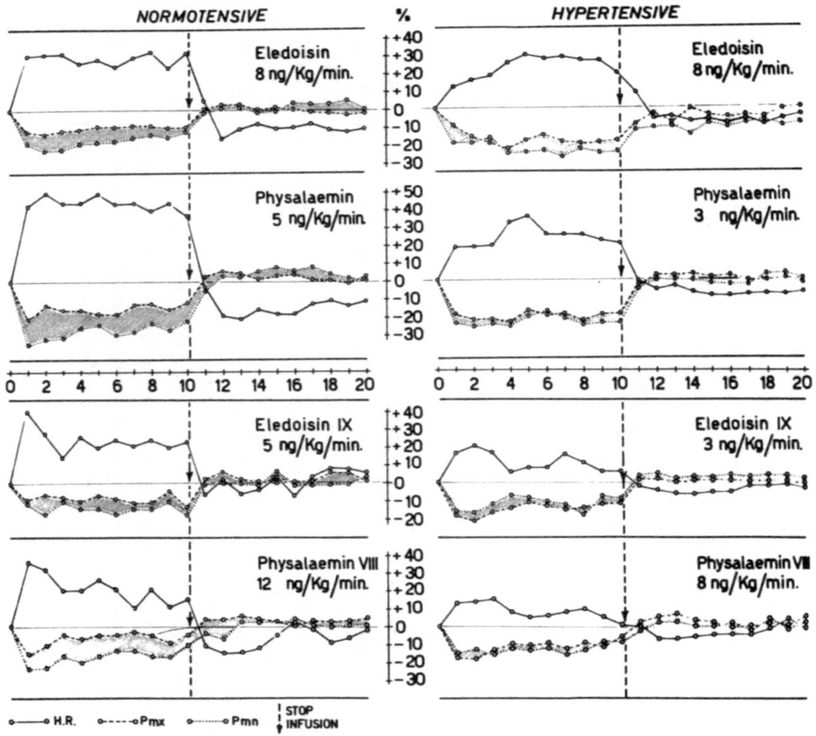

Fig. 1. H.R. = heart rate; Pmx = systolic blood pressure; Pmn = diastolic blood pressure. Effects of i.v. infusion of eledoisin, physalaemin, EI IX, and Ph VIII in normotensive and hypertensive subjects. Average values obtained from five subjects. Abscissa: time in minutes. Ordinate: per cent changes of blood pressure (starting values = 100). Dashed line at end of infusion.

in five subjects with evident circulatory insufficiency who had chronic arteriopathy of obstructive type.

Volume variations were converted to pressure variations by means of Sanborn transducers connected to a closed-air system and were recorded. Calibration was carried out at the beginning and end of the infusion by introducing known quantities of air into the system. Flow at minutes 1, 3, 5, 7, 9 during infusion, and at 2, 5, 8, 10 afterward was recorded. Venous occlusion pressure was 50 mm Hg.

RESULTS

Effects on systemic arterial pressure

Results are expressed in Fig. 1 as mean values of per cent changes with reference to starting values, in the two groups (normotensive and hypertensive subjects). A greater sensitivity to hypotensive action was observed in hypertensive subjects. Larger variations of diastolic pressure changes were observed in normotensive subjects. A relationship between heart rate and pressure changes was seen only in normotensive subjects ($P < 0.01$). Furthermore, a greater tendency toward normalizing the blood pressure was observed in the same group. Heart-rate changes in the group of normotensive subjects were likely to express a more effective reflex regulation of hypotension in peripheral circulation.

Effects on cardiac output

The literature does not offer conclusive data on the clinical use of eledoisin. Output and stroke volume increase (12), simple output increase (13), and output decrease (14), were observed during continuous infusion, while cyclic variations occurred during fast infusion (15).

In our previous studies, cardiac output (Fick method) during the hypotension showed no variations in normotensive subjects receiving eledoisin infusion (8 ng/kg/min). At the same time no significant alveolar ventilation (VA) variations, and consequently of the VA/Q (perfusion) ratio, were observed.

In normotensive subjects a polygraphic method also was adopted (11) to check the eventual phasic variations of cardiac output.

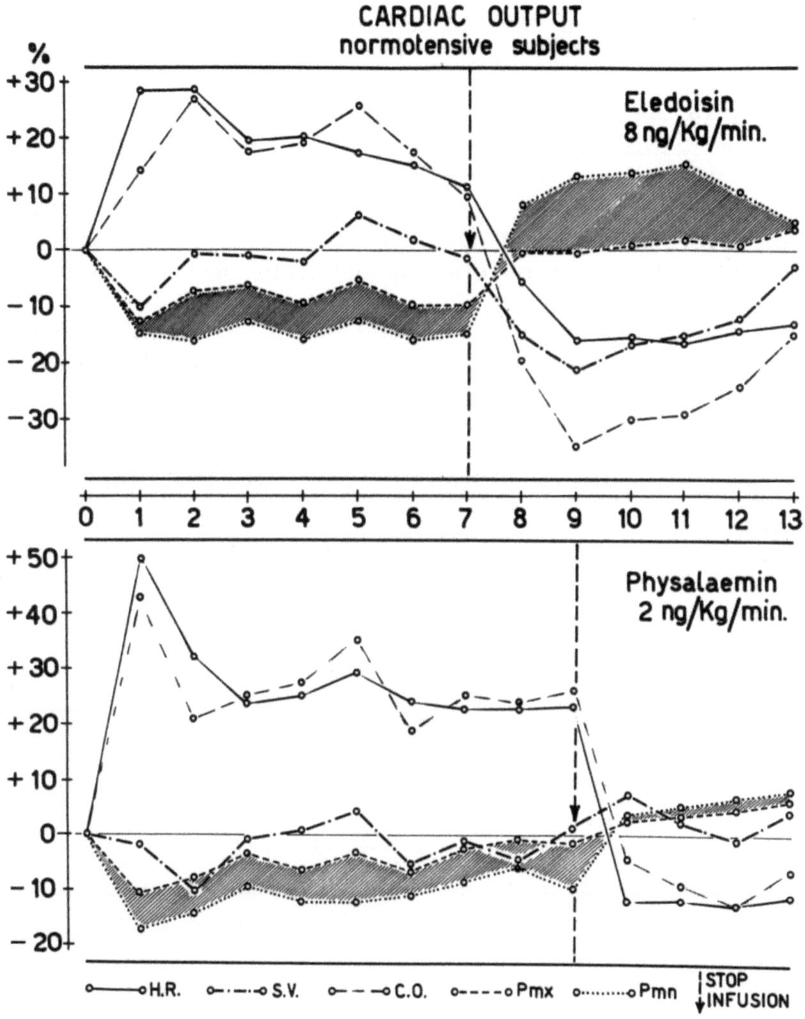

Fig. 2. H.R. = heart rate; S.V. = systolic volume; C.O. = cardiac output; Pmx = systolic blood pressure; Pmn = diastolic blood pressure. Abscissa: time in min. Ordinate per cent changes (starting values arbitrarily taken as 100).

During infusion of eledoisin or physalaemin a small decrease in the stroke volume together with an immediate heart-rate increase was observed during the first 2 min. During the whole infusion period the cardiac output increased in correlation with the heart rate. At the same time the diastolic pressure dropped more than the systolic. After the infusion stopped, a decrease of stroke volume, heart rate, and cardiac output was observed.

It is evident, therefore, that the hypotensive effect depended almost entirely on the peripheral vasodilatation; however, the cardiac-regulating mechanism was also important. At the end of the infusion the effects of arteriolar counterregulation were predominant: systolic blood pressure came back to normal values, while reflex adjustments of heart stopped abruptly (Fig. 2).

Hypotensive effects and cardiac output modifications were studied in dog (direct Fick method) during treatment with eledoisin, physalaemin, and the various polypeptides marked with asterisks in Table 1.

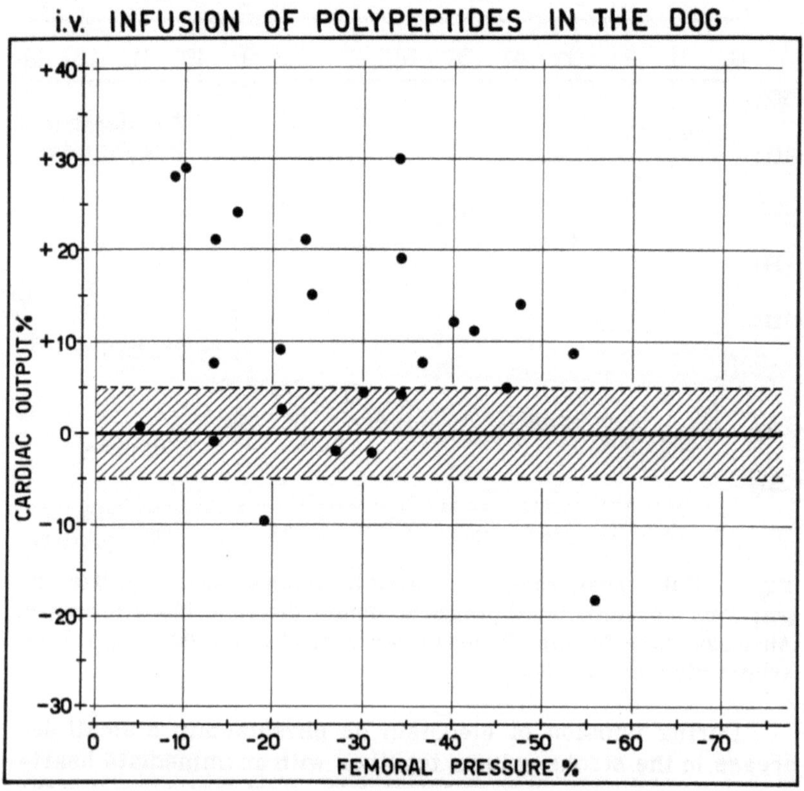

Fig. 3. Effects of i.v. infusion of different doses of eledoisin, physalaemin, and polypeptides marked with asterisk (in Table 1), on cardiac output. Hatched area corresponds to nonsignificant variations (± 5%).

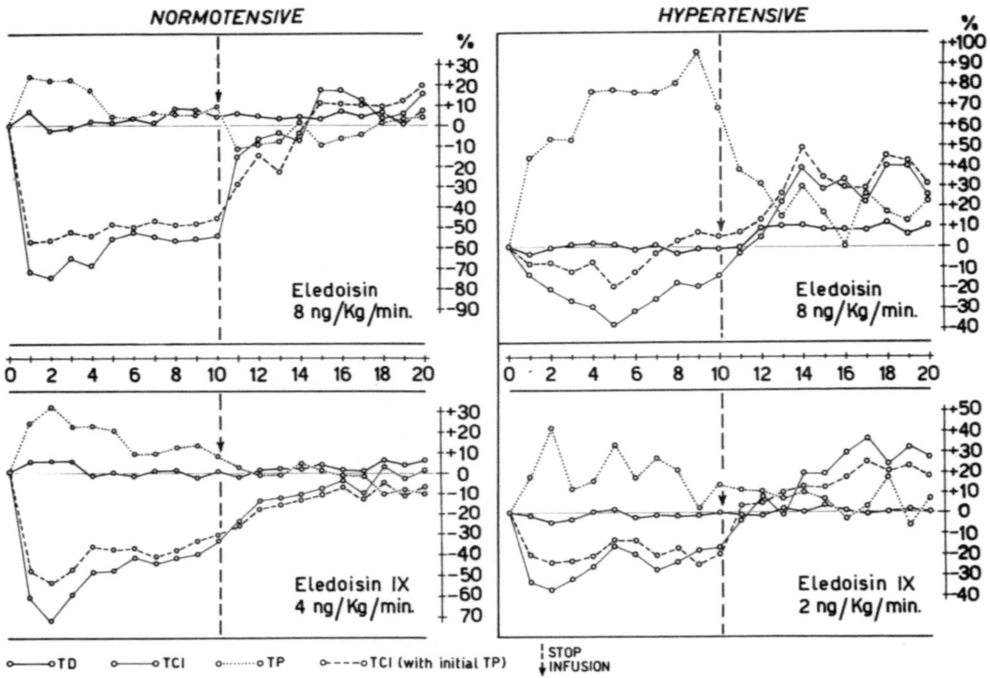

Fig. 4. Polygraphic findings. TD = deformation time; TCI = isovolumetric contraction time; TP = propagation time. Effects of i.v. infusion of eledoisin and EI IX in five normotensive and five hypertensive subjects (average values). Abscissa and ordinate as in Fig. 1.

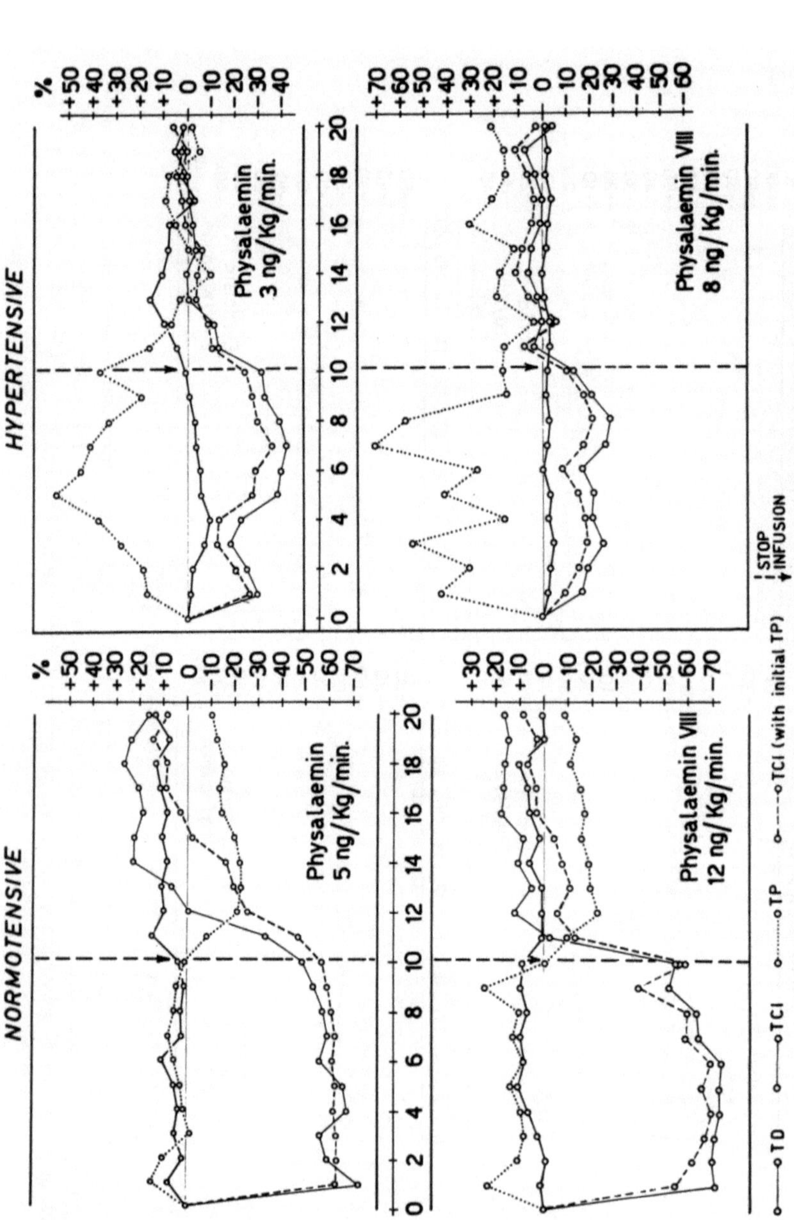

Fig. 5. Explanations as in Fig. 4; i.v. infusion of physalaemin and Ph VIII.

An output increase was observed, even in the presence of remarkable hypotensive effects. Only in sporadic cases (one was obtained with doses about 1,000 times the threshold value) was an output decrease seen (Fig. 3). Variations of A-V difference were very scarce.

Effects on the left ventricular dynamics (polygraphic method)

In two groups of normotensive and hypertensive subjects we observed the following parameters during eledoisin and physalaemin administration. Deformation time (TD); propagation time (TP); isovolumetric contraction time (TCI); isovolumetric contraction time calculated with basal propagation time. The results (Fig. 4 and 5) show a certain stability of deformation time, an increase of the propagation time, and a reduction of the calculated isovolumetric contraction time. Quantitative differences exist between the two groups regarding the PT and TCI indexes.

The greater per cent increase of the propagation time in hypertensive subjects was probably due to the hypertonia of the vessel walls and to the consequent greater reduction of the sphygmic wave speed related to the sudden peripheral resistance variations. The greater reduction of the calculated TCI in normotensive subjects, in the presence of similar pressure variations, probably showed a greater contraction of the vessels because of the interaction of catecholamines.

It is improbable that changes in heart rate caused these phenomena, since it was impossible to perceive parallel variations in the deformation time.

Effects on regional circulation

EKG findings. The evaluation of the coronary blood flow in man is difficult, owing to the extravasal influences. Experimentally, the effects already observed with one of the polypeptides (eledoisin) proved to be controversial (16-19). Therefore in man we considered only the indirect data obtained from electrocardiographic changes induced by eledoisin in selected cases (Table 2). The positive modifications (white circles) were strictly connected to the primary ischemic alterations of the terminal phase.

Similar results have been observed also for physalaemin and for two physalaemin and eledoisin derivatives in one subject (Table 3).

TABLE 2

Electrocardiographic findings after i.v. infusion of eledoisin in normal and pathological subjects*

ELECTROCARDIOGRAPHIC FINDINGS

N. of subjects		SEX	AGE	T-WAVES PRIMARY CHANGES				INFARCTION										T-WAVE SECONDARY CHANGES										NORMAL				
								INJURY						ISCHEMIA				VENTRICULAR LOADING					BUNDLE BRANCH BLOCK									
ng/Kg/min				4	8	16	4	10	16	20	4	8	10	16	24	4	8	8	4	8	12	16		4	6	8		4	8	12	16	
1	F.P.	♂	58		O																											
2	B.R.	♂	58		O																											
3	M.G.	♂	38		O																											
4	C.I.	♀	41		O		●																									
5	S.G.	♂	64		O																											
6	Z.E.	♂	43		O	O	●																									
7	M.C.	♀	64		O	O																										
8	B.P.	♀	62	O	O																											
9	S.S.	♂	61													●	●															
10	B.A.	♂	65													●	●															
11	G.M.	♂	50											O	O																	
12	C.N.	♂	52											O		●																
13	P.T.	♂	57								●	●	●																			
14	F.F.	♂	54						●	●		●																				
15	B.E.	♀	66								●																					
16	G.F.	♂	58					●	●																							
17	B.I.	♂	61									●	●																			
18	S.M.	♀	63															●														
19	C.A.	♂	46																●	●												
20	M.V.	♂	60																	●		●										
21	R.M.	♀	60															●	●													
22	M.H.	♂	52																●	●												
23	B.C.	♀	68																	●												
24	S.R.	♂	14																	●		●										
25	B.A.	♂	73																			●										
26	F.V.	♂	75																		●	●										
27	M.D.	♂	38																		●	●										
28	B.D.	♂	69																	●		●										
29	V.U.	♂	56																	●												
30	S.F.	♂	65																					●	●	●						
31	B.L.	♂	75																						●							
32	F.T.	♂	40																										●		●	
33	R.L.	♂	27																									●	●			
34	B.A.	♂	36																										●		●	
35	L.G.	♂	29																										●	●	●	
36	A.N.	♂	25																										●	●	●	
37	A.D.	♀	17																										●		●	
38	L.G.	♂	33																										●		●	
39	A.I.D.	♂	56																										●		●	
40	P.D.	♂	51																										●			
41	M.C.	♂	39																										●		●	
42	V.A.	♂	26																											●		
43	R.P.	♂	36																											●		
44	T.A.	♂	39																										●		●	
45	S.M.	♂	41																									●	●			

*0 = significant changes; ● = no changes.

<u>Plethysmography.</u> We observed in a group of subjects without any obstructive or vasospastic arteriopathy in the skin or muscle area a flow increase during the administration of eledoisin or physalaemin (Fig. 6). The cutaneous blood flow showed greater increases, probably because of the higher variability. When the hypotensive activity was equal, the intravenous

TABLE 3
EKG findings (lead V5) before, during, and after i.v. infusion of eledoisin, physalaemin, EI IX, and Ph VIII*

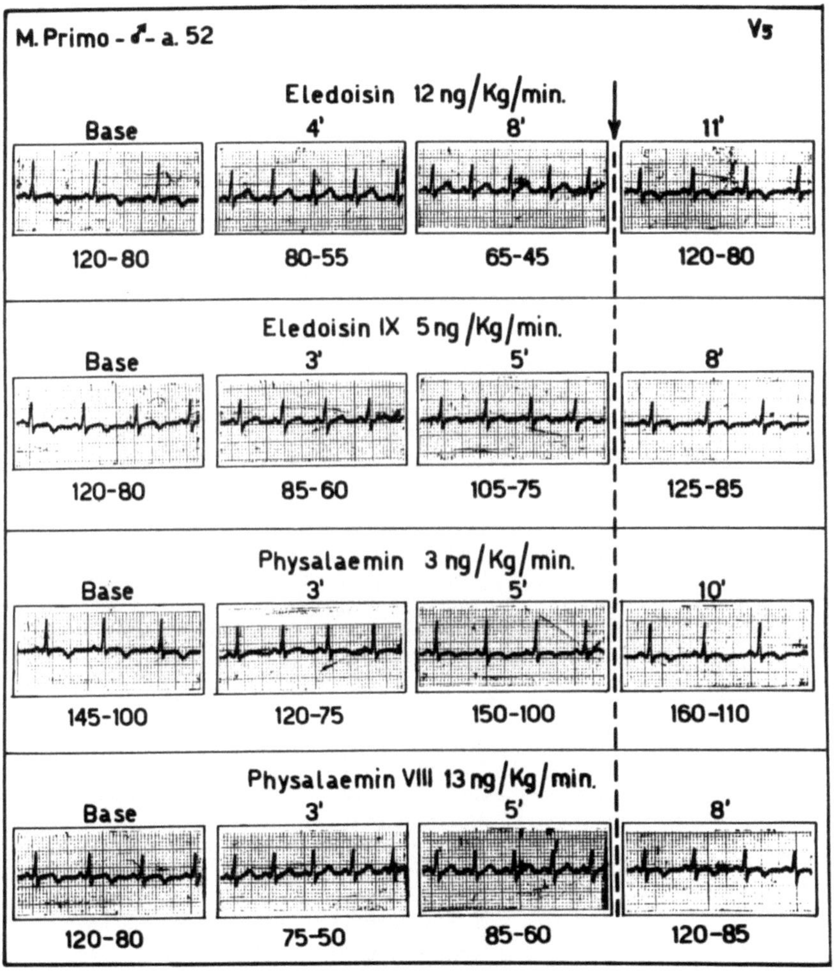

*Figures below EKG tracings represent systolic and diastolic blood pressure valves. End of infusion is at dashed line.

infusion of physalaemin caused greater effects, although with more evident phasic variations in the cutaneous and muscular regions.

A study was also carried out in a group of subjects with chronic obstructive arteriopathies; in these cases the vasodilatory effects were more strictly connected to the perfusion

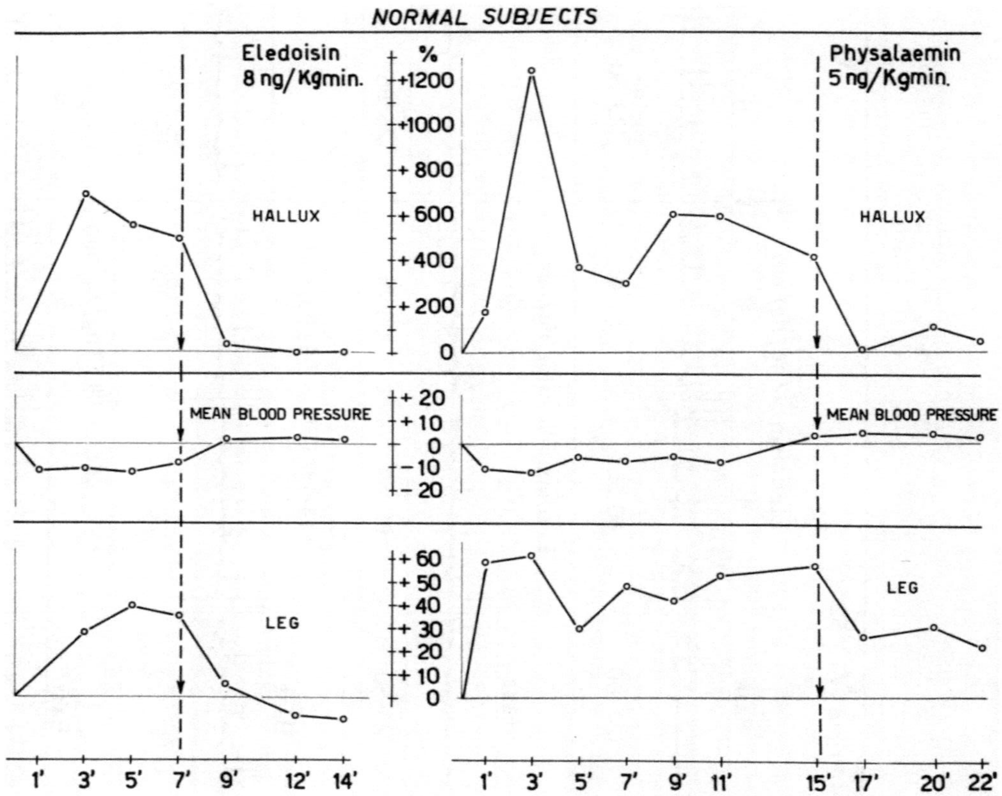

Fig. 6. Plethysmographic changes after i.v. infusion of eledoisin and physalaemin in five normal subjects (average values). Abscissa and ordinate as in Fig. 1.

pressure variations, even if they might have been concealed by greater hypotension, as previously remarked by Gersmeyer et al. (20).

DISCUSSION

The above findings prove that the peptides tested have a higher vasodilator activity than all other known biological substances. The cardiovascular activity of the polypeptides, is qualitatively not different in man. The hypotension is caused directly by the peripheral vasodilatation, which is evident mainly in particular regions (muscolocutaneous area). The different sensitivity between normotensive and hypertensive subjects may also depend on the different effectiveness of some homeostatic regulation mechanisms.

The hypotension, first determined by the vasodilatation, certainly causes sympathetic reflexes acting on the myocardium with increase of heart rate and, probably, of myocardial contraction. In fact, the cardiac output during infusion in the normotensive subject remains constant or even increases; which means a tendency to normalize the pressure. This activity may not always occur in hypertensive subjects or in cases of overdosage. The EKG variations indicate a vasodilatory effect in this region as well.

The blood-flow variation in the leg shows a different dependence on pressure changes in the healthy subject and in chronic obstructive arteriopathy; in the latter, the effect probably depends on the favorable relation between pressure modifications and regional vasodilatation.

SUMMARY

The authors have studied systemic and regional hemodynamic modifications induced in man and dog by various vasoactive polypeptides.

Hypotension is a consequence of a very strong peripheral vasodilatation, especially in musculocutaneous vessels. During hypotension, sympathetic myocardial reflexes appear, acting by increasing heart rate, stroke volume, and cardiac output.

The findings in healthy, hypertensive, and arteriopathic subjects are discussed.

RIASSUNTO

Gli AA. hanno studiato le modificazioni emodinamiche generali e distrettuali indotte nell'uomo e nel cane dal trattamento con vari polipeptidi vasoattivi.

L'ipotensione risulta conseguente alla intensa vasodilatazione periferica, particolarmente evidente nel distretto muscolo-cutaneo. Durante l'effetto ipotensivo si rivelano meccanismi riflessi di origine simpatica agenti a livello cardiaco, responsabili del comportamento della frequenza cardiaca, della gittata sistolica e della portata circolatoria.

I reperti ottenuti in soggetti sani, ipertesi ed arteriopatici sono discussi in base a questi elementi.

REFERENCES

1. Borghetti, A.: Minerva Europa (Symposium su "L'ipotensione arteriosa") Atti Convegno Farmitalia, Milano, Maggio 1964.
2. Ferrari, S., G. La Greca, E. Minardi, and A. Novarini: Minerva cardioangiol. In press.
3. La Greca, G., S. Ferrari, A. Novarini, and L. Chiampo: Minerva cardioangiol. In press.
4. Borghetti, A., V. Ferioli, S. Ferrari, G. La Greca, and A. Novarini: Minerva cardioangiol. In press.
5. Erspamer, V.: Experientia (Basel) 5, 79 (1949).
6. Erspamer, V., and A. Anastasi: Experientia (Basel), 18, 98 (1962).
7. Anastasi, A., and V. Erspamer: Brit. J. Pharmacol. 19, 326 (1962).
8. Erspamer, V., and G. Falconieri-Erspamer: Brit. J. Pharmacol. 19, 329 (1963).
9. Erspamer, V., and A. H. Glaesser: Brit. J. Pharmacol. 20, 516 (1963).
10. Erspamer, V., A. Anastasi, G. Bertaccini, and J. M. Cei: Experientia (Basel) 20, 489 (1964).
11. Meyer, cf. H. Waremburg, J. Desruelles, J. F. Merlen, and J. J. Waucampt: Actual. cardiol. 14, 5 (1965).
12. Kontos, H. A., W. Shapiro, H. Page, K. P. Mauck, and J. R. Patterson: J. appl. Physiol. 19, 113 (1964).
13. Ruol, A., G. d'Amelio, P. Donaggio, and I. dal Zotto: Minerva Europa (see Reference 1).

14. Ciampolini, E., A. Gerola, S. Grazi, and A. Libretti: B.S.I.B.S. XL, 72 (1964).
15. Broghammer, H. Klin. Wschr. 41, 1097 (1963).
16. Nakano, J.: Life Sci. 3, 291 (1964).
17. Beretta-Anguissola, A., and F. S. Feruglio: Atti LXIV cong. Soc. It. Med. Int., Padova, Oct. 1963, Ed. Pozzi, Roma.
18. Bergamaschi, M., and A. H. Glaesser: Circulat. Res. 13, 329 (1963).
19. Alela, A., L. Granata, and O. Pinotti: B.S.I.B.S. XL, 1915 (1964).
20. Gersmeyer, E. F., A. Castenholz, and M. Nicolay: Klin. Wschr. 43, 309 (1965).

THE EFFECTS OF ELEDOISIN AND BRADYKININ ON THE GENERAL AND VISCERAL CIRCULATION

A. BERETTA ANGUISSOLA, F. S. FERUGLIO, S. CAMPUS, L. CHIANDUSSI, G. PANDOLFO, and G. BERT

Institute of Medical Pathology
of the University of Turin, Italy

The quite numerous studies of the hemodynamic action of eledoisin and bradykinin have not come to a univocal conclusion. We intended, therefore, to carry out a comparative study of the cardiovascular activity of these vasoactive polypeptides, using them in doses which cause a similar hypotensive effect.

MATERIALS AND METHODS

The experiments have been carried out on medium- and large-sized dogs. The effect of bradykinin on hepatosplanchnic circulation was studied in healthy men. Experiments have been done in the dog under general anesthesia (chloralose and sodium pentothal) and in man receiving morphine sedation.

Eledoisin was given by means of a constant i.v. infusion of 0.01 $\mu g/kg/min$, and bradykinin was administered at the dose level of 0.5 $\mu g/kg/min$. In man, an initial dose of bradykinin was infused (0.5 $\mu g/kg/min$); then the rate of infusion was increased to maintain a constant hypotensive effect. The blood pressure and other variables were recorded before and 8 to 10 min after the beginning of the infusion, i.e. when the hemodynamic conditions had reached a stable stage.

In the animal, the following parameters have been determined: systemic and pulmonary arterial pressure, recorded manometrically (Sanborn) from femoral and pulmonary arteries; consumption of total oxygen; arteriovenous difference in oxygen (systemic and the single districts) (Van Slyke-Neill); the cardiac output (Fick); and the electrocardiogram. Cerebral and coronary flow (N_2O method), the renal flow (clearance of PAH), and the hepatosplanchnic flow (indocyenine green method) have also been determined. Finally, the following data have been calculated: the general and regional vascular resistances, the oxygen consumption of the various organs, and the work and efficiency of the left ventricle.

In man, the cardiac output was determined with albumin, labeled with I^{131}. The hepatic flow was measured by the indocyanine green method.

RESULTS

General hemodynamics. Eledoisin in the doses indicated on the following page (Table 1) increased the cardiac rate of dog by 8.3% ($P < 0.05$), whereas the mean arterial pressure decreased by 8.8% ($P < 0.001$); the cardiac output increased by 19.8% ($P < 0.01$), and the total vascular resistance was clearly reduced (−23.9%; $P < 0.001$). These last three changes are statistically highly significant.

We may add that the pulmonary vascular resistance decreased considerably, whereas the pulmonary arterial pressure did not drop significantly.

With bradykinin, in the doses indicated, the mean arterial blood pressure of the dog decreased by 10% (Table 2) ($P<0.001$). In contrast, the changes in the cardiac rate, the cardiac output, and the vascular resistance were not significant. In man, the situation is quite different. Here bradykinin caused a conspicuous fall in the mean arterial pressure (−25.84%; $P < 0.001$), a moderate increase in the cardiac output (+9.04%; $P < 0.05$), and a definite decrease in the total vascular resistance (−34.06%; $P < 0.001$).

Finally we may add that bradykinin, like eledoisin, did not induce significant variations in the dog's pulmonary pressure and pulmonary resistance.

The hemodynamics and the O_2 consumption of the single visceral regions are as follows.

Eledoisin in the cerebral circulation (Table 3) caused a

TABLE 1
Systemic and Pulmonary Hemodynamics

Eledoisin	No.	(mean ± S.D.)	(M.D. ± S.E.)	(dm%)	P
Cardiac rate (beat/min)	21	b. 165.2 ± 22.5 a. 178.9 ± 38.9	+13.7 ± 6.63	+8.3	< 0.05
Mean art. pressure (mm Hg)	21	b. 140.5 ± 17.6 a. 128.1 ± 19.6	− 12.4 ± 2.03	− 8.8	< 0.001
Cardiac output (cc/min)	19	b. 2,814 ± 1,060 a. 3,371 ± 1,372	+556.7 ± 148.6	+19.8	< 0.01
O_2 diff. (A-V) (vol %)	19	b. 5.52 ± 1.96 a. 5.23 ± 1.95	− 0.29 ± 0.27	− 5.3	0.3-0.2
O_2 Consumption (cc/min)	19	b. 140.8 ± 27.3 a. 155.2 ± 29.3	+14.4 ± 5.29	+10.2	0.02-0.01
Total vasc. resist. (dyne/sec/cm^5)	19	b. 4,621 ± 1,944 a. 3,517 ± 1,495	− 1,104 ± 194	− 23.9	< 0.001
Left ventric. work (kg-M/min)	19	b. 5.5 ± 2.40 a. 6.0 ± 2.79	+0.5 ± 0.27	+9.1	0.1-0.05
Mean pulm. pressure (mm Hg)	18	b. 11.4 ± 2.38 a. 10.4 ± 2.33	− 1.0 ± 0.54	− 8.8	0.1-0.05
Pulmon. resist. (dyne/sec/cm^5)	16	b. 361.1 ± 140.4 a. 271.5 ± 119	− 89.6 ± 20.05	− 24.8	< 0.001

*S.D. = standard deviation; S.E. = standard error. M.D. = difference between the means. Dm% = % difference between the means.

TABLE 2
Systemic and Pulmonary Hemodynamics

Bradykinin (dog)	No.	(mean ± S.D.)	(M.D. ± S.E.)	(dm%)	P
Cardiac rate (beat/min)	15	b. 192 ± 36.7 a. 188 ± 35.5	− 4 ± 5.76	− 2.08	0.5–0.4
Mean art. pressure (mm Hg)	15	b. 134 ± 22.8 a. 122 ± 20.4	− 12 ± 1.32	− 9.95	< 0.001
Cardiac output (cc/min)	12	b. 2,950 ± 1,448 a. 2,833 ± 1,322	− 117 ± 186	− 3.78	0.6–0.5
O_2 Diff (A–V) (vol %)	12	b. 4.70 ± 1.27 a. 4.94 ± 1.15	+0.24 ± 0.29	+5.10	0.5–0.4
O_2 Consumption (cc/min)	12	b. 127.5 ± 36.2 a. 128.9 ± 34.5	+1.4 ± 2.64	+1.09	0.7–0.6
Total vasc. resist. (dyne/ sec/cm^5)	12	b. 4,460 ± 2,175 a. 4,134 ± 1,823	+326 ± 389	− 7.30	0.5–0.4
Left ventric. work (kg-M/min)	12	b. 5.19 ± 2.54 a. 4.74 ± 2.57	− 0.45 ± 0.29	− 8.67	0.2–0.1
Mean pulm. pressure (mm Hg)	12	b. 9.6 ± 3.62 a. 9.1 ± 3.32	0.5 ± 0.48	− 5.20	0.4–0.3
Pulm. resist. (dyne/sec/cm^{-5})	12	b. 337 ± 229 a. 350 ±284	+13 ± 159	+3.85	> 0.9
Bradykinin (man)					
Mean arterial pressure (mm Hg)	11	b. 82.8 ± 10.5 a. 61.4 ± 9.3	− 21.4 ± 2.50	− 25.84	< 0.001
Cardiac output (cc/min)	11	b. 5,175 ± 1,198 a. 5,643 ± 1,243	+408 ± 198	+9.04	< 0.05
Total vasc. resist. (dyne/sec/cm^5	11	b. 1,364 ± 443 a. 897 ± 177	− 467	− 34.06	< 0.001

TABLE 3
Cerebral Circulation

Eledoisin	No.	(mean ± S.D.)	(M.D. ± S.E.)	(dm%)	P
Blood flow (cc/min/100 g)	7	b. 48.7 ± 8.96 a. 37.9 ± 8.13	– 10.8 ± 2.53	– 22.2	< 0.01
Resistance (units)	7	b. 2.8 ± 0.46 a. 3.3 ± 0.71	+0.5 ± 0.15	+17.8	0.05–0.02
O_2 Diff. (A–V) (vol %)	7	b. 6.97 ± 2.26 a. 8.07 ± 2.20	+1.10 ± 0.67	+15.8	0.2–0.1
O_2 Consumption (cc/min/100 g)	7	b. 3.39 ± 1.32 a. 2.97 ± 0.80	– 0.42 ± 0.45	– 12.4	0.4–0.3
Bradykinin					
Blood flow (cc/min/100 g)	7	b. 46.2 ± 11.3 a. 44.8 ± 8.28	– 1.4 ± 2.62	– 3.03	> 0.9
Resistance (units)	7	b. 2.8 ± 1.05 a. 2.5 ± 0.69	– 0.3 ± 0.21	– 10.71	0.3–0.2
O_2 Diff. (A–V) (vol %)	7	b. 3.02 ± 0.30 a. 2.79 ± 0.20	+0.23 ± 0.20	+3.10	0.2–0.1
O_2 Consumption (cc/min/100 g)	7	b. 4.43 ± 1.63 a. 4.37 ± 1.05	– 0.6 ± 0.35	– 13.54	0.2–0.1

TABLE 4
Coronary Circulation

Eledoisin	No.	(mean ± S.D.)	(M.D. ± S.E.)	(dm%)	P
Blood flow (cc/min/100 g)	7	b. 85.3 ± 8.12 a. 100.5 ± 20.1	+15.3 ± 5.67	+17.9	0.05–0.02
Resistance (units)	7	b. 1.67 ± 0.141 a. 1.34 ± 0.277	− 0.33 ± 0.113	− 19.8	0.05–0.02
0_2 Diff. (A–V) (vol %)	7	b. 15.93 ± 1.42 a. 15.04 ± 1.69	− 0.89 ± 0.51	− 5.6	0.2–0.1
Myocard. O_2 consumption (cc/min/100g)	7	b. 13.5 ± 0.938 a. 14.9 ± 2.46	+1.45 ± 0.97	+10.7	0.2–0.1
Left ventric. efficiency (work/O_2 consumpt.)	7	b. 0.38 ± 0.245 a. 0.41 ± 0.240	+0.038 ± 0.041	+10.0	0.5–0.4
Bradykinin					
Blood flow (cc/min/100g)	6	b. 95 ± 13.0 a. 114 ± 12.3	+19 ± 2.58	+20.00	< 0.001
Resistance (units)	6	b. 1.48 ± 0.279 a. 1.15 ± 0.168	− 0.33 ± 0.087	− 22.29	< 0.05
O_2 Diff. (A–V) (vol %)	6	b. 13.20 ± 1.95 a. 14.58 ± 2.00	+1.38 ± 0.71	+10.45	0.2–0.1
Myocard. O_2 consumption (cc/min/100g)	6	b. 12.41 ± 1.22 a. 16.55 ± 1.92	+4.14 ± 0.83	+33.36	< 0.01
Left ventricle efficiency (work/O_2 consumpt.)	6	b. 0.37 ± 0.934 a. 0.27 ± 0.068	− 0.10 ± 0.265	− 27.03	< 0.05

TABLE 5
Hepatosplanchnic Circulation

Eledoisin (dog)	No.	(mean ± S.D.)	(M.D. ±S.E.)	(dm%)	P
Hepatic blood flow (cc/min)	7	b. 545 ± 101 a. 741 ± 173	196 ± 32.6	+36.0	< 0.001
Hepatic wedged pressure (mm Hg)	7	b. 8.4 ± 1.42 a. 9.8 ± 2.11	+1.4 ± 0.427	+16.7	0.05–0.02
Splanchnic resistance (dyne/sec/cm^{-5})	7	b. 21,586 ± 4,183 a. 15,736 ± 5,958	– 5,850 ± 777	– 27.1	< 0.001
Bradykinin (man)					
Hepatic blood flow (cc/min)	11	b. 1,229 ± 230 a. 1,345 ± 306	– 116 ± 79.5	+9.43	0.2–0.1
Hepatic wedged pressure (mm Hg)	11	b. 5.1 ± 1.8 a. 5.4 ± 2.1	+0.3 ± 0.194	+5.88	0.2–0.1
Splanchnic resistance (dyne/sec/cm^{-5})	11	b. 5,246 ± 1,362 a. 3,459 ± 856	– 1787 ± 258	– 34.06	< 0.001

TABLE 6
Renal Circulation

Eledoisin	No.	(mean ± S.D.)	(M.D. ± S.E.)	(dm%)	P
Plasma flow (cc/min)	7	b. 367.1 ± 164 a. 403.3 ± 177	+ 36.1 ± 47.2	+ 9.8	0.5–0.4
Blood flow (cc/min)	7	b. 700.3 ± 316 a. 733.8 ± 362	+ 32.6 ± 99.4	+ 4.6	0.8–0.7
Resistance (units)	7	b. 0.22 ± 0.083 a. 0.19 ± 0.062	− 0.03 ± 0.0192	− 13.6	0.2–0.1
O_2 Diff. (A–V) (vol %)	6	b. 1.60 ± 0.60 a. 1.76 ± 0.54	+ 0.16 ± 0.199	+ 10.0	0.5–0.4
O_2 Consumption (cc/min)	6	b. 9.75 ± 4.18 a. 12.77 ± 6.44	+ 3.02 ± 2.65	+ 31.0	0.3–0.2
Bradykinin					
Plasma flow (cc/min)	6	b. 435 ± 95 a. 430 ± 217	− 5 ± 66.9	− 1.14	>0.9
Resistance (units)	6	b. 0.163 ± 0.04 a. 0.179 ± 0.09	− 0.016 ± 0.030	− 9.81	0.7–0.6
O_2 Diff. (A–V) (vol %)	6	b. 2.42 ± 0.975 a. 1.98 ± 1.22	− 0.44 ± 0.34	− 18.18	0.3–0.2
O_2 Consumption (cc/min)	6	b. 20.92 ± 8.28 a. 12.87 ± 6.13	− 8.05 ± 3.35	− 38.47	0.1–0.05

significant decrease in the flow (−22.2%; $P < 0.01$) with a corresponding equally significant increase in vascular resistance (17.8%; $0.05 > P > 0.02$), whereas the oxygen consumption did not change signficantly.

Bradykinin (Table 3) did not cause statistically significant changes in the cerebral flow, the cerebral vascular resistance, or the oxygen consumption of the brain.

The coronary circulation (Table 4) showed after the administration of eledoisin an increase of 17.9% ($0.05 > P > 0.02$) in the flow, whereas the vascular resistance decreased by 19.8% ($0.05 > P > 0.02$). On the other hand, the oxygen consumption and the cardiac efficiency were not altered significantly.

Administration of bradykinin (Table 4) increased the coronary flow by 20% ($P < 0.001$), while the vascular resistance decreased by 22.29% ($P < 0.05$). The myocardial oxygen consumption showed a significant increase (+33.36%; $P < 0.01$), and the cardiac efficiency was reduced (−27.03%; $P < 0.05$).

The hepatosplanchnic circulation (Table 5) of the dog showed after the infusion of eledoisin a conspicuous increase in the flow (36%; $P < 0.001$) and a slight mean increase in the wedged pressure (+16.7%; $0.05 > P > 0.02$) whereas the splanchnic vascular resistance decreased markedly (−27.1%; $P < 0.001$).

The results obtained with infusion of bradykinin in man, however, appear to be absolutely different. While the splanchnic vascular resistance decreased (34.6%; $P < 0.001$), the flow and wedged pressure did not change remarkably. It should be pointed out, however, that while in man the decrease in the mean arterial pressure reached 25%, in the dog it was only 9%.

In the renal circulation (Table 6) of the dog, eledoisin and bradykinin induced no statistically significant changes in the flow or in the resistance. Nor did the renal oxygen consumption change significantly.

CONCLUSIONS

In conclusion, with the reservations necessary for this kind of experiments in which so many variables can interfere (dose, species, method of administration, type of anesthesia, etc.) it can be said that eledoisin and bradykinin lower the arterial blood pressure markedly. Eledoisin is much more potent than bradykinin, as shown by the different amounts of the peptide employed. There are, however, some qualitative differences — for example, in the effect of the peptides on the cardiac output. Finally, some additional differences between the two substances were observed with regard to the regional circulatory changes (cerebral, coronary, and hepato-splanchnic circulation).

DISCUSSION

Erspamer: Non stupisce che il Prof. Beretta abbia trovato nei suoi esperimenti non solo differenze quantitative ma anche differenze qualitative fra eledoisina e bradichinina. Si tratta infatti di due polipeptidi che, all'infuori dell'azione ipotensiva, non molto hanno in comune. Le differenze nei risultati ottenuti dai vari Autori che si sono occupati dell'azione della eledoisina sui vari distretti vascolari sono in large misura ascrivibili, come giustamente ha affermato il Prof. Beretta, a differenza di metodica e di dose.

P. F. MANNAIONI

Department of Pharmacology, University of Florence

My question is whether, in the heart, the effect of bradykinin is mediated through the release of catecholamines or is accounted for by an interaction with specific receptors.

It is known that bradykinin releases catecholamines from adrenal medulla (J. Stassewska-Barczak, J. R. Vane: J. Physiol. (Lond.) 57P 177, 1965). Since bradykinin stimulates the isolated perfused guinea pig heart, it seemed conceivable that this effect is mediated through a release of catecholamines. In our experiments (Fig. 1), the stimulating activity of bradykinin on the heart could only partially be accounted for by a release of catecholamines. In the presence of a β-receptor blocking agent (dichloroisoproterenol), in a concentration high enough to inhibit the effect of norepinephrine completely, the stimulating activity of bradykinin, although partially blocked, is still evident.

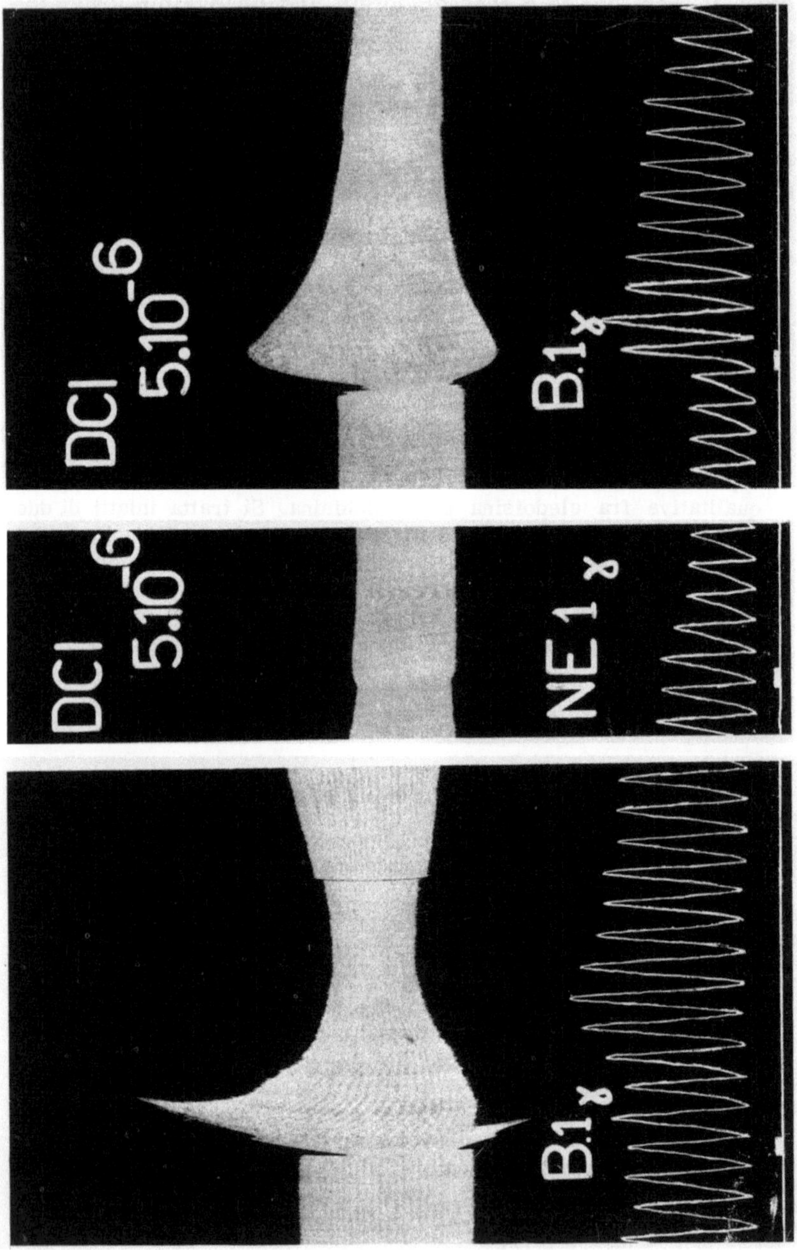

Figure 1

THE EFFECTS OF BRADYKININ AND SUBSTANCE P ON THE SPECIALIZED VENTRICULAR CONDUCTING SYSTEM OF DOG HEART

R. L. VICK, W. KRIVOY, and D. C. KROEGER

Departments of Physiology and Pharmacology, Baylor University College of Medicine, and Department of Pharmacology, University of Texas Dental Branch, Houston, Texas, U.S.A.

The wide range of effects attributed to the biologically active polypeptides indicates fundamental actions upon excitable cells, and thus suggests the possibility that some of these substances might affect cardiac tissue. With this in mind, we studied the effects of bradykinin and of 6-glycine bradykinin on hydrocarbon-epinephrine cardiac arrhythmia in dogs. It was found that bigeminy, or coupled ventricular beats, evoked in this situation could temporarily be restored to normal sinus rhythm, apparently by direct action on the heart (Vick et al., 1965).

This communication is a report of progress on a study of the effects of several polypeptides on the transmembrane potential of single cardiac cells. It was undertaken partly to find the mechanism of the antiarrhythmic action of bradykinin, but also to determine the effects of these compounds on the excitable process in heart tissue. It is a report of the failure to find marked actions of bradykinin on normal cardiac cells after it had appeared that these existed, and of the discovery of a new cardiotropic action of crude substance P.

MATERIALS AND METHODS

Isolated cardiac tissues from mongrel dogs were used in these experiments. After anesthesia with pentobarbital sodium

(30 mg/kg i.v.), the chest was opened and the heart rapidly removed. The right ventricle was opened and the anterior papillary muscle and attached free-running strand of Purkinje tissue (false tendon) was dissected out under a steady flow of oxygenated physiological solution. (Composition of the solution in m moles per liter: NaCl 122, KCl 2.7, $CaCl_2$ 2.0, $MgSO_4$ 1.2, NaH_2PO_4 1.2, glucose 5.5. Saturation with 95% O_2-5% CO_2 resulted in a pH of 7.4.) The tissue was mounted in 15 ml of the same solution, which was oxygenated and maintained at 37°C. In some experiments the tissues were separated and either the false tendon or the papillary muscle was studied.

KCl-filled microcapillary pipettes, selected for impedances between 10 and 20 megohms, were mounted on a micromanipulator and connected by chlorided silver wire to a cathode-follower input. A silver-silver chloride connection to ground was placed in the bath. Penetration of single cells was accomplished with the aid of a 20× binocular dissecting microscope. The signals were amplified by a Grass P6 preamplifier and a Tektronix 502 cathode-ray oscilloscope. Records were obtained by photographing the face of the cathode-ray tube.

In those experiments in which the configuration of the action potential was studied, the preparation was stimulated repetitively with single shocks through bipolar platinum electrodes insulated except at the tips. Biphasic stimuli of one msec duration and 1.5 times threshold were used at a frequency sufficient in each experiment to prevent spontaneous activity in the false tendon. The frequency of stimulation varied among different experiments from about 12 to 60/min, but was kept constant within a given experiment.

In other experiments the preparation was permitted to undergo spontaneous excitation, and effects upon the generation of excitation were studied.

The drugs used were: synthetic bradykinin (Sandoz), obtained in solutions of 80 or 100 μg/ml in sealed ampules; synthetic bradykinin triacetate (California Biochem.), obtained as a powder and dissolved in water at a concentration of 1 mg/ml before being used. The stock solutions were not diluted, and unused portions, kept in the cold, were discarded after 2 weeks. Bioassay on rat uterus showed no decrease in potency during that period. Chlorobutanol, USP, obtained in crystalline form, was used in two types of solutions at concentrations of 5 mg/ml: (1) dissolved in water, and (2) dissolved in a buffer solution containing acetic acid, 0.5 g/liter; sodium acetate $.3H_2O$,

2 g/liter; and NaCl, 7 g/liter. Crude substance P, extracted from hog intestine and assayed at 10 u/mg, was obtained as a powder and dissolved in the physiological solution at a concentration of 10 mg/ml immediately before use. The powder was stored at 0°C, and unused solutions were discarded.

RESULTS AND DISCUSSION

Bradykinin. Initially, only the synthetic bradykinin obtained in solution was used. This solution produced marked effects on specialized ventricular conducting (Purkinje) tissue. In preparations undergoing rhythmic stimulation, the duration of the action potential was decreased so that the tissue became fully re-excitable after a shorter period of refractoriness. In cells undergoing pronounced decline of the potential during diastole (pacemaker prepotential), the transmembrane potential immediately after the action potential was decreased in magnitude, and the pacemaker prepotential was reduced. In spontaneously active Purkinje cells this effect was seen as stabilization of the diastolic potential so that automaticity was reduced. These effects were seen after addition of the bradykinin solution sufficient to yield bath concentrations as low as 5×10^{-8} g/liter, but were more commonly seen with concentrations of 10^{-7} to 10^{-6} g/liter.

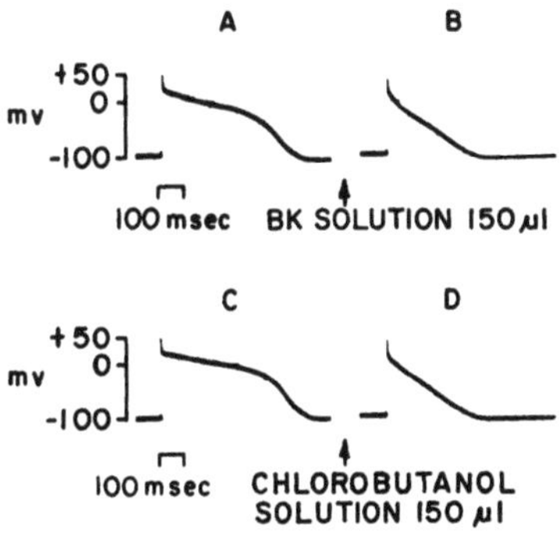

Fig. 1. Effects of bradykinin solution containing chlorobutanol (5 mg/ml) and of a chlorobutanol solution (5 mg/ml) on transmembrane potential of dog Purkinje cells. See text for discussion.

At this time it was pointed out that the vehicle in which bradykinin was contained might not be pharmacologically inactive in the amounts being used. A buffer solution identical with that used as a vehicle for the bradykinin, when added to the bath, appeared to produce all the effects that the bradykinin solution did. Since chlorobutanol, present in the vehicle as a preservative, seemed most likely to be an active constituent, a solution of that compound in water (5 mg/ml) was also tested. This solution also produced all the effects that the bradykinin solution did. An example of this action is shown in Fig. 1. After (A) 150 μ liters of the bradykinin solution was added to the bath, resulting in a concentration of 10^{-6} g/liter, and producing (B) a 71% reduction in action potential duration (at the level of 50% repolarization), and an average decrease in diastolic potential of 4 mV. After the bath was washed out, bradykinin triacetate, 10^{-6} g/liter, was added to the bath without effect. Again after washout, 150 μ liters of chlorobutanol solution produced effects (D) similar to those of the bradykinin solution: 73% reduction of action potential duration, and 3-mV average decrease in diastolic potential. Each of the two active solutions contained in common 5 mg chlorobutanol/ml.

In additional experiments with spontaneously active prep-

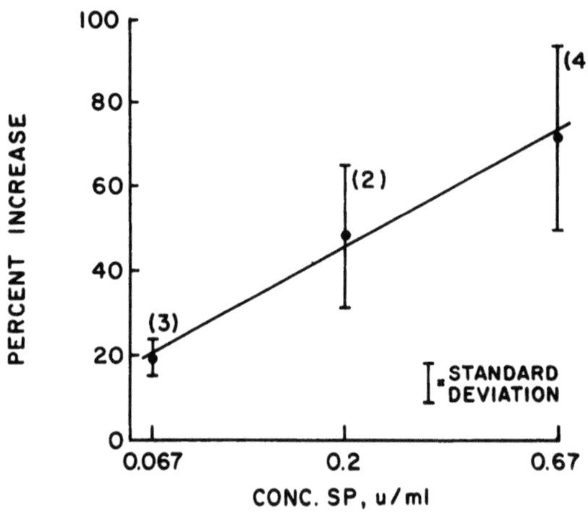

Fig. 2. Effects of crude substance P (10 u/mg) on frequency of spontaneous excitation in dog Purkinje tissue. The numbers in parentheses give the number of experiments at each point.

arations of dog Purkinje tissue, it was demonstrated that most or all of the effects in that preparation, also previously attributed to bradykinin, were actually produced by the chlorobutanol present in the vehicle.

Substance P. Initial investigation of the effects of another substance has been more fruitful. The effects of a sample of crude substance P (SP) have been studied on the specialized ventricular conducting (Purkinje) tissue. SP, in concentrations from 0.067 to 0.67 u/ml, increased the frequency of spontaneous activation in Purkinje tissue. The effect was rapid in onset, and it persisted in the tissue bath for more than 15 min. The effect appears to be concentration related, with average increases in frequency of about 20% to 70% in these concentrations (Fig. 2). Typical data recorded from a cell with a high degree of automaticity are shown in Fig. 3. (A) and (B) are controls, (C) and (D) were taken after the addition of SP. It can be seen that the rate of decline of the potential during diastole is increased (comparing A and C). (B) and (D), taken at faster sweep speeds immediately after (A) and (C), respectively, show more clearly the changes in action potential configuration. A possible mechanism of the increased rate is indicated. The diastolic potential is decreased (107 mV maximum in B, to 98 mV maximum in D), and this may be the principal contributor to the increased pacemaker activity. Not only is the diastolic potential nearer to threshold, and hence requires less depolarization to reach threshold but, if no other factor is involved, it is thereby made less stable, i.e. will more readily undergo spontaneous depolarization (Hoffman and Cranefield, 1960).

The data in Fig. 4 were taken from an experiment in which a higher concentration of SP was used. They give indication of

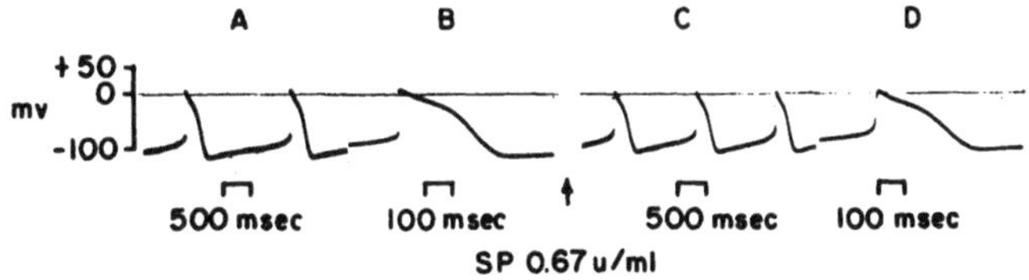

Fig. 3. Effects of substance P on transmembrane potential of spontaneously active dog Purkinje cells. See text for discussion.

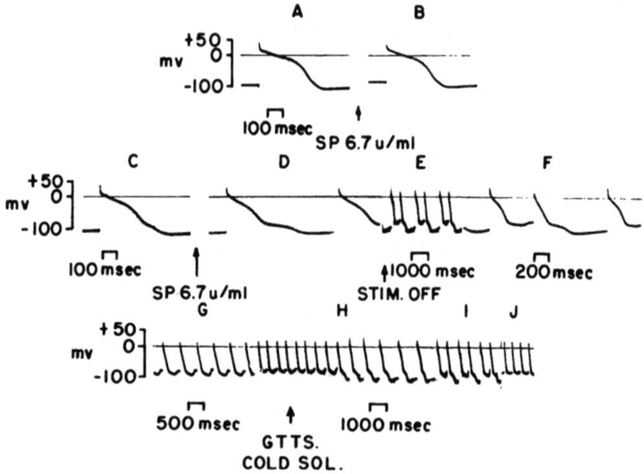

Fig. 4. Effects of a higher concentration of substance P on transmembrane potential of dog Purkinje cells. See text for discussion.

what might be another mode of action of SP on the excitable membrane in automatic cells of the specialized ventricular conducting system of dog. This experiment was begun in an attempt to study changes in action potential caused by SP independently of changes in rate. Hence, the preparation was stimulated regularly at a frequency exceeding the intrinsic frequency. Addition of SP (between A and B) decreased the maximal diastolic potential (from 112 to 106 mV), the diastolic potential immediately before excitation (from 100 to 91 mV), and the action potential duration slightly. Even though the preparation was being driven, increased automaticity could be detected by the occasional loss of control of the rate of excitation. After 15 min no additional changes had occurred, and the bath was washed out. Upon repenetration of a cell in the same immediate area, an altered configuration of the action potential was seen, which had not been recorded during the preceding period when SP was in the bath. This consisted of a small hump near the completion of repolarization, which persisted throughout a 17-min control period (C). Within 2 min after restoring SP to the bath, in the same concentration as before, the spontaneous rate exceeded that of the stimulator, and control could be maintained only by increasing the rate of stimulation. Fifteen minutes after the addition of SP the hump was more prominent

(D), and then an action potential arose from it (E, F), causing coupled beats. Note the differences in the time scale in the remaining sequences. The data were recorded at whatever speed seemed most appropriate to the events taking place.

At this point attempts to control the frequency were abandoned, and the preparation was allowed to beat spontaneously (E, F). Note in (F) that the negative wave follows each action potential, the regular as well as the coupled one, although, as yet, only the basic beat is followed by a coupled one. Within another 2 min the preparation was being excited regularly, with each after-potential wave proceeding into re-excitation (G). Eventually, each action potential followed so closely upon the preceding one that the after-potential from which it arose was not evident, and a typical pacemaker prepotential appears to have developed (H). Repolarization was interrupted so early in the course of recovery from the preceding excitation that a maximal potential of only 76 mV was reached during diastole, as compared with a maximum of 113 mV before the SP was readministered (C), and of 104 mV after the bigeminal rhythm first began (E).

The continued origin of the action potentials from after-potential waves was confirmed by adding a few drops of cold (0°C) physiological solution near the tissue, producing localized cooling and sufficient slowing of the frequency that repolarization was completed before re-excitation ensured (H). Under these conditions a maximal diastolic potential of 104 mV was reached, and the after-potential wave reappeared. In a few seconds, excitation again occurred with every second after-potential wave (I), and finally, the regular, rapid rhythm returned with a maximal repolarization during diastole of 76 mV (J). This maneuver demonstrated that the low diastolic potential was not caused by persistent depolarization, but that repolarization was interrupted by the depolarization process associated with the after-potential.

This effect of SP appears to be consistent with the findings of Caspers (1961), and of Krivoy (1961) and Krivoy and Kroeger (1963), of enhancement of residual negativity in brain and spinal cord, respectively. It is reminiscent of the veratrine effect in nerve and muscle (Kuffler, 1945), but differs in that with SP the after-potential is more clearly a second wave, whereas veratrine delays repolarization in Purkinje cells by prolonging the entire action potential until re-excitation interrupts it (Vick, unpublished results).

It cannot now be determined whether the appearance of a discrete after-potential, eventually progressing through a phase of coupled beats to a constant fast rate, is related to the effect of a smaller concentration of SP to produce a smooth rate increase. At this time we can only report this evidence of a profound effect upon the excitable membrane of cardiac Purkinje tissue by relatively crude SP. Additional study, particularly with higher potency SP, is needed to further elucidate the nature of this action.

SUMMARY

After the demonstration of temporary antagonism of chloroform-epinephrine-induced cardiac bigeminy by bradykinin (BK) and 6-glycine BK, a study of the effects of several polypeptides on the transmembrane potential of single cardiac cells was begun, which was undertaken to find the mechanism of the antiarrthythmic action of BK, as well as to determine the effects of the peptides on the excitable process in cardiac tissues.

Using only a solution of BK supplied in sealed ampules (Sandoz), marked effects were found in specialized ventricular conducting (Purkinje) tissue, which could be interpreted in terms of current theories to account for the antiarrhythmic action of BK. However, the observed effects were found to be attributable to the chlorobutanol (5 mg/ml) used in the solution as a preservative, and the effects of BK upon isolated cardiac tissues under normal conditions and in concentrations up to 10^{-6} g/liter are not marked. Further work is planned on isolated cardiac tissue under laboratory conditions more closely resembling those under which chloroform-epinephrine arrhythmias develop <u>in vivo</u>.

A marked effect of crude substance P (SP) on the generation of spontaneous activity in cardiac Purkinje tissue was observed in concentrations of 0.067 to 0.67 u/ml. Increases in spontaneous frequency of 20% to 70% occurred. In cells with marked pacemaker activity (spontaneous depolarization during diastole) the diastolic potential was reduced and the rate of spontaneous depolarization increased. At a concentration of 6.7 u/ml SP a prominent negative after-potential wave developed and increased in magnitude until it gave rise to excitation coupled with the preceding beat. Eventually, every action potential arose in this fashion, from the after-potential of the preceding action potential. The result was a rapid, spontaneous discharge with a low diastolic value of the membrane potential as re-excitation occurred before repolarization was completed.

Further study of this phenomenon with higher potency SP is planned.

ACKNOWLEDGMENTS

This research was supported in part by United States Public Health Service Grants HE 08372, HE 05435, and MH 3477. Synthetic bradykinin was kindly supplied by Dr. R. P. Bircher of the Sandoz Pharmaceutical Co., and substance P by Mr. D. Broida of the Sigma Chemical Co.

REFERENCES

Caspers, H: Some actions of substance P on the cerebral cortex and the brain stem reticular formation. Proc. sci. Soc. Bosnia and Herzegovina 1, 29 (1961).

Hoffman, B. F., and P. F. Cranefield: Electrophysiology of the heart. New York: McGraw-Hill (1960).

Krivoy, W.: A comparison of the action of substance P and other naturally occurring polypeptides on spinal cord. Proc. sci. Soc. Bosnia and Herzegovina 1, 131 (1961).

Krivoy, W., and D. C. Kroeger: The neurogenic activity of high potency substance P. Experientia (Basel) 19, 366 (1963).

Kuffler, S. W.: Action of veratrine on nerve-muscle preparations. J. Neurophysiol. 8, 113 (1945).

Vick, R. L., W. Krivoy, and D. C. Kroeger: The antagonism of chloroform-epinephrine induced cardiac arrhythmia by bradykinin and 6-glycine bradykinin. Arch. int. Pharmacodyn. 156, 100 (1965).

DISCUSSION

Zetler: Are these actions of substance P resistant to proteases? In unpublished experiments I have found that crude substance P, made from cattle brain, has positive inotropic and chronotropic actions on the isolated guinea pig atrium. These actions were resistant to trypsin and chymotrypsin, and are therefore due to unspecific impurities.

Vick: We have found in limited study that exposure of the material to α-chymotrypsin did not remove the activity. However, the results of our assay of substance P in the solution after enzymatic treatment have not been clear, and we cannot yet be sure if substance P activity remained. We are interested in this and will continue the study, for, to an electrophysiologist of the heart, this is a very exciting effect, and I would like to find the material that causes it, whether it be substance P or something else.

Collier: The authors have provided a good example of an effect that might have been attributed to bradykinin but was actually due to the chlorobutanol present at 50 times the concentration of bradykinin in the ampules used. We have had a somewhat similar experience in studying writhing response of mice to intraperitoneal injection of bradykinin. We found that chlorobutanol itself elicits a writhing response after a long latency. Bradykinin itself and not chlorobutanol, however, elicits writhing of very short latency when injected intraperitoneally into mice.

MICROCIRCULATORY EFFECTS OF POLYPEPTIDES

BENJAMIN W. ZWEIFACH

Department of Pathology, New York University Medical Center, New York, N. Y., U.S.A.

It has long been appreciated that tissue homeostasis is mediated by chemical agents which arise in the blood and the tissues proper and serve to increase or decrease blood flow locally in accordance with the metabolic needs of the tissue. The blood vascular elements respond to a wide variety of biologic materials that exist either in a free-form state in tissues or are produced as a result of chemical activity. These range from substances that dilate or constrict vessels, to those that potentiate or inhibit the response to other stimuli, and finally to those that affect the permeability of the barrier that separates the blood and extravascular compartments (Spector, 1958). Among the many chemical agents in the blood that exhibit vasoactive properties are polypeptides such as vasopressin and angiotensin. There is some question whether these agents have a particular function in normal homeostasis. Other vasoactive polypeptides appear to be present in tissues in extremely small amounts under normal circumstances and are detected in body fluids in substantial concentrations only after tissue injury, systemic shock-like reactions, immune responses, etc. Among these substances is bradykinin, which is known to have a direct action on vascular smooth muscle in vitro and to lower the blood pressure when injected intravenously into animals.

Local injection of bradykinin into the skin produces a typical hyperemia and tissue edema associated with an increased permeability to plasma proteins (Lewis, 1960). No systemic study of the vasomotor action of bradykinin has been carried out on the level of the microcirculation. The present study was undertaken with this in mind and to establish the extent to which vascular sequelae observed in the intact animal were the result of a direct or an indirect action of the polypeptide.

The method of study was direct microscopy of mesentery and skin vessels in anesthetized animals (Zweifach and Metz 1956). Mesenteric observations were carried out on the rat, guinea pig, and rabbit. It was also possible to study the responses of the skin vessels in the unanesthetized mouse, through a modified transparent skin chamber installed on the back of the animal. In order to provide a basis of comparison, the studies included the pattern of response to a series of smooth-muscle constrictor or dilator polypeptides in the same type of preparations.

VASOACTIVE POLYPEPTIDES

As can be seen in Table 1, polypeptides act both as constrictor and dilator agents. As will be indicated later, vasoactive polypeptides are biologically active in extremely small concentrations. In addition to their effect on vascular smooth muscle, agents of this type produce a significant increase in permeability to plasma proteins when they are injected into the skin of the rabbit. It should be noted that the permeability-

TABLE 1.
Vasoactive Polypeptides

Substance	Vasomotor	Permeability* local
Angiotensin II	vcs	++
Pitressin	vcs	0
PLV[2]	vcs	0
Bradykinin	vdil	+++
Substance P	vdil	+
Eledoisin	vdil	+
Lysosomal basic protein	vdil	++

*Intradermal response in rabbit; vcs = vasoconstriction, vdil = vasodilation.

increasing capacity of these materials does not appear to be related to their vasomotor activity, and, as will be shown later, the two effects can develop separately. Included in the table is a basic cationic protein obtained from polymorphonuclear leukocytes, which is believed to be a polypeptide (Janoff and Zweifach, 1964). Its vasomotor and permeability effects were produced by as little as 2 to 5 μg, despite the fact that the product used was still relatively impure.

BRADYKININ—MICROCIRCULATORY EFFECTS

An extensive study was made of the vascular effects of bradykinin in view of its proposed role in tissue injury (Rocha e Silva, 1964). As indicated in Table 2, topical application of bradykinin produces within a few seconds a pronounced increase in flow through the capillary vessels. Diameter measurements of the inflow and outflow channels show that the muscular venules are dilated to a proportionately greater extent than are the feeding arterioles; for example, a venule with a cross-sectional diameter of 50 μ under control conditions may

TABLE 2.
Vascular Effects of Bradykinin in Mesentery of Rat

Pretreatment (i.v.)	B'kinin topical (μg)	Vasomotor*		Permeability† local
		Arteriole	Venule	
Control	0.01	0	00	+++
Phenoxybenzamine	0.01	0	00	++
Chlorpromazine	0.001	00	000	++
Chlorpheniramine	0.01	0	0	+
48/80 acute	0.01	00	00	++
chronic	0.001	00	000	++
Adrenx + NaCl	0.01	0	00	+++
Cortisone	0.10	0	0	+

*Dilation graded from 0 (moderate) to 000 (severe).
†Intradermal blueing, 10^{-4} μg i.d.

increase in size to 70 to 75 μ. Arterioles, on the other hand, show only about a 10% increase in diameter. The dilation is accompanied by an increase in the number of capillaries with an active blood flow through the opening of additional precapillary sphincters. The precapillary effect appears after the rate and volume of blood flow have increased and hence may be secondary in nature. In a small number of experiments the polypeptide was applied directly into the tissue in the vicinity of the precapillary, by means of a micropipette, and amounts as small as 10^{-4} μg produced a dilation of the precapillary. Bradykinin, therefore, is an effective dilator of vascular smooth muscle in different parts of the capillary bed. It is not clear why muscular venules are affected earlier and to a greater extent than are the arterioles and precapillaries. On the other hand, other dilator polypeptides exhibit a similar predilection for the muscular venules.

In view of the fact that blood flow will change as the fourth power of the radius of the vessel, even slight changes in the caliber of the venules have a disproportionately large effect on capillary circulation. When amounts of bradykinin below 10^{-3} to 10^{-4} μg are applied topically, only vasomotor changes are observed. These persist for some 3 to 5 min and then wear off. When higher concentrations are used (10^{-1} - 10^{-2} μg), the vasomotor effects are much more protracted, lasting up to 20 to 30 min or longer, and indications of increased permeability develop in the postcapillaries and collecting venules. This is manifest by a tendency for red blood cells to clump together and to form large masses which eventually result in complete stasis in such vessels. The stasis then builds up in a retrograde manner until many of the capillaries proper are affected. Increased sticking of leukocytes is seen in the mesentery only when comparatively high concentrations of bradykinin are applied, for example, 10 to 10^{-1} μg. Such amounts of bradykinin lead to vascular stasis and impaired local blood flow for periods of an hour or longer in the mesentery. Although the data listed in Table 2 represent observations in the rat, comparable findings were obtained in the rabbit and guinea pig mesentery. In experiments in which bradykinin was applied to the exposed undersurface of the skin in the mouse, by the transparent chamber method, as little as 10^{-5} to 10^{-6} μg produced both vasomotor effects and changes in capillary permeability. Sticking of leukocytes and their emigration could be observed with concentrations of 10^{-2} to 10^{-4} μg.

BRADYKININ—PERMEABILITY EFFECTS

The last column in Table 2 lists the effects on capillary permeability as measured by the systemic injection of Evans blue dye and noting the blued area of the skin. The response was graded according to the size of the lesion after a standard 15-min period. The most striking permeability changes were obtained in the rabbit and guinea pig; the rat skin was the least susceptible to this type of reaction.

Selected pharmacologic blocking agents were used as a means of establishing whether bradykinin acted through recognizable receptor sites or possibly had an indirect action through the release of other vasoactive agents. As can be seen in Table 2, phenoxybenzamine, which effectively blocks the constrictor response to catecholamines and serotonin in this type of preparation, did not affect the vasomotor response, although there was some indication that the increased permeability in the skin was modified and somewhat reduced. A ganglionic blocker, such as chlorpromazine, actually increased the response of the vascular smooth muscle, and comparable dilator effects could be obtained with much smaller concentrations. Here again, there was no major interference with the local increase in permeability in the skin after bradykinin. Systemic administration of the antihistamine chlorpheniramine did significantly attenuate any of the reactions produced by bradykinin, although a positive reaction was still present even when amounts of chlorpheniramine were given which completely blocked the local effects of histamine. This seems to indicate that polypeptides of this kind, in addition to a direct action, probably bring about the release of histamine from tissue sources. In this regard, it should be pointed out that the mesentery of the rat is especially rich in mast cells. In the rabbit mesentery, chlorpheniramine had only a moderate suppressive action so that the response to bradykinin was essentially similar to that observed under control conditions.

BRADYKININ—HISTAMINE-DEPLETED ANIMALS

Various experiments were also carried out in animals which had been pretreated with compound 48/80, which is known to deplete the tissue stores of histamine, presumably by releasing material from mast cells (Riley and West, 1955).

Both acute and chronic experiments were done. In the acute preparations, 500 μg of 48/80 was given, and some 2 to 3 hours later the vascular response of the mesentery to bradykinin was observed. No differences from the control were noted. In several experiments the 48/80 was given twice daily for 3 days and the reactions to bradykinin observed on the fourth day. In these preparations the vasomotor responses were considerably exaggerated, again pointing to the fact that this polypeptide has as its major mode of action a direct effect on vascular smooth muscle.

ADRENALECTOMY

Results essentially the same as in controls were obtained in rats which had been subjected to bilateral adrenalectomy and maintained on 1% salt in their drinking water. There was some indication that in adrenal-insufficient animals the local increase in permeability was more pronounced, as evidenced by a much more extensive blueing of the local skin site.

CORTISONE

The administration of cortisone in amounts ranging from 1 to 5 mg for from 1 to 3 days served to blunt markedly the dilator aspect of the response to topical bradykinin. It should be pointed out that cortisone-treated animals are more responsive to the vasoconstrictor action of catecholamines and that the blood flow in such preparations is actually reduced below control levels (Zweifach, 1961). The response of the skin vessels to intradermal bradykinin was likewise reduced, with only a small area showing a comparatively light blue color.

In view of the possible influence of an increased constrictor tone on the dilator response to bradykinin, experiments were carried out in which norepinephrine was applied locally to produce an extensive closure of most of the precapillaries and a marked ischemia which, with amounts of 4-5 μg norepinephrine, persisted for some 4 to 5 min. When bradykinin was then applied to the mesentery during the period of maximal ischemia and vasoconstriction, a vasodilatation developed within 20 to 30 sec, indicating that bradykinin apparently could produce its effects even when the receptor sites were being stimulated by a catecholamine.

SYNERGISM—BRADYKININ AND SEROTONIN

In addition to its direct action, bradykinin was found to exhibit an interesting synergism with the amine serotonin or 5-hydroxytryptamine (5-HT). Table 3 shows that this synergism was manifest in terms of the vasomotor response of the arterioles and venules as well as in damage to the capillaries and venules which developed increased permeability and stasis. An especially prominent feature was the appearance of petechial hemorrhages in the collecting venules when mixtures of serotonin and bradykinin were present. This type of damage could be produced even when mixtures of the two substances were introduced with a micropipette locally, adjacent to a capillary or venule, suggesting a direct action. Furthermore, the two substances seemed to act additively, since one could introduce first serotonin and after an interval of some 30 to 60 sec, bradykinin; the reverse sequence, bradykinin and then serotonin, was also effective. The reaction seems to be a combination of venular spasm (which presumably leads to an increased capillary pressure) and a direct effect on the endothelial barrier. When one studies the endothelial response under high magnification, it can be seen that such mixtures produce

TABLE 3.
Vasoactive Amines and Bradykinin—Synergistic Action

Mixture applied locally (μg)			Rat mesentery*			
Nor.	5-HT	B'kinin	Arteriole	Venule	Petechiae	Stasis
1.0			++	+	none	none
	0.1			+	none	none
		0.001	0	00	none	none
1.0 +	0.1		++	+++	many	some
1.0		+ 0.001	+	0	none	some
	0.1 +	0.001	0	++	many	extensive

* +, ++, +++ = degree of vasoconstriction; 0, 00, 000 = degree of vasodilatation.

a swelling of the cell. This synergistic interaction could not be suppressed with systemic administration of the antihistamine chlorpheniramine.

BRADYKININ—SHWARTZMAN PHENOMENON

In view of the fact that a reaction analogous to the local Shwartzman phenomenon can be produced in the skin of rabbits by a combination of vasoactive amines and endotoxin (Thomas, 1956), a similar set of studies was carried out with various combinations of bradykinin. These data are summarized in Table 4. When amounts of epinephrine as small as 1 to 10 μg are injected into the skin, extensive hemorrhagic necrosis develops within 3 to 4 hours in animals which have received endotoxin either intravenously just prior to the skin test or directly into the skin site. To some extent serotonin can reproduce this phenomenon, but only in a small percentage of the rabbits. Where serotonin gave a positive response, the hemorrhagic interaction was just as marked and as extensive as that with the catecholamines. As can be seen in Table 4, bradykinin could not be used in combination with endotoxin to elicit

TABLE 4.
Skin Reaction—Rabbit

Intradermal (μg)	Hemorrhagic necrosis
Endotoxin + epinephrine (100) (10)	+++
Endotoxin + 5-HT (100) (10)	±
Endotoxin + B'kinin (100) (1.0)	0
Epinephrine + 5-HT (10) (10)	+
Epinephrine + B'kinin (10) (1.0)	0
5-HT + B'kinin (10) (1.0)	+

a similar type of increased permeability and local hemorrhage. However, mixtures of serotonin and bradykinin injected into the skin not only produced an extensive area of blue dye escape, but in 20% to 30% of the animals gave rise to local hemorrhage and central necrosis.

PLV_2

In view of the current interest in other polypeptides which have constrictor properties, some of these data are included in the present report. The polypeptide most extensively studied was PLV_2, an analog of vasopressin (2-phenylalanine-8-lysine vasopressin) (Boissonnas et al., 1961). The vascular responses in the mesentery of the rat are shown in Table 5. It can be seen that PLV_2 also has an unusual propensity for venules. The same set of pharmacologic blocking agents was used against both peptides, with no evidence of any suppression of the constrictor action of PLV_2. In fact, animals treated with ganglionic

TABLE 5.
Effect of Blocking Agents on
Action of Polypeptides of Vasopressin Type

	Vasopressin (0.01 p.u.)		PLV_2 (0.05 μg)	
Pretreatment*	Arteriole	Venule	Arteriole	Venule
Control	++	++	+	++
Chlorpromazine (2.5 mg)	+++	+++	++	+++
Phenoxybenzamine (40 μg)	++	++	+	++
Chlorpheniramine (5.0 mg)	++	++	+	++
Cortisone (5 mg)	+	+	±	+

*Per 1,500 g body weight; +, ++, +++ = vasoconstriction (diameter changes).

blocking agents showed an enhanced venular constriction and ischemia when PLV_2 was applied to the mesentery. Pretreatment of the rat with cortisone suppressed the constrictor action of PLV_2 and markedly reduced the constriction in the venules.

PLV_2, when injected into the skin, has only a minimal effect on permeability, as is evidenced by blue dye escape, and no synergistic interaction in terms of vascular damage or increased permeability could be demonstrated when the polypeptide was used in combination with either serotonin or norepinephrine.

SUMMARY

Evidence has thus been presented that vasoactive polypeptides have two types of effect — an action on vascular smooth muscle and a permeability-increasing effect on the endothelial barrier. The vasomotor action appears to be largely a direct one, although there is some evidence that vasoactive polypeptides of this type may also release histamine. The permeability-increasing action of bradykinin is independent of histamine release. This is in contrast to the permeability-increasing action of other polypeptides, such as that derived from the lysosomes of leukocytes, which is the direct result of histamine release (Janoff and Zweifach 1964). The polypeptides appear to act on the endothelial cell, since swelling is manifest during the period of permeability changes and probably is accompanied by a separation of cells from one another. It was not possible to determine whether these polypeptides have a direct action on the basement membrane proper. An effect on leukocyte and platelet sticking was present only when comparatively high concentrations of the polypeptides were used.

ACKNOWLEDGMENTS

Supported by Grant HE 02267 from the United States Public Health Service, and a Career Investigatorship Award from the Health Research Council of the City of New York (I-141).

REFERENCES

Boissonas, R. A. , St. Gutman, B. Berde, and H. Konzett: Relationship between the chemical structures and biological properties of the posterior pituitary hormones and their synthetic analogues. Experientia (Basel) 17, 377 (1961).

Janoff, A., and B. W. Zweifach: Production of inflammatory changes in the microcirculation by cationic proteins extracted from lysosomes. J. exp. Med. 120, 747 (1964).

Lewis, G. P.: Active polypeptides derived from plasma proteins. Physiol. Rev. 40, 647 (1960).

Riley, J. F., and G. B. West: Tissue mast cells. Studies with a histamine liberator of low toxicity (Compound 48/80). J. Path. Bact. 69, 269 (1955).

Rocha e Silva, M.: The participation of substances of low molecular weight with special reference to histamine and bradykinin, in Symposium on Injury, Inflammation and Immunity, ed. L. Thomas, J. W. Uhr, and L. Grant, p. 220. Williams & Wilkins, Baltimore (1964).

Spector, W. G.: Substances which affect capillary permeability. Pharmacol. Rev. 10, 475 (1958).

Thomas, L.: The role of epinephrine in the reactions produced by the endotoxins of Gram-negative bacteria. I. Hemorrhagic necrosis produced by epinephrine in the skin of endotoxin-treated rabbits. J. exp. Med. 104, 865 (1956).

Zweifach, B. W.: Functional Behavior of the Microcirculation. Thomas, Springfield (1961).

Zweifach, B. W., and D. Metz: Rat mesoappendix procedure for bioassay of humoral substances acting on peripheral blood vessels. Ergebn. Anat. Entwickl. — Gesch. 95, 176 (1956).

DISCUSSION

Guth: Did you have the opportunity to observe the large veins in the course of bradykinin application to mesenteric vasculature?

Zweifach: In the mesentery of the rat, veins from 150-500 μm in diameter can be observed. These vessels are narrowed by the local application of 1×10^{-5} to 1×10^{-6} μg per ml.

Keele: I agree wholeheartedly with Dr. Zweifach's statements concerning the multiplicity of factors which act upon the vascular bed. With respect to pain, my colleagues and I have shown that many naturally occurring substances e.g., potassium, serotonin, acetylcholine, and kinins are active by themselves. Prof. Sicuteri and his colleagues have given a beautiful demonstration of the potentiation by 5-HT of the algogenic action of bradykinin. This problem of interaction of kinins with other algogenic agents deserves much more study.

Lembeck: We compared the capillary permeability-increasing effect and the action on the smooth muscle of substance P extracts from various parts of the brain, and we found a close correlation. The highest activity was in the substantia nigra with both methods. Fractionation of this extract on an aluminum oxide column revealed exactly the same distribution for the factor acting on the guinea pig ileum and causing blueing effect upon intradermal injection into the rabbit skin. Amounts corresponding to 40 mg substanti nigra showed a pronounced effect on capillary permeability.

THE PROTEOLYTIC PROCESS IN PANCREATIC DISEASE

V. M. SARDESAI and ALAN P. THAL

The Robert S. Marx Laboratories of the Department of Surgery, Wayne State University School of Medicine, Detroit, Mich., U.S.A.

The diagnostic value of measuring proteolytic activity in the peripheral blood of patients with pancreatic disease has been questioned by several authors (Babson et al., 1962; Floch and Groisser, 1960). However, the fact remains that other investigators (Brown, 1959; Nardi, 1958; 1959) have demonstrated an increase in the proteolytic activity of blood taken from such patients. The purpose of this paper is to report yet another series of patients with pancreatic disease in which increases in plasma proteolytic activity is demonstrated and to inquire into the nature of the enzymes responsible for this activity.

METHOD

Determination of TAMe esterase activity of plasma. Plasma proteolytic activity was determined by its action on the synthetic substrate TAMe,* (p-tosyl arginine methyl ester hydrochloride), whose structural formula is given in Fig. 1. This compound is hydrolyzed by trypsin to yield tosyl arginine and methyl alcohol, the rate of reaction being proportional to the enzyme concentration. The methyl alcohol formed by hydrolysis can be measured fluorometrically (Sardesai and

*TAMe and lyophilized trypsin were obtained from Mann Research Laboratories, New York 6, N. Y.

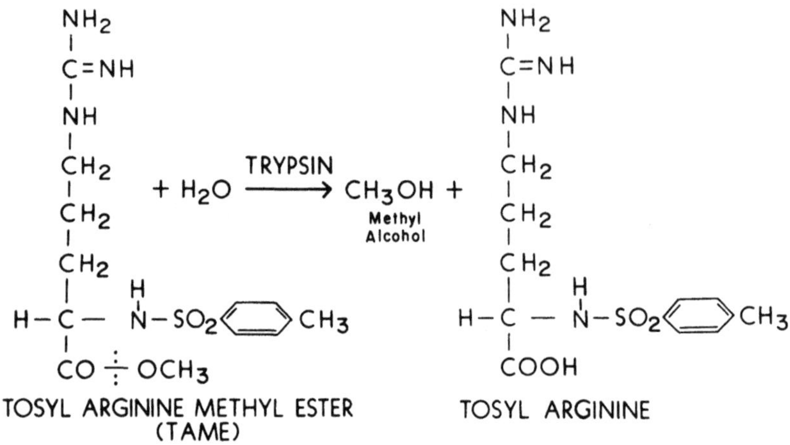

Fig. 1. The action of trypsin on TAMe.

Provido, 1964). Since other plasma enzymes such as plasmin (Ronwin, 1956). thrombin (Sherry and Troll, 1954), kallikrein (Webster and Pierce, 1961), C'1 component of complement (Donaldson, 1961), etc. besides trypsin can hydrolyze TAMe, the general term TAMe esterase activity has been used to indicate the proteolytic activity of plasma in view of the broad susceptibility of this amino acid ester to hydrolysis. The activity is expressed in terms of micrograms of trypsin, the concentration of which is known to cause an equivalent amount of hydrolysis of the substrate. All determinations were done in duplicate. A detailed account of the procedure for the determination of plasma TAMe esterase activity is described elsewhere (Sardesai and Provido, 1965).

Determination of the trypsin inhibitory capacity of plasma. Increasing quantities of pure trypsin are added to 1-ml samples of plasma. The TAMe esterase activities in these samples are then determined, and a curve of fluorescence intensity with respect to added trypsin to samples is plotted. The inhibitory capacity is expressed in terms of micrograms of added trypsin inhibited by 1 ml plasma.

RESULTS

In order to see the effect of Trasylol, a known proteolytic inhibitor (Forrell, 1963), on the action of trypsin, thrombin,

TABLE 1
Comparison of the Activity of Normal Plasma and Some Plasma Proteolytic Enzymes on the Synthetic Substrate TAMe, Before and After Incubation with Trasylol

Enzyme	Before Trasylol* (μg/cc)	After Trasylol (μg/cc)
Trypsin (lyophilized)	6.0	0.036
Normal plasma	3.6	3.6
Plasmin (234 NIH units/cc)	83.5	85.0
Thrombin (20.5 RPMI units/cc)	11.8	11.4

*Trasylol in concentration of 10 units/cc. Units are in terms of micrograms of trypsin.

TABLE 2
Plasma TAMe Esterase Activity in 14 Patients With Nonpancreatic Abdominal Disease

Patient	Diagnosis	Esterase Activity
		μg/cc
N.H.	Diverticulitis	11.6*
W.H.	Gastritis	7.0
M.U.	Cholecystitis	11.4
M.L.	Cholecystitis	10.9
R.L.	Cholecystitis	4.0
I.S.	Peptic Ulcer	6.4
L.C.	Gastritis	3.4
E.G.	Gastritis	5.6
J.H.	Gastritis	4.8
Y.B.	Pyelonephritis	7.5
F.A.	Hydronephrosis	8.8
J.B.	Peptic Ulcer	4.0
L.McC.	Appendiceal Abcess	4.0
G.C.	Peptic Ulcer	5.0

*Note: Patient N.H. had unexplained elevations of serum amylase.

and plasmin on TAMe, known amounts of these enzymes were incubated with the substrate and with and without Trasylol, and the fluorescence intensity was determined in each case. Table 1 represents the results obtained in this study. As is seen, trypsin, thrombin, plasmin, and normal plasma all have TAMe esterase activity. The action of trypsin on TAMe is inhibited by Trasylol, but this inhibitor has no effect on thrombin, plasmin, and normal plasma.

Three groups of human subjects were examined for their plasma TAMe esterase activity: (a) normal humans (nonpatients); (b) patients with nonpancreatic abdominal disease; and (c) patients with documented pancreatic disease.

Of the normal humans studied, all had measurable TAMe esterase activity in their plasma with a range of values from 3.35 to 8.5 μg/ml. Addition of Trasylol (10 units) to this plasma before incubation did not decrease the activity, suggesting that in the normal plasma, TAMe esterase activity may not be due to trypsin. Plasma obtained from 14 patients in group (b), with nonpancreatic abdominal disease, had an activity with a range of values from 4.0 to 11.6 μg/ml (Table 2). It is of interest to note that of the three false positives in this group, two were patients with cholecystitis and one with diverticulitis

TABLE 3
Plasma TAMe Esterase Activity in
9 Patients with Pancreatic Disease

Patient	Diagnosis	Esterase Activity
		μg/cc
A.S.	Chronic Recurrent	15.0
W.R.	Acute Hemorrhagic	30.0*
J.S.	Acute Necrotizing	46.0*
E.C.	Chronic Recurrent	23.4
L.R.	Carcinoma	26.5
O.F.	Chronic Recurrent	19.0*
M.M.	Chronic Recurrent	14.0
G.A.	Carcinoma	16.2
E.B.	Chronic Recurrent	19.0*

*Note: Serial samples taken from several patients during the course of their disease revealed some levels of activity which were within normal limits.

who had an unexplained elevation of serum amylase. Of the eight patients with pancreatic disease, all had elevation in plasma TAMe esterase activity ranging from 14.0 to 46.0 μg/ml, as seen in Table 3. All plasma samples obtained from this group were taken during the acute stage of the disease, and it should be noted in three patients that serial determinations resulted in an occasional falling of values within the normal range. Addition of Trasylol to plasma from patients with pancreatic disease resulted in fall in TAMe esterase activity to within normal ranges, suggesting that the elevated activity in this group was due to trypsin and/or kallikrein in their plasma.

Figure 2 summarizes the data on the in-vivo effect of Trasylol on the trypsin inhibitory capacity of plasma in a

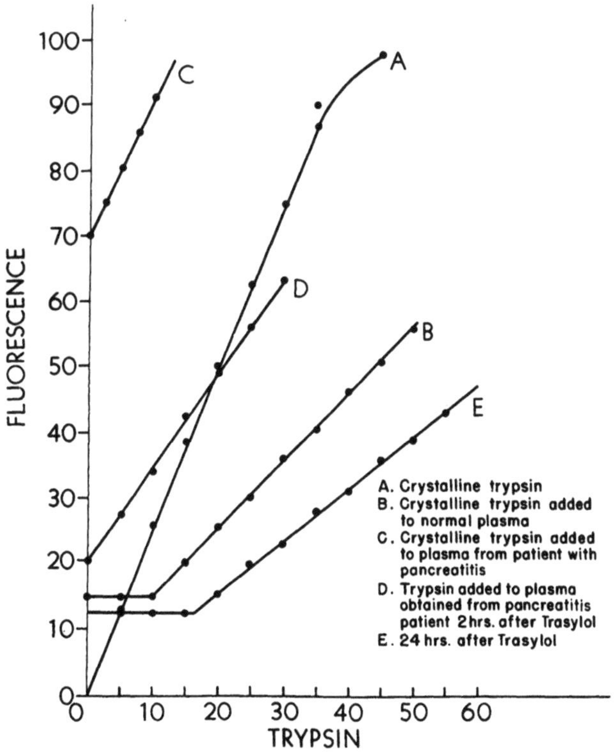

Fig. 2. Relationship between fluorescence intensity and amount of trypsin (μg/ml) added to the sample.

patient with pancreatitis. Curve A shows the linear relationship between the amount of trypsin added to $CaCl_2$ solution and the fluorescence intensity. Curve B represents the relationship between the amount of trypsin added to normal human plasma and the fluorescence intensity. As can be seen, addition of up to 10 μg trypsin to 1 ml normal plasma does not increase the fluorescence intensity, indicating that 1 ml normal plasma is capable of inhibiting up to 10 μg of added trypsin. Further addition of trypsin to the normal plasma increased the fluorescence intensity of samples gradually but at a rate less than that of pure trypsin in $CaCl_2$ solution (see slopes of curves A and B). The pattern of the trypsin inhibitory capacity in plasma from a patient with pancreatitis is shown by curve C. It can be seen that there is negligible trypsin inhibitory capacity in this plasma, since addition of trypsin to the plasma increased the fluorescence intensity at a rate similar to that of crystalline trypsin in $CaCl_2$ solution. This patient was then given 150,000 units of Trasylol, and 2 hours later a blood sample was drawn. Although the TAMe esterase activity in the plasma from this sample was considerably decreased, there was still no appreciable trypsin inhibitor capacity in this plasma (curve D). A sample drawn 24 hours after Trasylol administration had a lowered TAMe esterase activity and an increased plasma trypsin inhibitory capacity, as shown by curve E. The results suggest that one of the actions of Trasylol, in addition to lowering the proteolytic activity, is to increase the trypsin inhibitory capacity level in plasma in patients with pancreatic disease.

DISCUSSION

This series of experiments has demonstrated that in normal plasma there is at least one enzyme system present which hydrolyzes the synthetic substrate TAMe. In addition, it has been shown that there appears to be relatively significant elevation of plasma proteolytic activity in some cases of pancreatic disease and that similar elevations in activity are not seen in nonpancreatic abdominal diseases. These facts support the previous work of Nardi (1960) and Brown (1959).

Of additional interest is the observation that, whereas the proteolytic activity of normal plasma is not inhibited by Trasylol, a potent trypsin and kallikrein inhibitor, the abnormal elevations in plasma proteolytic activity seen in pancreatic disease could be reduced to within normal ranges by this inhibitor, suggesting that trypsin and/or kallikrein may be the enzyme(s) responsible for the increased TAMe esterase activity reported in this paper and by others.

The proteolytic enzyme systems exhibiting TAMe esterase activity known to be present in human plasma include trypsin, kallikrein, plasmin, thrombin, and C′1 component of complement. Trypsin and kallikrein are known to be inhibited by Trasylol (Forrell, 1963). The TAMe esterase activity of thrombin is not affected by this inhibitor as is seen under Results. There are reports in the literature claiming the inhibition of plasmin by Trasylol in vivo (Kazmers, 1964), but our data suggest that plasmin is not inhibited by this proteolytic enzyme inhibitor in vitro. It is possible, however, that plasmin has two separate actions: proteolytic and esterolytic and that the latter is not inhibited by Trasylol. There is no information currently available on C′1 component of complement.

These plasma proteolytic enzymes represent extremely complex and poorly understood systems, involving specific and nonspecific inhibitors and activators. Any explanation of the observations which have been made in this series of experiments must be highly speculative. However, it appears quite evident that plasma proteolytic enzyme activity does increase in pancreatic disease and that the increase is at least in part due to an enzyme system which is inhibited by Trasylol.

Addition of up to 10 μg trypsin to 1-ml samples of normal plasma does not increase the TAMe esterase activity apparently due to the presence of plasma trypsin inhibitors (Bundy and Mehl, 1958; Haverback et al., 1962). Once this inhibitory activity is overcome, further addition of trypsin to the plasma increases the esterase activity. In pancreatic disease, it appears that because of the excessive release of trypsin and/or kallikrein, the inhibitory activity is overcome and the plasma trypsin inhibitory capacity is negligible, as can be seen by the results presented in this report. Trasylol has been shown to decrease the TAMe esterase activity and to increase the inhibitory capacity in plasma from patients with pancreatic disease.

There have been several reports (Grözinger et al., 1964; Hermann and Knowles, 1962) indicating that Trasylol significantly reduced the mortality rate associated with experimentally produced acute pancreatitis in dogs. Our experimental data indicate that this proteolytic inhibitor would decrease the plasma proteolytic activity and increase the plasma trypsin inhibitor capacity in pancreatitis. Further studies of the trypsin inhibitory capacity of plasma in experimental pancreatitis are in progress.

On the basis of these results, it is evident that activation of proteolytic enzymes occurs in human pancreatic disease. Moreover, other studies in this laboratory suggest that in experimental endotoxic shock in dogs (Thal and Sardesai, 1965) and rats and in clinical septic shock, plasma proteolytic activity is increased, and plasma polypeptides such as bradykinin are released. The significance of these substances in the pathogenesis of these diseases remains to be determined.

SUMMARY

Plasma proteolytic activity was determined by a procedure wherein the methanol released by the action of plasma enzyme on TAMe was measured fluorometrically. This activity has been referred to as TAMe esterase activity and was determined in three groups of human subjects: (a) normal humans (nonpatients); (b) patients with nonpancreatic abdominal disease; and (c) patients with documented pancreatic disease. The TAMe esterase activity in normal humans and in patients with nonpancreatic abdominal disease was found to be in the range equivalent to 3.5 to 8.5 μg trypsin/ml plasma. Patients with documented pancreatic disease had plasma TAMe esterase activities in the range of 15 to 43 μg trypsin/ml. Whereas the activity present in the first two groups was not inhibited by Trasylol, the addition of this inhibitor to plasma from patients with pancreatic disease decreased its activity to within normal ranges.

Administration of Trasylol to a patient with pancreatitis decreased the plasma TAMe esterase activity and increased the plasma trypsin inhibitory capacity. The significance of these results is discussed.

ACKNOWLEDGMENTS

We are indebted to Dr. L. M. Wheeler of Parke, Davis & Co. for supplying us with bovine plasmin, human plasmin, and bovine thrombin. This research was supported by Grant AM-06385-04 from the National Institutes of Health, the Michigan

Heart Association, and The Receiving Hospital Research Corporation.

REFERENCES

Babson, A. L., P. A. R. Williams, and G. E. Philips: An evaluation of serum trypsin tests. Clin. Chem. 8, 62 (1962).
Brown, M. E.: Serum exopeptidase activity in diseases of the pancreas. New Engl. J. Med. 260, 331 (1959).
Bundy, H. F., and J. W. Mehl: Trypsin inhibitors of human serum — I. Standardization, mechanism of reaction, and normal values. J. clin. Invest. 37, 945 (1958).
Donaldson, V. H.: Studies on the activation of a serum esterase with ether and its relationship to C'1 esterase. J. clin. Invest. 40, 673 (1961).
Floch, M. H., and V. W. Groisser: Serum proteolytic enzyme activity in pancreatic disease. New Engl. J. Med. 263, 1129 (1960).
Forrell, M. M.: Therapy with kallikrein and protease inhibitors. Ann. N.Y. Acad. Sci. 104, 368 (1963).
Grözinger, K. H., C. P. Artz, A. V. Hollis, and R. L. Wesson: Evaluation of Trasylol in experimental acute pancreatitis. Surgery 56, 400 (1964).
Haverback, B. J., B. Dyce, H. F. Bundy, S. K. Wirtschafter, and H. H. Edmondson: Protein binding of pancreatic proteolytic enzymes. J. clin. Invest. 41, 972 (1962).
Hermann, R. E., and R. C. Knowles: Effect of a new trypsin inhibitor in the therapy of experimental pancreatitis. Surg. Forum 13, 306 (1962).
Kazmers, N.: Preliminary report on treatment and prophylaxis of acute pancreatitis with a kallikrein-trypsin inhibitor. Amer. J. Gastroent. 41, 28 (1964).
Nardi, G. L.: Serum trypsin determination in pancreatic disease. J. Lab. clin. Med. 52, 66 (1958).
Nardi, G. L.: Specific polypeptide substrates in enzymatic determination of pancreatic disease. Surgery 46, 30 (1959).
Nardi, G. L.: Serum trypsin (or arginine exopeptidase) screening test for cancer of the pancreas. Gastroenterology 38, 50 (1960).
Ronwin, E.: Enzymatic properties of bovine plasmin preparation; Evidence for similarity to but non-identity with trypsin. Canad. J. Biochem. 34, 1169 (1956).
Sardesai, V. M., and H. S. Provido: The determination of

methanol in biological fluids. J. Lab. clin. Med. 64, 977 (1964).

Sardesai, V. M., and H. S. Provido: A fluorometric method for the determination of TAMe esterase (tryptic) activity of plasma. J. Lab. clin. Med. 65, 1023 (1965).

Sherry, S., and W. Troll: The action of thrombin on synthetic substrates. J. biol. Chem. 208, 95 (1954).

Thal, A. P., and V. M. Sardesai: Shock and the circulating polypeptides. Amer. J. Surg. 110, 308 (1965).

Webster, M. E., and J. V. Pierce: Action of the kallikreins on synthetic ester substrates. Proc. Soc. exp. Biol. (N.Y.) 107, 186 (1961).

DISCUSSION

Trautschold: According to you, kinin liberation in acute pancreatitis is also due to activated trypsin. TAMe substrate, however, is a very unspecific one, which is attacked by many other enzymes besides trypsin. Thus the hydrolysis of this substrate does not reflect at all a tryptic activity. With a specific trypsin substrate (BAPA) we were not able, as I reported, to detect any tryptic activity during acute pancreatitits. In addition, there is the specific trypsin inhibitor in mamalian pancreas, which will inhibit about 5% of the total trypsin that can be activated. We found, however, a proteolytic and esterolytic activity in pancreas which is not identical with trypsin and which cannot liberate kinins. You mentioned that TAMe activity would be identical with this proteolytic activity.

Thal: The activity measured could be due to other proteolytic enzymes in addition to trypsin. BAPA is also not specific for trypsin just as is TAMe. We use Trasylol to distinguish between trypsin-like and non trypsin-like activities.

Back: You showed that Trasylol does not inhibit plasmin. Many investigators have reported, as our group has, the strong antifibrinolytic activity of Trasylol. This activity of Trasylol is employed clinically in controlling fibrinolytic hemorrhage due to a variety of conditions. We also have shown Trasylol to be effective in reducing mortality of mice shocked by scalding. High doses of Trasylol were required and had to be administered just prior to exposing the mice to the 10-sec burn-shock stimulus.

Thal: Our studies suggest that plasmin may have two actions, fibrinolytic and esterolytic as measured by the TAMe method. The esterolytic activity of plasmin is not inhibited by Trasylol whereas the fibrinolytic is.

Edery: I would like to raise a technical point. Is there any particular reason for the use of isolated aorta to determine kinin activity? This preparation has a rather poor sensitivity toward plasma kinin. It would be more advisable to use rat isolated uterus or guinea pig isolated ileum.

Thal: The preparation used originally was the ox carotid strip. While not extremely sensitive to kinin, it is sensitive to histamine and serotonin. Accordingly, with appropriate blocking agents we may determine the total vasoactivity due to several agents as well as the separate activities of these substances. More recently, all assays of the kinin in our laboratory have been carried out on the rat uterus.

Diniz: In shock produced by trypsin in collaboration with Dr. Corrado, we observed after the injection of small doses of trypsin (3 mg/kg) to dogs, an activation of BAMe-esterase trace activity and immediate release of histamine. Bradykinin release and bradykininogen depletion was observed only after the injection of large doses of trypsin (12 mg/kg).

Thal: Our results using trypsin would confirm your finding that the initial release following trypsin is histamine. When larger doses are used this is followed by the release of plasma kinin.

KININ FORMATION AND DESTRUCTION IN FLUID PRODUCED BY A STRANGULATED SEGMENT OF ILEUM IN THE RAT

EGIL AMUNDSEN

Institute of Physiology, University of Oslo, Norway

Intestinal obstruction is still a serious ailment with relatively high mortality (Troell, 1954; Höyer and Solheim, 1959). There are two main forms of this ailment. One is the simple obstruction of the intestine, the other the so-called strangulation obstruction in which the obstruction to intestinal passage is caused by strangulation of a segment of intestine. The two forms are illustrated in Fig. 1.

The mortality of patients with simple obstruction is now well below 10% in most hospitals, whereas the mortality of patients with strangulation obstruction still varies between 20% and 30%. (For survey, see Cohn, 1961.) The difference in mortality is related to processes that occur in the strangulated segment of the intestine and result in the formation of fluid that passes from it into the peritoneal cavity. The strangulation fluid collected from animals during the late stages of experimental strangulation obstruction is very toxic. Small volumes of it, injected intraperitoneally into healthy animals, has a lethal effect (Barnett, 1959). Microbes are the main cause of the toxicity, but do not fully explain it (Amundsen and Midtvedt, 1964).

This investigation deals with the content of plasma kinins and plasma kinin precursor in strangulation fluid and its ability to form and to destroy plasma kinins. The characteristics of strangulation fluid and the conditions under which it is formed are such that local formation of plasma kinins could be expected to take place.

In the present study, strangulation obstruction was carried

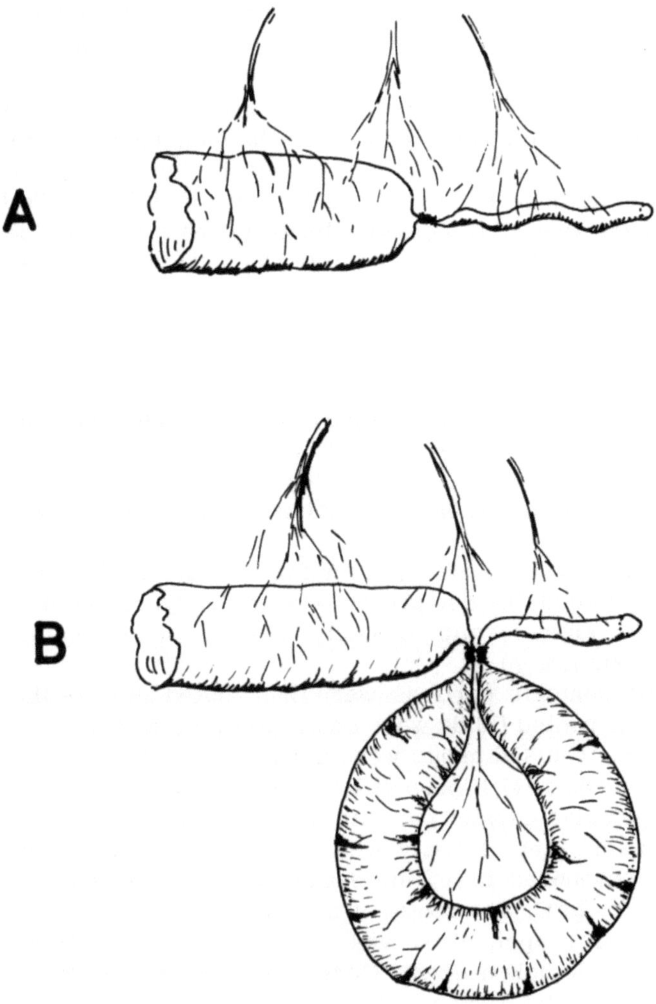

Fig. 1. Semischematic drawings of the two main forms of mechanical ileus. Both drawings illustrate findings made several hours after the conditions started. At A a condition of simple mechanical obstruction is illustrated. Note the dilated intestine cephalad to, and the collapsed intestine caudad to, the obstruction. At B a condition of intestinal strangulation obstruction is illustrated. The dilated strangulated loop of intestine with its mesentery can be seen in addition to the structure seen under A.

out in a loop of ileum 5 cm long in 15 rats. The distal end of the strangulation site was placed 5 cm orally to the ileocecal junction. Rats with this condition have an average survival time of about 90 hours. Each rat produces about 20 ml of strangulation fluid before it dies. The techniques to effect the strangulation and to drain and collect the fluid have been published elsewhere (Amundsen and Midtvedt, 1964). I shall only point here that it is possible to drain the strangulation fluid for collection in portions outside the body at +4°C over a 24-hour period.

All the fluid portions collected were tested for toxicity on mice; 0.5 ml of each fluid portion was injected i.p. into healthy male mice weighing 23-25 g. The 25-hour survival rate was registered. None of the fluid portions collected during the first 24-hour period was toxic, but an increasing number of the portions collected during the subsequent 24-hour periods were toxic, The strangulation fluid obtained during the terminal period was invariably toxic.

After having been tested for toxicity, all the fluid portions were examined for free plasma kinin activity, for kininase activity, for kinin-forming capacity, and for content of kinin precursor (kininogen).

The nontoxic portions were then pooled as were the toxic ones. The pooled fluids were centrifuged for 60 min at 100,000 g and 4°C. The sediments were discarded and the supernatants passed through Millipore filters with pore size 1 μ. Assays were then also carried out with these filtrates.

Plasma kinin activity was determined on the isolated rat uterus suspended in a 5-ml organ bath with aerated de Jalons solution at 29°C. The contact time was 45 sec, and the interval between each application was 5 min. The registrations were made on a smoked drum. Contractions were compared with those caused by synthetic bradykinin (Sandoz A. G.) The rats had been given i.p. injections of 0.3 ml of a 0.01% solution of stilbestrol 24 hours before being killed. Prior to some of the assays, Deseril (Sandoz A. G.), 0.2 μg/liter, and mepyramine maleate, 0.1 mg/liter were added to the de Jalons solution to exclude possible influence of serotonin and histamine.

Free plasma kinin activity. This could not be demonstrated in any of the 24-hour portions of strangulation fluid. When the kininase activity had been eliminated by acid treatment (Horton, 1959), some kinin activity would develop in the strangulation fluid collected during the first 24 hours and also in some fluid portions collected from subsequent periods, but never in any of those collected during the terminal period. When fluid portions were pooled, centrifuged, and filtered as described, and then exposed to acid treatment, considerable kinin activity appeared in the filtrate from nontoxic fluids, whereas no such activity could be detected in filtrate from toxic fluids. This is evident from Fig. 2.

Kininase activity. If a fluid contains kininase, it may be difficult to assess the possible content of kinin precursor and the kinin-forming capacity of the fluid. The kininase activity

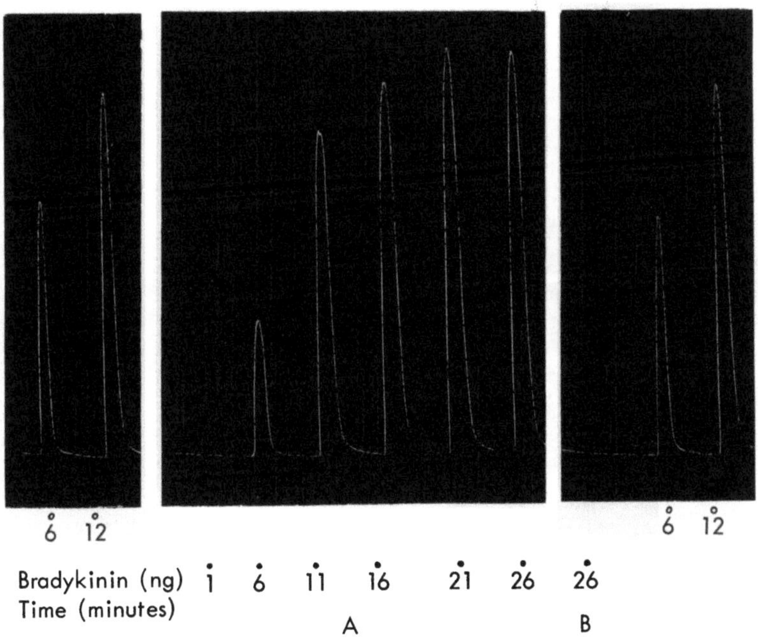

Fig. 2. Endogenous kinin formation in filtrates of pooled nontoxic (A) and pooled toxic (B) strangulation fluid portions after acid treatment of the filtrates. Portions of 0.2 ml were removed from the acid-treated filtrates at intervals of 5 min and tested for plasma kinin activity on the isolated rat uterus. Contractions were compared to those caused by known amounts of synthetic bradykinin.

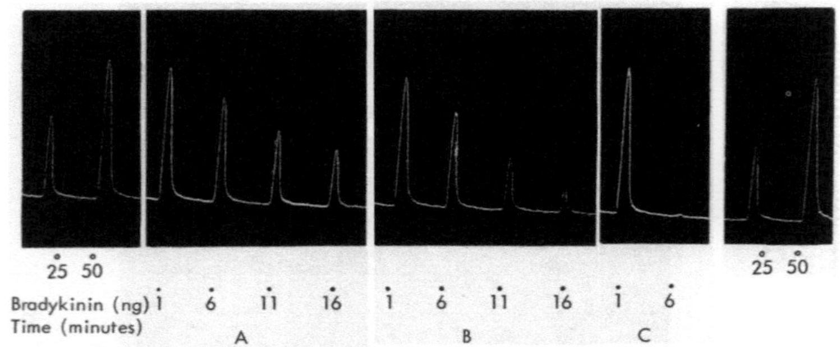

Fig. 3. Kininase activity of three consecutively collected portions of strangulation fluid from one rat. Collection periods: A, 0-24 hours after arrangement of strangulation obstruction; B, 25-48 hours after arrangement of strangulation obstruction; C, 73-96 hours after arrangement of strangulation obstruction. All fluids were diluted 1:10 with saline. To 0.9 ml of this diluted fluid was added 0.5 μg synthetic bradykinin (Sandoz A.G.) in 0.1 ml saline. The mixture was incubated in ice-cooled water. At 5-min intervals 0.1 ml of the incubated mixture was removed and added to the organ bath with the isolated rat uterus. Standard doses of bradykinin were added to the organ bath before and after tests.

of the different fluids was therefore evaluated before the other assays were performed.

Kininase activity was found in all the 24-hour portions of strangulation fluid collected from all 15 animals used. Fluid portions collected late in the course of the strangulation obstruction had, however, much more marked kininase activity than had those collected during the first 24 hours (Fig. 3). Filtrate of pooled nontoxic fluids portions had a less pronounced kininase activity than had filtrate of pooled toxic portions (Fig. 4). The kininase activity of the latter filtrate appeared to be about 8 times as strong as that of the former, as judged from the effect of dilution.

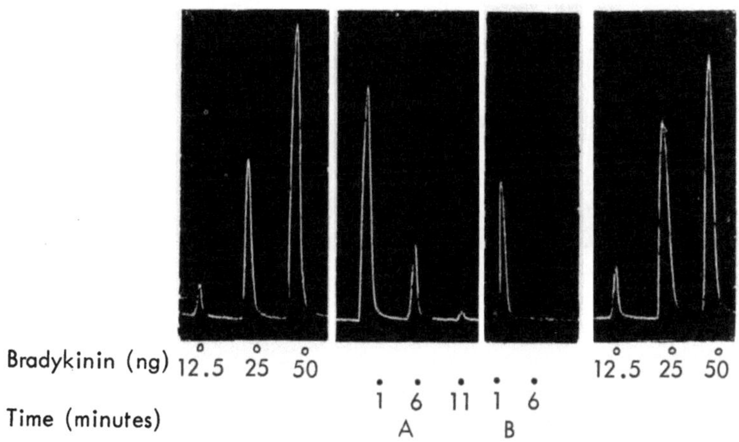

Fig. 4. Kininase activity of filtrate of pooled nontoxic (A) and pooled toxic fluid portions (B). The assay for kininase activity was carried out as described in Fig. 3.

Kinin-forming activity. No kinin-forming capacity could be detected in any portion of strangulation fluid upon direct incubation with a special plasma substrate (Amundsen et al., 1963). After acid treatment of the fluids, however, kinin-forming activity could be detected in several fluid portions. This activity was less pronounced in fluid portions collected late in the course of the obstruction than in those collected during the first 24-hour period. Figure 5 illustrates the

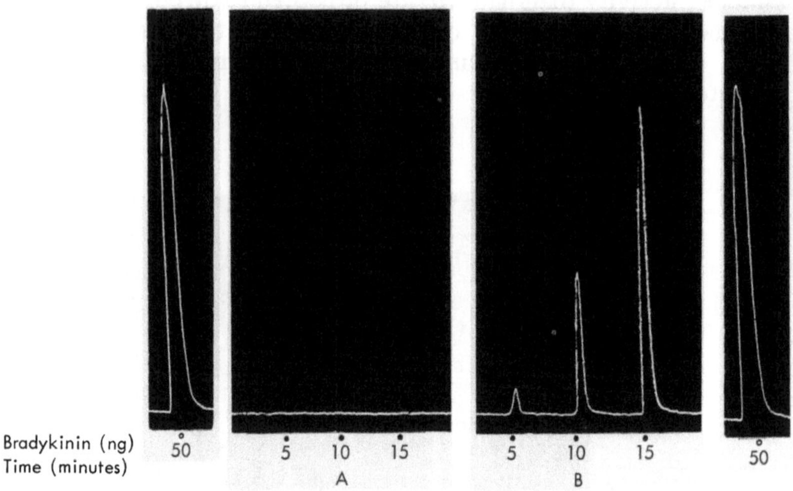

Fig. 5. Kinin-forming capacity of acid-treated filtrates of pooled toxic (A) and pooled nontoxic fluid portions (B); 0.2 ml of acid-treated filtrates was incubated with 0.8 ml of a stable plasma substrate at 37°C; 0.2 ml of the incubated mixture was removed at intervals and tested for plasma kinin activity on the isolated rat uterus. Contractions were compared to those caused by known amounts of synthetic bradykinin.

marked difference between filtrate of pooled early fluid portions (nontoxic) and filtrate of the late portions (toxic). The latter contained no kinin-forming capacity at all.

Plasma kinin precursor. The relative amount of precursor was determined in all the individual fluid portions, and also in the filtrates of pooled nontoxic and toxic fluids. The individual or pooled fluid portions collected during the terminal stages contained little or no plasma kinin precursor, whereas those collected during the first 24 hours always contained considerable amounts of precursor. The different contents of plasma kinin precursor in acid-treated filtrates of pooled nontoxic and toxic fluids is demonstrated in Fig. 6.

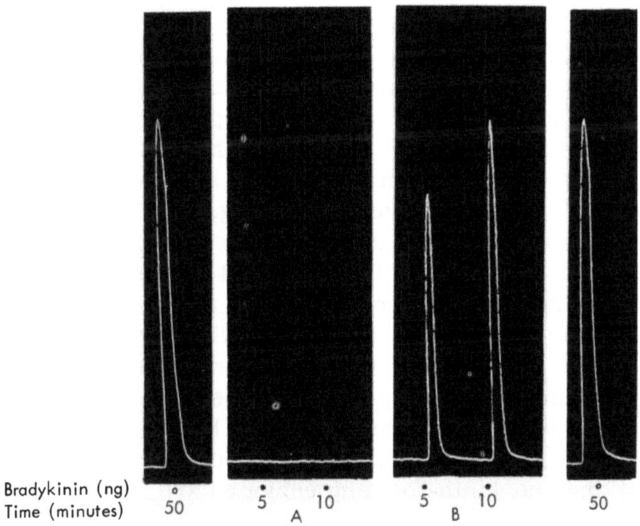

Fig. 6. Content of plasma kinin precursor in acid-treated filtrate of pooled toxic (A) and pooled nontoxic fluid portions (B); 0.8 ml of the acid-treated filtrates was incubated with 0.2 ml saliva at 37°C; 0.2 ml of the incubated mixture was removed at intervals and tested for plasma kinin activity on the isolated rat uterus. Contractions were compared to those caused by known amounts of synthetic bradykinin.

DISCUSSION

The findings presented indicate that plasma kinins are probably formed and broken down in the so-called strangulation fluid. In the fluid produced during the early stage of the ailment, kinin-forming activity, kinin precursor, and kininase activity are all present. The fluid produced during the terminal stages of strangulation obstruction, however, has a much stronger kininase activity but contains neither kinin precursor nor kinin-forming activity.

It is well known that plasma contains carboxypeptidases which break down bradykinin (Erdös and Sloane, 1962). Homogenates of a number of different types of cells also show a marked kininase activity (Amundsen and Nustad, 1965). Severeral strains of pathogenic microbes are also capable of producing kininases (Amundsen and Rugstad, 1965). The strong kininase activity found in terminally produced strangulation fluid is probably derived from several sources such as plasma, disintegrating cells, and perhaps also microorganisms.

It is more difficult to explain why both kinin precursor and kinin-forming activity are present in strangulation fluid produced during the early stages of the ailment, but not in that produced later. One explanation could be that while formation of plasma kinins is going on, the precursor is gradually consumed and the kinin-forming enzymes are subsequently inactivated or destroyed.

As regards the kinin-forming activity, it should be emphasized that the plasma substrate used will probably yield kinin only with the glandular type of kallikrein and not with plasma kallikrein (Jacobsen, 1965; personal communication). Kinin-forming activity of the fluids therefore must be re-evaluated with another substrate preparation.

The implications of these findings for the pathophysiology of strangulation obstruction are still obscure. It has been well established that microorganisms are the main cause of the toxicity of strangulation fluid (Amundsen and Gustafsson, 1963). The role of the plasma kinins in this connection is quite unknown. It seems worth examining, however, whether the lack of kinin activity, of kinin precursor, and of kinin-forming activity together with the high kininase activity of toxic strangulation fluid may interfere with the response of the organism to the massive peritoneal infection.

REFERENCES

Amundsen, E., and B. E. Gustafsson: Results of experimental intestinal strangulation obstruction in germfree rats. J. exp. Med. 117, 823 (1963).

Amundsen, E., and T. Midtvedt: The toxicity of fluid from experimentally strangulated intestinal loops in the rat. J. surg. Res. 4, 306 (1964).

Amundsen, E., and K. Nustad: Kinin forming and destroying activities of cell homogenates. J. Physiol. (Lond.) 179, 479 (1965).

Amundsen, E., and H. E. Rugstad: Influence of some pathogenic bacteria on kinin formation and destruction. Brit. J. Pharmacol. 25, 67 (1965).

Amundsen, E., K. Nustad, and B. A. Waaler: A stable substrate for the assay of plasma kinin forming enzymes. Brit. J. Pharmacol. 21, 500 (1963).

Barnett, W. O.: Lethal factors in intestinal obstruction. Surg. Gynec. Obstet. 109, 769 (1959).

Cohn, I., Jr.: Strangulation obstruction. Springfield, Thomas (1961).

Erdös, E. G., and E. M. Sloane: An enzyme in human plasma that inactivates bradykinin and kallidins. Biochem. Pharmacol. 11, 585 (1962).

Höyer, A. J., and Solheim, K.: The therapy of acute small bowel obstruction. Surg. Gynec. Obstet. 109, 555 (1959).

Horton, E. W.: The estimation of urinary kallikrein. J. Physiol. (Lond.) 148, 267 (1959).

Troell, L.: The treatment of acute intestinal obstruction. Acta chir. scand. Suppl. 189 (1954).

DISCUSSION

Habermann: How did you measure toxicity, and what were the symptoms causing death in mice?

Amundsen: A fluid was called toxic when all the mice injected intra peritoneally with 0.5 ml of it died within 24 hours. We did not record symptoms in the injected mice during the period elapsing between the time of injection and the time of death.

FIBRINOLYSIS AND VASOACTIVE PEPTIDES IN ANAPHYLAXIS*

N. BACK, H. WILKENS, and R. STEGER
Department of Biochemical Pharmacology, School of Pharmacy, State University of New York at Buffalo, and The Roswell Park Memorial Institute, Buffalo, N.Y., U.S.A.

During the complex sequence of events in allergic and anaphylactic states, proteases, anticoagulants, histamine, 5-hydroxytryptamine, catecholamines, hypotensive polypeptides, and many other biologically active principles may appear in different animal species. The mechanism of activation and/or release of these factors is understood poorly, as are their respective roles in the biochemical chain of events in different types of shock.

Protease activation during anaphylactic and anaphylactoid shock was first observed by Nolf (1908). Ungar and Hayashi (1958) have ascribed primary importance to protease activation in the pathogenesis of the allergic response. Activation of the fibrinolysin system has been reported during anaphylactic shock in the rabbit (Gans and Krivit, 1961; Salmon and Lecomte, 1962), guinea pig (Burdon et al., 1961), and dog (Rocha e Silva, 1950; Back, et al., 1963b). A link also has been established between the allergic stimulus and the release of hypotensive peptides (Dawson and West, 1965; Brocklehurst and Lahiri, 1962; Lecomte, 1961; Beraldo, 1950; Scharnagel et al., 1965; Diniz and Carvalho, 1963; Back et al., 1963a). In view of the ability of plasmin to form vasoactive peptides, both in vitro (Back and Steger, 1965) and in vivo (Horton and Lewis, 1960; Mackay,

*This investigation was supported by Public Health Service Research Grants HE-08633 and HE-03784, National Heart Institute.

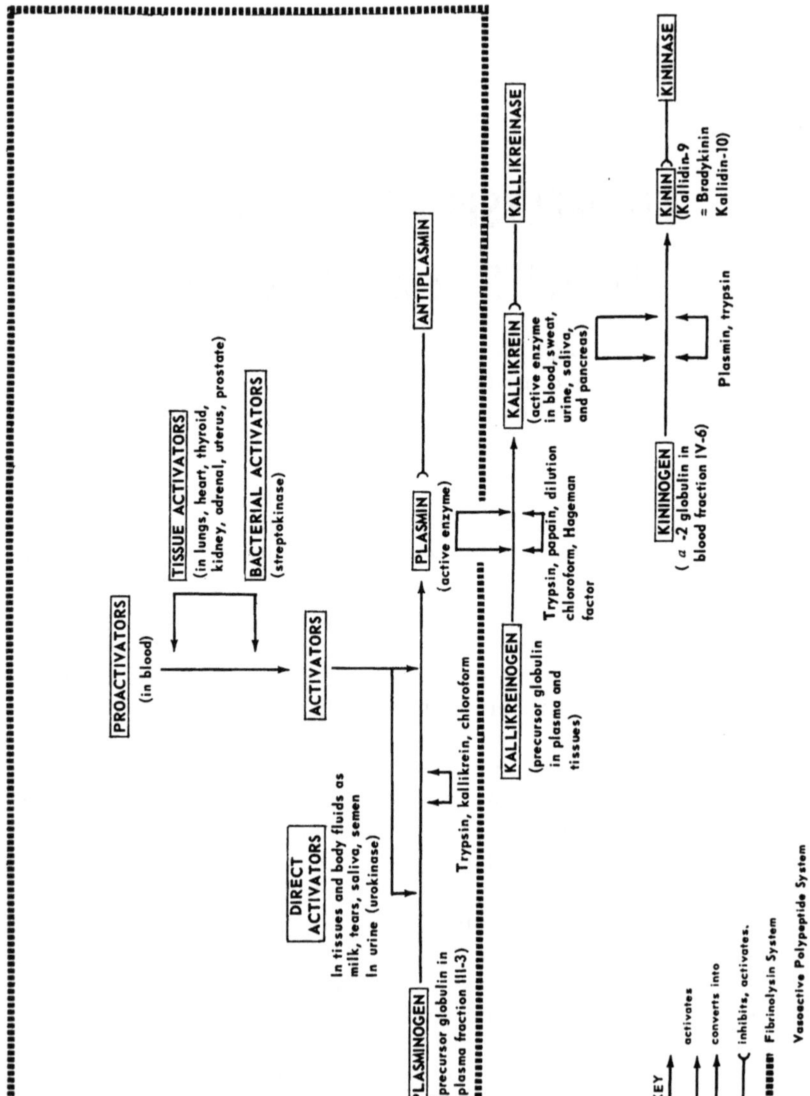

Fig. 1. Proposed relationship between the fibrinolysin and vasoactive polypeptide systems.

1963; Back et al., 1963a; Back, 1960), a possible relationship between the fibrinolysin and vasoactive polypeptide systems has been suggested (Back, 1966) (Fig. 1.) It is indicated that plasmin can form kinin both directly from kininogen and indirectly through a kallikrein-releasing mechanism.

This report summarizes initial studies designed to explore possible interrelationships among some of the various systems presumed to operate during anaphylaxis in the dog, which included the fibrinolysin, blood coagulation, and vasoactive polypeptide systems. An attempt to attribute a respective role to each system was made with the aid of appropriate pharmacologic agents.

MATERIALS AND METHODS

Initiation of anaphylaxis. Healthy mongrel dogs of either sex were sensitized by a single administration of 5.0 ml horse serum i.v. and 5.0 ml s.c. with 1.0 ml Freund's complete adjuvant. Fourteen to twenty days after initial sensitization, the dogs were anesthetized with sodium pentobarbital, 30 mg/kg, and challenged with 10 ml horse serum i.v. Femoral arterial blood pressure was recorded via a transducer attached to a multichannel polygraph. Venous pressure was measured by a cannula inserted into the inferior vena cava. Electrocardiograms, intestinal motility, and rate and depth of respiration were also recorded. In some instances, the thoracic duct was either ligated or cannulated for collection of lymph.

After pretreatment of the animals with the pharmacologic agents under study, sufficient blood was removed prior to and at frequent intervals after antigenic challenge for hematologic analysis including hematocrit, platelets, and white blood cells. Components of the blood coagulation, fibrinolysin, and vasoactive polypeptide systems also were analyzed. Artificial respiration was used at times to allow continued sequential analysis of blood changes during fatal anaphylaxis; 63 dogs have been studied thus far.

Blood analysis

1. Blood coagulation system. Twenty-five-ml samples of blood were withdrawn with siliconized syringes and 19-gauge hemorepellant needles. Fifteen-ml portions were distributed

either into 3.8% sodium citrate or 0.1 M sodium oxalate for the preparation of suitable plasma. The remainder of the blood was used for the preparation of serum and determination of hematocrit, platelets, whole-blood clotting time (Quick et al., 1948) and whole-blood thrombelastography (Nicola, 1957). The following factors or tests were studied: thromboplastic generation screening test (Hicks and Pitney, 1957), circulating anticoagulants (Stefanini and Dameshak, 1962), factor VII (Owren and Aas, 1951), factor V (Stormorken and Newcomb, 1957), real prothrombin and prothrombin consumption (Owren and Aas, 1951), partial thromboplastin time (Rodman et al., 1958) and fibrinogen (Wycoff, 1956; Andersch, 1944).

2. Fibrinolysin system. Plasma levels of plasmin, antiplasmin, plasminogen, and activator are being determined currently by methods previously described (Ambrus et al., 1960). Assay of proteolytic activity, with casein as a substrate (Astrup and Mullertz, 1955) also is in progress, as is estimation of esterase activity with *p*-toluene sulfonyl-1-arginine methyl ester (TAMe) as substrate. Initial estimate of plasma fibrinolytic activity was made by a modification of the fibrin plate method previously published (Ambrus et al., 1963).

3. Vasoactive polypeptide system: Kinin. Whole-blood kinin levels were determined by methods described previously (Back et al., 1963a). Five ml of freshly drawn venous blood was placed immediately in flasks containing 15 ml boiling methanol and boiled for 5 min. The mixture was centrifuged, the precipitate washed twice in 66% methanol, centrifuged once again, and the combined supernatants dried in an oven at 50°C. The residues were dissolved in 15 ml Tyrode's solution and assayed on the isolated rat uterus and duodenum perfused in a 10-ml bath with a modified Tyrode's solution (calcium concentration 10 mg/liter). Atropine, 1 mg/liter, and tripelennamine, 10 mg/liter, were added to the bath. Smooth-muscle-stimulating or -relaxing activity was compared with synthetic bradykinin.* Carboxypeptidase B† was added to those preparations showing uterine muscle-stimulating activity to determine whether kinin was involved.

* Bradykinin kindly supplied by Dr. Bircher, Sandoz Laboratories, Hanover, N. J.

† Carboxypeptides B, a product of Worthington Biochemical Corp., Freehold, N. J.

Kininogen. Plasma kininogen levels were estimated by a modification of a method described by Diniz and Carvalho (1963). Plasma (0.05 ml) was placed in siliconized glass tubes containing 0.45 ml of 0.2 M acetic acid. After 30-min boiling in water, the mixture was buffered with 0.4 ml of a 0.4 M Tris buffer at pH 7.8. The mixture was incubated for 30 min at 37°C with 50 μg crystalline trypsin in a volume of 0.2 ml. The reaction was stopped with 2 ml boiling ethanol. The contents were dried in vacuo at 35°C. The dried residue was suspended in 1.0 ml physiologic saline and assayed on the isolated rat uterus.

Kininase. A mixture of 0.3 ml plasma and 1.0 ml solution containing 1×10^{-6} g synthetic bradykinin was incubated for 5 min at 37°C. The kininase activity was then inhibited by the addition of 0.1 ml dimercaprol (BAL), 10 mg/ml. The remaining bradykinin was assayed on the isolated rat uterus preparation and the activity compared with a bradykinin reference standard containing 1×10^{-6} g bradykinin/ml and 10 mg BAL/ml. Data are expressed as the per cent change in activity with the base line taken as 100% enzyme activity.

Drugs. The following drugs were studied for their effect on the course of anaphylaxis: Trasylol,* heparin, protamine sulfate, diphenhydramine, ϵ-amino caproic acid,† and phenylbutazone. The agents were administered intravenously into sensitized dogs immediately prior to injection of the challenging dose of antigen. Drug dosages are as indicated in Table 1. In some dogs Trasylol was given together with the antigen.

RESULTS

Physiology. The typical physiologic responses recorded in sensitized saline control dogs following antigenic challenge are shown in a representative experiment in Fig. 2. An immediate fall in arterial blood pressure occurred with a concomitant decrease in pulse pressure and heart rate. The mean per

* Trasylol, a bovine kallikrein inhibitor, was generously supplied by Dr. R. Goesswald, Director of Research, FBA-Division of Metachem, New York, N. Y.

† ϵ-Amino caproic acid, a product of Lederle Laboratories, Pearl River, N. Y.

TABLE 1

Summary of Effect of Saline, Thoracic Duct Ligation, and Various Pharmacologic Agents on the Course of Anaphylactic Shock in the Dog

Drug or Procedure		Dose	Number of Dogs	Mean % Blood Pressure Fall	Mean Time for Maximum Blood Pressure Fall (min.)	Mean Time for 50% Recovery of Blood Pressure (min.)	Mortality # dead/# treated	
Saline	No Respirator	15 mls.	18	84	6.2	27.2	8/18	12/22
	Respirator	15 mls.	4	91	12.4	22.0	2/4	
TRASYLOL		12,500 units/kg.	4	66	2.3	12.5	0/4	3/10
		12,500 units/kg. followed by an additional 12,500 units/kg.	6	89	3.4	70.0	3/6	
Thoracic Duct Ligation		-	7	82	7.3	15.5	2/7	
Protamine Sulfate		8.5 mg/kg.	2	93	9.0	60.0	1/2	
Diphenhydramine (Benadryl)		20 mg/kg.	1	100	10.0	-	1/1	
Epsilon Amino Caproic Acid		250 mg/kg.	2	97	-	-	2/2	
Phenylbutazone		100 mg/kg.	4	90	3.8	23.3	1/4	

cent decrease in blood pressure from preinjection values was 84%. The mean time for maximal hypotension to occur was approximately 6 min without artificial respiration and 12 min with respiratory support. More than 50% of the dogs died acutely during respiratory paralysis within 3-20 min after antigen administration. All dogs experienced transient periods of apnea. At times, mean pressure and pulse rate increased transiently. Caval venous pressure increased slightly during shock in approximately 20% of the dogs and decreased significantly in the remaining dogs. Electrocardiograms revealed no abnormalities apart from a delayed tachycardia. Gastrointestinal hypermotility always occurred. Thoracic duct lymph flow increased significantly after onset of the shock, reaching peak values of 900% greater lymph flow 10 min after the shock. The lymph flow was still increased even after return of blood pressure to near control levels in surviving dogs.

The effect of various agents administered prior to antigenic challenge on the course of the shock syndrome is sum-

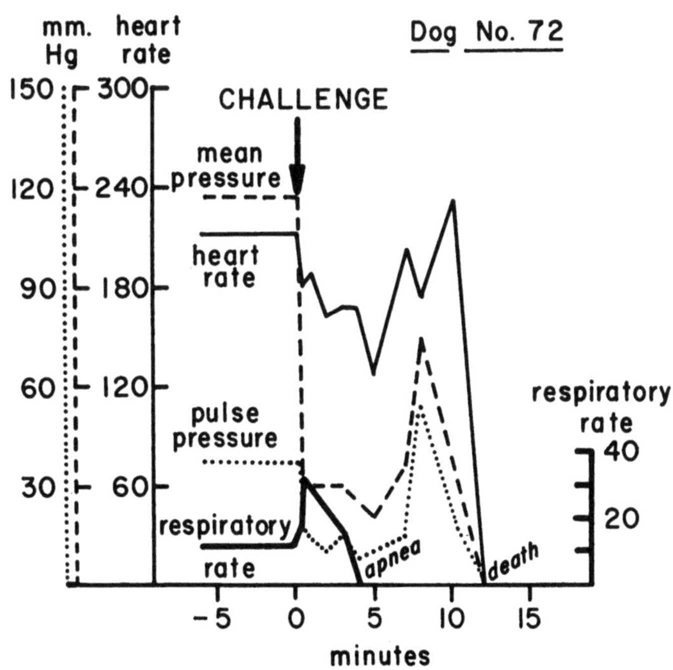

Fig. 2. Course of physiologic changes during anaphylaxis in saline-treated control dog.

marized on Table 1. Initial studies show that the following agents altered only slightly the picture of clinical shock: protamine sulfate, 8.5 mg/kg; diphenhydramine, 20 mg/kg; ϵ-aminocaproic acid, 250 mg/kg. On the other hand, phenylbutazone, 100 mg/kg, and the bovine protease inhibitor Trasylol, 12,000-25,000 units/kg, although not altering the clinical shock picture apart from decreasing the mean time for maximal hypotension to 2-4 min, did increase the survival rate by 25%. All agents were given intravenously just prior to initiation of the shock. Thoracic duct ligation also significantly increased the survival rate.

Typical altered shock response in a dog pretreated with Trasylol, 25,000 U/kg, is represented in Fig. 3. Transient hypotension and decreased respiratory rate were seen. The return to preshock levels followed within 20 min after onset of the shock. The cardiac rate, however, did not return fully to control values.

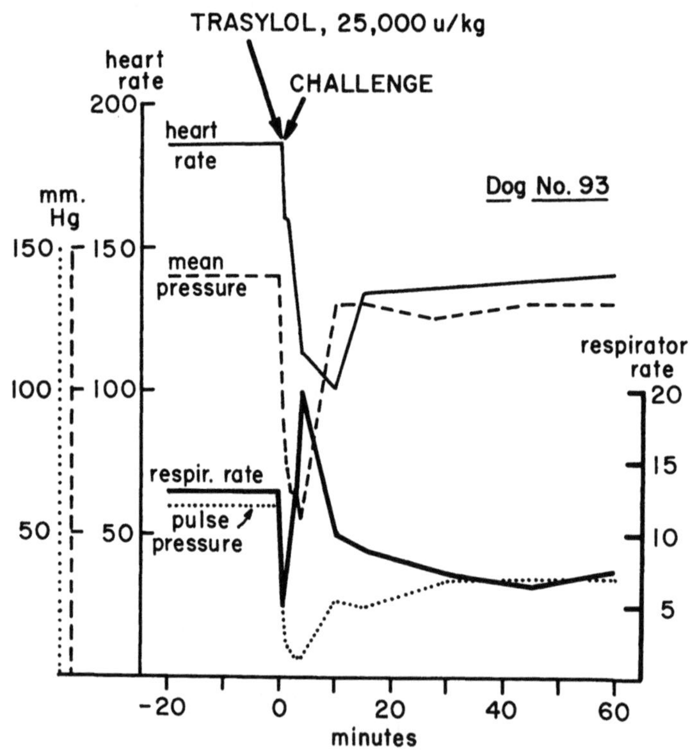

Fig. 3. Course of physiologic changes during anaphylaxis in dog pretreated with Trasylol, 25,000 units/kg i.v.

TABLE 2

Summary of Hematologic and Coagulation Changes During Anaphylaxis in the Dog After Pretreatment with Saline, Various Pharmacologic Agents, and Thoracic Duct Ligation

Drug or Procedure	Dose Range	Number of Dogs	% Showing Heparin Release	COAGULATION CHANGES								
				Platelets	WBC*	Hemato-crit	Factor V	Real Prothrom-bin	Factor VII	PTT**	Fibrinogen	TGT***
Saline	15 mls.	20	80	−	−	+	− o	o −	− o	+	−	−
TRASYLOL	12,500 – 25,000 u/kg	5	75	−	−	+ o	o +	− o	−	+	−	
Heparin	2.5 – 5.0 mg./kg.	2	-				−				−	
Thoracic Duct Ligation	-	5	100	−	−	+	+ o	−	o −	o	−	
Protamine Sulfate	8 – 18 ml. of 1% Solution	2	50	−	−	+	− o	− o	+ o	+ o	− o	o
Epsilon Amino Caproic Acid	250 – 500 mg./kg.	2	100	−	−	+	−	−	o		−	−
Diphenhydramine	20 mg./kg.	1	100				o	−	−	o	−	

− = decrease
+ = increase
o = no change

* WBC = White Blood Cell
** PTT = Partial Thromboplastin Test
*** TGT = Thromboplastin Generation Test

Blood analysis: Coagulation studies. Table 2 summarizes the hematologic and blood coagulation changes in saline-treated and drug-treated dogs. Changes in control dogs included thrombocytopenia, leukopenia, hemoconcentration, a decrease or no change in prothrombin, factors V and VII, an increase in partial thromboplastin test, and a decrease in thromboplastin generation time and fibrinogen. Heparin was also released frequently. Hypercoagulability (intravascular coagulation) usually occurred immediately and was followed by a prolonged state of hypocoagulability, presumably due to heparin and protease release. None of the agents in the doses employed or procedures described in Table 2 affected the hematologic or coagulation changes significantly. Heparin release was inhibited in one of the two protamine sulfate-pretreated animals. All other drugs did not influence in a consistent manner the changes seen in the control dogs.

Three representative coagulograms are seen in Fig. 4. Pretreatment with neither protamine sulfate nor Trasylo

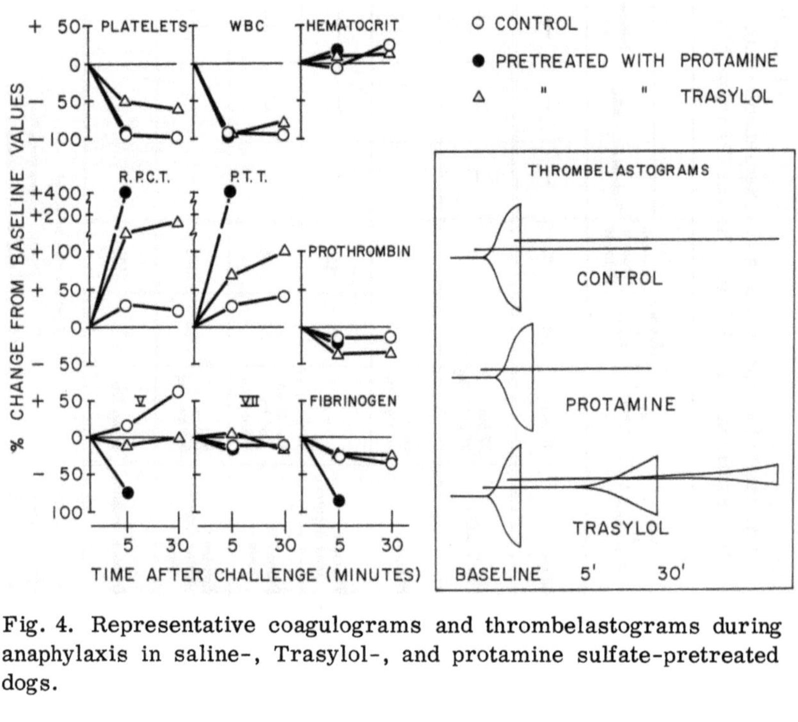

Fig. 4. Representative coagulograms and thrombelastograms during anaphylaxis in saline-, Trasylol-, and protamine sulfate-pretreated dogs.

influenced the coagulation changes signficantly apart from an initial decrease in factor V levels observed with protamine sulfate. This decrease was seen at times in the saline-treated shock dogs. Thrombelastograms also revealed anticoagulation in the blood samples of the saline and protamine sulfate-treated shocked dogs. Trasylol depressed the appearance of circulating anticoagulants until 5 min had elapsed after the shock, when anticoagulation was observed.

Fibrinolysis studies. While detailed analysis of changes in components of the fibrinolysin system are in progress, initial studies show a decrease in plasma plasminogen levels, free plasmin, and an increase in activator activity. Fibrin plate lysis studies showed free circulating plasmin and activator activity in the samples after shock, regardless of pharmacologic treatment. Plasmin was demonstrated in samples drawn immediately (1 min) and up to 1 hour after shock. The assay did not depend on inhibiting the anticoagulants in vitro.

Vasoactive polypeptide studies. Lyophilized alcohol extracts of whole blood stimulated the rat uterus and relaxed rat duodenum. Activity always was present in both preshock and postshock samples. Because in some instances the "kinin" activity in postshock samples was less than in the preshock samples, the possibility of an attentuated kinin effect due to histamine release during shock was considered. The response of the isolated rat uterus to a given concentration of bradykinin (1×10^{-9} g/liter), alone and in the presence of increasing concentrations of histamine (10^{-9} to 10^{-3} g/liter), was studied. Histamine was found to abolish completely the effect of bradykinin, even in the presence of antihistamines. Attempts to destroy any released histamine with diamine oxidase proved unsuccessful. The purest diamine oxidase preparation available destroyed synthetic bradykinin also, presumably owing to a carboxypeptidase contaminant.

Incubation of our test substance with carboxypeptidase did not affect its smooth-muscle-stimulating activity. A systematic study was undertaken to identify the factor responsible for the activity. Table 3 summarizes a comparison of the test substance (alcohol blood extract) with known vasoactive agents. Oxytocin and adenosine triphosphate (ATP) were the only two agents that shared practically all activities common to the test substance. Oxytocin was eliminated because it differed by its susceptibility

TABLE 3
Comparison of Activity of Whole-Blood Alcohol Extract (Test Substance) with Known Vasoactive Agents

Test	Test Substance	Bradykinin	Oxytocin	ATP	Histamine	Serotonin	Acetylcholine	Slow Reacting Substance	Substance P	Angiotensin	*Vaso-Pressin
Rat Uterus	contracts	contracts	contracts	contracts	relaxes or no effect	contracts	contracts	no effect	contracts	contracts	contracts
Rat Duodenum	relaxes	relaxes	relaxes	relaxes	contracts	contracts	contracts		contracts	contracts	contracts
Guinea Pig Ileum	relaxes	contracts	relaxes	relaxes	contracts	contracts	contracts	contracts	contracts	contracts	contracts
Guinea Pig Uterus	contracts	contracts	contracts	contracts	contracts		contracts	no effect	contracts		
Uterine Tachyphylaxis	no	no	no	no	no	yes			yes		
Dog Blood Pressure	lowers	lowers	lowers	lowers	lowers	raises	lowers	lowers	lowers	raises	raises
Dog Salivary Gland	dilates	dilates	dilates		dilates						
Carboxypeptidase Incub.	not destroy	destroys	not destroy	not destroy	not destroy	not destroy	not destroy				
Chymotrypsin Incub.	not destroy	destroys	destroys	not destroy					destroys		
Guinea Pig Plasma Incub.	not destroy	destroys	not destroy	not destroy			destroys		destroys		destroys
Acid Phosphatase Incub.	destroys	not destroy	not destroy	destroys	not destroy	not destroy	not destroy				
0.1 N NAOH	destroys	destroys	destroys	destroys	destroys				destroys		
0.1 N HCL (boil for 30')	destroys	not destroy	destroys	destroys	not destroy				not destroy	destroys	
Dialysis	dialyzable	dialyzable	dialyzable	dialyzable	dialyzable				dialyzable		dialyzable
Boiling Methanol	not destroy	not destroy	not destroy	not destroy	not destroy						

[ ] = lack of correlation with test substance.

to chymotrypsin and resitance to acid phosphatase action. By all criteria studied, the activity of our test substance was identical with that of ATP. Treatment of whole-blood extracts with acid phosphatase destroyed all smooth-muscle-stimulating activity. The use of plasma instead of whole blood for the extraction of kinin failed to demonstrate kinin levels in either the pre- or postshock samples.

Plasma kininogen levels were determined to study any possible substrate utilization during the initial phases of the shock. Figure 5 summarizes some of the initial kininogen data together with kininase levels in shocked dogs pretreated with saline (control), Trasylol, or phenylbutazone, and in dogs with their thoracic ducts ligated. Gradual significant increases in kininogen levels occurred in control dogs as well as in those pretreated with Trasylol or whose lymph ducts were ligated. Decreases occurred in phenylbutazone-treated dogs. Kininogen levels and plasma protein concentrations correlated, reflecting hemoconcentration with a shift of fluids and proteins during shock. In all instances, with the exception of the initial time period in the Trasylol-treated dogs, a concomitant increase in kininase activity occurred. In the exception noted above, a decrease in kininase activity was observed with no change in kininogen levels immediately after initiation of the shock in the Trasylol-treated dogs. This observation is under further study.

The adequacy of our assay method to detect kininogen changes was demonstrated in studies showing plasma kinino-

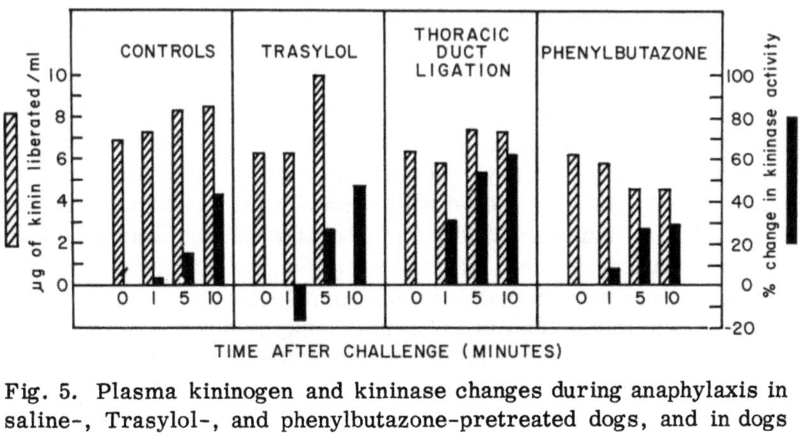

Fig. 5. Plasma kininogen and kininase changes during anaphylaxis in saline-, Trasylol-, and phenylbutazone-pretreated dogs, and in dogs with the thoracic duct ligated.

gen depletion in animals made tachyphylactic to human urinary kallikrein. Many plasma samples were assayed for kininogen, with plasmin and kallikrein as well as trypsin. Trypsin was able to activate at least 70% more kininogen than human urinary kallikrein or human plasmin.

DISCUSSION

Despite the use of a number of pharmacologic inhibitors of the systems shown to be activated during anaphylactic shock, the clinical picture of canine anaphylaxis remained essentially unaltered. The decreased mortality in the group with the thoracic duct ligated suggests the involvement of the lymphatics as a possible route through which factors liberated during shock may enter the systemic circulation (Mayerson, 1963). The proteinase inhibitor Trasylol also reduced mortality. This agent has been shown to reduce mortality in rodent anaphylaxis (Back et al., in press), and in shock due to scalding (Back et al., submitted for publication). The use of Trasylol in controlling fibrinolytic hemorrhage has been reviewed (Back and Ambrus, in press).

These initial studies with various pharmacologic agents have shed little light on the sequence of biochemical events during shock. Intravascular coagulation followed by hypocoagulability appears to be a consistent finding. Although histamine release is considered a major mediator of the shock syndrome, studies with antihistamines and compound 48/80 would involve other mediators. Others (Austen and Humphrey, 1963) have concluded that histamine is not the sole mediator in anaphylaxis.

The hypotensive action of plasmin (Back et al., 1956), its ability to activate kininogen (Back and Steger, 1965), and its formation during many shock states (Back and Ambrus, in press), suggest the possibility that kinin is formed during anaphylaxis. Earlier studies of Beraldo (1950) showed kinin release during shock. Recent studies by Scharnagel et al. (1965), Diniz and Carvalho (1963), and Lecomte (1961) demonstrated kininogen depletion during anaphylactic shock in the rabbit and guinea pig, endotoxin shock in the dog, and heat and dextran shock in rats. Cirstea et al. (1965), however, were unable to detect decreases in bradykininogen levels during canine and guinea pig anaphylaxis, although significant decreases were demonstrated in rabbits experiencing lethal anaphylaxis. Diniz and Carvalho (1963) also reported on bradykininogen increases during the acute phase of hemorrhagic shock in dogs, suggesting substrate release into the general circulation in response to the shock insult. Trypsin-induced shock, however, did not alter kininogen levels in the majority of the dogs studied.

The data reported here are in agreement generally with those of Cirstea et al. (1965) and Diniz and Carvalho (1963). Analysis of kininogen changes during the acute shock phase failed to reveal substrate utilization, nor were any vasoactive polypeptide factors detectable by the assay techniques used. If kinin is formed by the action of proteases shown by us and others (Rocha e Silva, 1960; Ungar et al., 1961) to be activated during anaphylactic shock, it is possible that rapid hydrolysis by the increased kininase prevents its assay. Even in the face of increased kininogen levels, the possibility cannot be excluded that kinin release during both the acute and recovery phase of anaphylaxis may contribute toward the severity of the shock. Whether the role of kinin is essential or major is still undetermined.

SUMMARY

Physiological and biochemical changes occurring during anaphylactic shock in the dog have been studied. The clinical picture of shock (hypotension, bradycardia, apnea, intestinal hypermotility) was unaffected by pretreatment with proteinase inhibitors, coagulants and anticoagulants, antihistamines, kinin inhibitors, and thoracic duct ligation. Survival rate was increased by the proteinase inhibitor Trasylol, phenylbutazone, and thoracic duct ligation.

During the acute phase of anaphylaxis, intravascular coagulation occurred, followed by hypocoagulability. Heparin release and activation of the fibrinolysin system also took place. Attempts to detect circulating kinin during this acute phase were unsuccessful. Kininogen level increases concomitant with increases in kininase activity were seen. The significance of these findings is discussed.

ACKNOWLEDGMENT

The authors acknowledge with appreciation the critical review of this manuscript by Dr. Julian L. Ambrus, Roswell Park Memorial Institute. The technical assistance of I. B. Mink and A. E. Munson also is greatly appreciated, as is the assistance of G. Braverman, Dickinson College, Carlisle, Pa., an undergraduate college participant in the Research Participation Program of the Roswell Park Memorial Institute supported by Grant GE-9380, National Science Foundation.

REFERENCES

Ambrus, J. L., C. M. Ambrus, J. E. Sokal, G. Markus, N. Back, L. Stutzman, R. Razis, C. A. Ross, B. H. Smith, A. C. Rekate, G. L. Collins, D. L. Kline, and J. B. Fishman: Clinical pharmacology of various types of fibrinolytic enzyme preparations. Amer . J. Cardiol. 6, 462 (1960).

Ambrus, C. M., D. H. Weintraub, D. Dunphy, J. E. Dowd, J. W. Pickren, K. R. Neiswander, and J. L. Ambrus: Studies on hyaline membrane disease. I. The fibrinolysin system in pathogenesis and therapy. Pediatrics 32, 10 (1963).

Andersch, M. A.: Determination of proteins in cerebrospinal fluid by biuret method. Amer. J. clin. Path. Tech. Suppl. 8, 43 (1944).

Astrup, T., and S. Mullertz: Fibrin plate methodfor estimating fibrinolytic activity. Arch. Biochem. 40, 346 (1952).

Austen, K. F., and J. H. Humphrey: In vitro studies of the mechanism of anaphylaxis. II. The role of pharmacological materials in in vivo anaphylaxis. Advanc. Immunol. 3, 18 (1963).

Back, N.: Acquired resistance to the pharmacologic effects of plasmin. Study on tachyphylactogenic proteins. J. Pharmacol. exp. Ther. 130, 474 (1960).

Back, N.: Fibrinolysin system and vasoactive kinins. Fed. Proc. 25:77 (1966).

Back, N., and R. Steger: Activation of bovine bradykininogen by human plasmin. Life Sci. 4, 153 (1965).

Back, N., and J. L. Ambrus: Fibrinolytic hemorrhage in surgical bleeding, in Handbook for Surgery and Specialties. New York: McGraw Hill-Blakiston Division, in press.

Back, N., J. L. Ambrus, S. Goldstein, and J. W. E. Harrisson: In vivo fibrinolytic activity and pharmacology of various plasmin (fibrinolysin) preparations. Circulat. Res. 4, 440 (1956).

Back, N., P. S. Guth, and A. E. Munson: On the relationship between plasmin and kinin. Ann. N. Y. Acad. Sci. 104, 53 (1963a).

Back, N., A. E. Munson, and P. S. Guth: Anaphylactic shock in dogs. Role of fibrinolysin and vasoactive polypeptide systems. J. Amer. med. Ass. 183, 260 (1963b).

Back, N., M. Jainchill, H. Wilkens, and J. L. Ambrus: Effect of kallikrein, kinin and plasmin inhibitors on experimental burn shock. Submitted for publication.

Back, N., H. Wilkens, A. E. Munson, and R. Steger: Trasylol in experimental shock states; a preliminary report. Germany: Schattauer Verlag, in press.

Beraldo, W. T.: Formation of bradykinin in anaphylactic and peptone shock. Amer. J. Physiol. 163, 283 (1950).

Brocklehurst, W. E., and S. C. Lahiri: The production of bradykinin in anaphylaxis. J. Physiol. (Lond.) 160, 15P (1962).

Burdon, K. L., J. P. McGovern, G. D. Barkin, and W. M. Meyers: Fibrinolysis and anaphylaxis I. Activation of blood fibrinolysin during acute anaphylactic shock in guinea pigs. J. Allergy 32, 55 (1961).

Cirstea, M., G. Suhaciu, and I. Butulescu: Bradykinin and anaphylactic shock in dogs, guinea-pigs and rabbits. Arch. int. Physiol. 73, 231 (1965).

Dawson, W., and G. B. West: The importance of bradykinin in anaphylactic shock. J. Pharm. Pharmacol. 17, 246 (1965).

Diniz, C. R., and I. F. Carvalho: A micromethod for determination of bradykininogen under several conditions. Ann. N. Y. Acad. Sci. 104, 77 (1963).

Gans, H., and W. Krivit: Study of fibrinogen and plasminogen concentrations in rabbits during anaphylactic shock. J. Lab. clin. Med. 58, 259 (1961).

Hicks, N. D., and W. R. Pitney: A rapid screening test for disorders of thromboplastin generation. Brit. J. Haemat. 3, 227 (1957).

Horton, E. W., and G. P. Lewis: A comparison of the plasma kinin forming enzymes in urine, kallikrein and urokinase. J. Physiol. (Lond.). 149, 477 (1960).

Lecomte, J.: Consommation des kininogenes plasmatiques au cours du choc anaphylactique de lapin. C. R. Soc. Biol. (Paris) 155, 1411 (1961).

Mayerson, H. S.: The physiologic importance of lymph, in Handbook of Physiology. Washington: Amer. Physiol. Soc. (1963).

Mackay, M.: The activation of endogenous plasma proteolytic enzymes. Vox Sang. (Basel) 8, 489 (1963).

Nicola, P. de: Thrombelastography. Springfield: Thomas (1957).

Nolf, P.: Contribution a l'etude de la coagulation. V. Fibrinolyse. Arch. int. Physiol. 6, 306 (1908).

Owren, P. A., and K. Aas: Control of dicumarol therapy and quantitative determinations of prothrombin and proconvertin. Scand. J. clin. Lab. Invest. 3, 201 (1951).

Quick, A. J., C. R. Honorato, and M. Stefanini. Value and limitations of coagulation times in study of hemorrhagic diseases. Blood 3, 1120 (1948).

Rocha e Silva, M.: The role played by leukocytes and platelets in anaphylactic and peptone shock. Ann. N. Y. Acad. Sci. 50, 1045 (1950).

Rocha e Silva, M.: Biochemistry and Pharmacology of Bradykinin in Polypeptides Which Affect Smooth Muscles and Blood Vessels. p. 210. New York: Pergamon (1960).

Rodman, N. F., E. M. Barrow, and J. B. Graham: Diagnosis and control of the hemophiloid states with the partial thromboplastin time (PTT) test. Amer. J. clin. Path. 29, 525 (1958).

Salmon, J., and J. Lecomte: Role de la fibrinolyse dans le choc anaphylactique du lapin. Int. Arch. Allergy 22, 378 (1962).

Scharnagel, K. von, K. Greeff, R. Lühr and H. Strobach: Freisetzung von bradykinin beim toxischen, anaphylaktischen, und anaphylaktoiden shock. Naunyn-Schmiedeberg's Arch. exp. Path. Pharmak. 250, 176 (1965).

Stefanini, M., and W. Dameshek: The Hemorrhagic Disorders. New York: Grune and Stratton, 1962.

Stormorken, H., and T. F. Newcomb: The use of buffers in blood coagulation studies. Scand. J. clin. Lab. Invest. 9, 237 (1957).

Ungar, G., and H. Hayashi: Enzymatic mechanisms in allergy. Ann. Allergy 16, 542 (1958).

Ungar, G., T. Yamura, J. B. Isola, and S. Kobrin: Further studies on the role of proteases in the allergic reaction. J. exp. Med. 113, 359 (1961).

Wycoff, H. D.: A microassay for plasma fibrinogen. J. Lab. clin. Med. 47, 645 (1956).

DISCUSSION

Austen: Systemic anaphylaxis in the actively immunized animal is a syndrome involving several immunoglobulins and several mechanisms; therefore, the mechanism can be studied only in passively sensitized dog, with single immunoglobulins. Based on work in vitro (Austen, and Becker and Austen) and in vivo (Bindon et al.) there is evidence that activation of the fibrinolysin system is not causal in anaphylactic syndrome. Presumably activation is secondary to consequences of the anaphylactic syndrome in vivo.

Zweifach: I would question the use of active anaphylaxis as a test for the involvement of the kinin system in shock, since tissue reactions may occur locally which are not reflected in blood levels. Passive anaphylaxis would be preferable as a test model.

Back: I quite agree that using passive anaphylaxis as a model may simplify the study of the role of the many systems presumed to be involved in anaphylaxis, particularly the kininogenase-kinin system. We are using the isolated perfused hind limb prepared from sensitized dogs in the hope of learning something of the possible local tissue release of kinins.

Waaler: In the complex picture of a special type of shock you showed, I should like to draw attention to the marked fall in the circulating plasma. If ATP is present one would expect ADP also to be present, and ADP will specifically cause thrombocyte aggregation and clumping. One would expect that these animals might have multiple thrombocyte-thrombi in several of their organs, including the central nervous system. Have you any indications that such thrombocyte-thrombi occurred? They could have contributed to the death of the animals.

ATP (and ADP) also causes a marked pulmonary vasoconstriction in several species. Have you any evidence of such a constriction occurring here?

Back: On autopsy we found no evidence of disseminated thrombi. In view of the high fibrinolytic activity, it is quite possible that any thrombi formed (as reported by other investigators) may have dissolved during the course of the experiment. The hypercoagulable state noted during the early phase of anaphylaxis would favor thrombus formation.

The brief period of apnea occurring immediately after initiation of the shock has been shown to be due to pulmonary vasoconstriction. This observation presently is under study. We also are studying the role histamine plays in this pulmonary effect.

Erdös: We have carried out similar experiments with Dr. Massion. In the dog i.v. injection of phenylbutazone (or acetylsalicylic acid) prevented or greatly reduced both the primary and the secondary drop in systemic blood pressure which follows the i.v. injection of E. coli endotoxin (CARI Symposium, Bacteremic Shock Mechanisms and Therapy; in press).

Back: We have also studied the effect of phenylbutazone in anaphylactic shock in mice, produced by scalding; traumatic shock (Noble-Collip) in rats; anaphylactic shock in mice; and aerosol shock in guinea pigs. Phenylbutazone increased survival in anaphylactic shock in the guinea pig only.

Melmon: We have found that Trasylol, while a good inhibitor of plasma and pancreatic kallikrein, requires preincubation with the activated enzyme before it can produce this inhibition. Do you think that the lack of early effect of Trasylol may be related to the phenomena observed? The decrease in mortality observed in your dogs may be due to the fact that Trasylol has no time to inhibit the enzyme released.

Back: We also have made similar observations. Trasylol pretreatment did not inhibit the hypotensive effect of kallikrein, kinin, or plasmin (Fed. Proc., in press). The pharmacologic effects of kallikrein, however, could be inhibited if the enzyme were incubated in vitro with Trasylol.

It is possible that the mechanism by which Trasylol reduces mortality due to anaphylaxis involves inactivation of proteases released. Because of the reported short biologic half-life of Trasylol (10 min), we have administered the inhibitor either immediately or together with the antigenic challenge followed by an infusion of the Trasylol. Reduced mortality was demonstrated under these conditions despite the inability to demonstrate any inhibition by Trasylol of the biochemic and physiologic events during anaphylaxis.

Diniz: In experimental shock produced by proteolytic enzymes, in the Department of Pharmacology of the Faculty of Medicine in Ribeirao Preto, we observed in collaboration with Dr. Corrado that kinin could be detected only in the first minute or two after protease injection.

We have made several attempts to show bradykininogen depletion in anaphylactic shock in dogs — without success, however.

Back: Our data from these experiments, correlated with those reported by us in the JAMA (183, 260, 1963) would suggest that kinin activity appears late in the shock syndrome. This may represent the ability of kinin-forming enzymes to activate kininogen substrate only slowly, not only in vitro but in vivo.

Melmon: We have found some situations which may apply to your inability to detect kinins in the blood plasma. (1) Heparin appears to bind bradykinin and will mask positive biologic assay. (2). The time needed for destruction of kinin in plasma is extremely short and therefore kinin may be destroyed while the blood is being centrifuged. Have you any recovery data regarding bradykinin added to the syringe used to draw your samples from the dog? Of course, in the shock animal you have also found an increase in kininase activity which would add to the problem of a time factor in preparation of the plasma before biologic assay.

Back: In some experiments we have drawn blood with a syringe containing the kininase inhibitor DFP. Under this condition we still were unable to detect plasma kinin activity in animals in shock. The increase in kininase activity resulting from the generalized insult certainly renders more difficult the demonstration of free plasma kinin if kinin release were to occur. The possibility of recovering kinin later during the shock sequence is greater, for kininase levels would be lower, and the proteolytic enzymes released would have more opportunity to act.

We also have attempted to isolate kinin in blood with heparin to which has been added the heparin inhibitor protamine sulfate. Even under these circumstances, no kinin was detected. It is possible that the protamine sulfate does not interfere with bradykinin-binding capacity of heparin.

Zweifach: I would like to bring to the attention of the audience the fact that leukocytes release from their lysosomal granules a basic cationic protein which can reproduce all the manifestations of local tissue inflammation despite the fact that these fractions possess no enzymatic activity of the type under discussion — cathepsins.

TREATMENT OF EXPERIMENTAL AND HUMAN ATHEROSCLEROSIS WITH BRADYKININ-ANTAGONIST, PYRIDINOLCARBAMATE: A PRELIMINARY REPORT

T. SHIMAMOTO, F. NUMANO, T. FUJITA, T. ISHIOKA, and T. ATSUMI

Institute for Cardiovascular Diseases and Department of Medicine, Tokyo Medical and Dental University, Tokyo, Japan

At the beginning of this century there were three incurable lesions having dead spaces with an accumulation of dead subtances: gummatous in syphilis, caseous in tuberculosis, and atheromatous in atherosclerosis. The discovery of modern chemotherapeutic agents has enabled cures for the first two, but the problem of atherosclerosis remains to be solved.

Recently the authors have found a compound which competitively and reversibly antagonizes the venoconstrictor effect of bradykinin and lysyl-bradykinin (kallidin) *in vitro* and also *in vivo*. With this compound we have demonstrated definite success in treating experimental and human atherosclerosis.

PART 1. TREATMENT OF ATHEROSCLEROSIS OF CHOLESTEROL-FED RABBITS

The curative effect of pyridinolcarbamate was tested in atherosclerosis of cholesterol-fed rabbits. The data for this report were obtained in 28 animals sacrificed recently. Large numbers of animals are still in experiment.

Materials and methods. Twenty-eight albino rabbits from the Takeda farm (6 females and 22 males, 8 months old.) with starting body weight of 2.2 0.3 kg were used. In order to produce atherosclerosis (Shimamoto *et al.*, 1966 b) all animals were kept on pellets produced by Oriental East Co., containing 1% cholesterol, for 15 weeks. These 28 atheromatous rabbits were divided into the placebo control group and the treated group. The treated group received 10 mg pyridinolcarbamate/kg daily in a gelatin capsule by mouth and the placebo control group received potato starch in a gelatin capsule.

Three animals in each group were sacrificed on days 4 and 7 during the cholesterol feeding. In the remaining 16 animals the administration of cholesterol pellets was discontinued after 15 weeks of the cholesterol feeding, and they were kept on RC-5 pellets which contain no cholesterol. Two animals were sacrificed after 3, 6, 10, and 15 weeks of the treatment with pyridinolcarbamate or with placebo.

In order to see the morphological changes, including edematous changes in arterial walls, the rapid sampling and fixation of aortic specimens in ice-cold glutaraldehyde solution (5%) or in 1% osmic tetraoxide were employed, and the procedure was finished within 30 sec after the animals were stunned by a blow on the head. For the embedding, paraffin and celloidin were used; stains were hematoxylin-eosin, Elastica van Gieson's, Heidenhains' Azan variant, Sudan III, and Sudan IV. For the identification of myoblast an electron microscopic technique was used.

Pyridinolcarbamate (Anginin) was supplied by Banyu Co.

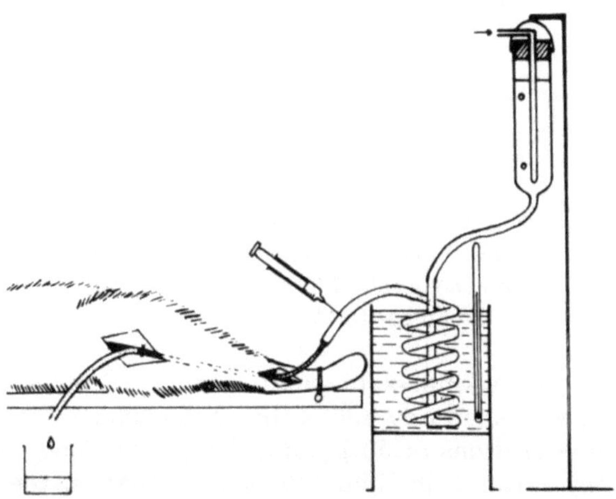

Fig. 1. Technique for measuring venoconstrictor effect.

Ltd. Since the compound is new, it seems to be desirable to present a brief summary of its pharmacological characteristics, including a technique for detecting the venoconstrictor effect of bradykinin (Rowley, 1964).

1) The bioassay technique for measuring the venoconstrictor effect of bradykinin and lysyl-bradykinin in vitro: The authors have established a new technique, using a perfused saphenous vein of amputated rabbit hind leg with Locke-Ringer's solution (37°C) under the pressure of 150 mm H_2O, as shown in Fig. 1. In this preparation the venoconstrictor effect of bradykinin is dose dependent and reproducible, and no tachyphylaxis was observed. A dose-response curve is shown in Fig. 2. The precision index was 0.103.

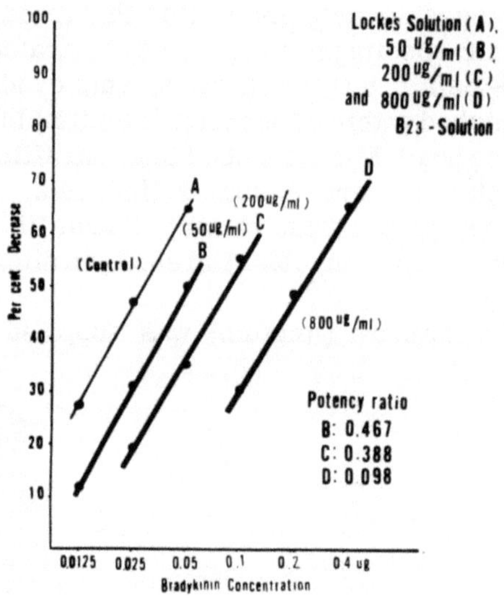

Fig. 2. Dose-response curve of bradykinin. Venoconstrictor effect on rabbit saphenous vein. B23 antagonizes bradykinin competitively.

2) The bradykinin-antagonistic effect of pyridinolcarbamate (B23) was shown by adding this compound to the perfusate in the concentrations of 50 μg/ml, 200 μg/ml, and 800 μg/ml. The potency ratios of pyridinolcarbamate mixed in the perfusate with bradykinin were 0.467, 0.388, and 0.098 respectively, as shown in Fig. 2. The antagonistic effect of pyridinolcarbamate is competitive and reversible and has been shown also in

marginal ear veins of the rabbit and in coronary veins of the dog in vitro. Similar observation had been made in situ as well.

The capillary dilation, exudation, leukocyte imigration, and hemorrhage induced by bradykinin were inhibited by pyridinolcarbamate in rabbits, rats, guinea pigs, and dogs. One of the striking effects of this compound is the prevention and efficient treatment of the edematous arterial reaction induced by cholesterol and epinephrine or by traumatization such as crushing of thigh muscles in rabbits. Oral pretreatment of animals with 5 to 10 mg of the compound/kg had a maximal preventive effect against the appearance of the edematous arterial reaction. The treatment with the same dose of the compound induced the disappearance of the established reaction within 2 hours. This compound exhibited a bradykinin-antagonistic effect on rat uterus and in the pseudoaffective response to bradykinin, but in guinea pig bronchus preparation almost no antagonistic effect to bradykinin was observed.

Otherwise, this compound appeared to have no pronounced effect on the cardiovascular, respiratory, digestive, autonomic nervous, and central nervous systems. Also, monoamine oxidase-inhibiting, antiserotonin, and antihistaminic effects were not seen.

Results. Aortas of the animals of the placebo control group exhibited severe atheromatous changes occupying over 50% of the whole internal surface. The plaques seemed white and slightly yellowish, and the lack of luster of their surfaces was noted.

The histological findings: The typical, well-known features of atheroma were observed in cholesterol-fed rabbits. The lesions exhibited some fibrous tendency in the superficial portion, accompanied by marked edematous features, as shown in Fig. 3-A.

In animals kept on a cholesterol-free diet for 3 to 15 weeks, minimal fibrous changes were seen, a few fibroblasts were found inside the atheroma, and a few smooth-muscle fibers appeared on the surface area of the atheroma (Fig. 3-B, 3-C). However, the amount of fibroblasts and muscle fibers was small, and the main parts of the atheroma were occupied either by an atheromatous mass showing severe fatty degeneration and necrotic foci or by an abundant accumulation of foam cells; the Sudan stain exhibited an abundant accumulation of sudanophilic substance in all the atheromatous lesions (Fig. 5-A).

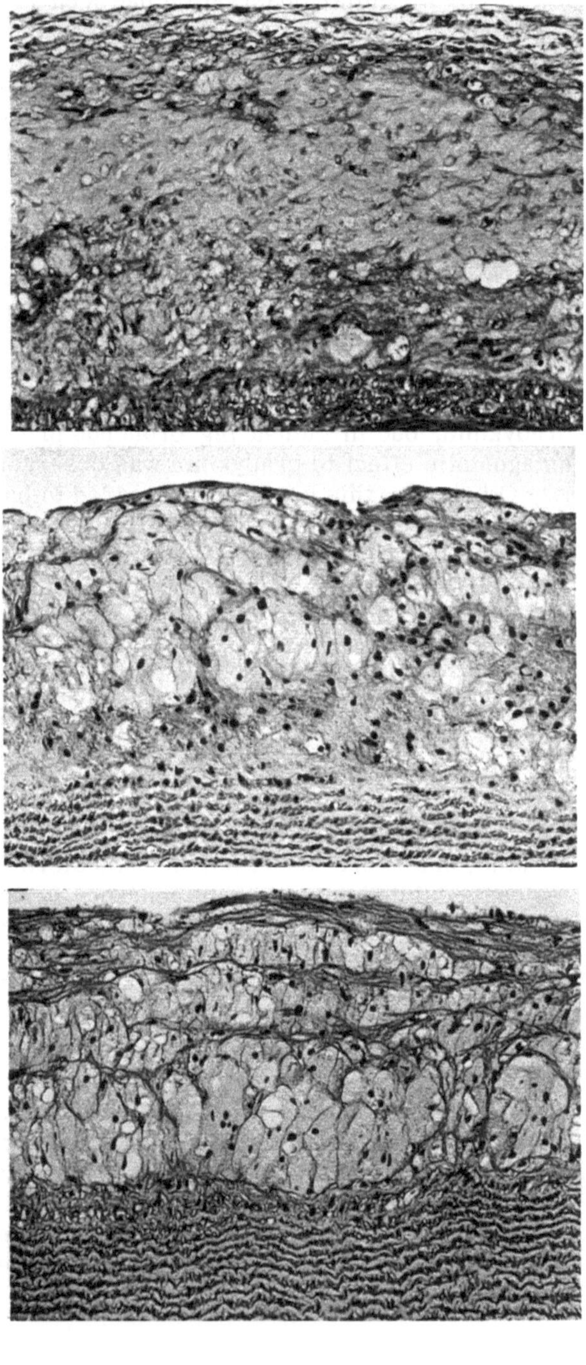

Fig. 3. Poor repair process in atheromatous lesion in placebo-treated group (A, for 4 days; B, for 6 weeks; C, for 10 weeks). H.E. stain, ×200. A: A typical atheroma; edema, acellular spaces with an accumulation of atheromatous material and degenerated foam cells in the intima. B and C: Large amounts of foam cells in the intima and few (B) or a small (C) amount of elastic and smooth-muscle fibers.

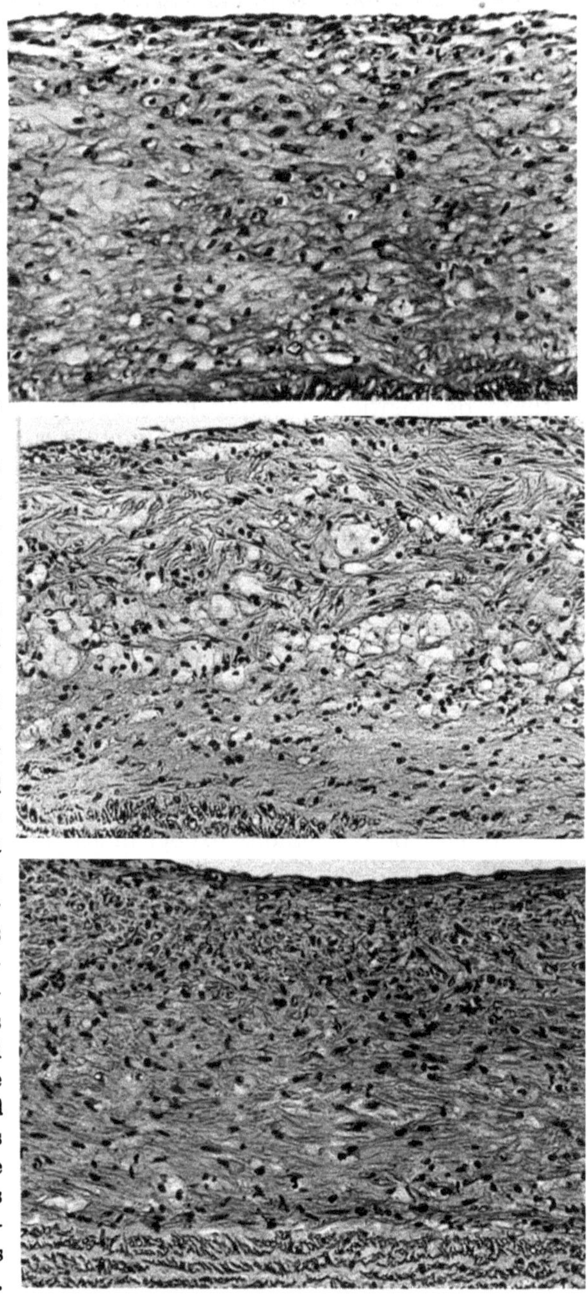

Fig. 4. Repair process in atheromatous lesions by smooth-muscle regeneration in pyridinolcarbamate-treated group (A for 4 days; B, for 6 weeks; C, for 10 weeks). H.E. stain, ×200. A: An appearance of fibroblasts, myoblasts (center) and some repair cells took place inside the hyalinous mass of atheroma, and the disappearance of edema is also noted. B: The smooth-muscle fibers appeared, separating the atheromatous mass into small pieces. C: The regenerated muscle fibers replaced the atheromatous mass; little atheromatous mass remains.

The grade of stainable fat content seemed increased, not decreased, in animals which were kept on the cholesterol-free basic feed especially in the period of 6 to 15 weeks.

The aortas of animals treated with pyridinolcarbamate for 4 to 7 days while they were on cholesterol diet, exhibited grossly almost the same features as aortas of the placebo control group. However, there was a striking difference in the histological findings. In animals which received the compound for 4 days, the atheroma already showed myoblasts, fibroblasts, and fibrocytes inside the atheromatous mass, as seen in Fig. 4-A, and a slight extracellular deposition of collagenous substances was found. The edematous feature had disappeared entirely, and the foam cells exhibited already a slight but definite shrinkage.

In animals which received the compound for one week, a definite desposition of collagen fibers was found in atheromatous lesions. Young smooth-muscle cells with their extremely find and short fibers and elastic fibers appeared in the middle part of atheromatous lesions.

In animals receiving the compound for 3 and 6 weeks, the surface of atheromatous plaques seemed to be somewhat lustrous. The most striking histological finding was abundant mature smooth-muscle fibers inside the atheroma, invading the atheromatous mass occupying the hyalinous foci, and dividing the hyalinous substances or degenerated foam cells into smaller portions, as shown in Fig. 4-B. The smooth-muscle fibers were also accompanied by elastic fibers. In animals receiving the substance for 6 to 10 weeks, the foam cells exhibited further absorption and almost disappeared in the majority of the lesions except in atheromas located at the point of bifurcation showing definite degenerative features (Fig. 4-C). The sudanophilic substance in the majority of the atheromatous lesions was definitely diminished, especially in the areas adjacent to medial layers (Fig. 5-B).

In animals receiving the compound for 15 weeks the percentage of surface area involved in atheromatous lesions was definitely smaller than in the control group. Histological examination showed that regenerated smooth muscle fibers occupied almost fully about half the atheromatous lesions in the thoracic aorta, and elastic fibers were also found among the smooth-muscle fibers. No foam cells were seen, and the Sudan-positive substance was minimal and scattered as small particles in such lesions. In the remaining lesions a few degenerated foam cells were divided among extremely small

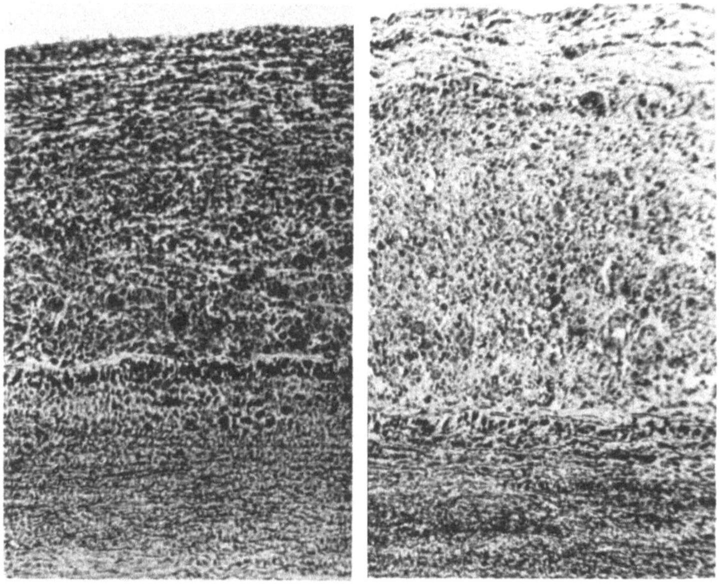

Fig. 5. Sudan III stain of atheromatous lesions, ×100. A: Treated with pyridinolcarbamate for 10 weeks. B: Treated by placebo for 10 weeks. A marked reduction in sudanophilic substances was noted in atheromatous lesion of treated animals.

areas surrounded entirely by regenerated smooth-muscle fibers. A sudanophilic substance was still found in these lesions, however; it exhibited a striking further reduction as compared to that in the animals treated for shorter periods. There were no necrotic or hyalinous foci (Table 1).

At the beginning of this experiment the mean serum cholesterol level of test animals was 1,651.3 ± 68.5 mg/dl and 2, 4, 6, 10, and 15 weeks after the withdrawal of cholesterol diet it was still high, amounting to 1,101, 651, 428, 191, and 113 mg/dl respectively. There was no significant difference between the control and the treated group.

Discussion. The early appearance of repair process in the atheroma was the most striking evidence in the treatment of experimental atherosclerosis with pyridinolcarbamate. It is well known that repair of common aseptic wounds is fast and is usually completed within 8 to 10 days; it comprises two processes: the production of fibrous material filling the dead

TABLE 1

Pyridinolcarbamate Treatment in Atherosclerosis of Cholesterol-fed Rabbits

Histological Findings in Atheromatous Lesions	Placebo Control Group						Pyridinolcarbamate Group					
	Treatment with Placebo for						Treatment with Pyridinolcarbamate for					
	4 d.s.*	1 wks*	3 wks	6 wks	10 wks	15 wks	4 d.s.*	1 wks*	3 wks	6 wks	10 wks	15 wks
Hyalinous & Necrotic Foci	╫ ╫ +	+ + +	+ +	+ +	+ +	+ ±	+ − ±	− − −	− −	− −	− −	− −
Edema	╫ ╫ +	+ + +	+ +	+ ±	+ +	+ ±	+ ± +	+ + +	+ +	− −	− −	− −
Foam Cells	⧻ ╫ ⧻	⧻ ╫ ╫	⧻ ╫	╫ ╫	+ ╫	╫ +	+ + +	± ± +	± +	± ±	± ±	± ±
Collagen Fibers	± ± ±	± ± ±	+ +	+ +	+ +	+ +	+ + ╫	╫ ╫ ╫	╫ ╫	⧻ ╫	⧻ ⧻	⧻ ⧻
Elastic Fibers	− − −	− − −	± +	+ +	+ +	+ +	± ± ±	+ + +	+ +	╫ ⧻	╫ ⧻	⧻ ⧻
Smooth Muscle Cells	− − −	− − −	± +	+ +	+ +	+ ╫	± ± ±	╫ ╫ +	╫ ╫	⧻ ⧻	⧻ ⧻	⧻ ⧻

*1% Cholesterol-Diet was kept on.

spaces and the regeneration of local tissues. In the repair of atheromatous lesions the rapid and striking regeneration of smooth-muscle fibers replacing the atheromatous mass is the most important finding in the treatment of atheroma with pyridinolcarbamate.

Since Virchow (1871), the appearance of edema, small round cells, and polymorphonuclear leukocytes has been described by many morphologists in atheromatous lesions, and Rössle (1934) proposed his "serous inflammation of artery" as a causative mechanism involved in atherogenesis. Shimamoto and Sunaga (1962) and Shimamoto (1963) found that so-called atherogenic substances such as cholesterol or epinephrine or high-molecular weight substances induce the edematous arterial changes as an immediate reaction, and the pretreatment of animals with bradykinin-forming enzyme inhibitor, Trasylol, or with bradykinin antagonists such as pyridinolcarbamate or cyproheptadine (Rocha e Silva, 1964) or acetylsalicylate prevents the edematous reaction induced by these atherogenic substances (Shimamoto, 1965a). These evidences obtained by the authors have indicated that venoconstriction (Rowley, 1964; Shimamoto and Sunaga, 1962; Shimamoto et al., 1966a), in the vasa vasorum induced by kinin may be the causative factor in the production of exudation in the arterial wall; that is, this compound may inhibit the excessive constriction of venous segments of vasa vasorum by bradykinin or by its active homologs which produce the exudation in the arterial wall leading to the production, maintenance, and progression of atheromatous changes and prevents the healing of atheromatous lesions.

The antagonistic effect of pyridinolcarbamate (Shimamoto, 1965a; Shimamoto et al., 1966 a), against the excessive venoconstriction caused by some substances such as, bradykinin may prevent the congestion and exudation and establish a smooth circulation in vasa vasorum, which contributes to the healing process of the hitherto incurable atheroma by regeneration of local structural elements even under the continuous high cholesterolemia.

2. TREATMENT OF HUMAN ATHEROSCLEROSIS

The first therapeutic attempt with pyridinolcarbamate in human atherosclerosis has been successful in patients suffering from atherosclerotic diseases of the extremities and of renal, coronary, and cerebral arteries (Shimamoto and Atsumi, 1965; Shimamoto et al., 1966 a).

This report is limited to the results obtained in 33 patients suffering from atherosclerotic diseases of the extremities, i.e. from arteriosclerosis obliterans, in which the spontaneous improvement is rare, the morbid condition is always progressive, and the improvement of the atheromatous lesions is related directly to the easily observable increase in the peripheral circulation (to the reappearance of the dead arterial pulsation or to the increase in the amplitude of peripheral arterial pulsation).

Materials and methods. Thirty-three male patients suffering from mild arteriosclerosis obliterans mainly in the lower extremities were hospitalized in our university hospital and treated with pyridinolcarbamate. Ages ranged from 30 to 62 years. 19 of all 23 cases had no cutaneous changes and remaining 4 had cutaneous ulcer and coloration. The drug was given in tablet form containing 250 mg of the compound, two tablets after breakfast and supper. The daily dose was 1 g. Besides the routine clinical and laboratory examinations, arteriography, electrical recording of arterial pulsation under controlled pressure in the affected dorsal pedal arteries, and plethysmography mainly in the second toe of the affected foot were performed at certain intervals before and after initiation of the treatment. Arteriography revealed either an occlusive change or a typical narrowing in the lower part of the femoral artery or in the upper part of the popliteal artery in all patients.

Results. The most striking effect was the early disappearance of cyanosis of the affected foot, which was observed in all 3 patients on days 3 to 20 after initiation of the treatment. At the same time the patients felt warmth in the affected foot; however, the skin temperature rose only 0.1 to 0.2°C in this initial stage of the improvement. The spontaneous painful sensation due to local circulatory insufficiency disappeared concomitantly; however, only a minimal improvement in intermittent claudication was observed at this stage. The definite increase in the amplitude of arterial pulsation or the reappearance of dead arterial pulsation of the affected dorsal pedal artery was often confirmed on days 4 to 8 after the initiation of the treatment; however, the reappearance of visible and recognizable arterial pulsation of affected arteries as judged by palpation took place on days 5 to 32 of the treatment in 20 arteries of all 24 affected arteries; 4 arteries exhibited no change (Table 2; Fig. 6).

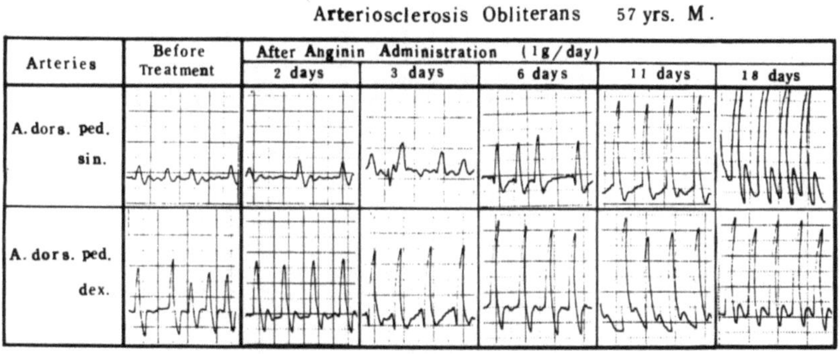

Fig. 6. Rapid improvement of affected arterial pulsation by treatment of atherosclerosis with Anginin.

In one patient aged 57 years the dead arterial pulsation of the right pedal artery reappeared at the end of the third week of treatment, while in the left pedal artery it took place at the end of the seventh month of treatment. The plethysmography recorded at the end of the fourth month of treatment exhibited a definite shortening in crest time in his bilateral second toe, and the arteriography revealed the reopening of the affected arteries. The intermittent claudication was improved concomitantly with the increase in the arterial pulsation recorded electrically, but in the stage of reappearance of palpable arterial pulsation, the intermittent claudication showed a marked improvement. For instance, before the treatment, 2 minutes was the upper limit of painless walking for one patient; during the treatment it increased gradually, especially during the third week, concomitantly with the reappearance of the dead dorsal artery of his right foot; at the end of five months, the patient could walk without pain for almost 20 minutes. The skin temperature was normal at this stage. Cutaneous wounds of affected toes and amputated ends of toes exhibited a rapid and dramatic healing in the second to fifth weeks of treatment in all four patients who had cutaneous wounds.

All patients have been taking the compound for 1 to 13

TABLE 2
Pyridinolcarbamate Treatment in Arteriosclerosis Obliterans
Reappearance of Dead Arterial Pulsation

Age & Sex	Arteries	Pyridinolcarbamate Treatment (1.0 gm/day p.o.) 10 · 20 · 30	150 days
62 y. M	A.dor.ped.d.	- -	- - - - - - - - - - - - - - -
	A.dor.ped.s.	± ± ± ± ± ± ± ± ± ± ± ± ± +	+ + + + + + + + + + + + + + + +
	A.pop.d.	± ±	± ± ± ± ± ± ± ± ± ± ± ± ± ± ±
60 y. M	A.dor.ped.s.	± ± ± ± ± ± ± +	+ + + + + + + + + + + + + + + +
57 y. M	A.dor.ped.d.	- - - - - - - - - - - - - - - - - - - + + + + + + + + + + + + + + + + + +	+ + + + + + + + + + + + + + + +
	s.	- -	- - + + + + + + + + + + + + + +
57 y. M	A.dor.ped.d.	± ± ± ± ± ± +	+ + + + + + + + + + + + + + + +
	s.	- - - - - - ± ± ± ± ± +	+ + + + + + + + + + + + + + + +
57 y. M	A.dor.ped.d.	- + + + + + + +	+ + + + + + + + + + + + + + + +
	s.	- + + + + + + +	+ + + + + + + + + + + + + + + +
56 y. M	A.dor.ped.d.	- + + + + + + + + +	+ + + + + + + + + + + + + + + +
55 y. M	d.	- - - - - - - - - - - - - - - - +	+ + + + + + + + + + + + + + + +
42 y. M	d.	- - - - +	+ + + + + + + + + + + + + + + +
39 y. M	s.	- - - - - - - - - - +	+ + + + + + + + + + + + + + + +
37 y. M	d.	- + + + + + +	+ + + + + + + + + + + + + + + +
33 y. M	A.dor.ped.d.	- - - - - - - - - - - - - +	+ + + + + + + + + + + + + + + +
	A.pop.d.	- - - - - - - - - - - - - - - ± ± ± ± ± ± ± ± + + + + + + + + + + + + + +	+ + + + + + + + + + + + + + + +
	A.tib.pos.d.	- + + + + + + + + + + + +	+ + + + + + + + + + + + + + + +
30 y. M	A.dor.ped.d.	± ± ± ± ± ± ± ± ± ± ± ± ± ± ± ± +	+ + + + + + + + + + + + + + + +

Arterial Pulsation {(+) (±) (−)}

months and the improvement continues. Neither relapse nor untoward effect has been observed.

Discussion. The prompt improvement of local circulatory disturbance such as the disappearance of local cyanosis or pain of affected feet, which takes place on days 3 to 10, corresponded to the rapid disappearance of edema and the appearance of repair process in the experimental atherosclerosis by pyridinolcarbamate described in the first part of this paper. It also corresponds to the prompt disappearance or definite amelioration of anginal attack by pyridinolcarbamate treatment, which occurs mainly on days 3 to 5 of treatment of patients suffering from coronary sclerosis (Shimamoto, 1965b; Shimamoto et al., 1966a).

The reappearance of dead arterial pulsation of affected feet was often slow, occurring in the course of 1 to 4 weeks in the majority of cases and in some cases during the seventh month of treatment. Such characteristic features in the course of clinical improvement by pyridinolcarbamate seem also to correspond to the pathological signs of improvement in experimental atherosclerosis, as shown in the first part of this paper.

Summary. Pyridinolcarbamate, a bradykinin antagonist, exhibited a definitely curative effect in atheroma of cholesterol-fed rabbits. The appearance of the effect was prompt, and the leading process was the replacement of atheromatous mass with regenerated smooth-muscle fibers; this begins early, within the first week of the treatment and progresses gradually over 15 weeks. Pyridinolcarbamate treatment with a daily dose of 1 g was definitely successful in 33 patients suffering from mild arteriosclerosis obliterans. These preliminary data warrant further clinical investigations of the therapeutic effectiveness of pyridinolcarbamate in human atherosclerosis.

REFERENCES

Rocha e Silva, M.: Ann. N.Y. Acad. Sci. 105, 899 (1964).

Rössle, R.: Verh. dtsch. path. Ges. 27, 152 (1934).

Rowley, D. A.: Brit. J. exp. Path. 45, 56 (1964).

Shimamoto, T.: J. Atheroscler Res. 3, 87 (1963).

Shimamoto, T.: in Methods and Achievements in Experimental Pathology, E. Bajusz and G. Jasmin, eds., Vol. 1. Basel and New York: Karger (1965a).

Shimamoto, T.: Jap. Med. 4, 230 (1965b).
Shimamoto, T., and T. J. Sunaga: Jap. Heart J. 3, 521 (1962).
Shimamoto, T., and T. Atsumi: Jap. Heart J. 6, 57 (1965).
Shimamoto, T., H. Maezawa, H. Yamazaki, T. Atsumi, T. Fujita, T. Ishioka, and T. Sunaga: Amer. Heart J. 71, 297 (1966a).
Shimamoto, T., T. Fujita, and F. Numano: Amer. Heart J. 71, 216 (1966b).
Virchow, R.: Die zellulär Pathologie, und ihre Begründung auf physiologische und pathologische Gewebelehre (1871).

DISCUSSION

Collier: Does pyridinolcarbamate share with antipyretics the ability to antagonize bradykinin in the smooth muscle and blood vessels of guinea pig lung? Conversely, do antipyretics antagonize bradykinin-induced venoconstriction in your preparation of rabbit hind leg? If so, then does pyridinolcarbamate show antipyretic action and fit into the group of antipyretics?

Shimamoto: In the guinea pig the bronchoconstriction caused by bradykinin was not significantly antagonized by pyridinolcarbamate. I have no data on blood vessels of guinea pig lung. However, the blueing lesion induced by local application of bradykinin was antagonized by this substance. In saphenous vein preparation of rabbit aspirin (also nialamide and cyproheptadine) did antagonize bradykinin-induced venoconstriction. This was of noncompetitive type, differing from the competitive-type block by pyridinolcarbamate. However, pyridinolcarbamate showed no antipyretic action like that of aspirin. This effect might be extremely minimal, if any. I have no idea whether it is good or not to classify this compound as antipyretic.

Oates and Melmon: If there actually is a therapeutic action of your compound, it is most important to perform controlled clinical studies which will clearly demonstrate this. Arterio-occlusio obliterans is far from a static disease, and therapeutic effect can be established only by a controlled, double-blind study using placebo. We also suggest a crossover design in the studies on patients with angina, because continued administration of placebo alone will produce progressive improvement in angina.

Pratesi: What is the proof of the assumption that the pyridinolcarbamate treatment acts by means of the canalization of the occluded artery? What is the importance of the collateral circulation?

Shimamoto: Arteriographical analyses showed the partial or complete reopening of the occluded main trunk of the artery after the treatment. The collateral circulation seems not to be important. The opening of the trunk simply made its local branches available for circulation.

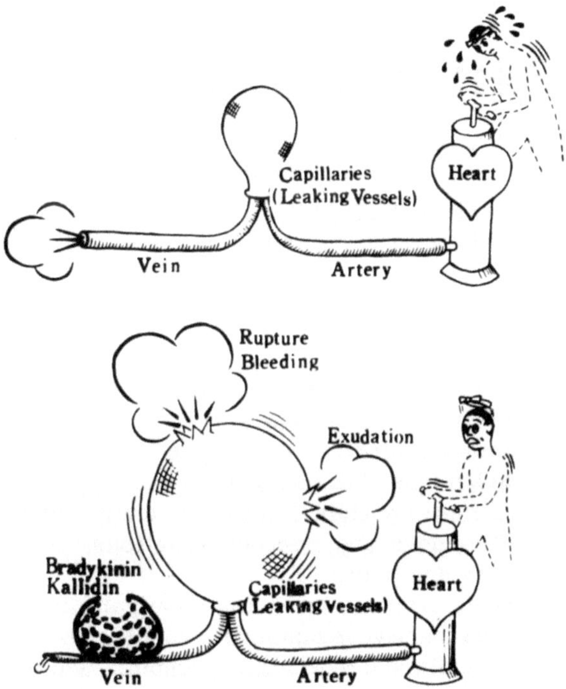

Significance of venous tone in exudation and bleeding, showing the need of bradykinin antagonist in these morbid conditions.

SOME PHYSIOLOGICAL AND PATHOLOGICAL ROLES OF KININOGEN AND KININS

F. SICUTERI, G. FRANCHI, P. L. DEL BIANCO, and M. FANCIULLACCI

Center for the Study of Headaches, General Medical Clinic of the University of Florence, Italy

INTRODUCTION

Some biological substances may be called "vasoneuroactive" since their target organs are the blood vessels and the nervous system (Sicuteri, 1965b). This is an arbitrary way to group these substances but useful for handling the pathogenetic problems of some important syndromes characterized by local circulatory disorders and nervous reactions, such as neuropsychic troubles and pain.

Polypeptides such as bradykinin and kallidin may be considered vasoneuroactive substances, since they certainly act on the microvessels and on the pain receptors and perhaps also on the central nervous structures. Our investigations in man deal with physiological and pharmacological aspects first and then with the pathological aspects of this action.

Role of the Kinin System in Relation to Normal Human Vascular Tissue

Control of the microcirculation. The first question is whether capillary dilatation represents under normal conditions an active phenomenon, and whether vasodilating polypeptides such as kallidin and bradykinin take part in this regulation. Part of our research indicates that plasma kinins play a role in the physiologic control of microcirculation. For example, we have found that plasma kininogen, which is the kinin precursor, is

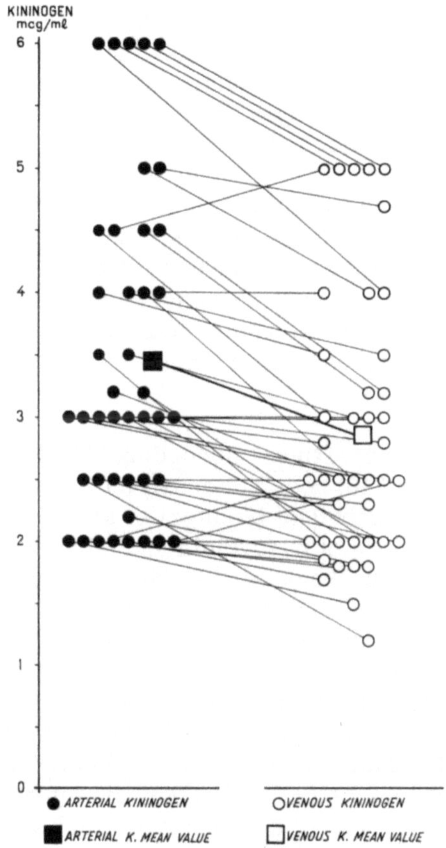

Fig. 1. Plasma kininogen values in 42 normal subjects. The arterial kininogen on the left; the venous kininogen on the right.

higher in the arterial than in venous blood (Sicuteri et al., 1963b) (Fig. 1). This suggests that the blood at capillary level may release some kinins. The difference between arterial and venous kininogen in some cases may be as great as 200 or even 1000 ng/ml. Evidently there exists at capillary level an enzymatic kinin-releasing activity, which is prolonged by circulatory arrest. In fact, after induced ischemia the venous blood contains less precursor than before.

Moreover, another interesting finding is the "clearing" power of the capillary bed for synthetic bradykinin (BRS-640).

If this peptide is injected intravenously or intra-arterially in such amounts as to induce a striking capillary dilatation, there will be no, or only a very slight, increase in the kinin level of the venous blood. This may mean that the microvessels destroy bradykinin immediately. These findings have been confirmed by the more extensive investigations of Allwood and Lewis (1964). It can be excluded that this phenomenon is due solely to plasma kininase (Erdös 1962) because of the very short time taken by the blood to pass from the capillaries to the veins. We have found no convincing differences between the activities of arterial and venous kininase. It may simply happen that microvessels "retain" the bradykinin and then destroy it.

Algogenic properties. According to our observations, bradykinin (BRS-640) does not constrict the veins in man. Using our original technique (Sicuteri et al., 1964), we have also studied the influence of bradykinin and other vasoactive agents on the veins of the back of the hand. This technique is very sensitive and shows, through an increase in the local venous pressure, even a very small constriction of the veins.

Whereas the vasoconstrictor actions of small doses (e.g. 200-500 ng or 50-100 ng respectively) of catecholamines (epinephrine and norepinephrine) or 5-hydroxytryptamine (5-HT) may be easily demonstrated on our "preparation," we find that bradykinin (BRS-640) does not constrict the veins in man even when it is used in high doses which affect the general circulation.

The other vasoactive peptides — angiotensin, oxytocin, arginine-oxytocin, kallidin, eledoisin, physalaemin — behave likewise. The slow local infusion of ergotamine induces a long-lasting venoconstriction; other substances such as indomethacin, which constricts intracranial vessels in man, and methysergide and bradykinin, even when slowly infused for a long time, do not induce venoconstriction in man.

Another of our pharmacological observations on the nervous action of the kinins is the potentiation of the pain-producing property of bradykinin by 5-HT (Sicuteri et al., 1965c). From the investigations of Armstrong et al., (1957) and Armstrong and Keele (1964), we know that bradykinin can produce "moderate" to "severe" pain and that its algogenic threshold concentration is very low. If nociceptors are sensitized previously by 5-HT, the pain-producing activity of bradykinin, kal-

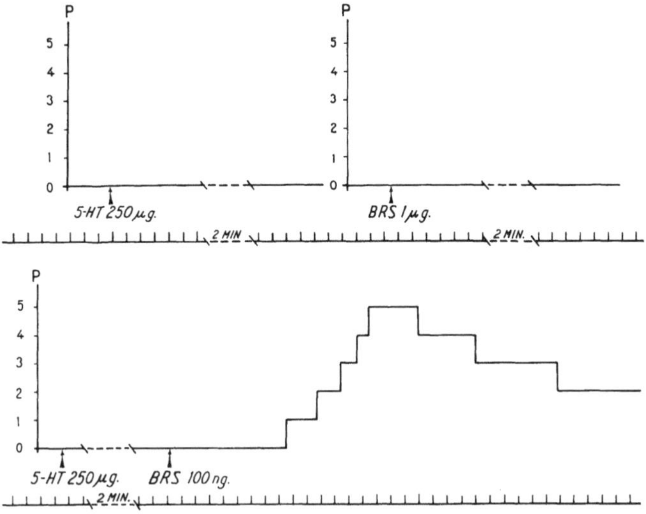

Fig. 2. 5-HT and synthetic bradykinin (BRS-640) given separately in different veins of the hand do not provoke pain (top). If BRS-640 is given in the same vein with 5-HT, pain will arise after a latent period (bottom).

lidin, and kallikrein will be greatly increased. This phenomenon which can be shown on dermal, subcutaneous, and muscular nociceptors can be very clearly demonstrated on the veins of the back of the hand (Fig. 2). There is a latent period between the contact of bradykinin with the 5-HT-sensitized nociceptors and the start of the pain. In some subjects there exists a reproducible inverse relation between the latency and the intensity and duration of the pain; the latency allows a semiobjective estimation of the pain and, by applying a simple formula, we can measure the pain in terms of the latent period (seconds). The 5-HT vein preparation only rarely shows tachyphylactic response to the pain-producing actions of bradykinin (Fig. 3).

Other vasoactive peptides such as angiotensin, eledoisin, physalaemin, vasopressin, and oxytocin do not induce pain on this substrate.

Kinins released by dilution of plasma also evoke pain. Vasoactive peptides released in vitro by incubation of trypsin with plasma, inexplicably do not provoke pain, even in doses which produce general reactions. Even trypsin itself does not cause pain on the 5-HT vein preparation.

There is chemical specificity in the polypeptides which induces the usually strong pain from these veins.

It can be suspected in thrombosis or embolism, where 5-HT is largely released from disintegrated platelets, that this 5-HT-kinin potentiation, which we have demonstrated to be so remarkable with bradykinin, may play an important part in the syndrome by sensitizing the vascular nociceptors.

Antiserotonin drugs such as methysergide, according to our previous observations, are able to inhibit the 5-HT venoconstriction but not to reduce the bradykinin pain. The bradykinin pain on the 5-HT vein preparation is not affected by antipyretic-analgesic drugs such as pyrazolones, salicylates, indomethacin (Sicuteri et al. 1965d), and morphine. A definite decrease of the pain-producing property of bradykinin is observed by previous local treatment of the vein with mono-

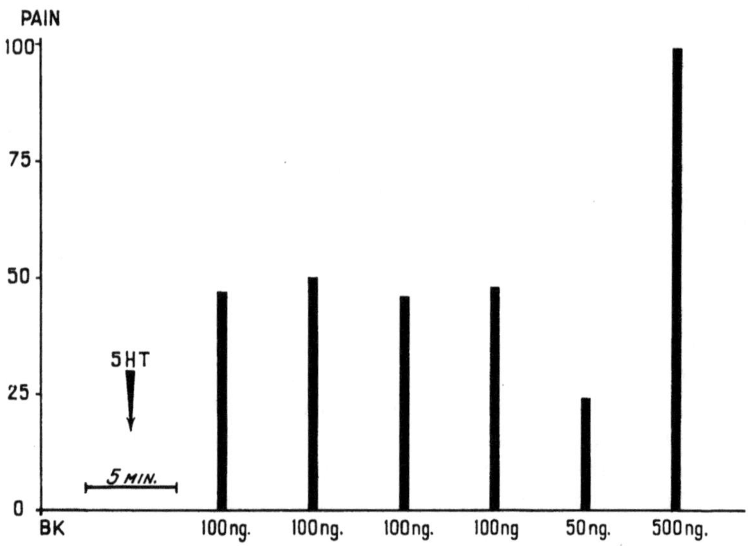

Fig. 3. Potentiation of the algogenic activity of bradykinin on the 5-HT vein preparation. After 5-HT treatment, very low doses of bradykinin provoke severe pain.

amine oxidase inhibitors, such as nialamide, pargyline, and phenelzine. Phenelzine is the most active drug and may completely inhibit the pain induced by kallikrein on the 5-HT vein preparation.

The Kinin System in Relation to the Pathology of Myocardial Infarction and Subarachnoid Hemorrhage

The implication of the kininogen-kinin system in pathological conditions has been extensively discussed. We would like to present briefly our observations on myocardial infarction and then describe our observations in the field of subarachnoid hemorrhage. These latter findings have led us to formulate a new hypothesis concerning the production and pathological importance of kinins.

Myocardial infarction. This seems to be a condition in which the sensitization of nociceptors by 5-HT is very important. During thrombotic or embolic vascular occlusions the disintegrated platelets undoubtedly release large amounts of 5-HT. This amine will sensitize to kinins the nociceptors of the vascular and perivascular structures.

The hypotension, shock, and pain during myocardial infarction remain as yet unexplained. It may be postulated that proteolytic kinin-releasing enzymes are activated by the necrotized myocardium. According to our preliminary observations, the plasma kininogen is decreased during myocardial infarction. Our investigation concerns 14 cases of myocardial infarction, three of which were fatal.

During the time of circulatory collapse, plasma kininogen is lowered, with a tendency to rise again to normal values in the subacute stage (Fig. 4). The patients who suffered a sudden aggravation during this stage also had a definite decrease in kininogen level.

In interpreting these findings it should be mentioned that a rapid decrease of kininogen level, as we observed during myocardial infarction, can hardly be related to a reduced synthesis of the kinin precursor. It is more likely that there is a kinin release which might partially explain the serious circulatory collapse in myocardial infarction. On the other hand, Trasylol seems to be able to improve pain and shock in myocardial infarction (Sotgiu and Artuso, 1965).

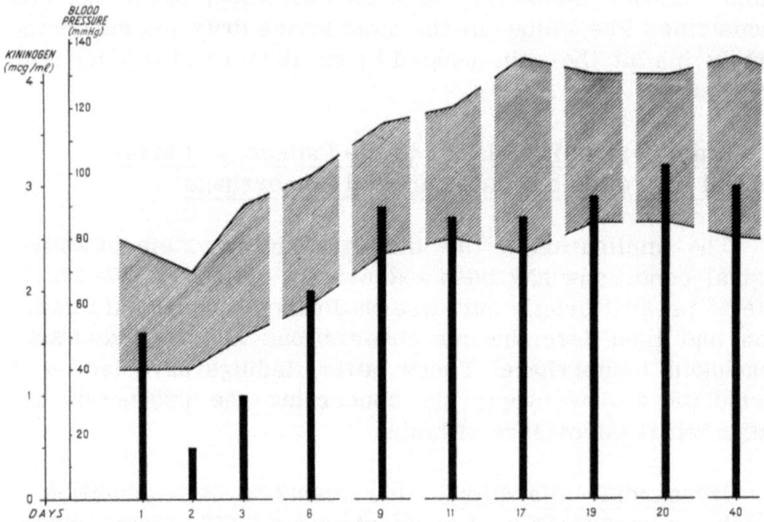

Fig. 4. Relationship between kininogen level and blood pressure in a typical myocardial infarction with shock.

Subarachnoid hemorrhage. According to our hypothesis the implication of the kinins in the most important phenomena of subarachnoid hemorrhage — headache and meningism — seems to be direct and convincing. We believe that the main pathogenic factor is the mixture of blood with the cerebrospinal fluid (csf). (Sicuteri, 1965a) (Fig. 5).

When blood, for instance from a bleeding aneurysm flows into the subarachnoid spaces, it is diluted with the cerebrospinal fluid, and large amounts of kinins are released into the intrathecal spaces. According to their characteristic properties, kinins will affect the intracranial structures and particularly blood vessels, sensitive neuroreceptors, and central nervous structures.

The following findings in vitro and in vivo support this hypothesis.

a) When plasma is diluted with either csf or Ringer's solution, kinin formation can be detected on the isolated rat uterus (Sicuteri et al., 1965a) (Fig. 6).
b) On the 5-HT vein preparation, plasma-csf, and Ringer's solution, if injected separately, do not induce pain. The in-

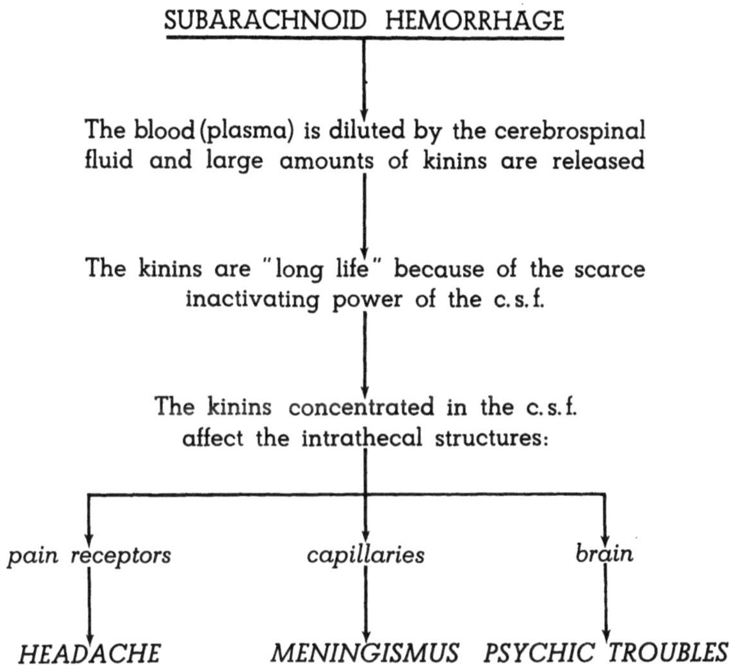

Fig. 5. Pathogenesis of the subarachnoid hemorrhage phenomena.

jection of plasma diluted with Ringer's solution or cerebrospinal fluid induces pain (Fig. 7).

c) We have observed that the csf contains only small amounts of kininase (Sicuteri et al., 1965b). As measured on the rat uterus, bradykinin is destroyed much more quickly when incubated with plasma than when incubated with cerebrospinal fluid (Fig. 8). The algogenic activity of bradykinin, when incubated with plasma, rapidly decreases, whereas it remains for a long time when incubated with csf (Fig. 9).

d) We have found in normal cerebrospinal fluid only trace amounts of kinins, whereas in recent subarachnoid hemorrhages the csf contains kinins in concentrations 20 to 50 times higher than normal (Sicuteri et al., 1963a).

e) The intrathecal injection of minimal doses of synthetic bradykinin provokes throbbing headache and meningism. The injection of physiological salt solution may or may not induce headache. The injection of 2 to 3 ml of "pre-active" plasma is followed by throbbing headache and later by meningism. These phenomena end in a benign way within 2 to 3 days.

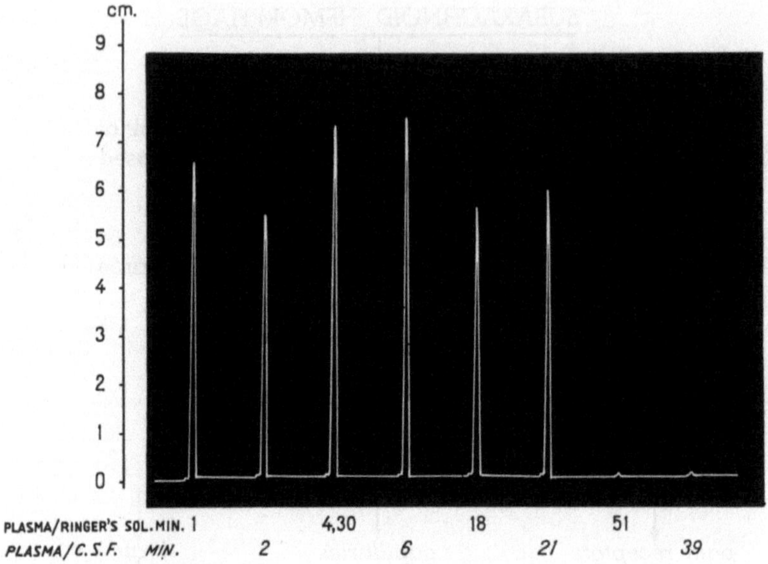

Fig. 6. The same amounts of kinins are released when the plasma is diluted with Ringer's solution or csf (isolated rat uterus in 10 ml Ringer's at 31°).

We feel there is much to commend the theory that the mechanism of the dilution of blood with cerebrospinal fluid is responsible for both headache and meningism in subarachnoid hemorrhage. It may also cause the meningeal re-

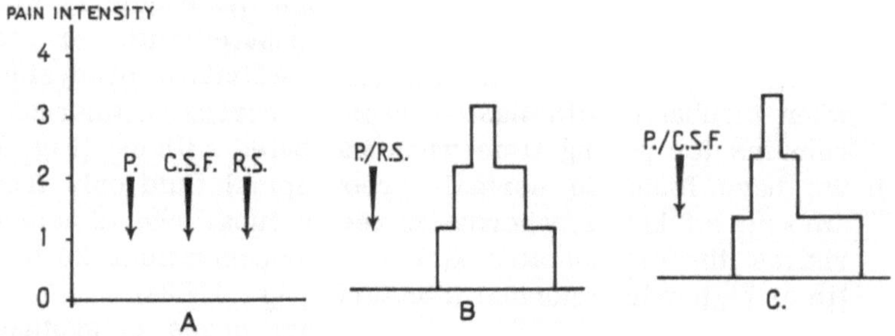

Fig. 7. Plasma diluted with csf (C) exibits the same pain-producing ability as with Ringer's solution (B). Plasma csf, and Ringer's given separately do not provoke pain (A) (5-HT-pretreated vein preparation).

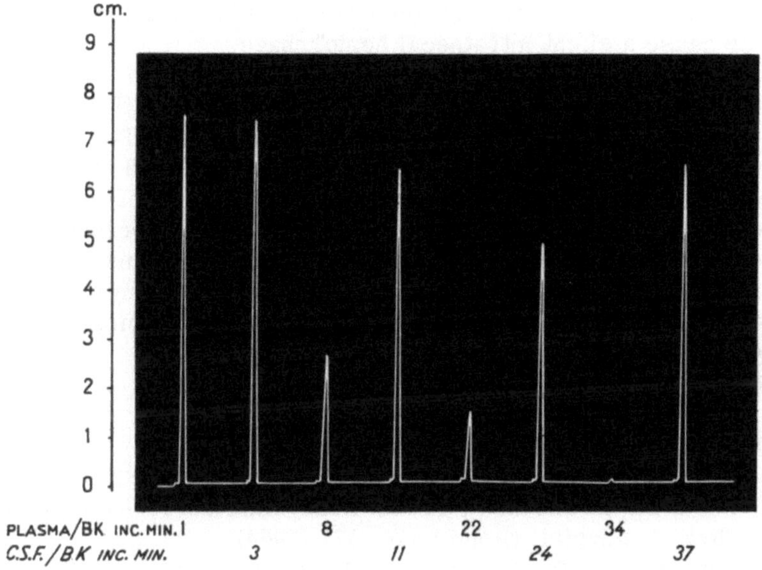

Fig. 8. Synthetic bradykinin incubated with human plasma loses its smooth-muscle-contracting activity more quickly than with csf.

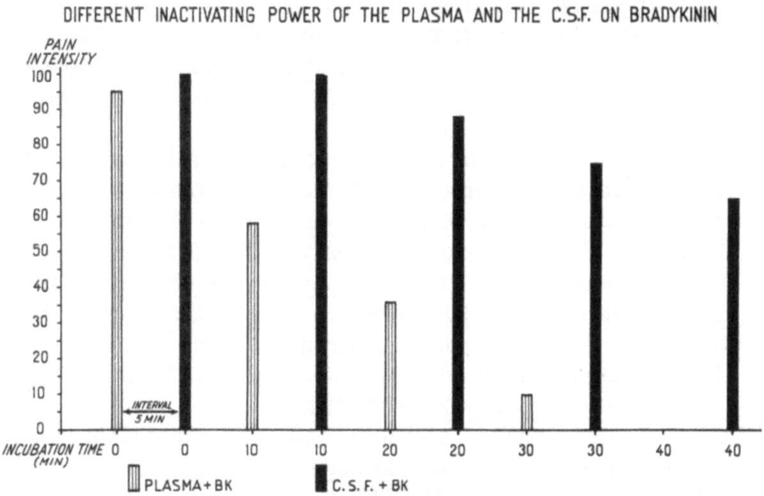

Fig. 9. Synthetic bradykinin incubated with human plasma loses its pain-producing activity more quickly than with csf (5-HT-pretreated vein; pain is evaluated with semiobjective criteria).

actions that sometimes follow simple lumbar puncture, which may cause a slight intrathecal hemorrhage.

We shall test our hypothesis still further, since it is the only one that gives a convincing explanation for the serious clinical phenomena of subarachnoid hemorrhage. The old hypothesis that the phenomena are due to intracranial hypertension is to be rejected: (1) because intracranial hypertension is not always observable during subarachnoid hemorrhage; (2) because it has been shown that intracranial hypertension by itself does not provoke headache; (3) because intracranial hypertension does not induce the meningitic picture, typical of this syndrome.

REFERENCES

Allwood, M. S., and G. P. Lewis: Bradykinin and forearm blood flow. J. Physiol. (Lond.) 170, 571 (1964).

Armstrong, D., C. A. Keele, J. B. Jepson, and J. W. Stuart: Pain producing substance in human inflammatory exudates and plasma. J. Physiol. (Lond.) 135, 350 (1957).

Armstrong, D., and C. A. Keele: Substances Which Produce Pain and Itch. London: Oxford University Press (1964).

Erdös, E. G.: Enzymes that inactivate polypeptides. Proceedings of the First International Pharmacological Meeting, Vol. 6. B. B. Brodie and E. G. Erdös, Eds., p. 159 Oxford: Pergamon Press (1962).

Sicuteri, F.: "La emodiluizione liquorale" quale fattore patogenetico della sindrome cefalalgica-meningeale nella emorragia subaracnoidale. Settim. med. 53, 173 (1965a).

Sicuteri, F.: Vasoneuroactive substances and their implication in vascular pain. Research and Clinical studies in Headache: An International Review. In press, Karger (1965b).

Sicuteri, F., P. Periti, and G. Franchi: Bradichininogeno e chinine libere nel liquido cefalo-rachidiano dell'uomo. Boll. Soc. ital. Biol. Sper. 40, 67 (1963a).

Sicuteri, F., P. Periti, G. Franchi, and D. Santini: Il quoziente artero-venoso del bradichininogeno dell'uomo. Boll. Soc. ital. Biol sper. 39, 310 (1963b).

Sicuteri, F., P. L. del Bianco, M. Fanciullacci, and G. Franchi: Il test della venocostrizione per la misura della sensibilita alla 5-idrossitriptamina e alle catecolamine nell'uomo. Boll. Soc. ital. Biol. sper. 40, 1148 (1964).

Sicuteri, F., P. L. del Bianco, M. Fanciullacci, and B. Anselmi: Liberazione di plasmachinine da diluizione del sangue nel liquido cefalo-rachidiano. Boll. Soc. ital. Biol. sper. 41, 462 (1965a).
Sicuteri, F., M. Fanciullacci, P. L. del Bianco, and B. Anselmi: Bradichininasi nel liquido cefalo-rachidiano nell'uomo. Boll. Soc. ital. Biol. sper. 41, 464 (1965b).
Sicuteri, F., M. Fanciullacci, G. Franchi, and P. L. del Bianco: Serotonin-bradykinin potentiation on the pain receptors in man. Life Sci. 4, 309 (1965c).
Sicuteri, F., S. Michelacci, and B. Anselmi: Termination of migraine headache by a new antiinflammatory vasoconstrictor agent. Clin. Pharmacol. Ther. 6, 336 (1965d).
Sotgiu, G., and P. Artuso: Action of protease inhibitors in myocardial infarction. Mal. cardiovasc. 6, 283 (1965).

DISCUSSION

R. Galletti, N. Marra, and L. Matassi: We studied the properties of some biologically active peptides such as bradykinin, kallidin, oxytocin, and angiotensin. Bradykinin and kallidin (Sandoz BRS 640 and KL 698) have been used in 10^{-6} µg/ml; Syntocinon (Sandoz) in 10^{-5} g/ml concentration; and angiotensin (Ciba) in 5.10^{-5} g/ml concentration. The peptides in these doses are known, from the published data, to produce pain.

These peptides have been injected into the skin of normal human subjects and the sensory effects they caused have been analyzed.

Among the four substances employed, bradykinin was the most painful. Bradykinin in 10^{-6} g/ml concentration causes a stinging pain of a considerable intensity during the injection, a delayed burning pain 18 sec after the injection, and then flare and wheal around the injection point. Finally it causes a hyperalgesic halo developing eccentrically until reaching its maximal diameter of 5 cm after 15 min.

In addition, the algogenous properties of glass-activated plasma and bradykinin in different concentrations from 10^{-9} to 5.10^{-4} g/ml in 0.9% NaCl solution were compared. The effects of plasma activated with glass for 12 min were very similar (10^{-5} g/ml).

In fact the delayed burning pain arises in both cases with a difference of 7 sec (the delay is 10 sec for the activated plasma and 17 sec for bradykinin) while the spreading of secondary hyperalgesia and its maximal diameter, after 15 min, are about the same.

The immediate pain is due to the chemical constitution of the solvent of bradykinin. The injection of solvent produces only an immediate pain, but not a burning one or a secondary hyperalgesia. It is possible that flare and wheal are caused by the solvent as well as by bradykinin.

In conclusion, the plasma factor acting in human skin (after a glass activation of 12 min) and bradykinin at 10^{-5} g/ml concentration equally stimulate "pain receptors" (burning pain) as well as cutaneous nerves (modified dermographic reaction and secondary hyperalgesia).

Arcangeli: I should like to present our data dealing with the effect of stasis on venous bradykininogen. Drs. Panerai, Ghetti, and I were able to show that plasma bradykininogen drops by roughly 60% in venous blood after 15-min stasis. Such changes are cancelled or sharply diminished if stasis follows a 20-min period of ischemia (1).

These phenomena could be related to the changes in skin mast cells during venous stasis; stasis is followed by diffuse degranulation of mast cells which is inhibited by previous ischemia (2). Since mast cell granules contain proteolytic enzymes which might be able to activate plasma bradykininogen directly or indirectly the two phenomena (decrease in bradykininogen and mast cell degranulation) could be related. We feel that the role bradykinin plays in causing ischemic pain and pain of myocardial infarction is of no primary importance. Pain-producing substances can induce tissue hyperalgesia of long duration. In this respect we saw in patients with intermittent claudication that the affected limb has a marked muscular hyperalgesia which lasts for several hours (3). Since hyperalgesia produced by bradykinin is of short duration (4), such reduction of muscular pain threshold can not be caused by such substances. We feel that it could be due to other algogenic substances, (e.g. some phospholipids) as suggested in our previous studies (5).

1. Panerai, A., A. Ghetti, and P. Arcangeli: Rass. Neurol. veg. 18, 131 (1964).
2. Arcangeli, P., V. Digiesi, and A. Arcangeli: Riv. crit. Clin. med. 65, 197 (1965).
3. Archangeli, P., F. Corradi, and G. D'Ayala Valva: Rass. Neurol. veg. 18, 167 (1964).
4. Arcangeli, P., A. Panerai, and A. Ghetti: Atti X Symposium della Soc. Internaz. Ricerche Neurovegetative. Basilea, 1964. Acta Neuroveg. (Wien) 1965.
5. Arcangeli, P., R. Galletti, and G. Rapi: Rass. Neurol. veg. 10, 1 (1963); L. Lorenzini, P. Arcangeli, and G. D'Ayala Valva: Riv. crit. Clin. med. 56, 897 (1956-57).

P. Panuccio, F. Cresci, and L. Viroli (General Medical Clinic of University of Florence): These investigations were done with the cooperation of Sicuteri, del Bianco, and Branchi.

Cardiac catheterism was performed in 7 patients; blood was sampled from the vena cava inferior and from the pulmonary capillaries. For comparison, samples were taken from the median cubital vein.

The amounts of kininogen found are shown in Table 1.

The kininogen level was markedly lower in the pulmonary capillaries than elsewhere. No significant difference was seen between the samples taken from the median cubital vein and from the inferior vena cava.

We believe that the slowing down of local circulation, caused by the catheter blocking a small pulmonary artery, would cause local release of active kinins together with a decrease in kininogen in plasma.

This seems to be in agreement with the findings of other investigators, that kininogen in venous blood is lower during ischemia or stasis in an arm.

TABLE 1

Median cubital vein	Inferior vena cava	Pulmonary capillaries
2,2	3	1,5
3	2	1
3	4	
4	3,5	1,2
3	2,5	
3	3	
2,8	3	

BRADYKININOGEN IN THE BLOOD OF WOMEN DURING PREGNANCY, LABOR, AND PUERPERIUM

P. PERITI and F. GASPARRI

Istituto di Farmacologia e Istituto di Clinica Ostetrica della Universita degli Studi di Firenze, Florence, Italy

The unexplained increase of oxytocic but thioglycolate-resistant activity in the blood during labor (Centaro et al., 1959, 1961; Hawker et al., 1961; Armstrong and Stewart, 1960) induced us (Periti et al., 1962) and also others (Martinez et al., 1962) to study the behavior of bradykininogen in plasma in this clinical condition. As this increase in oxytocic activity is due to polypeptides such as plasma kinin and slow-reacting substances in general, the behavior of the inactive blood precursor of some of these substances (i.e. bradykininogen) assumes particular significance. First it was thought that plasma kinin activated during labor would act synergistically with oxytocin on the uterus, which is the common target organ. Bradykinin and kinin derived from the plasma of a woman in labor, however, have no oxytocic effect on the human uterus in vitro (Berde and Saameli, 1961; Centaro et al., 1963), the oxytocic activity obtained in vivo with high doses of bradykinin or by means of pharmacological endogenous activation of the plasma kinin is certainly not to be attributed to a direct action on the uterus of these polypeptides (Berde and Saameli, 1961; Centaro and Massi, 1962). The increase of plasma kinins in blood during labor, therefore, very probably has physiological significance not due to the oxytocic activity, even if this increase appears in some way to be related to the dilatation of the uterus and expulsion during labor. We have investigated the changes in blood in relation to labor, assuming that a release of active kinins corresponds to the consumption of plasma bradykininogen, which is estimated in bradykinin equivalents. This research has been carried out during the last four years together with other studies on the behavior of bradykininogen in several physiological and pathological conditions (Sicuteri et al., 1963a,b; Centaro and Periti, 1963; Panconesi et al., 1963; Gasparri et al., 1964).

The conclusion that can be drawn from all these studies is that plasma reserves of bradykininogen are mobilized every time a functional or reactive hyperemia occurs in any area, large or small, of the organism. Pregnancy and labor appear to be similar physiological situations for a more or less localized involvement of the peripheral vascular bed.

The plasma bradykininogen of 44 healthy women admitted to the Institute of Obstetrics and Gynecology of the University of Florence for a delivery, have been assayed by the micromethod of Rocha e Silva (Diniz et al., 1961), the rat uterus in vitro (Periti and Sicuteri, 1962), and standard solutions of synthetic bradykinin, BRS/640 Sandoz (Boissonnas et al., 1960). As control, bradykininogen has been assayed in 135 healthy women with regular menstrual cycles, aged between 15 and 45 years.

The results calculated as mean values of bradykininogen in plasma equivalent to μg synthetic bradykinin per ml are shown in Tables 1, 2, 3, and 4; standard deviation and P (Student's "t") are indicated. During the last two months of pregnancy the bradykininogen in the venous blood increases significantly ($P = 0.001$).

The newborn has a decidedly lower venous bradykininogen ($P = 0.001$) level than normal. The arteriovenous ratio of plasma bradykininogen at the end of pregnancy (blood taken from femoral artery and saphena vein) is 1.25 ($P = 0.02$); in nonpregnant women this difference is not significant. At the beginning of labor the venous bradykininogen does not vary significantly,

TABLE 1
Bradykininogen in Venous Blood*

	Mean	S.D.	n	P
Nonpregnant women	2.77	1.10	79	
Pregnant women	4.55	1.60	16	0.001

*Expressed as μg synthetic bradykinin per ml, in all tables.

TABLE 2
Bradykininogen in Venous and Arterial Blood

	Blood	Mean	S.D.	n	P
Nonpregnant women	Venous	2.68	0.90	56	
	Arterial	2.91	1.00	56	0.30
Pregnant women	Venous	3.87	1.80	7	
	Arterial	4.84	1.60	7	0.02

TABLE 3
Bradykininogen in Venous Blood

	Mean	S.D.	n	P
At the beginning of labor	4.81	1.30	22	
Full dilatation and expulsion	3.74	1.20	25	0.01
Postpartum (few hours)	3.68	1.30	14	0.02

TABLE 4
Bradykininogen in Newborn

	Mean	S.D.	n	P
Umbilical blood	2.33	0.90	18	
Newborn child (1-5 days: venous blood)	1.78*	0.70	21	0.40

*This mean value is statistically significant by different (P = 0.001) from that of nonpregnant women (μg/ml 2.77: S.D. 1.10).

compared to the eighth and ninth months of pregnancy. During labor there is a significant decrease (P = 0.01) persisting for some hours after the delivery (P = 0.02) (Table 3).

The blood taken from the placental umbilical stump has a normal bradykininogen content which is remarkably lower than the maternal blood taken at the same time (Table 4). From blood samples taken from twelve women, starting a few hours after the beginning of the dilating stage and during the various

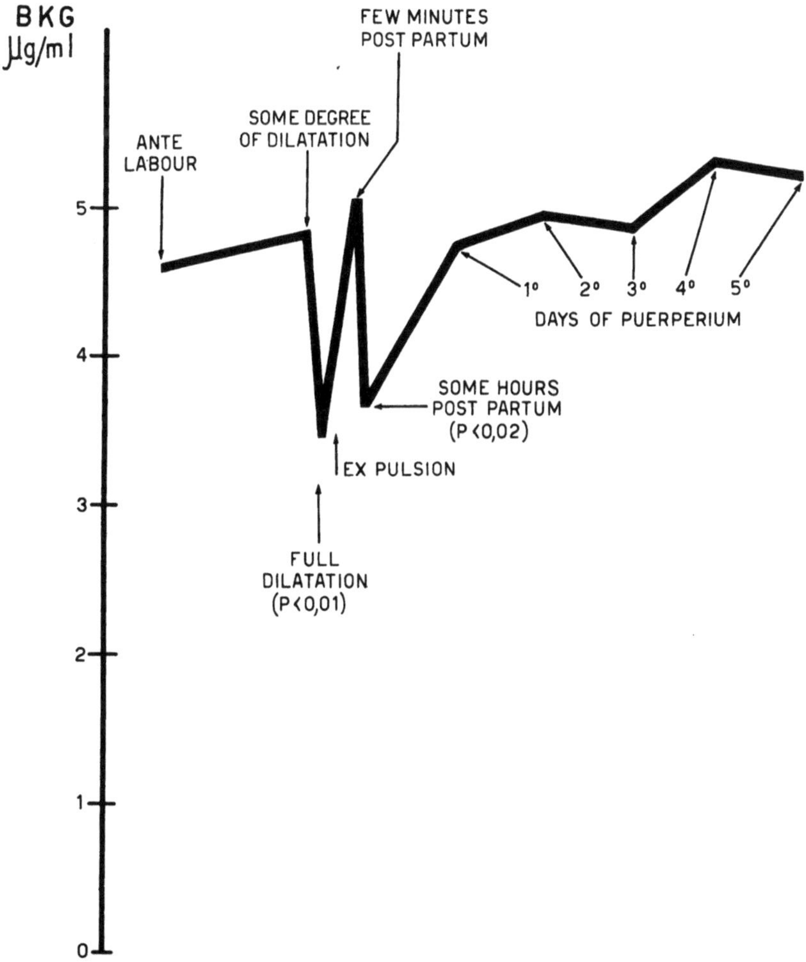

Fig. 1. Bradykininogen in blood of women in labor and puerperium. Each point represents a mean value. (Total = 132 samples).

phases of labor up to the puerperium, we have observed a decrease which reaches its lowest values at the very beginning of the fetal expulsion, followed immediately after the delivery by a rapid return to the initial level. Another decrease characterizes the first hours of puerperium, after which the bradykininogen reaches the starting figures again. The bradykininogen in different stages of labor between the prepartum and puerperium is presented in Fig. 1. The statistical significance (P) of variations is also shown. From these data we conclude that: (1) plasma bradykininogen in the blood of a woman during the last two months of pregnancy is decidedly higher than normal, and shows an arteriovenous difference which is statistically significant. (2) The plasma bradykininogen in the venous blood of a woman in labor undergoes a marked decrease which recedes rapidly in the first period of puerperium. (3) The venous blood of the newborn during the first week of life has a plasma bradykininogen content which is lower than normal and lower than the maternal blood level at the same time.

The interpretation of these results cannot be attempted unless the behavior of plasma bradykininogen in clinical conditions other than labor is taken into consideration. In Fig. 2 we show what we think is the most likely interpretation. It is based on the behavior of bradykininogen in the venous blood of the forearm submitted to ischemia with or without muscular activity of the ischemized limb (Sicuteri et al., 1964). Muscular exercise in ischemia is characterized by a notable consumption of bradykininogen during the reactive hyperemia which follows the decrease in blood supply.

On the other hand, the powerful uterine contractions of labor take place in ischemia, as in the contraction of the skeletal muscle, and have a functional hyperemia during the contractions themselves. This hyperemia certainly exceeds a liter of blood per minute in the uterus at the end of pregnancy and during labor (Spetz, 1965).

The consumption of circulating plasma bradykininogen could be attributed to the strong, prolonged, rhythmic uterine contractions of the dilating and expulsive stages of the delivery, especially because every 4 to 5 minutes all the blood of the parturient woman goes through the vascular bed of the myometrium. The increased flow of blood in different parts of the peripheral vascular bed of the woman in the second half of pregnancy, especially in the limbs, skin, kidneys, and uterus (Spetz, 1965), cannot be explained unless the existence of a

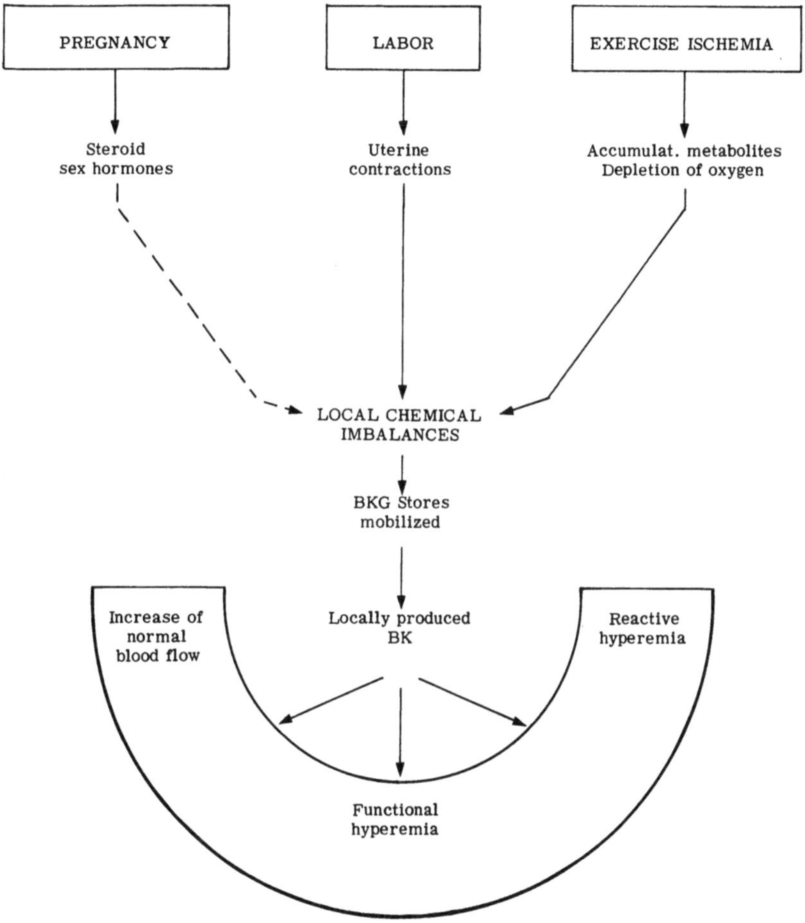

Fig. 2. Peripheral vascular bed.

vasodilator factor (Mason and Melmon, 1965) is considered (Spetz, 1965). The hypothesis that this factor could be plasma kinin is supported by the hyperkininogenemia with an arterio-venous quotient decidedly higher than the unity which characterizes the blood of a woman during the last months of pregnancy.

Indeed, an activation of plasma bradykininogen at a higher than normal rate may occur and result in local increase in free kinins. The consequence of this would be an increased blood flow in some organs and parts functionally involved during pregnancy.

The relatively low level of kininogen in the blood of the newborn should probably be attributed to a reduced hepatic capacity for synthesis of plasma bradykininogen.

REFERENCES

Armstrong, D. A. J., and J. W. Stewart: Observations on the formation of kinins in human plasma collected during labour. Nature (Lond.) 188, 1193 (1960).

Berde, B., and K. Saameli: Effect of bradykinin on uterine activity. Nature (Lond.) 191, 83 (1961).

Boissonnas, R. A., S. Guttmann, J. P. Jaquenoud, H. Konzett, and E. Stürmer: Synthesis and biological activity of peptides related to bradykinin. Experientia (Basel) 16, 326 (1960).

Centaro, A., and G. B. Massi: Aspetti della regolazione neuro-ormonale della contrazione uterina. Minerva ginec. 228 (1962).

Centaro, A., and P. Periti: Ricerche clinico-sperimentali su alcuni polipeptidi endogeni di interesse ostetrico e ginecologico. Riv. Ostet. Ginec. 18, 263 (1963).

Centaro, A., G. De Laurentis, and P. Periti: Oxitocina nel sangue di donna in travaglio di parto. Boll. Soc. ital. Biol. sper. 35, 1454 (1959).

Centaro, A., P. Periti, and G. De Laurentis: Ossitocinemia e attivita ossitocinasiche ematica e miometriale nella donna in travaglio di parto. Riv. Ostet. Ginec. 16, 297 (1961).

Centaro, A., F. Sicuteri, G. B. Battaglia, and G. B. Massi: Esperienze di induzione al parto con l'associazione plasma-chinine-ossitocina. Boll. Soc. ital. Biol. sper. 38, 670 (1962).

Centaro, A., P. Periti, and F. Sicuteri: Bradykininogen and bradykininase in blood of women during labour. Settim. Med. 51, Suppl. I, 70 (1963).

Diniz, C. R., I. F. Carvalho, J. Ryan, and M. Rocha e Silva: A micromethod for the determination of bradykininogen in blood plasma. Nature (Lond.) 192, 1194 (1961).

Gasparri, F., G. B. Battaglia, and P. Periti: Bradichininogeno e plasmachinine del sangue dopo interventi chirurgici ginecologici. Riv. Ostet. Ginec. 19, 194 (1964).

Hawker, R. W., C. F. Walmsley, V. S. Roberts, J. K. Blackshaw, and J. C. Downes: Oxytocic activity of blood in parturient and lactating women. J. clin. Endocr. 21, 985 (1961).

Martinez, A. R., I. F. De Carvalho, and C. R. Diniz: The bradykininogen content of plasma from women in labour. J. Obstet. Gynaec. Brit. Cwlth. 69, 1011 (1962).

Mason, T., and K. L. Melmon: Effects of bradykinin on forearm venous tone and vascular resistance in man. Circulat. Res. 17, 106 (1965).

Panconesi, E., F. Sicuteri, and P. Periti: Bradichininogeno e plasmachinine in dermatologia. Ricerche preliminari. Rass. Derm. Sif. 16, 19 (1963).

Periti, P., and F. Sicuteri: Rilievi sul substrato biologico per il dosaggio del bradichininogeno plasmatico e della bradichinina urinaria nell'uomo. Boll. Soc. ital. Biol. sper. 38, 316 (1962).

Periti, P., A. Centaro, F. Sicuteri, and B. Leocani: Bradichininogeno ematico durante il parto. Boll. Soc. ital. Biol. sper. 38, 672 (1962).

Sicuteri, F., A. Centaro, G. B. Massi, and P. Periti: Azione utero-tonica della bradichinina sintetica nella donna a termine di gravidanza. Boll. Soc. ital. Biol. sper. 38, 665 (1962).

Sicuteri, F., G. Franchi, and P. Periti: Physiopathological aspects of bradykinin and its precursor in man. Settim. Med. 51, Suppl. I, 53 (1963a).

Sicuteri, F., A. Comparini, and P. Periti: Comportamento del bradichininogeno ematico in corso di malattie infettive acute. Settim. Med. 51, 133 (1963b).

Sicuteri, F., P. Periti, M. Fanciullacci, and D. Santini: Bradichininogeno dopo lavoro muscolare in ischemia nell'uomo. Boll. Soc. ital. Biol. sper. 40, 955 (1964).

Spetz, S.: Studies on peripheral circulation in normal pregnancy and in pregnancy complicated by toxaemia and diabetes mellitus. Acta obstet. gynec. scand. 44, Suppl. I, 3 (1965).

DISCUSSION

Margolis: I would like to comment on the agreement between the normal kininogen values obtained in our laboratory and those reported here. These are of the order of 4 µg/ml BK equivalent which is much lower than those obtained by the Brazilian workers. It would be interesting to analyze the fall after labor in terms of the two kininogens I and II (component B). We have followed these in cord blood and found them proportionally depressed (approx. 50% of adult values). Qualitative data suggest that in parturient women there is a selective depression of "component B," but this is subject to quantitative confirmation.

Armstrong: We found that freshly withdrawn parturition plasma is not kininogen-depleted but has an exaggerated kinin-forming ability after a short time (e.g. 20 min). This is followed by increasing evidence of kininogen depletion. I agree with Dr. Margolis' comment that the first kininogen to disappear is the nontrypsin one. Thereafter, the trypsin substrate may or may not be attacked. A specific followed by a nonspecific action of the plasma kallikrein II is strongly indicated by our experiments.

Zeitlin: I would like to add a note on the "normal" levels of precursor as determined by our method. It is based on the concept behind the method of Brocklehurst and Lahiri (1962) and Diniz and Carvalho (1963), which is the incubation of trypsin with denatured blood proteins to release their potential kinin, but it differs enormously in details. This is not the moment to discuss these details. Using this method, I found that in convalescent elderly female patients, bedbound in hospital, the mean kininogen level was 7.1 μg bradykinin equivalent/ml plasma. (S.D. ± 1.1 μg/ml). For hospitalized elderly men it was a little lower, 6.2 μg/ml, with the same standard deviation, while young, nonhospitalized active adults had a mean level lower again (5.4 μ/ml), but more important, showing a far greater scatter, and ranging from 2.8 μg/ml to 12.0 μ/ml plasma (S.D. ± 2.8 μg/ml). It would seem necessary to state the conditions under which any "normal" control samples are obtained for comparison with abnormal samples; or preferably to obtain both the control and the test sample from the same individual if the experiment permits.

"Normal" Precursor Levels (as stated at the symposium)

Sicuteri	c. 2.8 μg/ml (Bk Eg)
Periti	c. 2.8
Thal	5.0
Margolis	c. 4.0
Diniz and Carvalho*	10.3

*Diniz, C. R.; Carvalho, I. F.; Ann. N.Y. Acad. Sci. 104, 77 (1963).
Brocklehurst, W. E.; Lahiri, S. C.; J. Physiol. (Lond.) 165, 39 (1962).
Brocklehurst, W. E.; Lahiri, S. C.; J. Physiol. (Lond.) 160, 15 (1961).
Gaddum, J. H.; Horton, E. W.; Brit J. Pharmaçol 14, 117 (1959).

PHARMACOLOGICALLY ACTIVE PEPTIDES IN TRYPANOSOME INFECTIONS

L. G. GOODWIN and P. F. L. BOREHAM

Nuffield Institute of Comparative Medicine, The Zoological Society of London, Regent's Park, London, England

The parasitic trypanosomes cover the whole range from harmless commensals to dangerous pathogens. They are motile, relatively large protozoa which are not difficult to separate from the blood of the host; they contain and produce antigens which provoke varying degrees of antibody response.

The most important African species are the *Trypanosoma brucei* group (*T. rhodesiense and T. gambiense*) which cause human sleeping sickness, and *T. vivax*, *T. congolense*, and *T. simiae* which, although apparently commensals in the local game animals, cause disease and death in domestic cattle and pigs. Parasitemia may be scanty and the exact cause of death is unknown. *T. brucei*, when inoculated into mice and rats, causes an acute infection; the organisms multiply in the peripheral blood until they are almost as numerous as the erythrocytes, and the animals die in 4 or 5 days.

Goodwin and Richards (1960) showed that in the course of the infection in mice, pharmacologically active peptides are found in the urine. These substances resemble kinins in their activity but have not yet been purified and identified. Richards (1965) has now shown that the kinin activity of the plasma steadily increases during the infection and that the activity of extracts of the tissues, especially of the ears, skin, and feet, increases even more than the plasma. Plasma kininogen decreases. It seems likely that the high kinin activity in the urine toward the end of the infection is derived from the tissues. Activity in the urine may, of course, be associated with renal damage or activity, but the presence of active peptide in plasma and skin indicates that other tissues are also involved. If infected animals are treated with a trypanocidal drug, the urinary output of peptides rapidly diminishes.

Pharmacologically active peptides are also excreted in the urine of patients with severe burns (Goodwin et al., 1963); recent analysis of the amino acid content of such urine suggests that the peptides are derived from collagen (Jackson, 1965). It is possible that the active peptides in the ears and skin of mice infected with trypanosomes are also derived from collagen.

The histamine content of urine, plasma, and tissues also increases during the first few days in infected mice but then decreases on the last day. Preparations made from the subcutaneous tissue and the ears show that the numbers of mast cells present and their degree of granulation do not change significantly at any stage of the infection. This means that the high histamine content of the ears of infected animals must be derived from sources other than the mast cells. So far we have made no determinations of histidine decarboxylase activity, but this would clearly be of interest.

In rabbits, *T. brucei* subgroup trypanosomes cause a chronic infection ending in death in about 30 days. Very few trypanosomes appear at any time in the peripheral blood, and Gray (1962) has shown by immunological techniques that the initial parasitemia is followed by the emergence of successive antigenic variants of the inoculated strain of organisms. Boreham (1965) has recently studied the kinin activity of the urine and plasma of infected rabbits. There was an initial output of activity in the urine which reached its peak in about 5 days; thereafter the activity varied but remained higher than that of control rabbits. The initial output of kinin activity may have been augmented by the products of the hemolysis that occurred early in the infection, but it is unlikely that hemolysis was responsible for the degree of activity observed. The subsidiary peaks of output of activity in the urine may perhaps be related to the emergence of antigenic variants of the original strain. This is a difficult problem to study in the rabbit. On the other hand, *T. lewisi* infection in rats produces a single peak of parasitemia; the trypanosomes then disappear and the animals are immune.

This seemed to be an ideal example in which to study the effects of the antigen-antibody reaction on output of active peptide. However, it was found that the peak of *T. lewisi* parasitemia in the rat was accompanied by no change at all in the normal level of urinary peptide (Boreham, 1965).

By contrast, the acute lethal infection produced in the rat by *T. brucei* leads, as in the mouse, to an increasing output of active peptide. It appears that *T. lewisi*, which is generally regarded as nonpathogenic in the rat, does not trigger the mechanism responsible for the production of pharmacologically active peptides.

This mechanism remains a mystery. In view of the results of the experiments with *T. leursi*, the physical presence of motile organisms in the blood stream does not appear to be the cause. Moreover, the addition of a concentrated suspension of washed *T. brucei* to normal rabbit plasma did not release kinin A similar lack of kinin-forming activity in vitro was observed by Amundsen and Rugstad (1965) among a variety of pathogenic bacteria. Estimation of plasminogen activator in the blood of normal and heavily parasitized mice showed no difference between them; whatever causes kinin release during infection seems not to be intimately associated with clotting mechanisms.

T. brucei is known to produce an exoantigen (Weitz, 1960), and it may be that this or other metabolic products of pathogenic trypanosomes cause peptide release. Another possibility is that pathogenic organisms may cause local occlusion of capillaries, resulting in the liberation of active substances from anoxic tissue. The ears of an infected mouse injected intravenously with Indiaink become darker and remain so for longer than the ears of an injected normal mouse.

Feldberg and Lewis (1964) have shown that bradykinin is a powerful releaser of catecholamines from the adrenal glands, and Maegraith et al. (1956, 1957) found that the fatal hypotension of *Plasmodium* infections in monkeys and *Babesia* infections in dogs could be alleviated for a time by injections of norepinephrine.

The release of peptides which we have observed is probably a general phenomenon that accompanies pathogenic infections. It seems likely that a mixture of substances, some physiologically active and others inactive, are released from damaged tissues and may play a part in causing fatal hypotension in infected animals. Perhaps the balance between plasma kinin and tissue stores suggested by Habermann (in this volume) is disturbed. Perhaps the vicious circle described here by Greenbaum is set up locally in suitably acid environments. The condition of an animal at the end of an acute infection certainly resembles that of patients with acute pancreatitis, described by Thal. The nature of the active substances clearly needs a degree of detailed chemical investigation which we have not so far been equipped to apply.

ACKNOWLEDGMENT

We wish to acknowledge the help and advice given by Dr. B. Weitz and the technical assistance of Miss S. V. M. Hook. The work relating to chronic infections in rabbits and rats was supported by Grant R.1524 of the Ministry of Overseas Development.

REFERENCES

Amundsen, E., and H. E. Rugstad: Influence of some pathogenic bacteria on kinin formation and destruction. Brit. J. Pharmacol. 25, 67 (1965).

Boreham, P. F. L.: Pharmacologically active peptides produced in the tissues of the host during chronic tyrpanosome infections. Nature (Lond.), in press (1965).

Feldberg, W., and G. P. Lewis: The action of peptides on the adrenal medulla. Release of adrenaline by bradykinin and angiotensin. J. Physiol. (Lond.) 171, 98 (1964).

Goodwin, L. G., and W. H. G. Richards: Pharmacologically active peptides in the blood and urine of animals infected with *Babesia rodhaini* and other pathogenic organisms. Brit. J. Pharmacol. 15, 152 (1960).

Goodwin, L. G., C. R. Jones, W. H. G. Richards, and J. Kohn: Pharmacologically active substances in the urine of burned patients. Brit. J. exp. Path. 44, 551 (1963).

Gray A. R.: The influence of antibody on serological variation in *Trypanosoma brucei*. Ann. trop. Med. Parsit. 56, 4 (1962).

Jackson, S. H.: The urinary excretion of free and peptide amino acids by a patient with thermal burns. Clin. chim. Acta 12, 389 (1965).

Maegraith, B. G., K. Devakul, and C. S. Leithead: Action of noradrenaline in medical shock in *Plasmodium knowlesi* malaria. Trans. roy. Soc. trop. Med. Hyg. 50, 311 (1956).

Maegraith, B. G., H. M. Gilles, and K. Devakul: Pathological processes in *Babesia canis* infection. Z. Tropenmed. Parasit. 8, 485 (1957).

Richards, W. H. G.: Pharmacologically active substances in the blood, tissues and urine of mice infected with *Trypanosoma brucei*. Brit. J. Pharmacol. 24, 124 (1965).

Weitz, B: The properties of some antigens of *Trypanosoma brucei*. J. gen. Microbiol. 23, 589 (1960).

DISCUSSION

Back: Is the infection characterized by a pyrogenic reaction? And if so, perhaps one should study the possible activation of the fibrinolysin system which is known to be activated by pyrogens? Thus, plasmin may be responsible, in part, for the kinin formed?

Goodwin: Yes, I agree that the function of plasmin in trypanosomiasis should be studied further, although so far we have been unable to correlate changes in plasminogen activator with degree of infection.

Erdös: In 1952, Professor Werle and I found a peptide which was probably a mixture of bradykinin and kallidin in human urine. Others continued the work but found no correlation between the level of urinary kinin and the pathological condition of the patients investigated.

Goodwin: Patients with severe burns have high urinary kinin activity; the more shocked they are, the more kinin is present and when they get better the activity in the urine decreases. This may well be correlated with proteinuria. However, the presence of raised kinin activity in the plasma and tissues suggests that this may have a part to play in the pathological condition.

Sicuteri: Intravenous injection of kinin does not increase the level of urinary kinin. Kinin is either formed in the kidney or it is liberated in the urine from the kininogen.

Goodwin: It is very probable that the kinin activity of the urine in infected animals results from renal processes. Nevertheless this does not explain the presence of high activity in the plasma and tissues, and we do not know how a kidney in an infected animal deals with substances in comparison with a normal kidney.

Oates: In our studies of urinary bradykinin (using an IRC-50 column for isolation of the peptide) we have found supernormal values only with associated proteinuria. This could be a result of the filtration of kininogen into the urine.

Eisen: You showed a rise in the urinary peptide concentration infected mice. These animals were probably oliguric, and I would like to know whether there was an increase in the amount of excreted peptide. Did these mice have proteinuria? Urinary protein is sometimes rich in angiotensinogen, and, as I have found, also kininogen. This would obviously affect peptide content. Can phagocytes ingest parts of trypanosomes? Phagocytosis could lead to the release of kinin-forming factors from lysosomes into body fluids.

Goodwin: Urine was adjusted to a constant 24-hour volume; the peptide increase did not depend on oliguria and high concentration but was an absolute increase. Proteinuria does sometimes occur, but the kinin activity is not related to the protein content of the urine. Toward the end of the infection a polymorphonuclear leukocytosis occurs, and polymorphs do ingest dead trypanosomes. It is possible that lysosomal enzymes are released.

POSSIBLE MODES OF KININ FORMATION IN SOME PATHOLOGICAL STATES IN MAN

V. EISEN and C. A. KEELE

Department of Pharmacology, The Middlesex Hospital Medical School, London, England

Reactions to transfusions of plasma fractions

The first of the clinical conditions to be discussed is what one may term a case of iatrogenically induced kinin formation. We are reporting studies which show that some clinically used plasma fractions contain significant amounts of active kinin-forming factors, which on rapid intravenous infusion sometimes produce unpleasant side effects in the patient. Our studies were mainly concerned with human fibrinogen preparations isolated by cold ether, by the method developed at the Lister Institute in London (Kekwick and Wolf, 1957). When prepared from freshly collected plasma, this fraction contains 5 to 8 times the plasma concentration of factor VIII or antihemophilic globulin. The fraction is described as AHG concentrate or HAHG, and is given to patients with hemophilia A. Fibrinogen from outdated transfusion plasma is called ETF and is used in various types of hypo- and afibrinogenemia.

In earlier studies (Maycock et al., 1963), side effects were reported in 43 out of 72 transfusions. In more recent series, reactions occurred less frequently (Biggs et al., 1965). The pattern of the clinical reactions suggested that they may be caused by plasma kinins (Table 1).

Members of the group which developed the plasma fractionation by cold ether set out to find the causes of these reactions. Mackay et al., 1965, found that HAHG and ETF con-

TABLE 1

Effects of Kinins	Reactions to AHG conc.
Pain	Pain at site of operation. Pain in the back, headache. Paraesthesiae
Vasodilatation	Flushing
Hypotension	Hypotension
Bronchoconstriction	Feeling of constriction in the chest
Increased pressure in pulmonary artery	
Increased capillary permeability	

tained small amounts of plasminogen and kininogen. On incubation at 37°C, the two fractions formed 2-30 ng bradykinin/ml. Activation of the plasminogen by streptokinase enhanced the kinin formation only in HAGH (30-117 ng/ml), but not in ETF. HAHG also contained 6-12% of the plasma content of factor XII.

The observed kinin formation seemed too slow and small to account for the prompt clinical reactions. Besides, there was no evidence that the contaminating plasminogen was activated in the patient. We tried to find out therefore whether the fractions contained already active kinin-forming factors that would produce kinins by interacting with the appropriate proteins in the patients' plasma (Eisen, 1965).

HAHG is usually transfused in amounts of 200-800 ml and at rates of 3-10 ml/min. When fresh human plasma was incubated with HAHG or ETF in proportions that could be expected to occur during such transfusions, kinin levels of 50-400 ng were promptly formed in 1 ml of plasma. Different fresh plasmas varied in their responsiveness. This could be seen when different batches were incubated with the same fresh plasma.

The active factor in the fractions which was mainly responsible for this kinin formation had the characteristics of plasma kallikrein. It acted directly on the substrate in the patients' plasma. This could be demonstrated by incubating HAHG and ETF with plasma whose intrinsic kinin-forming factors had been destroyed by heating (Fig. 1). Comparison with purified human plasma kallikrein (Webster and Pierce, 1961) showed that six batches of HAHG and ETF attacked heated plasma at rates which corresponded to plasma kallikrein concentrations of 0.05-0.15 Frey units/ml. Figure 2 illustrates this for HAHG batch 373 and ETF batch 422.

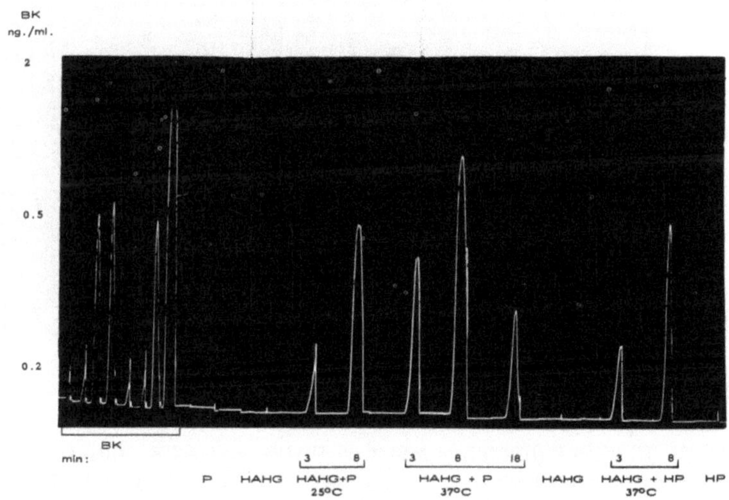

Fig. 1. Kinin formation by HAHG in fresh plasma (P) and in plasma heated at 61°C for 30 min (HP). Rat uterus. Bk = responses to bradykinin 0.2, 0.5, and 2.0 ng/ml (drum running at half speed). One ml HAHG 268, 1 ml Tris buffer (pH 7.4), and 2 ml plasma were incubated for the specified period, and 0.1 ml of the mixtures tested. The kinin levels formed in 8 min were in 1 ml of plasma, 50 ng at 25°C and 100 ng at 37°C, and in 1 ml of heated plasma, 50 ng.

The enzyme was inhibited by 0.05-0.5 mg SBTI/ml and by 250-500 kallikrein inhibitor units Trasylol/ml, but not by the antiheparin agent hexadimethrine bromide in concentrations up to 0.5 mg/ml.

Previous incubation with streptokinase enhanced the direct kinin-forming action of the fractions.

A factor that could activate the precursors of the kinin-forming enzymes in human plasma was also present in HAHG and ETF. Figure 3 shows that it could be readily and firmly adsorbed on to glass, and it thus resembled factor XII. Such

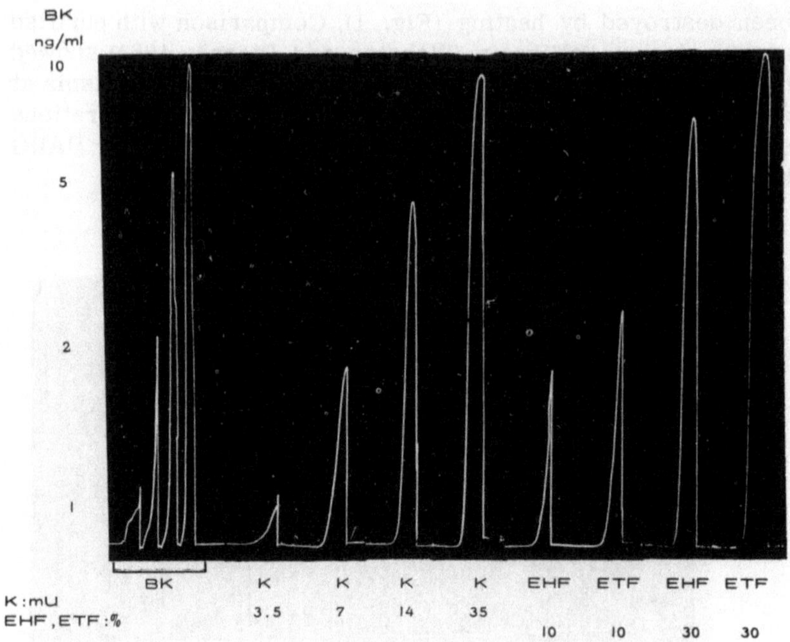

Fig. 2. Plasma kallikrein content of HAHG and ETF. Rat uterus. Bk = responses to bradykinin 1, 2, 5, 10 ng/ml (drum running at half speed). One ml substrate plasma (61°C for 60 min, then pH 2 for 15 min) was incubated at pH 7.4 and 37°C with 3.5, 7, 14, and 35 Frey milliunits of purified human plasma kallikrein (K), and with 10 and 30% of the transfusion concentration of HAHG 373 or ETF 422. Aliquots of 0.1 ml were tested after 3-min incubation. (After prolonged incubation, 1 ml of the mixtures formed 5 μg bradykinin.)

coated glass induced kinin formation in Hageman trait plasma, which does not respond to clean glass because it lacks factor XII although it contains kallikreinogen and kininogen. Heated plasma was not activated by this glass, which confirmed that the adsorbed factor was an activator of other enzymes and not itself a kinin-forming enzyme. Glass which had previously been immersed in hexadimethrine bromide and then thoroughly washed did not adsorb this factor. The factor could be adsorbed onto glass in the presence of SBTI. These findings were compatible with the view that the agent adsorbed onto glass from the fractions was factor XII (Eisen, 1964).

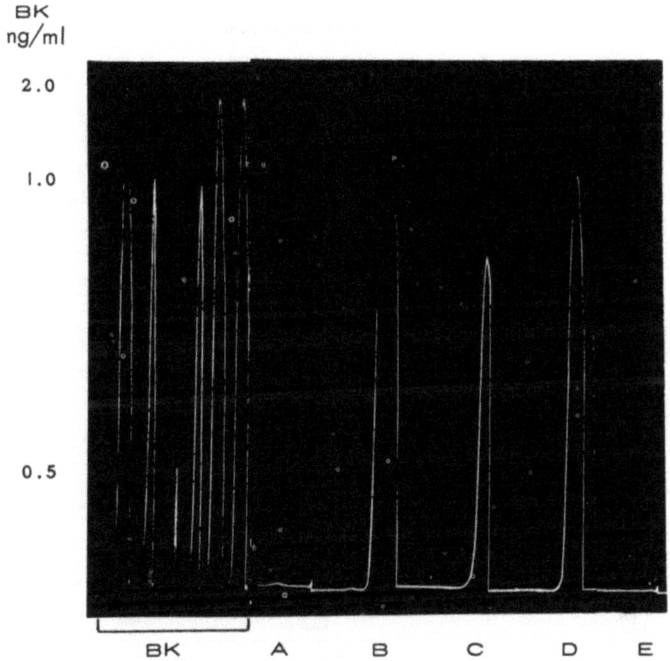

Fig. 3. Glass adsorbs factor XII-like activator from HAHG and ETF. Isolated rat uterus; 5-ml bath. Bk = responses to bradykinin 0.5, 1.0, and 2.0 ng/ml (drum running at half speed). Tests A-E = 0.1 ml. Glass powder was mixed with HAHG or ETF for 15 min, then separated and washed. Hageman-trait plasma was tested after 5-min rotation with clean glass powder (A), and with powder from HAHG 373 (B), ETF 422 (C), and ETF 417 (D). Powder from HAHG 373 had no effect on heated plasma (E).

Hydrolysis of amino acid esters

An attempt was made to determine the nature of the contaminant in HAHG and ETF which hydrolyzes p-tosyl-L-arginine methyl ester (TAMe) even without prior activation by streptokinase (Mackay et al., 1965). Table 2 shows that HAHG and ETF hydrolyzed TAMe and N-benzoyl-L-arginine ethyl ester (BAEe) at similar rates. The plasma kallikrein levels present in HAHG and ETF could account only for 1-5% of this esterolytic activity. The bulk of this esterolysis, which was not inhibited by SBTI, must have therefore been due to another

TABLE 2

Hydrolysis of esters by HAHG and ETF, and by purified plasma kallikrein, at 37° C and pH 7.8. Final concentration of the Tris buffer was 0.2 M, of TAMe, BAEe, or ATEe 0.02 M, and of HAHG or ETF 1/3-1/9 of the concentrations used for transfusion. The μ-moles of ester hydrolysed in 60 minutes by the transfusion concentrations of 1 ml of HAHG or ETF were calculated. Plasma kallikrein was incubated in the concentrations specified.

Batch	μ-moles ester/60 min/37 C		
	TAMe	BAEe	ATEe
HAHG 268	80	84	95
HAHG 353	120	142	160
HAHG 362	120	150	160
HAHG 373	100	110	106
ETF 417	112	120	84
ETF 422	150	150	140
Plasma kallikrein			
Frey U/ml			
0.35	11	10.5	
0.14	5	5	
0.07	2	2	

enzyme. Of the arginine esterases in plasma which are resistant to SBTI, thrombin and factor XI (plasma thromboplastin antecedent) had been excluded by the appropriate clotting tests (Mackay et al., 1965). The arginine esterase belonging to the first component of complement (C′1) may be activated by cold ether (Donaldson, 1961). This induced us to test the effect of the ether-precipitated fractions HAHG and ETF on N-acetyl-L-tyrosine ethyl ester (ATEe), the substrate for which C′1-esterase has the highest affinity. The hydrolysis of ATEe was of the same order as that of the arginine esters.

The pH optima of the hydrolysis of TAMe and ATEe by the fractions were at pH 8.3 and 7.8, respectively, and were thus similar to those found with highly purified preparations of C′1-esterase (Austen, 1960; Haines and Lepow, 1964).

If the ATEe hydrolysis by HAHG and ETF is indeed due to C′1-esterase, a large part of their effect on arginine esters may also be attributed to this enzyme. The ratio of TAMe/ATEe μmoles hydrolyzed by six batches of HAHG and ETF was 0.8-1.5. Ratnoff and Lepow (1957) found with crude C′1-esterase a TAMe/ATEe ratio of 0.79, and Haines and Lepow (1964), using a more purified preparation, a ratio of 0.34.

Studies with other therapeutic plasma fractions

Fibrinogen prepared by precipitation with ethanol, as for example in the method by Blombäck and Blombäck (1956), contains about one fifth to one sixth of the kinin-forming and arginine esterase activity found in fibrinogen precipitated with ether. No ATEe hydrolysis was detected. Small amounts of kinin-forming activity are present in therapeutic γ-globulin fractions (M. Mackay, personal communication).

In conclusion, the findings show that with a large transfusion, a patient may receive up to 100 Frey units of plasma kallikrein. The plasma kinin formation resulting from such an amount may well be the main cause of the clinical reactions to transfusions.

Since factor XII was detected by its adsorption and activation onto glass, the present study did not distinguish whether the fractions contained the factor in its active or preactive form. Mackay et al. (1965) believed that only the precursor was present. It is therefore likely that factor XII contributes little to the kinin-forming effect of HAHG and ETF in patients.

The levels of the presumed C′1-esterase in HAHG and ETF seemed sufficient to produce side effects. The fractions contained 2 to 4 times higher concentrations of this activity than did the sera from patients with hereditary angioneurotic edema, a syndrome which Donaldson and Evans (1963) attribute to circulating C′1-esterase. The pathological effects of C_1-esterase in this disease and in reactions to transfused HAHG and ETF, may be partly based on its property to liberate histamine and thus increase capillary permeability (Ratnoff and Lepow, 1963).

Kinin formation by urate crystals

Seegmiller et al. (1962) and Faires and McCarty (1962) have in recent years provided convincing support for the hypothesis proposed by Sir Alfred Garrod in 1876 that the urate crystals present in most gouty joint effusions contribute to the symptoms of acute gouty arthritis. The extremely painful inflammations of the spontaneous acute attack could be reproduced by injecting suspended microcrystals of monosodium urate into the knee joints of healthy or gouty volunteers. Numerous leukocytes appeared in the synovial fluid, and intense phagocytosis of the crystals ensued. Both groups of workers have reported that the intensity of the inflammation is correlated to the degree of phagocytosis of the crystals.

More recently, Seegmiller (1964) and his colleagues found in the synovial effusions of gout and of other types of arthritis, kinins in concentrations up to 50 ng/ml. The synovial kinin levels were more closely correlated to the tenderness of the joint than to its warmth or swelling; they showed no apparent correlation with the synovial leukocyte count or protein concentration. In these concentrations, kinins could produce most signs of inflammation. The pain-producing effect of the kinins may be potentiated by 5-hydroxytryptamine (Sicuteri et al., 1965) released from disintegrating platelets.

Our studies suggest two ways by which kinins can be formed in gouty arthritis:

1. Urate particles activate the kinin-forming enzymes in synovial effusions. Kellermeyer and Breckenridge (1965) have recently reported that urate crystals promote blood clotting by activating factor XII in plasma. As Fig. 4 shows, factor XII appears to be present in synovial effusions as well, and is capable of initiating kinin formation in these fluids. These experiments were performed with effusions from patients with rheumatoid arthritis. The protein concentration of these fluids is similar to that of effusions in acute gouty arthritis (Ropes and Bauer, 1953). Mixing with 5 mg monosodium urate microcrystals per ml induced the formation of a bradykinin-like activity. The activity corresponded in twenty of these fluids to that of 50 to 3,000 ng synthetic bradykinin/ml.

Several lines of evidence confirmed that factor XII was engaged in this kinin formation. (a) Brief immersion in the antiheparin agent hexadimethrine bromide abolishes the capacity of glass, Celite, kaolin, and other surfaces to activate factor XII (Eisen, 1964), and this was also true of the surface of urate crystals. (b) Like all activating surfaces, urate crystals firmly adsorbed factor XII. They were then capable of inducing kinin formation in Hageman-trait plasma, which did

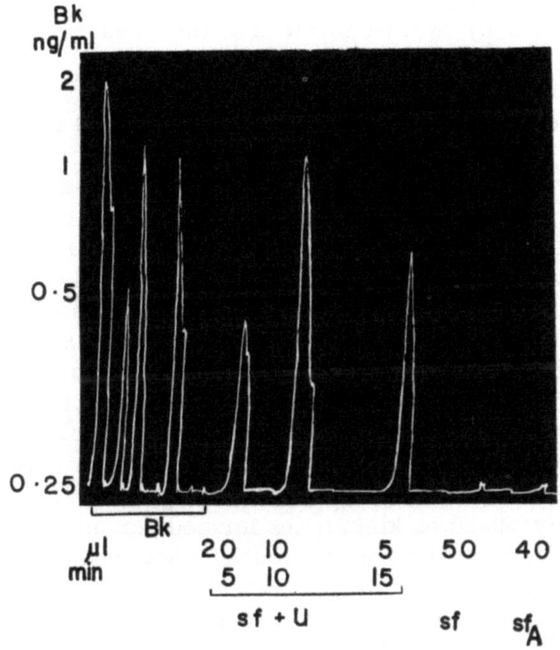

Fig. 4. Urate induces kinin formation in synovial fluid (sf). Rat uterus. Bk = responses to bradykinin 0.25, 0.5, 1, and 2 ng/ml. Volumes and times of tests as specified. Agitation at pH 7.4 and 37°C with 5 mg monosodium urate crystals (U)/ml led in this synovial fluid to a kinin level of 600 ng/ml at 15 min. No kinin was detected in synovial fluid after corresponding dilution and agitation with buffer (sf_A).

not respond to clean particles. (c) The presence of factor XII in synovial effusions was confirmed by their response to ellagic acid, a specific activator of factor XII (Ratnoff and Crum, 1964).

In general, the kinin-forming response in synovial effusions was greater than in normal plasma. Studies of various parameters of kinin formation, and of the relative responsiveness of effusions and of the corresponding plasmas, are now in progress. Preliminary results show that the TAMe hydrolysis induced by kaolin or urates is smaller than in normal plasma. The destruction of bradykinin is slower. Effusions mostly contain high levels of kininogen. The removal of hyaluronic acid by hyaluronidase and the resulting fall in viscosity do not affect the kinin-forming response.

2. Our findings suggest that the important role played by the phagocytosis of urate crystals by leukocytes in acute gouty arthritis (Seegmiller et al., 1962), may also be mediated – at least partly – by kinins. In the course of phagocytosis, enzymes are released from lysosomes into the cytoplasm and into the environment of the cell (Weissmann, 1964). The cell itself is often killed by the action of the released enzymes. One of the lysosomal enzymes that could induce kinin formation in synovial effusions is cathepsin B whose bond specificity resembles that of trypsin. Admittedly, the pH optimum of cathepsins is 5.5 or less, but kinins are effective in such small amounts that even very low forming activity would be significant. We found that lysosomal fractions from rat liver induced kinin formation in synovial fluid. The apparent formation was slight, only 30-50 ng kinin being formed per ml. However, the true formation must have been much higher because the lysosomal fraction also contained a potent kinin-destroying factor. It is interesting that the kinin-destroying factor was very active at pH values up to 7.8.

There is another pathway by which lysosomal enzymes may induce kinin formation in synovial fluid. The proteases among these enzymes digest the mucoproteins in cartilage and connective tissue (Lack, 1964). The breakdown of mucoproteins results in the release of soluble mucopolysaccharides, some of which have kinin-forming properties. Figure 5 illustrates this for purified chondroitin sulfate, one of the principal polysaccharides of cartilage, which appears to activate the kinin-forming enzymes in synovial fluid.

In Figure 6, the kinin-forming mechanisms have been summarized and added to the scheme of pathogenesis of acute gouty arthritis proposed by Seegmiller et al. (1963). Urate crystals precipitate from body fluids and tissues with raised uric acid levels. The crystals lead to phagocytosis and to activation of factor XII. Both processes bring about kinin formation. By their effects on the microcirculation, kinins enhance the influx into the joint space of fluid and proteins, which will promote further humoral kinin formation. Kinins also attract leukocytes, thus fostering phagocytosis. Kinins could therefore both produce and perpetuate inflammatory changes. The increased production of lactate during phagocytosis reduces the pH and

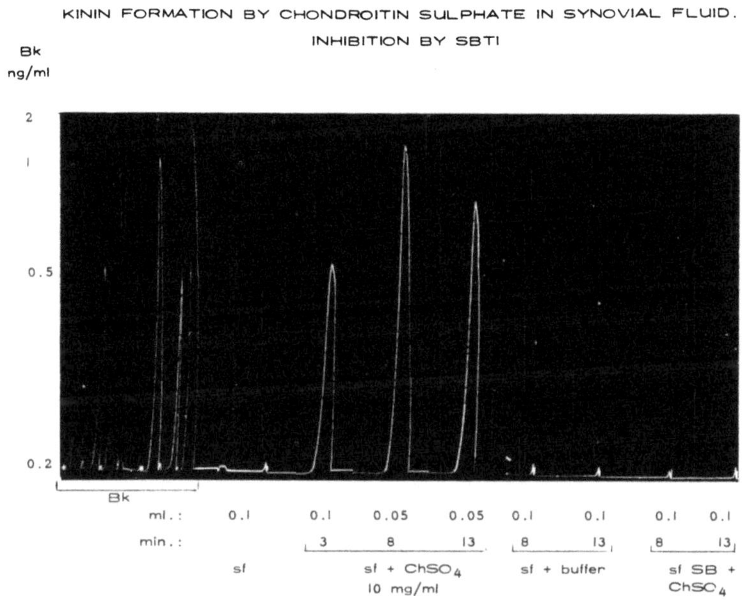

Fig. 5. Rat uterus. Bk = responses to bradykinin 0.2, 0.5, 1.0, and 2.0 ng/ml. Volumes and times of tests as specified. Incubation at pH 7.4 and 37°C with chondroitin sulfate ($ChSO_4$) led in this synovial fluid (sf) to a kinin level of 140 ng/ml at 13 min. The effect was inhibited by soybean trypsin inhibitor (SB).

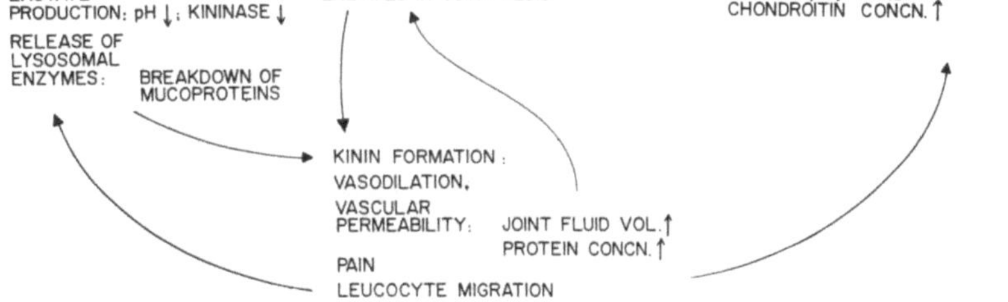

Fig. 6. Possible role of kinin formation in acute gouty arthritis.

thus causes further precipitation of urates. The more acid milieu should also depress kininase (Edery and Lewis, 1962), preserving in this way formed kinins. Released mucopolysaccharides may contribute to kinin formation and, according to Laurent (1964) lower the solubility of uric acid.

ACKNOWLEDGMENTS

Figures 1, 2, and 3 and Table 2 are published by permission of the Editor of the British Journal of Haematology, and Figures 5 and 6 by permission of the Honorary Editors of the Proceedings of the Royal Society of Medicine.

One of us (V.E.) was supported by the Medical Research Council.

REFERENCES

Austen, K. F.: The differentiation of the chloroform, peptone and antigen-antibody inducible esterase activities of human serum from plasmin. Immunology, 3, 152 (1960).

Biggs, R., J. M. Matthews, B. M. Rush, F. C. Johnstone, R. G. MacFarlane, and D. S. Hayton-Williams: Further experience in the use of human antihaemophilic globulin (H.A.H.G.) for the control of bleeding after dental extraction in haemophilic patients. Lancet 1, 969 (1965).

Blombäck, B., and M. Blombäck: Purification of human and bovine fibrinogen. Ark. Kemi. 10, 415 (1956).

Donaldson, V. H.: Studies on the activation of a serum esterase with ether and its relationship to C'1-esterase. J. clin. Invest. 40, 673 (1961).

Donaldson, V. H., and R. R. Evans: A biochemical abnormality in hereditary angioneurotic oedema. Amer. J. Med. 35, 37 (1963).

Edery, H., and G. P. Lewis: Inhibition of plasma kininase activity at slightly acid pH. Brit. J. Pharmacol. 19, 299 (1962).

Eisen, V., Effect of hexadimethrine bromide on plasma kinin formation, hydrolysis of p-tosyl-L-arginine methyl ester and fibrinolysis. Brit. J. Pharmacol. 22, 87 (1964).

Eisen, V., Studies on fibrinogen fractions isolated from human plasma by precipitation with cold ether. II. Enzymic activities resembling plasma kallikrein and C'1-esterase. Brit. J. Haemat. 11, 576 (1965).

Faires, J. S., and D. J. McCarty: Acute arthritis in man and dog after intrasynovial injections of sodium urate crystals. Lancet 2, 682 (1962).

Haines, A. L., and I. H. Lepow: Studies on human C′1-esterase. I. Purification and enzymatic properties. J. Immunol. 92, 456 (1964).

Kekwick, R. A., and W. P. Wolf: A concentrate of human antihaemophilic factor. Its use in six cases of haemophilia. Lancet, 1, 647 (1957).

Kellermeyer, R. W., and R. T. Breckenridge: The inflammatory process in acute gouty arthritis. I. Activation of Hageman factor by sodium urate crystals. J. Lab. clin. Med. 65, 307 (1965).

Lack, C. H.: Proteolytic activity and connective tissue. Brit. med. Bull. 20, 217 (1964).

Laurent, T. C.: Solubility of sodium urate in the presence of chondroitin-4-sulphate. Nature (Lond.) 202, 1334 (1964).

Mackay, M., W. d'A. Maycock, E. Silk, and B. S. Combridge: Studies on fibrinogen fractions isolated from human plasma by precipitation with cold ether. I. Plasma kinin formation by the activation of contained plasminogen. Brit. J. Haemat. 11, 563 (1965).

Maycock, W. d'A., S. Evans, L. Vallet, B. Combridge, P. Wolf, B. H. MacGibbon, E. E. French, L. H. Wallett, J. V. Dacie, R. Biggs, D. Handley, and R. G. MacFarlane. Further experience with a concentrate containing human antihaemophilic factor. Brit. J. Haemat. 9, 215 (1963).

Ratnoff, O. D., and J. D. Crum: Activation of Hageman factor by solutions of ellagic acid. J. Lab. clin. Med. 63, 359 (1964).

Ratnoff, O. D., and I. H. Lepow: Some properties of an esterase derived from preparations of the first component of complement. J. exp. Med. 106, 327 (1957).

Ratnoff, O. D., and I. H. Lepow: Complement as mediator of inflammation. J. exp. Med. 118, 681 (1963).

Ropes, M. W., and W. Bauer: Synovial Fluid Changes in Joint Disease. Cambridge, Mass.: Harvard University Press. (1963).

Seegmiller, J. E.: Lecture, Royal Society of Medicine, London (1964).

Seegmiller, J. E., R. R. Howell, and S. E. Malawista: The inflammatory reaction to sodium urate. J. Amer. Med. Ass. 180, 469 (1962).

Seegmiller, J. E., L. Laster, and R. R. Howell: Biochemistry of uric acid and its relation to gout. New Engl. J. Med. 268, 712 (1963).

Sicuteri, F., M. Fanciullaci, G. Franchi, and P. L. del Bianco: Serotonin-bradykinin potentiation on the pain receptors in man. Life Sci. 4, 309 (1965).

Webster, M. E., and J. V. Pierce: Action of the kallikreins on synthetic ester substrates. Proc. Soc. exp. Biol. (N.Y.) 107, 186 (1961).

Weissmann, G.: Lysosomes. Blood 24, 594 (1964).

DISCUSSION

Greenbaum: I was pleased to see that Dr. Eisen and I agree that proteases other than kallikreins may produce kinins. I refer to the lysosomal proteases.

Eisen: Yes, I think this is true. Of course, we still have to ascertain that the kinin-forming effect of lysosomal fractions can be attributed to one of the known lysosomal enzymes.

Melmon: While synovial and white cell lysosomes may be an important source of kallikrein in the pathogenesis of acute joint inflammation, there is poor correlation between synovial protein and white cell content and the degree of heat and pain production and concentration of synovial fluid kinin. Another source of kallikrein may be the synovial tissue itself. We have preliminary data suggesting that there is an enzyme with a pH optimum between 7 and 8, which may be adsorbed on DEAE-cellulose and may form kinin when incubated with crude bradykininogen. The synovial tissue, especially in acute rheumatoid arthritis where the synovium is inflamed, may be another source of kinin-forming enzyme.

BIOCHEMICAL AND PHYSIOLOGIC STUDIES OF THE KININS IN CARCINOID SYNDROME

JOHN A. OATES and KENNETH L. MELMON

Departments of Medicine and Pharmacology, The Division of Clinical Pharmacology, Vanderbilt University, Nashville, Tenn. and Departments of Medicine and Pharmacology, The Section of Clinical Pharmacology and Cardiovascular Research Institute, University of California, San Francisco, Calif., U.S.A.

Patients with carcinoid tumors metastatic to the liver may exhibit a syndrome which includes attacks of vasodilatation with cutaneous flushing, diarrhea, cardiac valvular lesions, and occasional bronchial constriction. Overproduction of serotonin is an almost uniform finding. Several studies, however, have shown poor correlation between serotonin release and flushing attacks (Robertson et al., 1962; Levine and Sjoerdsma, 1963; Oates et al., 1964), suggesting that other vasodilators must be present.

The carcinoid flush may be induced by small doses of catecholamines. The parallel between this and the observations of Hilton and Lewis (1956) that epinephrine releases kinin-forming enzymes from the salivary gland was intriguing and prompted investigation of the possible role of kinin peptides in the carcinoid syndrome.

The release of bradykinin into hepatic venous blood. To ascertain whether there was an increase in kinin peptide in the venous blood from the liver during flushes, hepatic vein catheterization was performed on eleven subjects with carcinoid syndrome and on fifteen control subjects. After obtaining blood in a controlled state, epinephrine was administered in doses averaging 5 μg; blood samples were then collected from the carcinoid patients at the beginning of the flush and at a comparable time after epinephrine in the control group. In one patient, the flush was induced with the oral administration of ethanol.

The kinin peptides were isolated by rapid precipitation of the whole blood with ethanol followed by an ion-exchange procedure, and then assayed on the estrus rat uterus (Oates, *et al.*, 1964).

TABLE 1

Kinin in Hepatic Venous Blood During Flush

	Hepatic venous kinin (μg/100 ml) (bradykinin equivalents)	
Subjects	Control	After epinephrine
Noncarcinoid	0.1 - 7.9 (avg. < 4) (n = 15)	0.1 - 4.6 (avg. < 2.5) (n = 11)
Carcinoid		
C.P.	4	27
	2	27
O.S.	4	61
	8	53, 50
E.S.	7	56
	4	48
L.C.	23	53, 39
F.S.	64	88
	77	120
C.S.	11	16
		41
F.G.	12	19
A.F.	9	9
	7	16
F.Ga.	10	24
J.B.	1	1
F.D.	5	6
	5	6

Epinephrine did not cause an increase in kinin levels in the noncarcinoid subjects (Table 1). In the controlled state before the induction of flushing, most of the patients with carcinoid syndrome did not have increased amounts of kinin activity. Patients L. C. and F. S. differed in that they appeared to have continuous release of kinin into the hepatic vein. After the injection of epinephrine, an increase in blood kinin activity was seen in most of the patients with carcinoid syndrome. In the first five patients shown in the table, there was a striking increase to levels of over 25 μg/100 ml. Four other patients showed either variable increases or slight increases to values less than 25 μg/100 ml, and two patients showed no increase in blood kinin activity.

The kinin obtained from the hepatic venous blood of four of these patients has been further examined. These studies on the physicochemical characteristics, enzymatic inactivation, and pharmacologic properties have all indicated that the substance measured in the hepatic venous blood is bradykinin. (Oates et al., in press). The kinin initially isolated from hepatic venous blood was additionally purified by a fractional elution from a weakly acidic cation-exchange resin and then applied to a small column of CM-Sephadex. By a gradient elution with increasing pH and increasing ionic strength, it was possible to separate bradykinin from lysyl-bradykinin (kallidin) on this column and to isolate as little as 1.5 μg bradykinin. It was shown that the kinin isolated from the blood of a patient with carcinoid syndrome had elution characteristics that were similar to bradykinin. Furthermore, when the kinin isolated from the blood of a carcinoid patient was chromatographed on the CM-Sephadex column together with authentic bradykinin, a single peak of activity was obtained, indicating that the elution characteristics of peptide were identical with those of bradykinin and different from those of lysyl-bradykinin.

Paper chromatography with a butanol:acetic acid:H_2O system was then carried out on the freeze-dried eluates from the CM-Sephadex column. It was found that both authentic bradykinin and the kinin obtained from carcinoid blood had Rf's of 0.35.

Electrophoresis of the kinin isolated from carcinoid blood and of the authentic bradykinin was carried out for 2 hours at 1,000 V in a pH 3.5 pyridine acetate buffer. The rates of migration were found to be similar; peak activity for bradykinin was found at 15 cm from the origin, whereas the carcinoid kinin migrated to 14 cm. The inactivation of vasoactive peptides by certain enzymes has been used as a means of characterizing small quantities of these peptides. In this study, added specificity was obtained by determining the rate of inactivation of the peptide by chymotrypsin and trypsin. Authentic bradykinin (0.5 μg) and an equivalent amount of bradykinin from carcinoid blood were incubated at pH 7.4 at 12°C for 12 min with five μg chymotrypsin. The rates of inactivation were similar, with 22% of the authentic bradykinin remaining at the end of the incubation period and 22.7% of the carcinoid kinin remaining. These mean values fall within one standard error of each other. In addition, 0.5 μg bradykinin and an equivalent amount of kinin from carcinoid blood were incubated separately with trypsin in Tyrode's solution for 60 min at 38°C. It was found that very little inactivation (less than 20%) of either authentic bradykinin or carcinoid kinin occurred during incubation with trypsin. Complete inactivation of substance P has been shown to occur during incubation with trypsin under these conditions (Pernow, 1955).

The pharmacologic properties of the kinin from carcinoid blood were compared qualitatively with those of bradykinin in several systems. Similar contractions of the rat uterus and guinea pig ileum were observed, and both substances had a hypotensive action in the rabbit. Both bradykinin and the kinin from carcinoid blood relaxed the rat duodenum, whereas substance P is known to contract this organ. In addition, the response of kinin from carcinoid blood was compared quantitatively with the response to submaximal doses of bradykinin on the rat uterus, guinea pig ileum, and rabbit blood pressure, and they were found to be equivalent within a range of 15%.

Studies of the kallikrein found in tumors and arterial blood of carcinoid patients. The content of kallikrein was investigated in liver tissue obtained at necropsy from five patients without carcinoid and in tumor metastasis from the livers of six patients with the carcinoid syndrome(two obtained at the time of surgery, four at necropsy). Kallikrein was isolated from tissue by adsorbtion onto diethylaminoethyl (DEAE)-cellulose and quantified by measuring the kinin formed on incubation with human kininogen (Oates et al., 1964). Kallikrein activity was expressed in Frey units. The results are summarized in Fig. 1. No kallikrein activity could be demonstrated in liver tissue from patients without carcinoid tumors. The hepatic metastasis from five of the six patients with the carcinoid syndrome, however, contained from 0.4 to 5.0 Frey units/g tissue. The highest values were found in the tumors obtained at laparotomy on two patients with severe flushes and hypotension. There was no detectable enzyme in one tumor taken from a patient without flushes.

The anionic kallikrein from carcinoid tumor can be adsorbed onto DEAE-cellulose at pH 6.4 and can thereby be isolated from the cationic, glass-activated kallikrein present

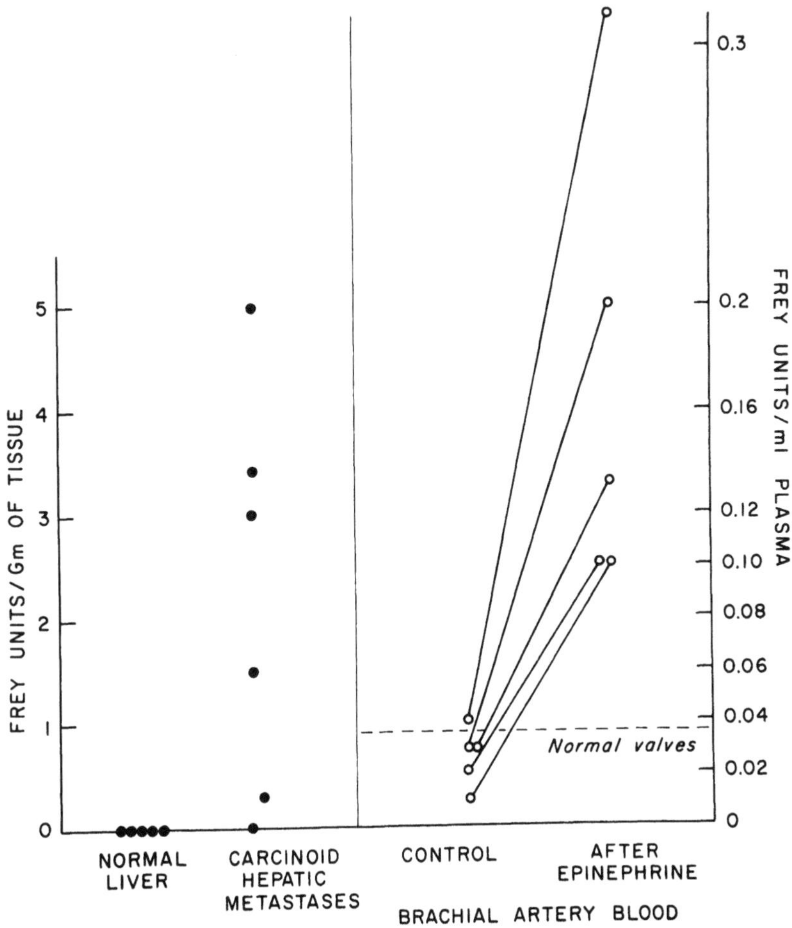

Fig. 1. Determination of anionic kallikrein of hepatic metastasis and brachial artery.

in human plasma. To determine whether an anionic kallikrein is released into the circulation, arterial blood from five normal and five patients with carcinoid flushes was sampled before and after intravenous injection of epinephrine (as was described above). The heparinized plasma was diluted with 8 volumes of 0.05 M acetic acid and stirred at 4°C for 12 hours. Then it was centrifuged and the supernatant adjusted to pH 6.4. The anionic kallikrein released into the arterial blood was selectively adsorbed onto DEAE-cellulose by the same procedure as for tumor extracts. Assays showed less than 0.038 Frey units/ml plasma in normals during both control and experimental periods. In the carcinoid patients, the arterial content of anionic kallikrein was usually below 0.04 Frey units/ml during control periods, but rose 2 to 8 times higher after epinephrine was administered.

Studies were then designed to characterize the peptide(s) produced by the action of tumor kallikrein on human plasma substrate. Fifteen Frey units of tumor kallikrein were added to 0.5 M phosphate buffer (pH 7.4) containing concentrated kininogen obtained from the equivalent of 100 ml human plasma. The mixture was incubated at 37°C for 2 hours. The peptide formed was adsorbed on 2 g of Amberlite CG 50 (H^+) by the method of Pierce and Webster (1961). Biologic assay on the rat uterus revealed activity equivalent to 700 μg of synthetic bradykinin. The product was susceptible to destruction by chymotrypsin and produced hypotension when injected into the rat. The peptides were characterized by chromatography on carboxymethyl (CM)-cellulose, Sephadex G-25, paper electrophoresis, and amino acid analysis after acid hydrolysis. These procedures (Melmon et al., 1965), identified lysyl-bradykinin as the major peptide produced by the incubation in vitro. There is at least one other peptide with biologic activity constituting less than 2% of the total biologic activity. This substance appears to be a larger molecule than bradykinin (by Sephadex G-25 chromatography) and migrates in the same area as bradykinin on CM-cellulose chromatography. Further identification is pending.

Conversion of lysyl-bradykinin (kallidin) to bradykinin by blood. The tumor enzyme initially splits lysyl-bradykinin from kininogen, whereas bradykinin was the peptide isolated from the hepatic venous blood. To determine whether rapid conversion of lysyl-bradykinin to bradykinin by enzymes in blood could account for this, 40 μg lysyl-bradykinin was incubated in fresh whole human blood in siliconized flasks at 37°C for 80 sec. This was the approximate time which was estimated to elapse between the release of kinin into hepatic capillary blood and the complete withdrawal from the catheter tip. The kinin isolated after the incubation had the same elution characteristics on CM-Sephadex as bradykinin and was not lysyl-bradykinin. This finding is consistent with a rapid conversion of lysyl-bradykinin to bradykinin in whole blood and is in agreement with the studies of Webster and Pierce, 1963, and Erdös et al., 1963; demonstrating the presence of an aminopeptidase in plasma that can bring about this conversion. The lysyl-bradykinin added to blood was incompletely recovered as bradykinin, probably because there are at least three enzymatic processes rapidly operating simultaneously: the conversion of lysyl-bradykinin to bradykinin, the inactivation of lysyl-bradykinin by a carboxypeptidase, and the inactivation of bradykinin by a carboxypeptidase (Erdös and Sloane, 1962).

The effects of bradykinin in carcinoid patients. The intravenous injection of epinephrine (1 to 8 μg) in a patient with carcinoid syndrome will produce tachycardia, hypotension, tachypnea, and a flush which resembles the patient's spontaneous flushes. In four patients with carcinoid syndrome, the injection of bradykinin produced similar changes as well as a flush resembling the one induced by epinephrine.

It is now known that the carcinoid tumor may elaborate a variety of biologically active substances simultaneously. Most patients with spontaneous or epinephrine-induced flushes experience a decrease in forearm vascular resistance, a rise in forearm blood flow, and unpredictable changes in venous tone. Studies have shown that the magnitude of change in any of these hemodynamic parameters is dependent upon a variety of factors. For instance, bradykinin produces direct venodilatation but increases venous tone in the presence of an intact autonomic nervous system (Mason and Melmon, 1965). Likewise, renal hemodynamics and function are variously affected by bradykinin, dependent upon intact sympathetic reflexes (Gill et al., 1965). That serotonin may also influence hemodynamic responses to bradykinin was suggested by the increase in mean arterial blood pressure in a patient whose serotonin synthesis was remarkably elevated (5-hydroxyindole acetic acid excretion 870 mg/day, normal less than 15; platelet serotonin content

3.54 μg/mg protein, normal less than 0.8). During her flush or bradykinin infusion, forearm blood flow rose (from 1.0 to 3.6 ml/100 g/min) and venous tone rose (from 10.2 to 35.0 mm Hg/ml) much higher than that seen in patients with less exaggerated serotonin synthesis (5HIAA excretion less than 400). The decrease in forearm vascular resistance was average for most carcinoid patients studied (15 mm Hg/ml/min). Similar

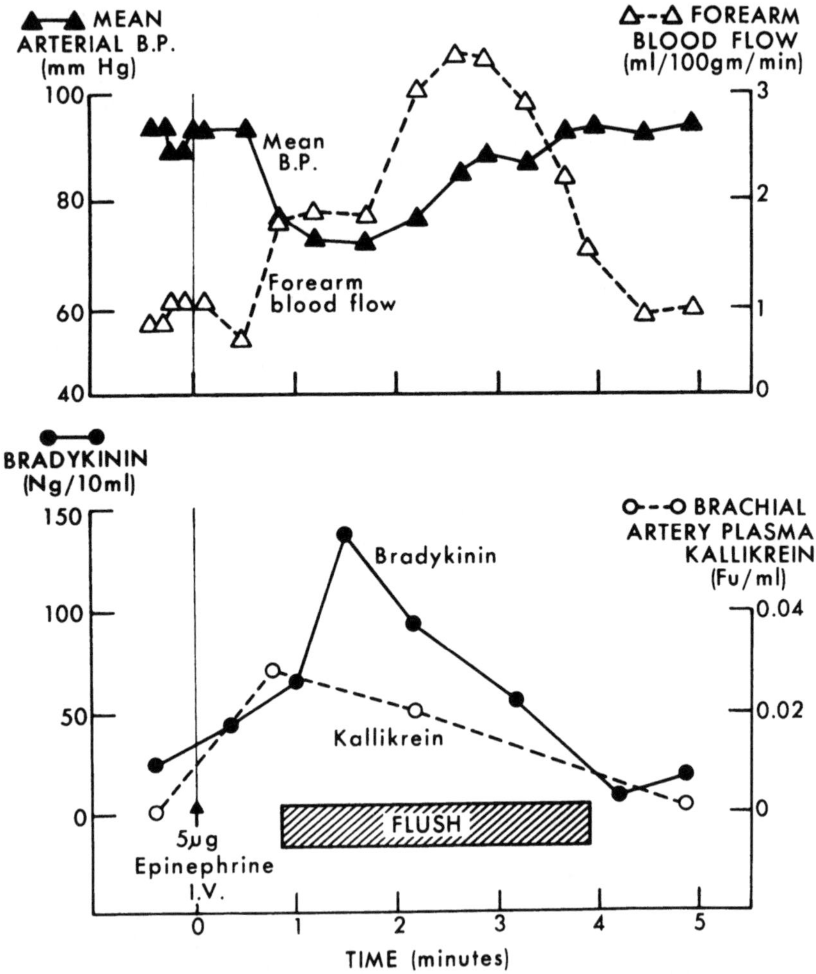

Fig. 2. Simultaneous hemodynamic and brachial artery blood bradykinin and anionic kallikrein measurements in a carcinoid patient. The flush was provoked by intravenous injection of 5 μg epinephrine.

studies of five patients with carcinoid syndrome showed that while hemodynamic measurements varied in separate patients, those produced by infusion of exogenous bradykinin mimicked the changes seen during spontaneous or epinephrine-induced flushes. Furthermore, the rise in the concentration of kinin peptide and anionic kallikrein in blood correlated with the hemodynamic changes seen in one patient during epinephrine-induced flushes (Fig. 2).

DISCUSSION

From the available data, it would appear that flush-provoking stimuli such as epinephrine act on the carcinoid tumor metastases to liberate the enzyme kallikrein into the blood. There it acts on the kininogen to liberate lysyl-bradykinin. This decapeptide is then rapidly converted to bradykinin by an aminopeptidase in plasma. In order to verify this proposed mechanism of bradykinin release, it would be desirable to have further characterization of the kallikrein found in the blood during flushes.

The release of bradykinin by these well-differentiated enterochromaffin cell tumors suggests that this mechanism for bradykinin release may exist also in normal enterochromaffin cells. Certainly, the mechanism of serotonin production is similar in normal enterochromaffin cells and carcinoid tumors. Further studies of the exaggerated release of bradykinin and serotonin in the carcinoid syndrome may provide clues to the mechanism by which these vasoactive compounds are released physiologically.

Several vasoactive substances have now been related to the carcinoid syndrome. Excessive synthesis of 5-hydroxyindoles by these tumors is the most constant biochemical indicator of the disease. Increased histamine excretion is found in a few of these patients, chiefly those with gastric carcinoid (Oates and Sjoerdsma, 1962). With the demonstration that bradykinin is released during flushes, consideration must be given to its possible role in the pathophysiology of the syndrome. Because bradykinin produces flushes resembling those occurring spontaneously in some carcinoid patients, it is likely that this peptide plays a role in the flushes of those patients who release large amounts into the circulation. Bradykinin concentrations of over 25 $\mu g/100$ ml hepatic venous blood have been found during flushes in five of the eleven patients studied to date; these values suggest that amounts are released into the circulation which are comparable to the amounts shown by us and by Fox et al. (1961) to produce vasodilatation in man. Because of the multihumoral potential of these tumors, it is not surprising that the appearance of the flushes seen in carcinoid syndrome may vary remarkably from patient to patient, dependent on which vasoactive agent or combination of agents is released.

There is an interesting phylogenetic analogy between the carcinoid tumors and the cells of the wasp venom sac, for they both can release serotonin, histamine, and kinin peptides. While the wasp venom sac has a highly specialized function, one might speculate that the enterochromaffin cells in man have evolved to the point where these agents are released locally to act singly or in concert to influence function of the gut, modulate neuron transmission, or alter the microcirculation.

There is an abundance of data which indicates that serotonin is not the sole mediator of the paroxysms of vasodilatation seen in carcinoid patients. Neither, however, is bradykinin the universal vasodilator in this syndrome. We have studied two patients who did not release appreciable amounts of bradykinin into the hepatic venous blood during hypotensive flushing attacks. Simultaneous measurement of the hepatic venous levels of both bradykinin and free plasma serotonin during flushing will be necessary to further elucidate the relationship between the various flushing patterns and specific vasoactive substances, and to explore the possibility that yet another vasoactive compound may be released during flushes.

ACKNOWLEDGMENTS

The authors gratefully acknowledge the cooperation of Drs. Albert Sjoerdsma, Walter Lovenberg, Louis Gillespie, Dean Mason, William Pettinger, and R. B. Doctor in these studies. This work was supported, in part, by Research Grants HE-5545 and HE-08668-03 and HE-09964 from the National Institutes of Health.

REFERENCES

Erdös, E. G., A. G. Renfrew, E. M. Sloane, and J. R. Wohler: Enzymatic studies on bradykinin and similar peptides. Ann. N.Y. Acad. Sci. 104, 222 (1963).

Erdös, E. G., and E. M. Sloane: An enzyme in human blood that inactivates bradykinin and kallidins. Biochem. Pharmacol. 11, 585 (1962).

Fox, R. H., R. Goldsmith, D. J. Kidd, and G. P. Lewis: Bradykinin as a vasodilator in man. J. Physiol. (Lond.) 157, 589 (1961).

Gill, J. R., K. L. Melmon, L. Gillespie, Jr., and F. C. Bartter: Bradykinin and renal function in normal man. The effects of adrenergic blockade. Amer. J. Physiol. 209, 844 (1965).

Hilton, S. M., and G. P. Lewis: The relationship between glandular activity, bradykinin formation, and functional vasodilatation in the submandibular salivary gland. J. Physiol. (Lond.) 134, 471 (1965).

Levine, R. J., and A. Sjoerdsma: Pressor amines and the carcinoid flush. Ann. int. Med. 58, 818 (1963).

Mason, D. T., and K. L. Melmon: Effects of bradykinin in forearm venous tone and vascular resistance in man. Circulat. Res. 17, 106 (1965).

Melmon, K. L., W. Lovenberg, and A. Sjoerdsma: Characteristics of carcinoid tumor kallikrein. Identification of lysylbradykinin as a peptide it produces in vitro. Clin. chim. Acta 12, 292 (1965).

Oates, J. A., and A. Sjoerdsma: A unique syndrome associated with secretion of 5-hydroxytryptophan by metastatic gastric carcinoids. Amer. J. Med. 32, 333 (1962).

Oates, J. A., W. A. Pettinger, and R. B. Doctor: Evidence for the release of bradykinin in carcinoid syndrome. J. clin. Invest. In press.

Oates, J. A., K. Melmon, A. Sjoerdsma, L. Gillespie, and D. T. Mason: Release of a kinin peptide in the carcinoid syndrome. Lancet 1, 514 (1964).

Pernow, B.: Inactivation of substance P by proteolytic enzymes. Acta physiol. scand. 34, 295 (1955).

Pierce, J. V., and M. E. Webster: Human plasma kallidins: Isolation and chemical studies. Biochem. biophys. Res. Commun. 5, 353 (1961).

Robertson, J. I. S., W. S. Peart, and T. M. Andrews: The mechanism of facial flushes in the carcinoid syndrome. Quart. J. Med. 31, 103 (1962).

Webster, M. E., and J. V. Pierce: The nature of the kallidins released from human plasma by kallikreins and other enzymes. Ann. N. Y. Acad. Sci. 104, 91 (1963).

DISCUSSION

Zeitlin: Working in Dr. Brocklehurst's laboratory in collaboration with the surgeon Mr. A. N. Smith, I have had an opportunity to measure the free kinins and kinin precursor occurring in the peripheral venous blood of three patients suffering from carcinoid tumor. The free kinin extraction used was not as described by Dr. Oates, but consisted of extraction of whole blood kinins into cold absolute ethanol, as described by Brocklehurst and Lahiri (1961), followed by a butanol extraction to remove interfering potassium and 5-HT, adapted from that described by Gaddum and Horton (1959). The activity was measured on estrus rat uterus in the presence of atropine and a 5-HT antagonist, BOL. Simultaneously, kinin precursor levels were assessed by trypsin incubation of denatured blood by an adaptation of the methods of Diniz and Carvalho (1963) and Brocklehurst and Lahiri (1962), which will be described in a paper to be submitted for publication. Both measurements could be carried out on the same 5-ml sample of blood.

Of the three patients examined, one was nonflushing, and the tumor was shown to be carcinoid-like by histological examination. The second showed the typical transient flushing of the carcinoid syndrome. The third showed continuous, very severe flushing. The nonflushing patient had a mean free kinin level of 3.5 ng bradykinin equivalent/ml plasma. He had a kinin precursor level of 5.5 μg/ml plasma. Both values are of the same order as normal controls. The patient showing transient flushing produced during flush a level of 26.3 ng/ml plasma, more than 5 times greater than normal controls. He too had normal levels of kinin precursor (4.8 μg/ml) plasma. It was not possible to obtain both nonflushing and flushing samples for comparison in this patient. The patient showing continuous flush had a mean free kinin level of 53.7 ng bradykinin equivalent/ml plasma which is about 10 times greater than normal. The kinin precursor level in this case, quite surprisingly, was 16.6 μg Bk equivalent/ml plasma, about 3 to 4 times the level of our normal controls.

The active material measured in the carcinoid samples contracted rat uterus in the presence of atropine and BOL; and it relaxed rat duodenum. In this way it was distinguished from acetylcholine, 5-HT, histamine, substance P, epinephrine, and gastrin, none of which produced the same combination of effects in these two tissues.

Chymotryptic incubation has not yet been carried out, owing to the small quantities involved in these samples. However, samples have been stored, and may be pooled and subjected to this test at a later date. Thus it would seem that these preliminary results support those of Dr. Oates and his colleagues.

Amundsen: Have you ever used any kind of kininase preparations in the treatment of your patients during their flush periods?

Oates: The only kinin-inactivating enzyme currently proved safe for human use is chymotrypsin; this is a weakly active "kininase", and we did not note amelioration of flushing after administration of this enzyme.

Shimamoto: Rowley and Guth and co-workers described the venoconstrictor effect of bradykinin. We also found that the mode of venoconstriction is either spastic or tonic according to the different types of veins. Cutaneous veins like saphenous vein or ear vein of the rabbit responded with closure of veins to increased doses of bradykinin. The coronary or liver vein exhibited a tonic constriction and never closed; there was merely prolonged constriction of a certain degree. Such data even suggest the existence of isometric constriction of some venous smooth muscles, caused by bradykinin. In vasodilatation the increase in the tone of venous smooth muscle of either common venous segments or arteriovenous shunt has to be taken into consideration. Recently, rosacea has disappeared after pyridinol carbamate treatment. Various hemorrahagic states such as thrombopenic or nonthrombopenic purpuras and various exudative inflammatory states responded to this compound. These have been confirmed in over 1,000 cases during the past 2 1/2 years in our University Hospital.

Lembeck: Your work clarifies some unexplained results I had years ago when I extracted the first carcinoid and wanted to identify the active compound as 5-HT. In paper chromatography I found, besides 5-HT, another active fraction which remained unexplained. I excluded histamine and substance P. In connection with your new finding I would remind you of Dengler's observations of a patient with a pancreatic carcinoma; there was a severe flush, but no 5-HT in the tumor.

Oates: Dr. Melmon and co-workers have found kinin in tumors, but have not drawn any conclusions from this because of the multiple other factors that can produce kinin in tissue. In studies with one such pancreatic tumor, we have not found kallikrein. This search should continue.

Collier: Herxheimer and Stresemann showed that in some chronic asthmatics ingestion of ethanol greatly potentiates bronchoconstriction induced by inhalation of bradykinin aerosol. Ethanol is also well known to induce flush in normal people. Dr. Oates has shown that epinephrine does not liberate kinin in normal people. Can he tell us whether ethanol liberates kinin in normal people?

Oates: We have not studied the effects of ethanol on kinin release in normal subjects.

Back: In collaboration with Dr. Gardner of Downstate Medical Center, New York, we have studied kininogen levels in a single patient with carcinoid syndrome. Whereas kininogen levels did not decrease during spontaneous flush, injection of epinephrine did result in a decrease in kininogen. Infusion of Trasylol (25,000 U/kg) did not improve the clinical symptoms seen in this patient, and none of the procedures caused any changes in plasma kallikrein or kininase levels. We are currently studying the localization of kallikrein and kininogen in normal and tumor tissue of this patient by the immunofluorescent techniques. Have you used Trasylol in any of your patients?

Oates: Yes; no amelioration of symptoms occurred. Studies in vitro suggest that preincubation of the inhibitor with enzyme is necessary for inhibition.

OCCURRENCE OF A POLYPEPTIDE BIOLOGICALLY NOT DISTINGUISHABLE FROM BRADYKININ IN HUMAN PULMONARY CARCINOMA

PIETRO DI MATTEI

Institute of Pharmacology, University of Rome, Italy

The presence of polypeptides with kinin-like activity was previously demonstrated in this laboratory in some pathological conditions in man and animal (Di Mattei, 1962).

Tissues capable of rapid and vigorous protein synthesis and rich in proteolytic enzymes, such as animal and human neoplastic tissues with very rapid cell multiplication, could be potential sources of biologically active peptides. In view of this possibility, we carried out investigations on a series of human adenocarcinomas from lung, stomach, kidney, and large intestine, made available through surgical intervention.

The tissues were frozen in liquid nitrogen immediately after removal and then preserved until sufficient material had been collected for the extraction of the polypeptides. The biological material was slowly brought to room temperature in about 2 hours by controlled raising of the temperature of the freezing room; it was then homogenized in glacial acetic acid at a constant temperature of 15°C.

The acetic acid extract was diluted to about 1 M concentration and dialyzed in Visking tubes against a solution of 1 M acetic acid for 48 hours, with twelve changes of the outside fluid.

The dialyzed material was lyophilized at 0°C. It was then recovered in absolute ethanol and transferred on a column of basic aluminum 80 cm in height and 2.5 cm in diameter, which had been activated previously and washed with absolute ethanol. After the extract had settled, the column was again washed with absolute ethanol (200 ml) and chromatography was started with decreasing concentrations of ethanol in water, from 95% to 30%. The fractions up to the concentration of 70% were discarded because they contained some biogenic amines. When the main biological activity of presumed peptides was assayed on the intestine of guinea pig ileum, it was found to be distributed mainly in a broad band in the fraction eluted with 60% and lower concentrations of alcohol, with maximal activity occurring in the fraction moving with 40% ethanol. The fractions with biological activity were then evaporated at low temperature, submitted again to chromatography on a column of DEAE-Sephadex and subsequently eluted with pyridine acetate (0.1 M at pH 4.5).

The fractions containing the biological activity and giving positive reaction with ninhydrin before and after hydrolysis were then collected and pooled. Subsequently these fractions were further fractionated by means of countercurrent distribution in the system n-butanol-ethanol-acetic acid-water with a total of 200 transfers. In this way we obtained fractions A_1, A_2, A_3, A_4, A_5, which were then tested for biological activity on isolated organs and on laboratory animals.

Of all the extracts of tumors tested, only those from lung adenocarcinomas were shown to possess peptides endowed with biological activity. This finding renders unlikely the hypothesis that the activity could be due to blood always present in the tissue. Bradykinin in the blood, on the other hand, does not seem to occur, although some kinin-like property in the circulating blood has been occasionally reported. By the same token, the possibility could be ruled out that the biological activity found in these tumors could be due to substances present everywhere.

In fractionation of lung tumors, the biological activity as tested on isolated organs was found almost entirely in fraction A_1, which was then submitted to further purification. The fractionation consisted of electrophoresis on starch gel column which was followed by a second electrophoresis on cellulose acetate. The fraction seemed to move as a unique band almost identical, with that of bradykinin. No difference between the substance and standard bradykinin was noticed when the terminal pipsylpeptides of the two peptides were synthesized by means of reaction with the I^{131}-phenylsulfonic acid chloride and then submitted to chromatography on a thin layer of silica gel with a mixture of ether-ethanol-acetic acid, with the technique previously described (Melchiorri, 1962).

TABLE 1
Biological Effects of Polypeptide Fractions from Human Pulmonary Carcinoma*

Counter-current fractions	Dog blood pressure (μg brad./ml)	Rabbit blood pressure (μg brad./ml)	Rat uterus (μg brad./ml)	Guinea pig ileum (μg brad./ml)	Rabbit large intestine (μg physal./ml)	Rat large intestine (μg eled./ml)	Rat writhing reflex (for pain measurement) (μg brad./ml)
Total extracts	+++	++	+++	+	–	–	+++
A_1	20	20–25	25	20	< 0.05	< 0.2	+
A_2	1–2	0.2	3	0.3	< 0.05	< 0.2	–
A_3	0	0.4	0.15	0.3	< 0.05	< 0.2	–
A_4	0	0.2	0.1	0.3	< 0.05	< 0.2	–
A_5	0	0.2	0.15	0.3	< 0.1	< 0.2	–

*Brad., bradykinin; physal., physalaemin; eled., eledoisin.

This procedure, when applied to about 3.500 g of lung tumor, led to the isolation of about 9 mg (as measured in total nitrogen) of peptide fraction A_1, which contained the maximal biological activity of the extract; the pain-producing activity was low in the fraction and much higher in the total extract. The biological data on the testing of this fraction, as well as the bioassay of the total extract of the other fraction, are reported in Table 1.

The amino acid content of this fraction is under investigation at present. A great deal of caution is needed in the identification of the polypeptide, and we can only indicate the great similarity between its properties and those of bradykinin.

REFERENCES

Di Mattei, P.: Arch. int. Pharmacodyn. 140, 368 (1962).
Melchiorri, P.: Biochim. appl. (Parma) 9, 309 (1962).

In Memory of Sir John Gaddum, F.R.S.

Chairman's remarks, F. Lembeck:

Before discussing substance P in this session, I feel that I should say a few words devoted to the memory of Sir John Gaddum. He discovered this peptide together with U. S. von Euler in 1931. He has contributed a great deal to our present knowledge during the past decades. Sir John designed the principles of parallel quantitative bioassays, which have been used so often in the work with these peptides. In 1961 when we had a symposium on substance P organized by Prof. Stern in Sarajevo, Sir John was the central figure in the research on substance P and kept this position during his last years of painful illness until his death on June 30th of this year. A few months ago he was still hoping to be able to participate in our present symposium, and he joined the honorary committee.

Here I need not recall all the work he has done. Behind the work stands a great and distinguished personality, whom it is hardly possible to honor in only a few words. He characterized best his unique ability to simplify things in Who's Who, where you will find his hobbies described as: "Reading, writing and arithmetic." Few of us know how much he read and how carefully he was in reading, but we all know how precise his writing was, in communications and in his textbook as well. His "arithmetic" resulted in the introduction of the term "lambda" into biometrics — now commonly used in this field as the terms chi-square or "t."

Now I think we should take with us into our discussions an aphorism Sir John Gaddum expressed some years ago. He said, "It will probably always be more important to try a thing out, than to argue about it."

PURIFICATION OF SUBSTANCE P

H. ZUBER

Laboratory for Molecular Biology, Federal Institute of Technology, Zurich, and Pharmaceutical Division of CIBA Ltd., Basle, Switzerland

In 1962 in New York I presented a paper on a method of isolating substance P from bovine brain (Zuber, 1963). Subsequently, I managed to improve and expand this method a little further at CIBA. Unfortunately, however, it was possible to continue this work for only one year – until May 1963. The results, therefore, are relatively old and may already have been superseded. Nevertheless, they might be of use to various groups who are interested in this problem.

The isolation of substance P is very difficult, chiefly for the following reasons. First, the principle is present only in small amounts in brain and intestine; it is estimated that approximately 1 mg peptide can be isolated from 100 kg of brain. Second, the basic substance displays a high degree of adsorption to carrier materials such as ion exchange resins, aluminum oxide, and cellulose powder, which leads to considerable losses during isolation. Third, other chemically similar peptides are present in great excess, and it is relatively difficult to separate these peptides from substance P. Finally, several active variants of substance P seem to exist. I shall return to this question later. When isolating the substance, therefore, we endeavored to keep the number of purification steps as low as possible and to find chromatographic methods that would involve only slight losses due to adsorption. As described previously (Zuber and Jaques, 1962; Zuber, 1963), Sephadex and CM-Sephadex proved to be very suitable for fractionation purposes. The method can also be used for isolating fairly large quantities of substance P. In order to simplify the isolation, the oxycellulose stage described previously was omitted.

First I should like to give two typical examples to show how the method, in its modified form, is used to isolate substance P from bovine brain, and in particular to demonstrate the difficulties that arise during the isolation. Later I shall discuss a few questions concerning the structure of the substance.

I have selected two examples of isolation which differ in respect of the amount of brain material used. Increasing the amount of brain from 100 to 1,000 kg does not yield an appreciably larger quantity of substance P. Owing chiefly to technical difficulties, a larger quantity of starting material leads to less pure products which are more difficult to purify and which lose a great deal of their activity during purification. For instance, in the 1,000-kg experiment the extract contains the same total activity as in the 100-kg experiment; this is probably due to large losses occurring during extraction. In addition, the ammonium sulfate precipitate contains a 20-fold greater amount of accompanying substances.

Isolation and characterization of substance P

In the first example (experiment 6) 100 kg bovine brain was extracted as described (Zuber and Jaques, 1962) and fractionated with ethanol and ammonium sulfate. Yield of the ammonium sulfate precipitation: 88 g crude product (25 μg/ml T.D.*; 100% = initial activity, cf. Table 1). For the purposes of calculating the yield, the active material used was assumed to have an activity of 100% (= initial activity) at this stage. It is, in fact, difficult to determine activity accurately in the crude extract, since the latter also includes other active components as well as inhibitory substances. Fractionation with glacial acetic acid and ether yielded 13.8 g of an active product (activity 5 μg/ml T.D., 30 units/mg, 78% of the initial activity). The active fraction of Sephadex G-25 fractionation (yield: 1.2 g) possessed an activity of 0.25 to 1.0 μg/ml T.D. (85% of the initial activity), and the fractions of CM-Sephadex fractionation with maximal activity (relatively broad distribution of the activity among many fractions) had an activity of 10 ng/ml T.D.†) (yield: 14 mg, 40% of the initial activity, 10,000 to 14,000 units/mg).

After fractionation by means of preparative high-voltage

*Contraction of the isolated guinea pig ileum in Tyrode's solution, 37°C, bath contents 20 ml, atropine and tripelennamine added: end concentration 2.10^{-3} g.T.D. = threshold dose/ml bath fluid.

†1 ng = 10^{-3} μg.

TABLE 1
Purification of Substance P

Experiment 6; 100 kg brain.

Stage	Yield: Weight	Yield: Activity (%)	Specific activity* (per ml T.D.)
1. Ammonium sulfate precipitation	88 g	100†	25 μg
2. Glacial acetic acid/ether fractionation	13.8 g	78	5 μg (30 U/mg)
3. Sephadex G-25	1.2 g	85	0.25–1.0 μg
4. CM-Sephadex	14 mg	40	10 ng (14,000 U/mg)
5. Prep. electrophoresis (pH 9.5)	7 mg	20	10 ng + 50 μg/ml Tween-80 2 ng/ml T.D.‡ (40,000 U/mg)
	= 0.7 mg peptide		1 ng + 50 μg/ml Tween-80 0.2 ng/ml T.D. (400,000 U/mg)

*T.D. = threshold dose.
†Activity assumed to be 100% (initial activity) at this stage.
‡Enriched (in relation to activity) 12,500 fold between stage 1 and stage 5.

electrophoresis on paper (pH 9.5, triethylammonium carbonate buffer, 180 min, 1,200 V) a fraction (B) (7 mg) was obtained with an activity of 10 ng/ml T.D. After the addition of Tween-80 the activity was 2 ng/ml T.D. = 40,000 units/mg in comparison with the Euler standard. According to Dr. Jaques of CIBA, the addition of Tween-80 stabilizes the purified fractions of the substance, resulting in an approximately five-fold increase in activity. Related to the peptide material (0.7 mg) in the preparation, the activity was 1 ng/ml T. D.; with the addition of Tween, 0.2 ng/ml T. D.; or, determined in Euler units, 400,000 units/mg. When submitted to chromatographic examination, the active fraction was found to be uniform in the solvent system sec. butanol: 3% NH_3 (100:44), but not homogeneous in in the system ethylacetate:pyridine:water (20:20:11).

In the second example (experiment 7), 1,000 kg of bovine brain was extracted. Yield of the ammonium sulfate precipitate: 1.9 kg (activity, 500 μg/ml T. D., 100% = initial activity; Table 2); 205 g of the glacial acetic acid ether fractionation (activity, 40-50 μg/ml T. D., 108% of the initial activity) yielded 49 g of the active fraction of Sephadex G-25 fractionation (activity, 10 μg/ml T. D., 120% of the initial activity). After CM-Sephadex fractionation, repeated four times, 66 mg substance with an activity of 35 ng/ml (55% of the initial activity; fractionation IV, Table 2) was obtained

Repeated preparative paper chromatography (solvent: sec. butanol: 3% NH_3, 120:44) yielded a fraction (11 mg) with an activity of 10 ng/ml T. D. (28% of the initial activity). Renewed purification with the aid of paper chromatography (Fig. 1; solvent, ethylacetate:pyridine:water, 20:

Fig. 1. Marginal strip of a preparative paper chromatogram (paper: Schleicher and Schüll 2043 b MGl; solvent, ethylacetate: pyridine: water, 20:20:11; color reagent, chlorotolidine). Fraction B has the highest activity (main component).

20:11) yielded 5 mg of an active fraction (B) (activity, 5 ng/ml T. D., related to peptide material 0.5 ng/ml; 25% of the initial activity). The substance P fraction was chromatographically uniform (solvent, ethylacetate-pyridine-water).

In the solvent ethylacetate-pyridine-water the R_f value was 0.5 with thin-layer chromatography and 0.32 with paper chromatography (Schleicher and Schüll paper 2043 b MG1, descending).

TABLE 2
Purification of Substance P

Experiment 7; 1,000 kg brain.

Stage		Yield: Weight	Yield: Activity (%)	Specific activity (per ml T.D.)
1. Ammonium sulfate precipitation		1,900 g	100*	500 μg (inhib. subst.)
2. Glacial acetic acid/ ether fractionation		205 g	108	50 μg
3. Sephadex G-25		49 g	120	10 μg
4. CM-Sephadex	I	900 mg	47	0.5 μg
	II	250 mg	65	0.1 μg
	III	142 mg	50	75 ng
	IV	66 mg	55	35 ng
5. Prep. paper chromatography	I and II	11 mg	28	10 ng
	III (fraction B)	5 mg	25	5 ng†
		0.5 mg peptide	25	0.5 ng

*Activity assumed to be equal to 100% (initial activity) at this stage.
†Enriched (in relation to activity) 100,000-fold between stage 1 and stage 5.

The reproducibility of the R_f values, especially in the thin-layer chromatogram, is not very good. In particular, we noticed that after being stored for some time — or even sooner when submitted to repeated evaporation — a chromatographically uniform fraction would suddenly display several components upon renewed thin-layer chromatography. We cannot at the moment decide whether this phenomenon is due to decomposition of substance P.

This observation, however, is probably closely linked to the fact that several active substance P fractions occur. In the examples of substance P isolation demonstrated above we deliberately followed only one component. Other fractions, however, possessing similar or less activity, are also present, fractions whose R_f values differ from that of the main component in the paper or thin-layer chromatogram. Upon total hydrolysis and end-group analysis, however, they yield the same amino acids. In an additional experiment (no. 8), we managed to separate a second active component from the main component on a cellulose column (ethylacetate-pyridine-water and 1 M pyridine acetate). The R_f value of this second component (paper chromatography) in the solvent system ethylacetate-pyridine-water was 0.5, as compared with an R_f of 0.3 for the main component. Both substances are composed of the same amino acids qualitatively and possess the same carboxyl end groups. We cannot say in what respects the two components differ chemically from each other since we do not know their structures. There may possibly be differences at the methionine (methionine sulfoxide; reduction of some fractions with thioglycollic acid produces a marked increase in activity), or at the aspartic acid or glutamic acid (free acid, amide, peptide bond linking the α- or β-carboxyl group at the aspartic acid, peptide bond linking the α- or γ-carboxyl group at the glutamic acid, or else formation of pyroglutamic acid).

Before passing on to studies of the structure of substance P, we must first ask ourselves the following question: Even if the substance behaves uniformly upon chromatographic examination, is it really a pure peptide? Even with the aid of enzymatic end-group determination, the results of which will be described later and which yield reasonable values, it is not possible to give a definitely affirmative answer to this question. The amino acids released by enzymatic hydrolysis at the carboxyl end (particularly since phenylalanine is the first amino acid to be split off) may also come from a mixture of different peptides. The actual substance P peptide would then contain other inactive peptides as impurities. Unfortunately, chemical methods of end-group determination cannot be employed, owing to the small amounts of substance available. The amount of peptide used for enzymatic degradation studies is 10 μg!

Studies of the chemical structure of substance P

a) Amino acid composition (qualitative). Total hydrolysis of the purified fractions (from experiments 6, 7, and 8) with 6 N HCl yields, as already described in part (Zuber and Jaques, 1962), the following amino acids: Lys, Arg, Asp, Glu, Gly, Ser, Ala, Thr, Pro, Val, Leu/Ileu, Phe. Tyrosine was demonstrated after acid hydrolysis when the accompanying substances eluted from the chromatography paper had been removed by renewed fractionation with CM-Sephadex (Tyr is destroyed by the paper impurities during hydrolysis). The tyrosine in the nonhydrolyzed peptide was also observed in the ultraviolet spectrum (differential curve, pH 6 and 0.2 N NaOH). Tyrosine is liberated by degradation with carboxypeptidase A, carboxypeptidase B, and carboxypeptidase C (see below). Methionine is more difficult to detect. After reduction with 1% thioglycollic acid, methionine was demonstrated both after total hydrolysis with 6 N HCl and by exhaustive degradation with carboxypeptidase B (see also Table 3 [quantitative amino acid analysis.]) Sometimes after enzymatic degradation a spot which might have been tryptophan appeared in the two-dimensional thin-layer chromatogram.

b) Amino acid composition (quantitative). The values obtained by quantitative analysis of the amino acids in various substance P preparations are not very consistent and display wide deviations. This is presumably due to the fact, first, that the losses incurred during the hydrolysis of such small amounts of substance are high; 20 μg peptide was hydrolyzed each time. The losses are particularly high when impurities eluted together with the peptide material from the chromatography paper destroy the amino acids during hydrolysis, as in the case of tyrosine. Second, analysis of such small amounts of amino acids does not yield very accurate results (0.001 μmoles/amino acid = approx. 0.15 μg/amino acid). Five analyses (hydrolysis with 6 N HCl in four cases, and combined degradation with trypsin, carbosypeptidase B, and leucine aminopeptidase in one case) yielded the round values found in Table 3.

The number of amino acid residues in the polypeptide chain can be estimated from the data on amino acid analysis and would seem to be 21 to 27. This corresponds to a molecular weight of 2,373 to 3,051 (mean weight of amino acid residues: 113).

TABLE 3
Amino Acid Composition*

Arg	1-2	Ser	2	Ileu	1
Lys	1-2	Ala	2-3	Phe	1-2
Asp or		Thr	1	Tyr	1
Asp (NH_2)	1	Pro	2	Met +	
Glu and/or		Val	1	MetSO	1
Glu (NH_2)	2	Leu	2-3		
Gly	2-3				

*Quantitative, rounded off mean values from five analyses of highly purified fractions from experiments 5, 6, and 7.

c) Enzymatic degradation experiments. The biological activity of substance P preparations (tested on guinea pig ileum) can be destroyed with the aid of chymotrypsin and trypsin (90 to 180 min, 37° C, ratio of enzyme to substrate = 1:20). When the ratio of enzyme to substrate is 1:200 (3 hours, 37°C), the activity is completely destroyed by chymotrypsin but only 50% destroyed by trypsin (splitting proceeds more slowly with trypsin). Strangely enough, activity is maintained after hydrolysis with carboxypeptidase A (DFP-treated, 2 hours and 20 hours, 37°C). The activity is destroyed by carboxypeptidase A that has not been treated with DFP (chymotrypsin content ?). Degradation with trypsin and chymotrypsin yields a series of cleavage products, the composition of which has not yet been studied in detail.

Hydrolysis with carboxypeptidase A (DFP-treated) releases one after the other Phe, Leu, Tyr, as well as traces of Lys, Gly, Ala, Ser, and Val, from the carboxyl end (degradation experiments of 10 min, 2 hours, and 20 hours). From this it can be inferred that the carboxyl end contains the sequence Phe-Leu-Tyr, provided, of course, that we are dealing with a uniform peptide. As already emphasized, however, it is not impossible – though it is very unlikely – that the amino acids Phe, Leu, Tyr, or the amino acids resulting from degradation with carboxypeptidases B and C, come from various peptides. This result was obtained with all fractions, even in the less active ones! In the degradation with carboxypeptidase B (DFP-treated), Lys, Arg, Gly, Ala, Ser, Val, Thr, and traces of amino acids (Met-SO ?) are released in addition to Phe, Leu, and Tyr. With carboxypeptidase C (Zuber, 1964), Phe, Leu, Tyr, Lys, Arg, Gly, Ala, Ser, Val, Thr, Pro, Asp(NH_2) + Asp, and Glu(NH_2) are split off from the carboxyl end.

With leucine aminopeptidase, on the other hand, no hydrolysis occurs from the amino end (only traces of amino acids are split off, largely owing to the carboxypeptidase A and B activity of leucine aminopeptidase). After prior hydrolysis for 20 hours with trypsin, leucineaminopeptidase releases the following amino acids: Phe, Leu, Tyr, Val, Ala, Gly, Ser, Lys, Arg, Thr(Met, Met-SO ?) [traces of Asp(NH_2) + Asp].

Prolidase, prepared by the method of Davis and Smith (1957) caused no clear-cut hydrolysis from the amino end (traces of Lys, Gly, Ala, Ser, Glu, and sometimes Arg occur). This enzyme preparation possesses virtually no endopeptidase activity but has some leucine aminopeptidase activity. This largely excludes the possibility of a NH_2-X-Pro . . . sequence at the amino end. Here, too, after prior hydrolysis with trypsin (20 hours), Phe, Leu, Tyr, Val, Gly, Ala, Ser, Lys, Arg, and Thr, are liberated, as well as Pro, Glu, and Asp. Similar findings were obtained after prior hydrolysis with chymotrypsin. Peculiarly enough, prior treatment with pepsin (48 hours, 37° C, 0.02 N HCl) followed by hydrolysis with prolidase, causes these amino acids to be liberated: Phe, Leu, Tyr, Val, Gly, Ala, Ser, Pro, Glu, Lys, and traces of Asp, Thr, and Met. After reaction with FDNB, no α-DNP amino acids were observed, only ϵ-DNP-Lys (O-DNP-Tyr ?).

In conclusion, I should like to mention an interesting experiment. In connection with the purity of highly purified fractions, it is in fact interesting to compare a preparation from brain with a preparation from intestine. Dr. Meinardi and Dr. Craig were kind enough to let us have a highly purified fraction obtained from horse intestine so that we could compare it with our substance P preparation. We degraded this fraction with carboxypeptidases A, B, and C and obtained in all three cases the same amino acids as with our own preparation. These findings make it seem most unlikely that the amino acids split off by the carboxypeptidases come from various peptides present in substance P in the form of impurities. For it is almost impossible that preparations from brain and intestine, submitted to entirely different methods of purification, should contain the same peptide impurities in addition to the substance P peptide. Moreover, these findings provide further evidence that the structure of substance P from brain must be identical with that of substance P from intestine.

ACKNOWLEDGMENT

I should like to thank Dr. Jaques of the CIBA Pharmacological Laboratories most warmly for the excellent tests he carried out on many of the fractions. Without these tests our attempts to purify substance P could not have succeeded.

REFERENCES

Davis, N. C., and E. L. Smith: Purification and some properties of prolidase of swine kidney. J. biol. Chem. 224, 261 (1957).

Zuber, H.: Isolation of substance P from bovine brain. Ann. N.Y. Acad. Sci. 104, 391 (1963).

Zuber, H.: Purification and properties of a new carboxypeptidase from citrus fruit. Nature (Lond.) 201, 613 (1964).

Zuber, H., and R. Jaques: Isolierung von Substanz P aus Rinderhirn. Angew. Chem. 74, 216 (1962).

DISCUSSION

Gladner: Is it possible that your carboxypeptidases A and B are contaminated each with a trace of the other, thus giving misleading results by yielding more materials than called for by the specificity?

Zuber: Carboxypeptidase A is not contaminated by carboxypeptidase B because only the amino acids phenylalanine, leucine, and tyrosine are liberated. Carboxypeptidase B is presumably contaminated by carboxypeptidase A, since in addition to lysine and arginine other amino acids are liberated. In the case of carboxypeptidase B is it not also possible that, as Neurath found, porcine carboxypeptidase B displays a higher specificity — i.e. releases only the amino acids arginine and lysine — but bovine carboxypeptidase B is not very specific; i.e. it releases other amino acids besides arginine and lysine?

STUDIES OF SUBSTANCE P

HARRY MEINARDI and LYMAN C. CRAIG

The Laboratories of the Rockefeller University,
New York, N. Y., U.S.A.

The physiological role of substance P (SP) is still largely a matter of conjecture. Its extremely high potency in pharmacological tests (stimulation of gut, depression of blood pressure, and vasodilation), its almost exclusive occurrence in the gut and certain areas of the brain (Pernow 1953; Lembeck and Zetler, 1962; McLennan, 1963), and its association with the subcellular particles containing nerve endings (Euler and Lishajko, 1961; Euler 1963; Ryall 1964), suggest an important role in nervous activity. However, this is a suggestion that can be substantiated only after the isolation of a pure chemical entity in sufficient quantity for characterization and pharmacological study. The pure material may after all not show all the biological properties heretofore ascribed to substance P.

The realization of a pure, highly active preparation has been complicated by the apparent properties of the principle which, in addition to being present in extremely small amounts in tissues difficult to work up, has been reported to have poor stability and to be adsorbed on the walls of the vessels used. This experience seemed to indicate that some new approach to the isolation procedure should be tried which would subject the material to milder treatment in solvents designed to inhibit adsorption on surfaces.

Not long after the conference in New York in 1962 (Erdös, 1963), discussions along this line with Dr. K. Vogler resulted in an arrangement whereby a collaborative effort would be carried out by investigators at the Rockefeller University and and Hoffmann-La Roche. The highly purified material from horse intestine described by Vogler et al. (1963) was made available to the Rockefeller University group for further study by the thin-film dialysis technique (Craig, 1960) and other suitable approaches including gel filtration on Sephadex.

These techniques soon indicated that the preparations received were grossly impure. When they were assayed again at the Hoffmann-LaRoche laboratories, results indicated the loss of the activity originally shown to be present. Moreover, fractionation by both gel filtration and thin-film dialysis suggested that the activity resided in a fraction that constituted less than 10% of the already precious material received from Hoffmann-La Roche. While the thin-film dialysis technique in dilute acetic acid showed that activity was present in a fraction in the 2,000-3,000 molecular weight range, quantitative amino acid analysis of several of these highly active fractions by the technique of Spackman et al. (1958) was still not consistent with the properties of a preparation approaching homogeneity.

At this point in the course of the investigation, Dr. H. Meinardi joined the group at the Rockefeller University and immediately set up the bioassay procedure so much needed for such work, particularly where loss of activity seemed to be a major problem. However, by this time it was obvious that the supply of material from Hoffman-La Roche would not be sufficient for completion of the original objective. It was accordingly set aside as a working standard for the bioassay to be used in connection with attempts to develop a more gentle procedure for the isolation of substance P from fresh tissue.

Speculation concerning the basis of the instability, and the recent experience of Guillemin (1964) in the isolation of the hypothalamic pituitary hormone-releasing factors, suggested the advisability of a procedure whereby the possibly destructive enzymes present in the tissue would be denatured as quickly as possible after the death of the experimental animal. This would not be easy to accomplish with intestine as source of material, because of the usual procedures in the slaughterhouse, but it seemed more feasible with brain. The isolation procedure reported here was therefore developed with the brains of calves or hogs. After removal of the cerebellum and both hemispheres, the remainder was blended in a Waring blendor with an equal volume of glacial acetic acid at room temperature, not more than 30 min after the death of the animal, and fractionated (Table 1).

After preliminary experiments to develop this procedure for isolation from tissues obtained from calves and hogs, another promising source of material came to our attention. Through

TABLE 1
Extraction and Purification Scheme

7 kg of tissue-acetic acid suspension stirred with 2 volumes of ethanol and centrifuged 90 min at 14,000 g.

Supernatant — Precipitate discarded

Evaporated under reduced pressure at 45°C
Residue washed twice with 20 times its volume of acetone and once with ether

224 g powder — Acetone and ether discarded

Stirred with 100 ml distilled water per g. and cetrifuged 15 min at 14,000 g

Supernatant — Precipitate discarded

Concentrated under reduced pressure at 45°C to smaller volume and lyophilized

39.2 g powder

Dissolved in 392 ml 0.1 N acetic acid and chromatographed on Sephadex G-25 in 50-ml batches (column 8 cm × 35 cm, solvent 0.1 N acetic acid throughout; rate 20 ml/10 min). Fractions tested for activity by pipetting 0.1 ml directly into isolated organ bath.

Activity in effluent fraction (450 ml-900 ml) concentrated under reduced pressure at 45°C; lyophilized; 9.6 g powder. Dissolved in 100 ml upper and 60 ml lower phase of countercurrent distribution system (water, n-butanol, pyridine, acetic acid, Tween 80; 675, 600, 150, 75, 0.5); 1,000 transfers (5 ml upper/3 ml lower phase).

Broad activity peak (tubes 425-625) K = 0.680 — Activity peak (tubes 160-390) Stored; K = 0.19

Evaporated under reduced pressure 50-60°C. Dissolved in 150 ml H_2O, lyophilized. Chromatographed on Sephadex G-25 (coarse) 2 cm × 200 cm. Flow rate 10 ml/10 min. Solvent 0.1 N acetic acid throughout.

Activity in band 600 ml-1,050 ml effluent. The effluent from 760 ml-820 ml was pooled, concentrated under reduced pressure at 45°C, and lyophilized; 0.424 mg residue; 1×10^{-8} g/ml caused half-maximal contraction of isolated guinea pig ileum.

the help of Dr. R. Guillemin of Baylor University, we learned that frozen goat hypothalamus could be obtained (Melton Provision Co., San Antonio, Texas), which had been removed from the freshly killed animal as quickly as possible according to Dr. Guillemin's procedure. (We are grateful to Dr. Guillemin for this assistance.)

There are available for comparison several crude preparations of substance P. They can be characterized in terms of their pharmacological properties, but only vaguely at present in terms of their chemical properties. In this paper we therefore propose to designate them as follows: Hoffmann-La Roche equine intestinal material = $HLRSP^e$; Hoffmann-La Roche bovine intestinal material = $HLRSP^b$; Ciba bovine brain material = CSP^b; Rockefeller University bovine brain material = $RUSP^b$; Rockefeller University hog brain material = $RUSP^h$; and Rockefeller University goat brain material = $RUSP^g$.

Sephadex chromatography of $HLRSP^e$. A pilot run on 10 mg with Sephadex G-25 in 0.1 N acetic acid in a column 0.9 × 150 cm indicated that all the material with activity emerged in a band with 40-80 ml of effluent. The inactive major part of the sample was retarded and easily separated from the active band. The shape of the active band was not that of a single component.

The active band from a larger sample (50 mg) was recovered and refractionated on a longer column of Sephadex G-25 (0.9 × 300 cm). This gave the pattern shown in Fig. 1.* Assay indicated the active material was in the region of 65-75 ml of effluent. An accurate weight on this material was not obtainable, but the band did not total more than a few per cent of the original $HLRSP^e$.

Goat hypothalamus material. The fractionation of this material will be described a little more in detail, since thus far it appears to be a more promising source than others tried. The frozen tissue was fractionated essentially as given in Table 1, with a few variations. Less material was used, and the original acetic acid suspension was not diluted with ethanol before centrifugation.

The first Sephadex chromatography effluent pattern is shown in Fig. 2. A small cut (38-46) was taken which included only material shown by bioassay to contain activity. This ma-

*We thank Dr. E. J. Harfenist for these results.

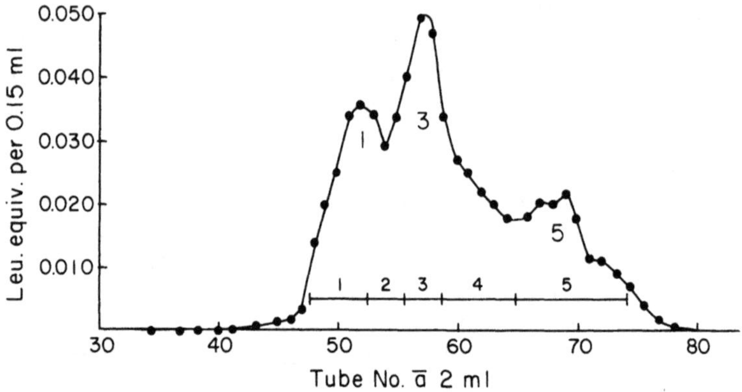

Fig. 1. Elution pattern of the active cut from a preliminary Sephadex separation of HLRSP[e].

terial was then distributed to 1,000 transfers. Since the total weight per tube of active principle is too small to be detected by any analytical procedure except bioassay, the pattern in terms of contraction units of the bioassay is given in Fig. 3.

The material from cuts 500-600 was recovered and fractionated on two columns in series (column dimensions 0.9 x 165 cm) of Biogel P 4. This gave the pattern shown in Fig 4, again in terms of the bioassay units. However, by this time sufficient concentration of the active material had been achieved so that a micro-weight analysis (Craig, 1960, p. 150) of certain of the cuts could be made. The values obtained are given on the pattern. A total of 85% of fraction 39 (Fig. 3) was used for the amino acid analysis.

Molecular weight measurements. Most highly purified SP preparations contain substantially more material than can be accounted for as amino acids (Boissonnas et al., 1963; Zuber, personal communication). It is therefore of the utmost importance to get an estimate of the molecular weight from data other than the amino acid analysis. For this purpose, the results of gel filtration (Sephadex G-25 and Biogel P 4) and thin-film dialysis have been used. As standards, comparative results with bacitracin (mol. wt. 1,420), tyrocidin (1,369) and bradykinin (1,060) have been obtained. The sample and the standards (with exception of bradykinin) were run separately, as there is no generally satisfactory way of distinguishing between overlapping bands in an effluent at such extremely low concen-

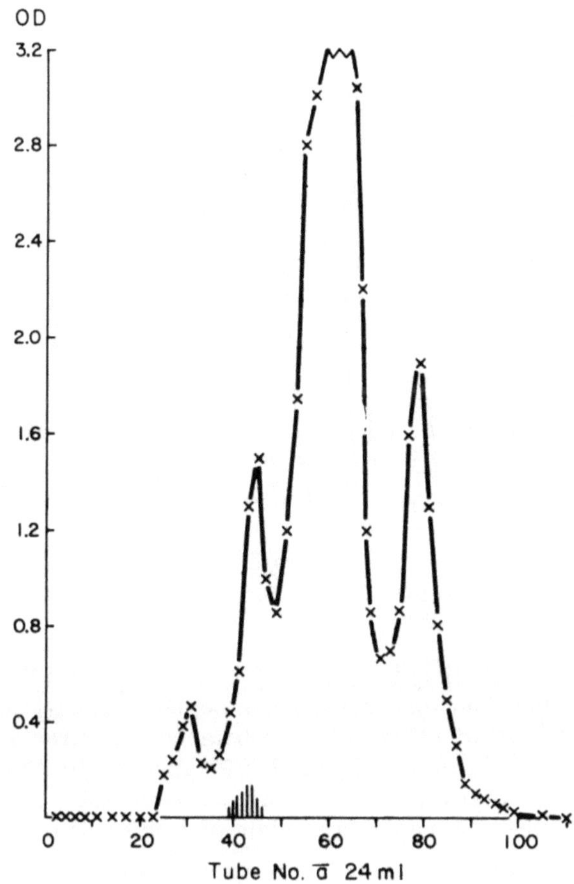

Fig. 2. Effluent pattern of Sephadex chromatography of crude extract.

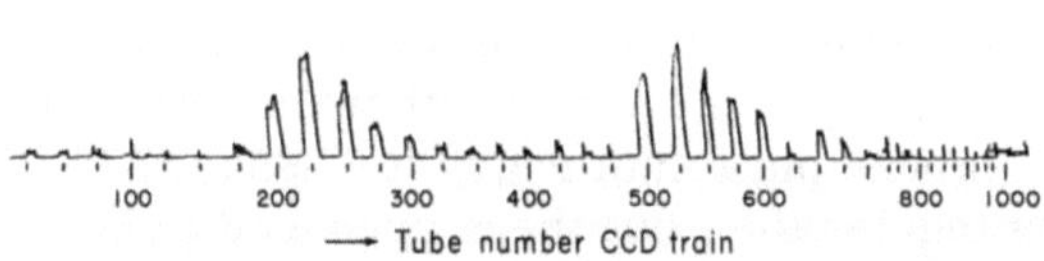

Fig. 3. Countercurrent distribution pattern in terms of bioassay after a 1,000-transfer distribution in the Vogler et al. system (1963). System = water, n-butanol, pyridine, acetic acid, Tween 80; ratios in ml 675, 600, 150, 75, 0.5. Phase volumes = 5/3.

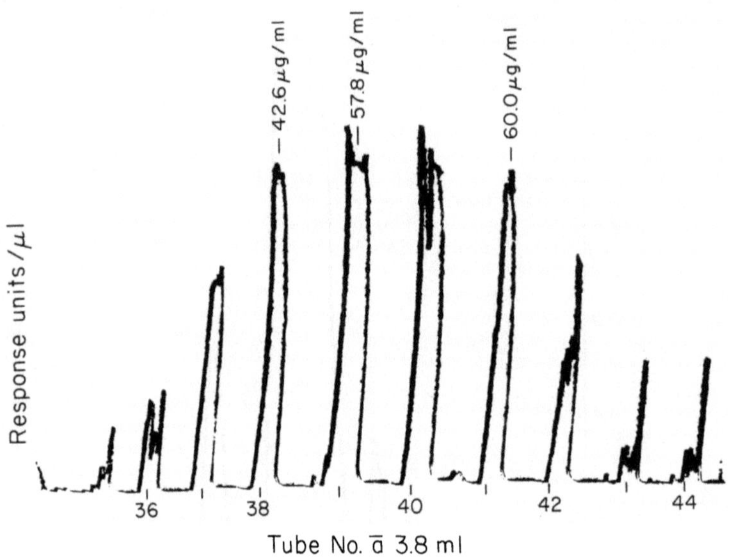

Fig. 4. Elution pattern from a Biogel 4 column of the material from cut 500-600 of Fig. 3.

trations. To distinguish between bradykinin and SP, however, the rat duodenum preparation can be used. Bradykinin causes rat duodenum to relax, while substance P contracts it. From the results, the molecule of $RUSP^h$ appears to be somewhat larger than bradykinin, tyrocidin, or bacitracin. Vogler et al. (1963) reported a molecular weight of $1,650 \pm 250$.

Fractionation by paper electrophoresis. This was performed on the most active preparations after the fractionation scheme in Table 1, or variations of it, and compared with $HLRSP^e$. The electrophoresis was carried out in a Wieland-Pfeiderer high-voltage Pherograph at 0°C with Nagel 214 chromatography paper and at several different pH's. The migration of the active principle was followed by cutting the paper in 0.5-cm wide strips and dipping them directly into the tissue bath with guinea pig ileum ready for assay. $RUSP^g$ and $RUSP^h$ were indistinguishable, but there appeared to be a definite difference between them and $HLRSP^e$. In general, the active material was found to move as one strongly tailing zone which was considerably wider than the zone for glutamic acid. At the low pH and higher load level, the goat and hog brain material seemed to be resolved into three bands, whereas the $HLRSP^e$

TABLE 2
Comparative Analytical Data on Purified Fractions

	HLRSP[e] *	HLRSP[e] (Sephadex)	RUSP[h] †	RUSP[g]
Lysine	1.65	1.62	1.58	1.84
Histidine	1.1	0.4	0.32	0.87
Arginine	0.85	1.5	0.64	2.21
Cystine	0.0	0.0		0.0
Aspartic acid	2.15	1.76	2.12	1.31
Threonine	1.33	1.00	0.65	0.83
Serine	2.7	1.6	1.73	1.01
Glutamic acid	0.87	1.72		2.95
Proline	2.25	3.25	1.6	3.35
Glycine	1.5	2.00		2.68
Alanine	2.15	2.2	1.0	1.0
Valine	2.52	1.5	1.12	1.02
Methionine + methionine sulfoxide	0.0	0.47	?	0.5-1
Isoleucine	0.6	1.00	0.76	0.49
Leucine	3.0	3.0	2.55	2.62
Tyrosine	0.0	0.7	0.0	0.0
Phenylalanine	1.6	1.95	0.82	3.28

*Analysis made by Dr. M. Brenner, University of Basel, on a sample of high activity sent by Dr. Vogler.
†We thank Dr. S. Moore for this analysis.

seemed to show two. This, however, apparently is a result caused by adsorption on the paper, since a rerun of a cut of the peak material recovered again gave a broad band or smear. A run of the original material at one fifth the load of the higher run gave a comparable smear.

Amino acid analysis. Numerous amino acid analyses were made by the method of Spackman et al. (1958) with the latest modifications for micro work. The most significant of these are presented in Table 2, even though the homogeneity of the preparations is in doubt. In Table 2, the results have been normalized to approach the nearest whole residue value for all the amino acids found.

DISCUSSION

In the attempt to isolate SP in a form sufficiently homogeneous for precise chemical characterization and structural study, one of the most important considerations concerns stability. When such extremely small amounts of material must be studied, the question of adsorption on surfaces is scarcely less important. The literature (Erdös, 1963, p. 389) seems to indicate that both these difficulties must be overcome.

The experience gained thus far with brain material extracted (Table 1) seems to suggest improved stability for the active principle isolated in this way. The further question of adsorption on surfaces does not seem insurmountable. In the earlier countercurrent distribution runs, a small amount of the detergent Tween was added† to the system in order to reduce adsorption. In later countercurrent distribution runs with goat hypothalamus material, the use of Tween was not necessary.

The first countercurrent distribution of the cut from a Sephadex column clearly showed two active components. Thus far, a serious attempt to isolate the low K (K = 0.19) material in pure form has not been made. Both bands are present in the material isolated from the brains of calves, swine, and goats, but the higher K material has usually been observed in higher amount. Zettler (1963) has described three components, but one of these does not meet the pharmacological criteria for SP.

Attempts to characterize chemically a preparation of the highest activity have been hampered by the very small amount of material obtainable. This has been of the order of a milligram or less from a single work-up. Since at best the quantitative amino acid analysis must be made on a residue obtained from an extremely dilute solution, contamination from a significant amount of foreign peptidic material is difficult to avoid with certainty. The recovery of amino acids calculated from the weight of $RUSP^h$ or $RUSP^g$ hydrolyzed has indicated either the presence of nonpeptidic residue in the active principle or its presence as a contamination. Hydrolysis of material after Sephadex fractionation gave low recovery, probably owing to the elution of carbohydrate material from the column. Biogel P 4 was better in this respect and gave a residue with $RUSP^g$ with a recovery of amino acids over 60%. When the hydrolysate of $RUSP^g$ was tested for carbohydrate material with the anthrone reagent, a

†A suggestion of Dr. H. Zuber.

positive test was not obtained. It does not appear likely that carbohydrate material is an integral part of the SP molecule.

Numerous amino acid analyses other than those reported in Table 2 have been made with less active fractions. In these, the relative amounts of the same residues shown in Table 2 do not vary as widely. These fractions all had been fractionated either by gel filtration or thin-film dialysis and they were therefore of similar size. The data suggest that SP is probably one of a family of closely related peptides, some of which may arise from the inactivation of the active principle. The differences shown in the last three columns of Table 2, however, are greater and are more likely a reflection of species differences. On the other hand, if this is true, an appreciable difference in the partition ratio in the countercurrent distribution system might be expected. Little or no difference was observed.

In an isolation procedure of the type undertaken here, accurate amino acid analyses have little meaning unless they are performed with a preparation of the highest biological activity. Those reported in the last two columns of Table 2 were done with a preparation of the highest activity we have been able to obtain. From the data in Fig. 4 it can be calculated that 0.043 μg, tube 38, is sufficient to give a half-maximal contraction. With a tissue bath of 8 ml this is a concentration approximating 0.005 μg/ml on a residue weight basis or 0.003 μg/ml when related to the peptide content of the residue. This is the same order of activity reported by Vogler et al. (1963) for their most active equine intestine material and Zuber (1963) for his most active bovine brain material CSP[b].

From the behavior of the sample of RUSP[g] in two separate preparations in amino acid analysis, one not given in Table 2, it seems certain that the active material contains methionine or methionine sulfoxide, almost certainly as a single residue. This would be in line with the apparent behavior, particularly with regard to stability. Tyrosine, tryptophan, and cysteine or cystine are definitely absent.

There can be little doubt but that SP is a peptide because of its susceptibility to hydrolytic enzymes. This is in accord with the results of others (Boissonnas et al., 1963). RUSP[h] is particularly sensitive to chymotrypsin but less sensitive to trypsin. It is also sensitive to pepsin. Carboxypeptidase B did not eliminate the activity of RUSP[b] but eliminated completely the activity of HLRSP[e].

A broad cut from the first active band with a K of 0.19 in

the countercurrent distribution run of Fig. 3 was purified further on a preparative column (8 × 36 cm) of Sephadex G-25 (fine). The active cut, though still impure, was studied with the pharmacological tests‡ used to characterize SP. It was indistinguishable from SP by these criteria except for the test on rat uterus. Here it produced a somewhat stronger contraction than the K = 0.68 material. The activity was eliminated by treatment with chymotrypsin and pepsin but not by trypsin. Thus it is most certainly a peptide. The apparent resistance to trypsin suggests that this form could be the form obtained by Euler (1963) when he treated SP extracts with a low concentration of trypsin.

From the results of Fig. 3 it can be said with certainty that there are two polypeptides with SP activity in goat hypothalamus, calf brain, and pig brain, and from the following reasoning almost certainly a third one in goat hypothalamus. The activity zone for the higher K material of Fig. 3 is much broader than a theoretical zone should be. If the substance behaved ideally, a zone of about 80 tubes in width would be expected. When a cut was taken for redistribution, only tubes 500-600 of the active band were taken, leaving behind the active material above 600 which was obviously there by virtue of a higher K. The active material in tubes 500-600 was redistributed in the same system to 2,000 transfers by the recycling procedure. On assaying the distribution train only activity in a band extending from 980 to 1,070 could be detected. This is the calculated spread of a single substance with an effective K a little above 1 but with an actual K of 0.68 when the volume corrections are made. Thus, there is a strong indication that the activity in the tubes above 600 in Fig. 3 is due to a third SP. The 2,000-transfer run also provided further assurance that the cut 500-600 of Fig. 3 contains only one entity as far as activity is concerned.

Certain doubts have been raised whether SP actually exists as such in the tissue (Euler, 1963) or is formed by the vigorous conditions used in the initial extraction process. These usually have involved boiling at pH 4 for a short time. Trypsin treatment (Euler, 1963) of crude extracts with a low concentration of the enzyme seemed to increase the activity. In the light of these reports it appeared of interest to see if

‡These results were obtained with the collaboration of David Forman.

SP could be extracted directly from tissue at room temperature with a strong concentration of acetic acid. Acetic acid is a strong dissociating solvent and, in addition, it would be expected to denature any destructive enzymes present in the tissue. The results shown in Table 1 leave little doubt that this is an improvement in the extraction of SP.

We were hoping that extraction of the active principle according to this procedure might prove it to be more stable and possibly also to be a more discrete entity from the chemical standpoint. Our impression is that improved stability is achieved up to the point given in Table 1. However, a considerable portion of the total activity seemed to be lost when material of this purity was redistributed to 2,000 transfers in the countercurrent distribution train in the same system used in Fig. 3. Whether or not this is due to the complete removal of an accompanying stabilizing impurity (Euler, 1963) cannot be stated for certain. Several different groups (Erdös, 1963, p. 389; (Cleugh et al., 1964) report instability of highly purified fractions.

SUMMARY

A new purification procedure for Substance P (SP) from brain is outlined. The treatment is milder than the one used previously leaving little doubt about the natural occurrence of SP. Two, possibly three, active principles have been isolated using a bio-assay on guinea pig ileum.

The amino-acid composition of an essentially pure active principle, obtained from goat hypothalami is given which corresponds closely to SP reported by Vogler (1963).

ACKNOWLEDGMENT

This work was supported in part by Grant A-02493 from the National Institutes of Health.

REFERENCES

Boissonnas, R. A., J. Franz, and E. Stürmer: Ann. N.Y. Acad. Sci. 104, 376 (1963).

Cleugh, J., J. H. Gaddum, A. A. Mitchell, M. W. Smith, and V. P. Whittaker: J. Physiol. (Lond.) 170, 69 (1964).

Craig, L. C.: in Analytical Methods of Protein Chemistry. P. Alexander and R. J. Block, Eds., Vol. I, p. 104. New York: Pergamon Press (1960).

Erdös, E. G.: Ann. N.Y. Acad. Sci. 104, 1 (1963).

Euler, U. S.von: Ann. N.Y. Acad. Sci. 104, 449 (1963).

Euler, U.S.von, and F. Lishajko: in "Symposium on Substance P," Proc. sci. Soc. of Bosnia and Herzegovina 1, 109 (1961).

Guillemin, R.: in Proteins and Polypeptides, E. G. Astood and J. C. Beck, Eds., p. 242. New York: Grune & Stratton (1964).

Lembeck, F., and G. Zettler: Int. Rev. Neurobiol. 4, 159 (1962).

McLennan, H.: Synaptic Transmission. Philadelphia: Saunders (1963).

Pernow, B.: Acta phys. scand. 29, Suppl. 105 (1953).

Ryall, R. W.: J. Neurochem. 11, 131 (1964).

Spackman, D. H., W. H. Stein, and S. Moore: Analyt. Chem. 30, 1190 (1958).

Vogler, K., W. Haefely, A. Hürlimann, R. O. Studer, W. Lergier, R. Strässle, and K. H. Bernais: Ann. N.Y. Acad. Sci. 104, 378 (1963).

Zettler, G.: Ann. N.Y. Acad. Sci. 104, 416 (1963).

Zuber, H.: Ann. N.Y. Acad. Sci. 104, 391 (1963).

DISCUSSION

Erdös: The question can be raised again, whether this smooth-muscle-stimulating material is identical with the original substance P, which was described as having an effect on the central nervous system.

Meinardi: There is no doubt that the original substance P contained smooth-muscle-stimulating properties, as this was how it was detected in the first place. Neither is there doubt that the original substance P was a quite crude extract, most likely containing several active peptides. If one is interested in a specific action one should use this as the bioassay during the isolation and purification of the active principle.

Stürmer: When we first isolated substance P we compared the activity of the purified material with that of the Euler-Gaddum standard. In this way it was easy to demonstrate the increase in purity. In the two papers we heard that bioactivity has been estimated in terms of threshold doses and 50% contraction doses respectively. I wonder what progress these estimations will provide; but this question should be answered by the pharmacologists.

Meinardi: We did use a standard, namely the Hoffmann-La Roche preparation. However, in chemical purification one is more interested in knowing whether one can increase the specific activity by further purification.

SUBSTANCE P IN LIPID EXTRACTS OF BRAIN TISSUE

F. LEMBECK, A. HEIZMANN, and G. SEIDEL

Department of Pharmacology, University of Tübingen, Germany

The common method of extraction of substance P from nervous tissue or gut employs a technique which was used by Starling for the extraction of secretin from the dog duodenum. Starling boiled the ground tissue in dilute acid, and followed this by neutralization to precipitate the proteins. Such an extract from brain or gut contains substance P and can be used for the bioassay.

Amin et al. (1954) first prepared a precipitate by adding 20 volumes of dry acetone to the homogenized tissue. Less acetone gave lower recoveries; the acetone precipitate was then suspended in N/10 hydrochloric acid and treated as mentioned above. Zetler and Ohnesorge (1957) applied a method similar to the one used by Euler and Gaddum in 1931. They suspended ground mouse brain in distilled water, added ethanol to a final concentration of about 75% and spun down the proteins. Fourfold repetition of this extraction yielded about 80% of substance P in the alcoholic extract ("a-extracts"). The rest of substance P could be extracted from the remaining precipitate by boiling it in dilute acid ("b-extracts"). All the usual methods that have been used for complete extraction of substance P include treatment of the homogenized tissue with dilute acids.

Further steps of purification, first initiated by Pernow (1953), led finally to preparations with a very high concentration of substance P, i.e. more than 100,000 units/mg, although different tissues and extraction procedures were used by three different research groups (Table 1). All of them recovered 13

TABLE 1
Extraction Procedures for Substance P

Author	Boissonnas et al. (1963)	Vogler et al. (1963)	Zuber (1963)
Animal	Horse	Cattle, horse	Cattle
Tissue	Gut	Gut	Brain
First extract.	Method: Pernow (1953)	Method: Pernow (1953)	Method: Pernow (1953)
Subsequent extract.	Ammonium sulfate precipitation	Ammonium sulfate precipitation	Ammonium sulfate precipitation
	Acetic acid-ether precipitation		Acetic acid-ether precipitation
	Amberlite IR-45		Sephadex G-25
	Amberlite IRC-50		Oxycellulose
	Al_2O_3		CM-Sephadex G-25
	Sephadex (desalt.)	Countercurrent distribution	High-voltage electrophoresis
	CM-cellulose		
Units/mg	140,000	120,000	150,000-300,000
Molecular wt.		1,650 ± 50	

TABLE 2
Chloroform-methanol Extraction

TISSUE HOMOGENATE
chloroform:methanol (2:1)
↓ → water phase
ORGANIC PHASE
washed with water
↓ → water phase
ORGANIC PHASE
evaporation
CM_1 ("slow contraction")
↓
suspension in N/10 HCl
boiled for 10 min
adjusted to pH 5.5
↓ → sediment
SUPERNATANT
freeze-dried (CM_2)
dissolved in acetic acid
↓ → sediment
SUPERNATANT
ether precipitation
↓ → supernatant
CM_3 ("fast contraction")

different amino acids from the final and presumably pure preparation. Almost all these amino acids have been found in physalaemin, the peptide that has much the same pharmacological action as substance P. Highly purified extracts have been reported to be extremely labile; this does not hold true for the unpurified extracts.

We tried to improve the quantitative extraction method for substance P from pig brain by previously defatting the homogenized tissue with chloroform:methanol (2:1, v/v). But substance P could no longer be extracted from the defatted residue of the tissue. However, when the chloroform-methanol extract was evaporated and suspended in saline, it caused a slow contraction of the guinea pig ileum. In further experiments the chloroform-methanol extract was washed with water to remove gangliosides and other water-extractable material from it. Thereafter the organic solvents were evaporated. The dry extract was then suspended in N/10 HCl, boiled, and neutralized with NaOH. This treatment yielded a white precipitate and a fairly clear supernatant fluid; the fluid produced the usual rather fast contraction on the guinea pig ileum in the presence of atropine, mepyramine, and LSD. The supernatant was freeze-dried, redissolved in glacial acetic acid, and precipitated with ether; this extract is called Cm_3 Table 2).

This chloroform-methanol extract was compared on several test preparations with a substance P extract prepared from pig brain according to the method of Gaddum et al.

TABLE 3

Extraction According to the Method of GADDUM et al. (1964)

TISSUE HOMOGENATE
boiled for 10 min, N/10 HCl
→ precipitate

↓

SUPERNATANT
concentration
ethanol precipitation
→ precipitate

↓

SUPERNATANT
freeze-dried

↓

GE_1

↓

ammonium sulfate precipitation
→ supernatant

↓

PRECIPITATE
dissolved in acetic acid
→ sediment

↓

SUPERNATANT
ether precipitation
→ supernatant

↓

GE_2

(1964), which included an acetic acid-ether precipitation after the previous ammonium sulfate precipitation. The extract is called GE_2 here (Table 3).

The two brain extracts were compared in various tests (Heizmann et al., 1966):

Total recovery

From 1 kg of pig brain 25,000 units of substance P could be recovered by the method of Gaddum (GE_2), 20,000 units by the chloroform-methanol method (CM_3). The activity of GE_2 was 10.2 units/mg (20 units/mg protein), the activity of CM_3 was 6.7 units/mg (100 units/mg protein).

Incubation with trypsin

CM_3 as well as GE_2 could be inactivated by incubation with trypsin, thus the peptide nature of CM_3 as well is established.

Discrimination test

Comparison of GE_2 in a discrimination test and a substance P standard preparation from gut indicated that the activities of two extracts (GE_2, CM_3) were about equal. But there was a remarkable difference between the extracts and the substance P standard preparation extracted from the cow gut (Table 4).

TABLE 4
Activity of CM_3, GE_2, and SP (gut standard) in Four Test Preparations*

	1 mg GE_2 = X mg CM_3	1 mg GE_2 = X mg SP	1 mg CM_3 = X mg SP
Guinea pig ileum	1.5	0.77	0.56
Rat uterus	1.4	0.20	0.13
Rabbit jejunum	1.5	1.10	0.76
Rabbit blood pressure	1.4	1.20	0.84

*Mean values of five experiments.

Extraction from areas of the brain

Four parts of the brain with different substance P contents have been extracted with both methods. A sample of each tissue was homogenized and the homogenate divided into two parts for the two extraction procedures. A close correlation could be seen (Fig. 1).

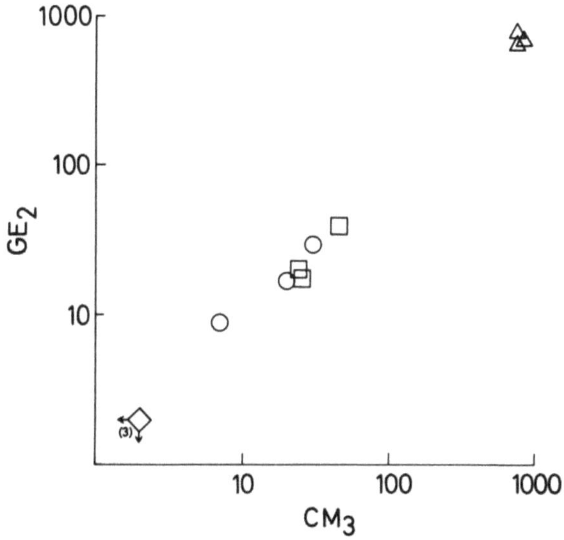

Fig. 1. Extraction of substantia nigra △, brain stem □, cortex ○, and cerebellum ◇ by the CM_3 (abscissa) and the GE_2 methods (ordinate).
Activity in substance P units on the guinea pig ileum.

Subcellular distribution

Whittaker (1959), Kataoka (1962), Ryall (1962), and Cleugh et al. (1964) found most of the substance P in a fraction of homogenized brain tissue which contained nerve endings (fraction B). In a recent investigation we used cow retina in subcellular fractionation, according to the methods of Whittaker (1959) and De Robertis et al. (1962), since retina contains no myelinated fibers but only afferent nerves and few glia cells. Here also, the highest amount of substance P, extracted by Gaddum's method (1964), was recovered from the nerve-ending fraction which also contained the highest cholinacetylase (Fig. 2). (Held et al., 1966).

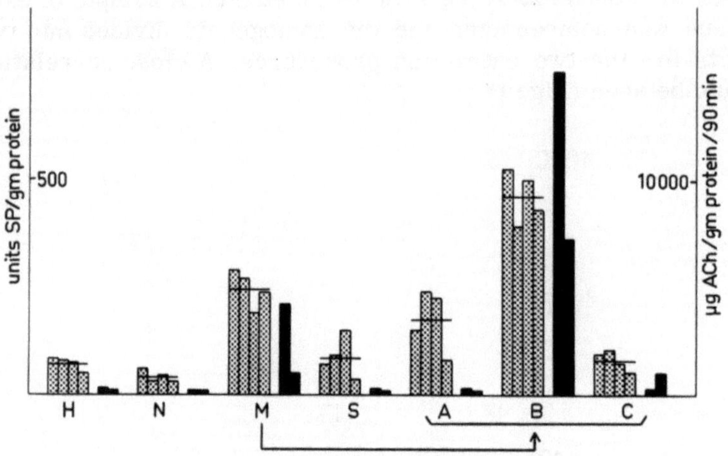

Fig. 2. Subcellular distribution of substance P in cow retina according to the method of Whittaker (1959). Dotted columns: substance P per mg protein: black columns: ACh/hr/g protein.

We studied the subcellular distribution of the substance P activity when the subcellular fractions of rat brain obtained by Whittaker's method (1959) were extracted by the GE_2 and the CM_3 methods. With either method the highest amount of substance P was recovered from fraction B (nerve endings) (Lembeck et al., 1965).

Column chromatography

Using Zetler's method (1963) of separation on an aluminum oxide column, the activity of CM_3 was found only in fraction F_a (extractable with 70% methanol); GE_2 also was extracted with fraction F_a and contained a small proportion of the water-extractable fraction F_b. The activity of the standard preparation of substance P of cow gut, however, was eluted mainly with fraction F_b besides a small percentage in F_a and F_c (Table 5).

TABLE 5
Fractionation of GE_2, CM_3, and SP (gut standard) on an Aluminum Oxide Column by Zetler's Method (1963)

	Fraction	Standard SP		GE_2		CM_3	
Eluting solvent		(units)	(%)	(units)	(%)	(units)	(%)
Methanol 70%	F_a	350	21	1,380	77	1,650	97
Water	F_b	1,375	81	240	13		
NaOH, 0.1 N	F_c			240	13		
Total recovery			102		103		97

Paper chromatography and paper electrophoresis

When two solvent systems in paper chromatography were used, GE_2 and CM_3 gave different R_f values. However, when GE_2 and CM_3 were purified by adsorbing them on an aluminum oxide column and eluted with 70% methanol, both gave identical R_f values in paper chromatography (Table 6).

TABLE 6
R_f Values of GE_2, CM_3, and the Substance P Standard Preparation from Gut*

	n	Unpurified fraction	n	After separation on aluminum oxide column
GE_2	5	0.73 (0.68-0.78)	3	0.15
CM_3	5	0.55 (0.53-0.60)	3	0.15
SP standard	5	0.42 (0.38-0.45)	3	0.24

*In n-propanol:acetic acid:water (4:2:1) before and after elution of fraction F_a (Zetler, 1961) from the aluminum oxide column.

In paper electrophoresis (0.05 M acetate buffer, pH 4.95, 440 V, 20 mA, 6 hours) GE_2 migrated twice as fast to the cathode as CM_3 on the aluminum oxide column (see above). Both extracts migrated with identical speed in paper electrophoresis.

These results support the assumption that substance P can easily be carried by other material in unpurified preparations, if such are used in physical separation methods, and can lead to the erroneous assumption of the existence of different compounds in two extracts.

Contaminants in GE_2 and CM_3

Both extracts, GE_2 and CM_3, were contaminated with small but roughly equal amounts of other active substances except AMP (Table 7), which was not present in CM_3. This is important, since AMP seriously interferes with bioassays of substance P (Laszlo, 1963).

TABLE 7
Contamination of CM_3 and GE_2 by
Other Biologically Active Compounds*

	(ng/mg CM_3)	(ng/mg GE_2)
Acetylcholine	100.0	100.0
Hydroxytryptamine	7.5	9.0
Catecholamines	3.4	6.8
ATP	230.0	290.0
ADP	90.0	120.0
AMP		710.0

*Heizman et al. (1965)

From these results it was concluded that the active material in GE_2 and CM_3 is identical. The extraction procedure of CM_3 led to the assumption that in the brain tissue, substance P is bound to a lipid, from which it can be separated during the usual extraction by acid hydrolysis. Leach (1959) has already suggested such a possibility, since she recovered greater amounts of substance P from tissues when the acetone extract (Amin et al., 1954) was treated with hot n-butanol, which is known to split lipoproteins.

In collaboration with Dr. Vogt we tried to adsorb the fraction CM_1 (using only the chloroform-soluble part of it) on a silicic acid column. After elution with chloroform-methanol

TABLE 8

Adsorption of CM_1 on a Silicic Acid Column*

	Elution with				
	Chloroform:methanol				Methanol:acetic acid
Activity	9:1	4:1	3:2	1:4	9:1
Phosphorus	present	present	present	present	trace
Fraction	(+)	(+)	(+)	(+)	+++
Fraction incubated with chymotrypsin	(+)	(+)	(+)	(+)	---
Fraction boiled in N/10 HCl					+++
Fraction boiled in N/10 HC1 and then incubated with chymotrypsin					---

*Elution first with chloroform-methanol mixtures and finally with methanol-acetic acid; (+), +++ activity on guinea pig ileum; chymotrypsin incubation (1 mg/ml chymotrypsin, 60 min, 37°C).

mixtures of increasing polarity we obtained four phosphorus-containing fractions. Suspended in Tyrode's solution and added to the bath, these fractions produced a rather slow and weak contraction of the guinea pig ileum. There was no change in the type of contraction after boiling the extract in acid. The contractions could not be abolished by preincubation of the four suspensions with chymotrypsin.

The effluent contained only a very small amount of phosphorus upon further elution with methanol:acetic acid (1:1). Suspended in Tyrode's solution it caused the typical fast contraction of the guinea pig ileum. When this fraction was evaporated, suspended in N/10 HCl, boiled, and neutralized, the activity, like CM_3, was found in the supernatant. It could no longer be transferred into chloroform. The suspension of this methanol-acetic acid fraction, as well as the subsequent N/10 HCl extract, was incubated with chymotrypsin. In both cases the activity was destroyed (Table 8).

The new extraction procedure which utilizes chloroform-methanol (CM_3) has the following advantages: (1) inactive proteins or peptides which do not enter the chloroform phase are excluded from the successive steps of extraction; (2) only fraction F_a is extracted by this method; (3) the recovery in the described steps of extraction was rather high; (4) the CM_3 extract is not contaminated with AMP, which interferes with bioassay.

The aim of our recent investigations was not yet to obtain highly purified extracts by the chloroform-methanol method. We first wanted to be sure that this extract contains the same active principle as the usual substance P extract. The results obtained with this extraction method lead us to assume that a "lipid-substance P complex" exists in brain tissue. We believe that the further purification of this complex offers a way to learn more about the nature and possible functions of substance P.

REFERENCES

Amin, A. H., T. B. B. Crawford, and J. H. Gaddum: The distribution of substance P and 5-hydroxytryptamine in the central nervous system of the dog. J. Physiol. (Lond.) 126, 596 (1954).

Boissonnas, R. A., J. Franz, and E. Stürmer: On the chemical characterization of substance P. Ann. N. Y. Acad. Sci. 104, 376 (1963).

Cleugh, J., J. H. Gaddum, A. A. Mitchell, M. W. Smith, and V. P. Whittaker: Substance P in brain extracts. J. Physiol. (Lond.) 170, 69 (1964).

De Robertis, E., A. P. Iraldi, G. R. De Lores Arnaiz, and I. Salganicoff: Cholinergic and non-cholinergic nerve endings in rat brain. J. Neurochem. 9, 23 (1962).

Euler, U. S. von, and J. H. Gaddum: An unidentified depressor substance in certain tissue extracts. J. Physiol. (Lond.) 72, 74 (1931).

Gaddum, J. H., M. Randic, and M. W. Smith: An antistrychnine extract from horse intestine. J. Physiol. (Lond.) 172, 207 (1964).

Heizmann, A., F. Lembeck, and G. Seidel: Die Gewinnung von Substanz P aus dem Lipidextrakt von Hirngewebe. Naunyn-Schmiedeberg's Arch. exp. Path. Pharmakol. 253, 246 (1966).

Held, H., A. Heizmann, F. Lembeck, and W. Schlote: Substanz P in der Retina. Naunyn-Schmiedeberg's Arch. exp. Path. Pharmakol. 253, 266 (1966).

Kataoka, K.: The subcellular distribution of substance P in the nervous tissue. Jap. J. Physiol. 12, 81 (1962).

Laszlo, I: Estimation of substance P in mouse brain. The identification of interfering nucelotides in the extract. Brit. J. Pharmacol. 20, 449 (1963).

Leach, E.: Extraction of substance P from nervous tissue. J. Physiol. (Lond.) 149, 34P (1959).

Lembeck, F., A. Heizmann, and H. Ochsenfahrt: Subzelluläre Verteilung der lipidextrahierbaren Substanz P. Unpublished Observations.

Pernow, B.: Studies on substance P. Acta physiol. scand. 29, Suppl. 105 (1953).

Ryall, R. W.: Sub-cellular distribution of pharmacologically active substances in guinea pig brain. Nature (Lond.) 196, 680 (1962).

Vogler, K., W. Haefely, A. Hürlimann, R. O. Studer, W. Lergier, R. Strässle, and K. H. Berneis: A new purification procedure and biological properties of substance P. Ann. N.Y. Acad. Sci. 104, 378 (1963).

Whittaker, V. P.: The isolation and characterization of acetylcholine-containing particles from brain. Biochem. J. 72, 694 (1959).

Zetler, G.: Zwei neue pharmakologisch aktive Polypeptide in einem Substanz P-haltigen Hirnextrakt. Naunyn-Schmiedeberg's Arch. exp. Path. Pharmak. 242, 330 (1961).

Zetler, G.: New pharmacologically active polypeptides present in impure preparations of substance P. Ann. N.Y. Acad. Sci. 104, 416 (1963).

Zetler, G., and G. Ohnesorge: Die Substanz P-Konzentration im Gehirn bei verschiedenen Funktionszuständen des Zentral nervensystems. Naunyn-Schmiedeberg's Arch. exp. Path. Pharmak. 231, 199 (1957).

Zuber, H.: Isolation of substance P from bovine brain. Ann. N.Y. Acad. Sci. 104, 391 (1963).

DISCUSSION

Walaszek: We recently found a competitive antagonist to ATP, ADP, AMP, and adenosine. This antagonist is caffeine. Theobromine and theophylline are also antagonists, but caffeine is best when investigating the vasodilatation produced by the adenosine compounds, because it causes very little change in base line blood pressure. Caffeine antagonizes the vasodilatation by adenosine compounds in dog, cat, rabbit, and chicken. Caffeine does not affect the vasodilatation caused by acetylcholine, histamine, oxytocin, and other vasodilators. Caffeine also antagonizes the effect of adenosine compounds on the isolated rabbit intestine. In this case it blocks the relaxation caused by ATP but not that by epinephrine.

In experiments in dogs, caffeine will antagonize the shock produced by 1mg ATP/kg and it will also reduce the LD_{50} of ATP in mice. In cats and rabbits, caffeine will reduce the vasodilatation produced by substance P by 20% to 30%. Thus we believe that in crude substance P extract at least 20% to 30% of the hypotensive effect is due to adenosine compounds. The dose of caffeine is 30-50 mg of the base/kg.

Meinardi: Did you have higher oxytocic activity in your chloroform-methanol extracts?

Lembeck: If you think of the action on the rat uterus as the oxytocic activity, this action was hardly different in both brain extracts, GE and CM . The relative activity on the guinea pig ileum was found to be several times higher in the substance P standard preparation than in the two brain extracts.

THE ROLE OF RELEASE OF ACETYLCHOLINE IN THE GUT-CONTRACTING ACTION OF A BRAIN POLYPEPTIDE

G. ZETLER

Pharmakologisches Institut, Medizinische
Akademie Lübeck, Lübeck, Germany

Chromatography on aluminum oxide columns of crude substance P preparations, purified from cattle brains, yielded three pharmacologically active fractions which were named Fa, Fb, and Fc, Fb being substance P (SP) itself (Zetler 1961, 1963, 1964). These three fractions are well characterized by their behavior during paper electrophoresis and paper chromatography in different solvent systems. Their active principles are very probably polypeptides because they are inactivated during incubation with proteases. According to biochemical and pharmacological findings, Fa and Fb must be chemically related, but Fc must belong to a chemically different group. The leading characteristics of Fc are: acidic nature, resistance against trypsin, a slow kinin-like action on the guinea pig ileum, and inactivity on the blood pressure of the atropinized rabbit. Thus, Fc contrasts clearly with Fa and Fb which are of basic nature and destroyed by trypsin, act much faster on the ileum, and are hypotensive in the rabbit.

This paper shows that, in contrast to Fa and Fb, Fc has an indirect action on the guinea pig ileum, liberating acetylcholine from postganglionic neurons in the intestinal wall.

METHODS

Segments about 15 mm long were taken from guinea pig ilea about 15 mm from the ileocaecal junction and immersed in an organ bath containing 3 ml Tyrode's solution maintained at 32°C and bubbled with a mixture of 95% oxygen and 5% carbon dioxide. The ilea were connected to an isotonic frontal-writing lever which magnified the responses tenfold. The contracting agents were left in the bath until the contraction stabilized, which took 40 to 60 sec. The interval between contracting doses was 5 min. In the first group of experiments the Tyrode's solution contained 10^{-7} g atropine sulfate ml, 10^{-6} g pyrilamine maleate/ml, and 2×10^{-5} g tryptamine hydrochloride/ml in order to keep the ilea insensitive to acetylcholine, histamine, and 5-hydroxytryptamine. All drugs that were tested as possible antagonists against the polypeptides in question were given 2 min before the agonists; exemptions will be mentioned. In the experiments which determined dose-response curves the differences in sensitivity between the ilea had to be allowed for. This was done by determining for each ileum the threshold dose of a polypeptide and by increasing this dose stepwise by a factor of 2 until a 64-fold (or if necessary in the presence of an antagonist, 128-fold) increase was reached. A dose was considered to be threshold strength if its effect was greater than that of a control dose of Tyrode's solution and again definitely greater after it was doubled. The ileum was ready for the experiment if increasing the threshold dose two- and four-fold and decreasing it again to its original value revealed a clear relation between dose and effect. Tachyphylaxis appeared only with Fc. Therefore it was possible with Fa and Fb to get from one piece of ileum two dose-response curves — one curve without and the next one with an antagonist. With Fc, however, a new piece of ileum had to be used for each dose-response curve.

The liberation of pharmacologically active material from the isolated guinea pig ileum was investigated in the following way. The ileum was suspended in 3.5 ml of Tyrode's solution and kept at 31°C. After contact of 10-min duration, the bath fluid was removed by means of a syringe, mixed with 30% distilled water to lower the osmotic pressure, and frozen until testing on the leech dorsal muscle, which was done the same afternoon. In these experiments the bath fluid of the ileum contained 10^{-7} g and that of the leech muscle 10^{-5} g eserine/ml. The distilled water to be added to the Tyrode sample from the ileum contained enough eserine to make a final concentration in the mixture of 10^{-5} g/ml. After the end of the experiment, the piece of guinea pig ileum under investigation was slightly pressed between filter paper to remove fluid and mucus, and weighed. The same Tyrode's solution as for the ileum was used for the leech dorsal muscle, but it was mixed with 30% distilled water to lower the osmotic pressure; this bath had a volume of 3 ml and was gassed with a mixture of 95% oxygen and 5% carbon dioxide. The unknown solution was assayed by "bracketing" with known doses of acetylcholine.

Fa, Fb, and Fc were prepared and standardized as described by Zetler (1964). Fa contained 12 units/mg, Fb 20 units/mg, and Fc 12 units/mg; some notes on the "true" activity of Fc will be found in Section B below. Bradykinin was kindly given to us by Dr. E. Stürmer of Sandoz A.-G./Basel. Drugs used were morphine hydrochloride, cocaine hydrochloride, atropine sulfate, eserine sulfate, hexamethonium bromide, d-tubocurarine chloride, pyrilamine maleate, tryptamine hydrochloride, and nicotine bitartrate. Doses or concentrations of these drugs are expressed in terms of the salts.

RESULTS

Experiments on ilea with blocked receptors for histamine, acetylcholine and 5-hydroxytryptamine

During our previous work on the isolation of Fa, Fb, and Fc, the isolated guinea pig ilea, in general, were kept insensitive to histamine, acetylcholine, and 5-hydroxytryptamine by the addition per ml of 10^{-6} g pyrilamine, 10^{-7} g atropine, and 2×10^{-5} g tryptamine to the bath fluid. It therefore seemed justified to begin the present experiments under the same conditions.

Figure 1 shows that 10^{-8} g morphine/ml antagonizes only Fc but not Fa, Fb, or bradykinin. This effect of morphine was easily reversible. Cocaine also diminished only the activity of Fc. In this case, however, a concentration of 10^{-5} g/ml was necessary, and the antagonism was much more stable. Another difference between the actions of morphine and cocaine was revealed by dose-response curves of Fc without and with antagonists. Under the influence of 10^{-8} and 10^{-7} g morphine/ml the curves ran parallel to the control curve but were shifted to the right, which hints of a competitive antagonism. Cocaine, however, led to a decrease of the slope of the curve, indicating a mechanism of action different from that of morphine. Experiments of this kind were also done with Fa, Fb, and bradykinin.

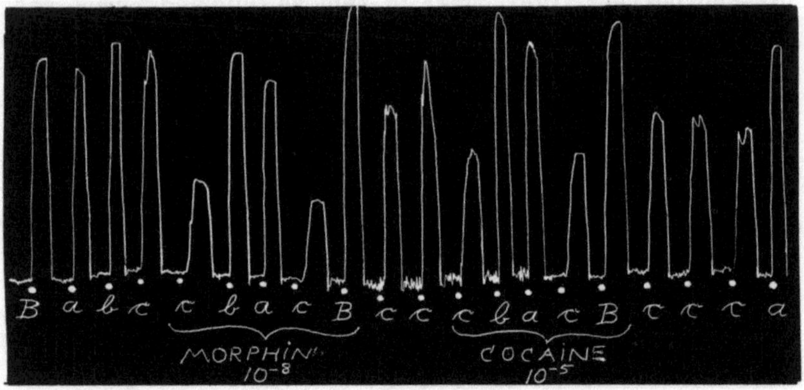

Fig. 1. Guinea pig ileum with receptors blocked to acetylcholine, histamine, and 5-hydroxytryptamine, in a 3-ml bath. B = 15 ng bradykinin; a = 50 μg Fa; b = 25 μg Fb; c = 50 μg Fc. Morphine or cocaine was given 2 min before the agonist.

There was a very weak antagonism of morphine (10^{-5} g/ml) against Fa, and cocaine (10^{-5} g/ml) against bradykinin and Fb. In each of these cases, the curves with and without antagonists ran parallel. Therefore, it was possible to establish from each experiment the dose-ratio (DR) (Gaddum and Picarelli, 1957) which is the distance between these two curves or the ratio of doses giving equal responses with and without an antagonist. For Fc, the DR with 10^{-8} g morphine/ml was 2.5 and with 10^{-7} g it was 6.5; for bradykinin, Fa, and Fb — each with 10^{-5} g morphine/ml — 1.0, 1.3, and 1.0 respectively were obtained.

Experiments on normal ilea

The results described in Section A were reminiscent of the findings of Paton (1957), Schaumann (1957), and Johnson (1963) on the influence of morphine on the release of acetylcholine from isolated guinea pig ilea. It was now necessary to repeat the experiments in absence of the antagonists routinely used before, especially without atropine. Figure 2 shows the results achieved with Fc. Morphine was now even more active, antagonizing Fc in a concentration as low as 10^{-9} g/ml. Atropine was also a clear antagonist but ten times weaker than morphine. Morphine and atropine were ineffective against Fa and Fb. The contractions caused by Fa and Fb were resistant even against 10^{-4} g morphine/ml and those of acetylcholine

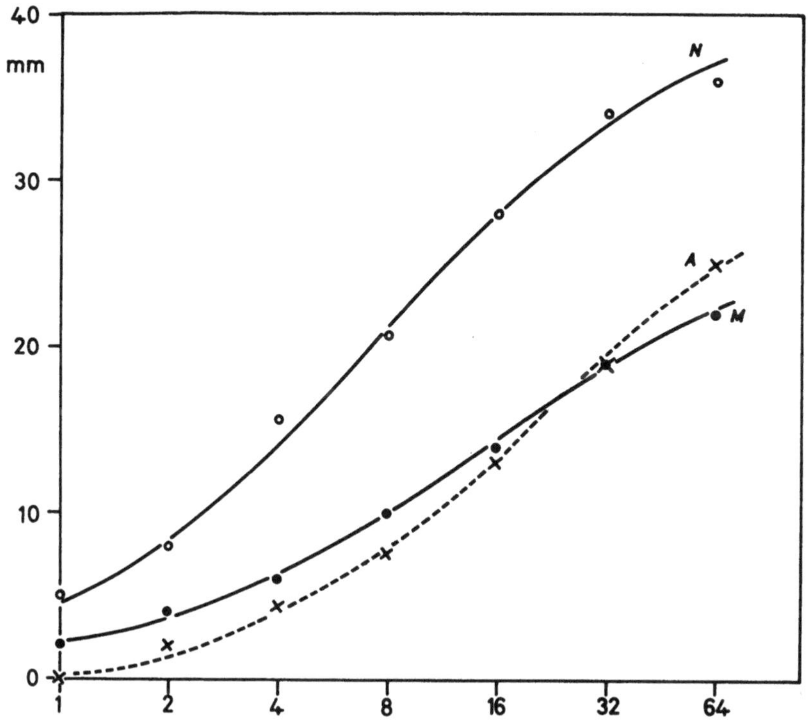

Fig. 2. Relation between increasing doses of Fc and height of contraction of normal guinea pig ilea. The abscissa indicates multiples of the threshold doses which were individually determined for each ileum, and the ordinate the height of contraction. The curves represent the mean values of 7 (N, normal controls), 5 (M, morphine 10^{-9} g/ml) and 5 (A, atropine 10^{-8} g/ml) experiments. Morphine and atropine were given 2 min before Fc.

against 10^{-5} g/ml. A weak antibradykinin effect was achieved with morphine concentrations of about 10^{-5} g/ml. These experiments permitted the conclusion that the antagonism of morphine against Fc is specific. This fact, together with the sensitivity to atropine, hinted at the release of acetylcholine as the basis of the mechanism of action of Fc. This view was corroborated by the action of 5×10^{-8} g eserine/ml, which enhanced the effects of acetylcholine and Fc to the same extent, but was without influence on Fa and Fb.

It clearly follows from these results that the biological standardization of Fc based on guinea pig ilea with blocked

TABLE 1

Release of Pharmacologically Active Material from Isolated Guinea Pig Ilea by 5.7 μg Fc/ml or 19 μg Fb/ml Within 10 Minutes, Expressed as Acetylcholine Equivalents on the Isolated Dorsal Muscle of the Leach

Experiment no.	Sequence of experimental steps a,b,c	Control period (ng/g/10 min)	Fb added: Fb added (ng/g/10 min)	Fb added: Change (%)	Fc added: Fc added (ng/g/10 min)	Fc added: Change (%)
1	a b c	20	14	−30	59	+195
2	b c a	37	35	− 5	67	+ 81
3	c a b	9	8	−11	16	+ 78
4	a c b	17	23	+35	48	+182
5	a b c	6	6	0	16	+167
6	a c b	20	13	−35	55	+175

acetylcholine receptors must give too low values. Indeed, all Fc standardizations of our foregoing work were performed with ilea kept insensitive to acetylcholine, histamine, and 5-hydroxytryptamine (see above). Consequently, Fc was now newly standardized in absence of any antagonist. Two four-point assays using Fb (20 units/mg) as reference standard yielded the same result: an activity of 33 units/mg for Fc to which formerly an activity of 12 units/mg was attributed. This change in activity of Fc has consequences for our knowledge of the quantitative composition of crude SP preparations, as will be discussed below.

The action of hexamethonium on Fa, Fb, and Fc

Hexamethonium was used to differentiate between preganglionic nerve fibers and/or intramural ganglia and postganglionic structures as possible points of attack for Fc; 2.5×10^{-5} g hexamethonium/ml was enough to diminish greatly the action of 3.3×10^{-6} g nicotine/ml but had no comparable influence on the activity of Fc. Fa was weakly antagonized by hexamethonium, but this effect was not investigated more closely. It must be concluded that Fc probably does not affect the preganglionic nerve fibers of the guinea pig ileum.

The release of active material by Fc

The dorsal muscle of the leech was used to test the quantities of active material released from the guinea pig ileum by Fc, since the amounts of Fb and Fc necessarily transferred with the bath fluid to be tested did not interfere with the acetylcholine assay of this preparation. The effect of 5.7 μg Fc/ml on the ileum was compared with that of 19 μ Fb/ml. The results of these experiments are summarized in Table 1 which also indicates the sequence of the three experimental steps a, b, and c. The spontaneous release of active material varied to a remarkable extent, which perhaps reflects the physiological differences in motor activity between the ilea. Nevertheless, independent of the individual basic release and of the sequence of experimental steps, Fc enhanced the release in each experiment by 80% to 200% of the control value. Fb, on the other hand, was not active in this respect. The decrease of release seen after Fb is possibly within the limits of experimental error.

The experiment shown in Fig. 3 throws some light on the

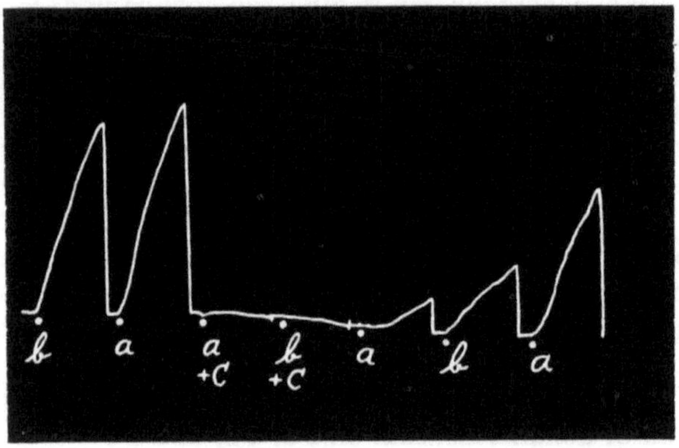

Fig. 3. Leech dorsal muscle in a 3-ml bath; a, 6 ng acetylcholine; b, 1 ml Tyrode's solution with active material released from an isolated guinea pig ileum by 5.7 μg Fc/ml within 10 min, "+C", 5×10^{-6} g d-tubocurarine/ml, added 2 min before.

nature of the principle released by Fc. On the leech dorsal muscle, d-tubocurarine blocked the action of the released material to the same extent as that of acetylcholine. During recovery from d-tubocurarine, the sensitivity of the preparation to both factors reappeared with about the same speed.

DISCUSSION

The results of our experiments indicate that Fc acts indirectly on the guinea pig ileum. In agreement with the views of Paton (1957) and Schaumann (1957), the pharmacological properties of Fc lead to the hypothesis that Fc acts by releasing acetylcholine from neural structures in the gut. This view is supported by the demonstration that Fc releases from the gut active material which contracts the leech dorsal muscle and is blocked by d-tubocurarine. Hexamethonium did not modify the action of Fc, which must therefore have its point of attack at the postganglionic structures. In a recent review article on vertebrate smooth muscle, Daniel (1964) reports with some regret an "unexpectedly large number of indirectly acting substances." It may therefore be pointed out that the results of my experiments clearly indicate a direct action of Fa, Fb (substanceP), and bradykinin, at least as far as acetylcholine is concerned. This can perhaps make up for another indirectly acting substance.

In our foregoing work the activity of extracts and fractions was tested on the isolated guinea pig ileum with blocked acetylcholine receptors. This means that all values of Fc activity published so far are too low. If atropine is omitted, the activity of the Fc standard preparation increases from 12 to 33 units/mg. Thus, the formerly published figures must be corrected by a factor of 2.75. Therefore, the composition of our crude SP preparation from cattle brain (Table 2 in Zetler 1961, and Table 4 in Zetler 1963) is now as described in Table 2.

TABLE 2
Corrected Values for Fa, Fb, and Fc, as Achieved by Column Chromatography on Aluminum Oxide of a Normal and a CCl_4-treated Crude Preparation of Substance P (SP), Prepared from Cattle Brains (further explanation in text)

	Total yield (units)	Percentages of total yield		
		Fa	Fb	Fc
Normal SP preparation	9,361	8	71	21
CCl_4-treated SP preparation	20,823	10	25	65

The purest Fc fraction achieved has now 36 units/mg dry weight or 59 units/mg protein. It is evident from Table 2 that crude SP from cattle brain contains considerable quantities of Fc, especially after pretreatment with CCl_4. It is still an open question whether Fc is a normal constituent of brain tissue. This deserves further investigation, since the presence of an acetylcholine-releasing, i.e. neurotropic, polypeptide in nervous tissue could be of physiological significance, linking the metabolism of brain proteins with that of acetylcholine. This possibility is corroborated by the fact that Fc is not present in SP preparations made from intestinal tissue (Zetler 1963, 1964).

SUMMARY

This paper describes some pharmacological characteristics of Fa, Fb, and Fc, which are active fractions of crude substance P of cattle brain and are polypeptidic in nature. On the isolated guinea pig ileum, Fc can easily be differentiated from Fa, Fb, and bradykinin, for its action is antagonized by morphine, cocaine, and atropine, but enhanced by eserine. Higher concentrations of these drugs have only a very weak influence, if any, on the actions of Fa, Fb, and bradykinin. In the presence of Fc but not of Fb, the isolated gut releases increased amounts of active material, which stimulates the dorsal muscle of the leech and is antagonized by d-tubocurarine. It is concluded that the polypeptide of fraction Fc releases acetylcholine from postganglionic neurons of the intestinal wall.

REFERENCES

Daniel, E. E.: Ann. Rev. Pharmacol. 4, 189 (1964).

Gaddum, J. H., and Z. P. Picarelli: Brit. J. Pharmacol. 12, 323 (1957).

Johnson, E. S.: Brit. J. Pharmacol. 21, 555 (1963).

Paton, W. D. M.: Brit. J. Pharmacol. 12, 119 (1957).

Schaumann, W.: Brit. J. Pharmacol. 12, 115 (1957).

Zetler, G.: Naunyn-Schmiedeberg's Arch. exp. Path. Pharmak. 242, 330 (1961).

Zetler, G.: Ann. N.Y. Acad. Sci. 104, 416 (1963).

Zetler, G.: Naunyn-Schmiedeberg's Arch. exp. Path. Pharmak. 246, 504 (1964).

DISCUSSION

Morley: It is of interest that action of gastrin on smooth muscle (isolated guinea pig ileum or whole stomach), similar to fraction Fc, seems also to be associated with the release of acetylcholine (antagonism by morphine, anoxia, etc.; stimulation by eserine). In the case of gastrin, the action is unaffected by hexamethonium and is therefore postganglionic. Do similar conclusions apply to fraction Fc?

Zetler: Yes, the action of Fc is also postganglionic. These are two more similarities between gastrin and Fc. Both principles have a strongly acidic nature and are inactive on rabbit's blood pressure.

G. PEPEU

Department of Pharmacology, University of Florence

The pharmacological properties of Dr. Zetler's fraction Fc remind me of an acetone extract of peripheral nerve described by Green and Carlini which causes a slow contraction of the guinea pig ileum. This extract has a phospholipid component and its action on the isolated intestine is blocked by atropine, by morphine and by a pretreatment with hemicholinium n. 3. Moreover the increase of an acetylcholine-like substance in the bath has been demonstrated. Did you consider the possibility that Fc and this extract might be related?

Furthermore, I wanted to know whether Fc may release acetylcholine (Ach) also from nervous tissues, for instance from the cerebral cortex. With the small amount of Fc kindly supplied by Dr. Zetler, Dr. Bartolini and I were able to perform two experiments applying Fc directly on the exposed somatosensory cortex of cats with midbrain transections. The method described by Mitchell was used. The Ach released from the cortex was collected into small perspex cups filled with eserinized Ringer solution. Every 20 min. the solution was changed and bioassayed on leech's dorsal muscle.

The results are shown in Fig. 1.

In cat No. 1 Fc has no effect on Ach release whereas in Cat No. 2 Fc causes a four fold increase. The two experiments seem at variance. On the other hand the two cats differ markedly since the first has a strongly activated EEG and a release of Ach higher than the second cat with an EEG characterized by high voltage slow waves and spindle bursts. From our previous work (Pepeu, Mantegazzini) we know that the cortex of cats with synchronized EEG has a higher Ach content than that of activated cats. Therefore, the experiments suggest that Fc releases Ach only when the stores of Ach in the neuronal terminals are fully replenished. Experiments are in progress in order to confirm this hypothesis. However, a finding on the isolated intestine seems to support it. We have, in fact, observed that tachyphylaxis occurs very rapidly when the administrations of Fc are repeated every 2 minutes whereas it is delayed remarkably when the administrations are made every 15 minutes. We may think that *in vitro* the Ach stores in the postganglionic neurons are not replenished in a 2 minute period.

Green, J. P. and A. Carlini. J. Pharmacol. Exper. Ther. 143, 96-106, 1964.

Mitchell, J. F. J. Physiol. 165, 98-116, 1963.

Pepeu, G. and P. Mantegazzini, Science 145.

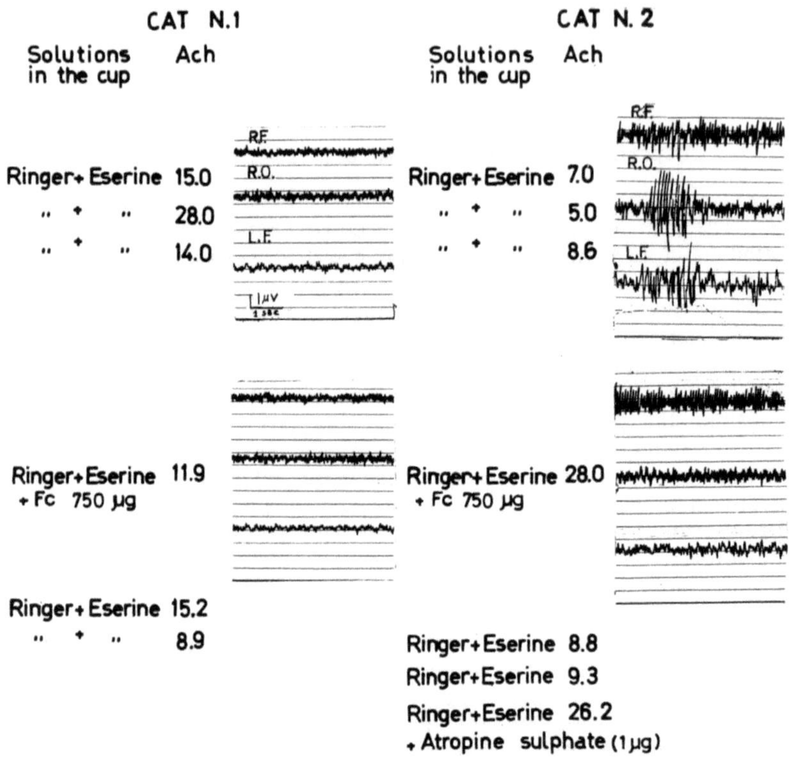

Figure 1. Cats with pontine pretrigeminal sections. Effect of Zetler's Fc fraction on acetylcholine release (ng in 20 minutes) from the cerebral cortex.

Zetler: I have, of course, considered the possible identity of the substance of Green and Carlini with Fc. However these authors consider their substance to be a phospholipid and not a peptide; its biological activity is greatly enhanced by boiling at pH 10, a treatment which does not affect Fc. Hemicholinium No 3 in concentration up to 10^{-4} is inactive toward Fc as well as against angiotensin which, according to Khairallah and Page also releases Ach from the gut. Pepeu pointed out that Fc and angiotensin act more on acetylcholine storage than on synthesis.

CONTRIBUTION TO THE SEDATIVE ACTION OF SUBSTANCE P

P. STERN

Department of Pharmacology, Medical Faculty, Sarajevo, Yugoslavia

We (Stern and Dobrić, 1957b), have put forward the hypothesis that substance P (SP) is a physiological tranquilizer. This was based particularly on the work of Zetler (1956a), and of Euler and Pernow (1956). Moreover, we were able to demonstrate some synergism between SP and mephenesin (Stern et al., 1957), and other central relaxants (Stern and Dobrić, 1957b). Some of our experiments support this hypothesis. Thus, SP tranquilizes *Betta splendens* fishes (Stern and Huković, 1958) and wild rabbits (Stern and Milin, 1957), and suppresses the psychomotor anxiety produced by iminodipropionitrile (Stern and Dobric, 1957a). This view is also supported by findings of Krivoy and Kroeger (1963). During this period, Franz et al. (1961) isolated SP of a purity of 30,000 μ/mg. Even purer SP was prepared by Zuber and Jaques (1962) and by Vogler et al. (1962).Purity was assessed by the contraction of the guinea pig ileum in the presence of antagonists of acetylcholine, histamine, and serotonin. Haefely et al. (1962) injected SP of a purity of 24,000 μ/mg into the ventricles of cats, and even purer SP into mice parenterally, without obtaining a sedative effect. They concluded that SP does not act on the central nervous system. Impure SP containing 100-300 μ/mg did produce sedation when injected into rabbit ventricles (Euler and Pernow, 1956).

First, the question arises whether SP passes the blood-brain barrier. Second, intraventricular administration of a drug may not produce the same effects as parenteral application, which Feldberg and Sherwood (1954) noted with various drugs. Third, it may not be possible to demonstrate depressive effects of drugs on normal animals, as was attempted in Haefely and Hürlimann's (1962) study. It is known that some depressive agents act only on aggressive animals (Horowitz et al., 1963). Zetler (1956a) reported that the sedative effect of SP given to mice i.p. was more pronounced in animals which were in a new environment and more active, than in animals which were used to their cages.

Before describing our own experiments, I should like to stress the following points. Today the nature of pure SP is being questioned. The only criterion of the purity of SP has been its effect on the guinea pig ileum. If SP loses during purification its principal effects described by Euler and Gaddum (1931), can one still regard it as SP? Lembeck (1957), Zetler (1956a), Stern et al. (1957), Stern and Huković (1958), Stern and Milim (1959), and Stern and Dobrić (1957a) obtained the central effects with crude SP. There is no guarantee that very pure SP preparations contain the same factors as does impure SP, even disregarding impurities like ammonium sulfate, adenine nucleotide, potassium, etc. (Laszlo, 1964). Relatively pure SP (275 μ/mg) does not posses the antistrychnine effect, but it still counteracts morphine analgesia (Stern and Huković, 1960). This relatively pure SP produced sedation, and the mice were definitely calmed. We suggest therefore that the purity of SP should not be determined only by its effect on the guinea pig ileum but also by some central or other specific effect of SP. The sedative effects could of course be due to impurities and not to the polypeptide SP itself. However, this has not been proved. On the other hand, it is not certain that pure SP has a sedative effect.

We have therefore investigated the sedative effect of impure and relatively pure SP on aggressive mice in order to see whether there is a parallel between the actions on smooth muscle and on the central nervous system. We have also found further evidence for central actions of SP, by using its sedative effect on aggressive mice.

METHOD

Male mice weighing 20-25 g were isolated in small cages for 21 days (Yen et al., 1959). After 21 days, two isolated mice or an isolated and a normal mouse were put together into one

TABLE 1
Sedation of Aggressive Mice by SP and Other Substances

	SP (6.5 U/mg)	SP (75 U/mg)	SP (330 U/mg)	SP inactivated by chymotrypsin (6.5 U/mg)	Meprobamate (100 mg/kg)	Bradykinin "Sandoz" (20 μg/kg)
Effect on aggressiveness in mice	Aggress. suppressed	Aggress. suppressed	Aggress. suppressed	No effect	Aggress. suppressed	No effect
Effect of desmethylimipramine on aggressive mice tranquilized with SP	Aggress. returns	Aggress. returns	Aggress. returns		Aggress. returns	

cage, and fighting ensued. Only mice still fighting after 2-3 min were used in tests. Test mice were given SP preparations of 6.5 U/mg, 75 U/mg, or 330 U/mg in a dose of 5,000 U/kg i.p. 15 min before exposure to another mouse, and the onset or absence of fighting noted. Animals in which SP suppressed fighting were then given the antidepressant desmethylimipramine (Sulser et al., 1964) and the reappearance of aggressiveness observed. Controls received SP of a purity of 6.5 or 330 U/mg which had been inactivated by chymotrypsin. A second group of aggressive mice was given bradykinin, 10 μg/kg i.p., instead of SP, and a third group the central relaxant meprobamate, 100 mg/kg i.p.

RESULTS

Table 1 shows that all three concentrations of SP produced sedation in aggressive mice 15 min after application. Desmethylimipramine restored aggressiveness. Bradykinin had no effect. Unlike SP, bradykinin does not influence harmine-induced tremor or bulbocapnine catatonia (Zetler, 1956b). Meprobamate acted like SP, and its tranquilizing effect could be abolished by desmethylimipramine. SP inactivated by chymotrypsin had no effect.

DISCUSSION

As mentioned earlier, the sedative effect of a substance is more easily assessed in excited animals than in animals behaving normally. SP has a pronounced sedative effect in excited animals. However, Haefely and Hürlimann (1962) observed decreased spontaneous activity also in normal mice given very pure SP (50,000 U/mg i.v.). We think that the sedative action of SP on aggressive mice may be attributed to an effect on the central nervous system, because it is abolished by desmethylimipramine which definitely acts on a central site. Furthermore, impure SP passes the blood-brain barrier (Serafimov and Stern, 1956), but so far as I know, the passage of pure SP has not yet been tested. The preparation of 330 U/mg which we used had lost its antistrychnine action, but retained certain sedative effects. The findings that bradykinin had no effect and that meprobamate behaved like SP support the view that SP is a sedative.

These experiments should be repeated with pure SP preparations in order to see whether the effects on smooth muscle and on the central nervous system are parallel. If this be the case, the term SP must be restricted to preparations in which purification concentrates both effects. Impurities should be examined in the same way. Accordingly, 20 mg ammonium sulfate/kg was given i.p. to aggressive mice and found to be without effect (Stern, unpublished). Injected into the ventricles, ammonium sulfate produces some sedation (Haefely et al., 1962). As mentioned earlier, Krivoy and Kroeger (1963) demonstrated in cats a neurotropic effect of very pure SP (10,000 U/mg) — an enhancement of the dorsal spinal root potential. It would be advisable to add this test and the one suggested by us, to smooth-muscle tests, when the purification of SP is being checked. The SP antagonists trimethaphan (Arfonad) and cystine-di-β-naphthylamide (Hukovic and Stern, 1963) could perhaps also be used for this purpose.

The mechanism of the sedative action of SP cannot be discussed here. Recent findings (Stern et al., to be published) suggest that SP increases in the serotonin of mouse brain, but not in other biogenic amines (acetylcholine, norepinephrine, dopamine, histamine).

Until this question is clarified, only the preparation described by Euler and Gaddum (1931) should be termed SP.

REFERENCES

Euler, U. S. von, and J. H. Gaddum: An unidentified depressor substance in certain tissue extracts. J. Physiol. (Lond.) 72, 74 (1931).

Euler, U. S. von, and B. Pernow: Neurotropic effects of Substance P. Acta physiol. scand. 36, 265 (1956).

Feldberg, W., and S. L. Sherwood: Injections of drugs into the lateral ventricle of the cat. J. Physiol. (Lond.) 123, 146 (1954).

Franz, J., R. A. Boissonnas, and E. Stürmer: Isolierung von Substanz P aus Pferdedarm und ihre biologische und chemische Abgrenzung gegenüber Bradykinin. Helv. chim. Acta 44, 881 (1961).

Haefely, W., and A. Hürlimann: Substance P, highly active naturally occurring polypeptide. Experientia (Basel) 18, 297 (1962).

Haefely, W., H. Thoenen, and A. Hürlimann: The effect of intraventricular application of Substance P preparations of different purity. Med. exp. (Basel) 7, 245 (1962).

Horovitz, Z., A. R. Furgiuele, L. J. Brannick, J. C. Burke, and B. N. Craver: A new chemical structure with specific depressant effects on the amygdala and on the hyperirritability of the "septal rat." Nature (Lond.) 200, 369 (1963).

Huković, S., and P. Stern: Antagonisten der Substanz P. Naunyn-Schmiedeberg's Arch. exp. Path. Pharmak. 245, 281 (1963).

Krivoy, W., and D. Kroeger: The neurogenic activity of high potency Substance P. Experientia (Basel) 119, 366 (1963).

Laszlo, I.: Interference by adenine nucleotides and potassium in the Substance P estimation in different parts of rat brain. J. Physiol. (Lond.) 172, 30P (1964).

Lembeck, F.: Untersuchung über die Auslösung afferenter Impulse. Naunyn-Schmiedeberg's Arch. exp. Path. Pharmak. 230, 1 (1957).

Serafimov, N., and P. Stern: Influence des Mediateurs chimiques sur la teneur en "Substances P" du cerveau du rat. Arch. int. Physiol. 66, 653 (1956).

Stern, P., and V. Dobrić: Beitrag zur sedativen Wirkung der Substanz P. Naturwissenschaften 44, 517 (1957a).

Stern, P., and V. Dobrić: Über die Wirkung der Substanz P im Zentralnervensystem, in Psychotropic Drugs. Amsterdam: Elsevier (1957b).

Stern, P., V. Dobrić, and D. Mitrović-Kocić: Synergistiche Wirkung von Substanz P und Mephenesin. Arch. int. Pharmacodyn. 112, 102 (1957).

Stern, P., and S. Huković: Über die inhibitorische Wirkung der Substanz P an kampflustigen Fischen Betta splendens. Naturwissenschaften 45, 626 (1958).

Stern, P., and S. Huković: Beziehungen zwischen zentraler und peripherer Wirkung der Substanz P. Med. exp. (Basel) 2, 1 (1960).

Stern, P., and R. Milim: Tranquilizing effect of "Substance P." Proc. Soc. exp. Biol. (N.Y.) 101, 298 (1959).

Sulser, F., M. H. Bickel, and B. B. Brodie: The action of desmethylimipramine in counteracting sedation and cholinergic effects of reserpine-like drugs. J. Pharmacol. exp. Ther. 144, 321 (1964).

Vogler, K., W. Haefely, A. Hürlimann, R. Studer, W. Lergier, R. Strässle, and K. Berneis: A new purification procedure and biological properties of Substance P. Ann. N.Y. Acad. Sci. 104, 378 (1962).

Yen, C. Y., L. Stanger, and N. Millman: Ataractic suppression of isolation-induced aggressive behavior. Arch. int. Pharmacodyn. 123, 179 (1959).

Zetler, G.: Substanz P, ein Polypeptid aus Darm und Gehirn mit depressiven, hyperalgetischen und morphinantagonistischen Wirkungen auf das Zentralnervensystem. Naunyn-Schmiedeberg's Arch. exp. Path. Pharmak. 228, 513 (1956a).

Zetler, G.: Wirkungsunterschied zwischen den Polypeptiden Bradykinin und Substanz P am Zentralnervensystem. Naunyn-Schmiedeberg's Arch. exp. Path. Pharmak. 229, 148 (1956b).

Zuber, H., and R. Jaques: Isolierung von Substanz P aus Rinderhirn. Angew. Chem. 74, 216 (1962).

DISCUSSION

Kroeger: We may have to accept the splitting of the two actions: the musculotropic and the neurotropic. Since there are very small quantities of the purified material available, I wish to suggest that a technique, such as a single cell-recording technique of Dr. Vick, might allow those working on the isolation of substance P to follow the neurotropic and muscolotropic actions.

Haefely: I would suggest reserving the name of substance P for the active material purified during recent years on the basis of its smooth-muscle-stimulating activity. Effects observed with so-called impure substance P preparations should preferably be associated with "impure tissue extracts."

We did experiments with highly purified substance P preparations not only on normal animals but also on animals excited with amphetamine. I was surprised to hear that DMI excited mice.

Stern: With regard to the nomenclature of substance P, I think we should use the name according to the discoverers (Euler and Gaddum), and make changes only in agreement with Professor von Euler. I did not know about the experiments with animals excited by amphetamine. This is another means of excitation. The action of imipramine is not the same on normal and excited animals.

Lembeck: When I try to conclude this discussion on substance P, I feel that substance P looks like a black sheep in the crowd of white, purified, and synthesized peptides. Its properties make this substance so attractive for the pharmacologist with regard to its high activity and so difficult for the biochemist because of the low amount in the tissue. The action on the smooth muscle may be regarded as a method to assay substance P. Just as, e.g., the action of bradykinin on the rat uterus, it may mean not very much with regard to its physiological functions. The matter is also confused by the finding of several active fractions or compounds. We shall, therefore have to decide whether we reserve the name substance P to one of the pure peptides, as Dr. Haefely suggested, or to use it in Sir John Gaddum's original sense for the substance P containing preparation of tissues.

THE HYPOTENSIVE ACTION OF ELEDOISIN IN SUBJECTS WITH ARTERIAL HYPERTENSION

A. RUOL, F. PALERMO, S. MINUTILLO, G. D'AMELIO
Clinica Medica of the University of Padova

Summary

We have investigated the reaction to eledoisin in 29 subjects with a high blood pressure of various kinds.

In 15 subjects with arterial hypertension, doses of eledoisin lower than 3γ/minute, corresponding to less than 45 ng/min/Kg. of body weight, lowered the systolic blood pressure by more than 20% just as in normotensive patients. In 14 other subjects with arterial hypertension, however, the doses of eledoisin required to lower the systolic blood pressure by more than 20% were much greater, always over 3γ/minute (corresponding to more than 50 ng/min/Kg.) (Fig. 1).

When the dose was increased (in order to lower the pressure further, very large quantities were given—8.7- 9.0- 10.5- 6.2- 7.2 γ min.), phenomena of marked intolerance to the drug appeared (tachycardia, dizziness, faintness, swelling of the face), but the blood pressure did not drop by 20%.

Both in the patients who responded readily to eledoisin and in those who responded to a lesser degree, the interruption of the infusion of the drug was followed by a return to initial pressure levels in one to seven minutes.

The reason for this varying individual response is not clear. Our patients with essential hypertension responded generally less than the patients with other kinds of arterial hypertension, but there are many exceptions to this rule.

The clinical evaluation of the functional damage of some peripheral circulatory districts does not supply any suitable criteria to explain the different response to the eledoisin of some patients with high blood pressure.

Table 1

	REDUCTION OF BLOOD PRESSURE OVER 20 PER CENT with doses of eledoisin	
	< 4γ/ minute (15 cases)	> 4γ/ minute (14 cases)
HYPERTENSION Essential	4 cases	10 cases
HYPERTENSION Nephrogenic	5 cases	4 cases
HYPERTENSION Other Types	6 cases	0 cases
FAMILIAL INCIDENCE	6 cases	9 cases
LENGTH OF ILLNESS (average)	3.5 years	7.0 years
AGE AT WHICH HYPERTENSION APPEARED (average)	43.0 years	34.9 years
CHOLESTEREMIA (>200 mg. %)	6 cases	7 cases
DAMAGE OF CIRCULATORY BEDS:		
aorta	4 cases	7 cases
kidneys	8 cases	7 cases
coronary arteries	5 cases	9 cases
limbs	3 cases	4 cases
brain	3 cases	3 cases
retina	11 cases	12 cases

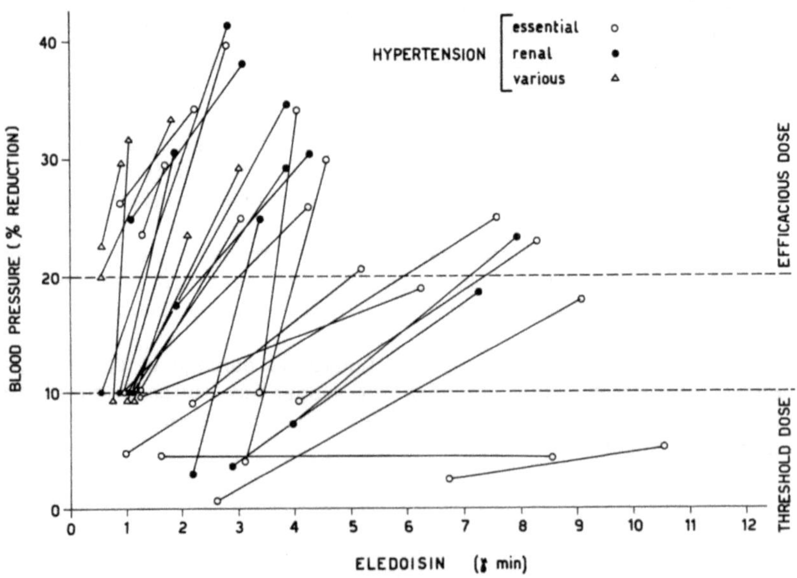

Figure 1

R. PANSINI

Semeitoica Medica dell'Universita di Ferrara

L'intervento in discussione riguarda un peptide ad azione ipertensiva: l'angiotensina. Collier ha sottolineato l'antagonismo tra bradichinina (broncocostrittrice) e angiotensina (broncodilatatrice) sui muscoli di Reissessen. Haefely e coll. hanno dimostrato una concorde azione inibitrice dei due polipeptidi sulla trasmissione nervosa ganglionare.

Il noto, opposto effetto sulla pressione arteriosa si realizzerebbe anche a livello di distretti vasali diversi per le due sostanze: le arterie (ed anche le vene) per quanto riguarda il potere vasocostrittorio dell'angiotensina; i più piccoli vasi arteriosi ed anche il letto capillare, per quanto riguarda l'azione vasodilatatoria della bradichinina.

Ma il dato che richiama maggiormente la nostra attenzione in questo momento in campo fisiopatologico è la comune capacità dei due polipeptidi sia di provocare liberazione di catecolamine adrenergiche dal surrene, sia di indurre una maggiore sensibilità a tutti quegli agenti ed a tutte quelle manovre che realizzano questo medesimo effetto: ed abbiamo sentito come possano risultarne, almeno in parte, mascherata l'azione broncocostrittrice della bradichinina e, per contro, potenziato l'effetto vasocostrittorio-ipertensiogeno dell'angiotensina.

Nel mio Istituto (con l'impiego di un metodo originale di dosaggio cromatografico-spettrofotometrico) si è potuta documentare la frequenza di un'iperangiotensinemia venosa in ipertensioni arteriose essenziali.

In questi stessi soggetti è stato dimostrato la concomitanza o di un'iper-aldosteronuria o di un'iper-catecolaminuria, pur non trattandosi di casi di feocromocitoma o di iperaldosteronismo primario.

La influenza che anche la bradichinina esercita sulla secrezione delle catecolamine ci fa prospettare qui l'interesse di due ipotesi di lavoro:

1) quella di ricercare il comportamento anche di questo polipeptide in ipertensioni arteriose come quelle suaccennate e cioè anche in forme morbose apparentemente con esso non correlabili sulla semplice base della vasomotilità: tanto più tenendo conto che l'atteggiamento vasomotorio sia nel suo equilibrio fisiologico sia nelle sue deviazioni patologiche viene ad essere sempre la resultante di effetti di segno opposto esplicantesi sul tono, sulla contrattilità e sulla reflettività vascolare.

2) Quella di accertare l'eventuale esistenza di rapporti più ampi tra bradichinina e surrene, nei confronti, ad esempio, anche della secrezione aldosteronica se non altro per analogia con quanto già si conosce, a tale proposito, proprio per l'angiotensina.

L. ROSA

Clinica Medica of Bologna

I wish to sum up briefly the preliminary results of research we are carrying out in the Clinica Medica of Bologna under the direction of Professor G. Sotgiu.

The consideration that necrotic damage to a highly differentiated tissue, such as the myocardium, may induce the liberation of bioactive amines and this process would be responsible, in part, for the syndrome following the necrotic damage, suggested the use of protease inhibitors in the treatment of acute myocardial infarction, although we still lack a full knowledge of the action of released peptides.

A protease inhibitor, Trasylol (Bayer) was given intravenously by slow injection. The treatment was started as soon as possible and lasted several days until the pain and the post-myocardial shock disappeared.

Preliminary results observed after Trasylol were as follows:

1. A pain-relieving action, which, in some cases, was dramatic;
2. A sense of well-being, unusual in patients with acute myocardial infarction;
3. A moderate action on the post-infarctual shock.

This treatment cannot be considered definitive in every case, having occasionally observed negative results in some severe cases.

Though several patients (70) have already been treated by Trasylol, we still consider our results to be of preliminary nature. Since no unpleasant effects have been observed, we think that clinical and laboratory tests should be carried out for further characterization.

Besides clinical observations, the bradykininogen blood levels have been measured according to Diniz's method (modified by Bayer's Lab., Wuppertal) in patients with acute myocardial infarction and in controls, before and after Trasylol treatment.

The following observations were made:

1. No significant changes in the bradykininogen level after giving Trasylol in controls;
2. Great variability of the bradykininogen level, with a general tendency to lower rates in patients with myocardial infarction (in some cases blood could be tested a few hours after necrosis, while in others after a longer delay).

Other studies were performed on patients suffering from arteriopathies in the legs. After Trasylol, a significant delay in the onset of pain was observed.

In the same patients, bradykininogen blood level was moderately but steadily reduced at the time of pain onset.

No changes of the bradykininogen blood level were observed in controls.

Further experiments on these subjects are being carried out in our department.

REFERENCES

Sotgiu, G. and P. Artuso: Azione degli inibitori delle proteasi nella sindrome da infarto miocardico. Bull. Sci. Med. Soc. Med. Chir. Bologna 137:135, 1965.

Sotgiu, G. and P. Artuso: Action of protease inhibitors in myocardial infarction. Mal. Cardiovasc. 6:283, 1965.

Diniz, C. R., I. F. Carvalho, Y. Ryan and M. Rocha e Silva: Nature (London) 192:1194, 1961.

CONCLUDING REMARKS

E. GREPPI

Clinica Medica dell'Universita di Firenze

In my capacity as an older fellow, and though I am well aware that my competence is confined to the clinical field, I should like to conclude this symposium with a brief critical review of the subject that has been discussed: a subject still fraught with many obscurities but equally attractive for the biochemist, the pharmacologist, and the student of physiopathology in the experimental animal as well as in man.

This being the case, I am obviously not qualified to give you a critical review, either extemporaneously or with appropriate meditation, of the kinins. All I can do is to offer informally a few comments, on these five days of meetings, simultaneously crammed with scientific activities and cheered by a most enjoyable hospitality.

It is possible that I feel more awed by the difficulties of the subject matter than other scientists here convened, precisely because I am old and because I am essentially a clinician. I remember a witty and elegant remark by Paul Valéry, that while he was examining some ponderous treatises of theoretical philosophy he felt like a candid farmer suddenly and unpreparedly dropped in the middle of the Louvre: confused and admiring at first, then increasingly awestruck and stunned, and at last running for dear life out of doors! I have experienced something of the sort, every now and again during the past few days: but I sat tight through the lectures, and let myself be persuaded to be the farewell speaker today because of my ties with individuals, and whole groups (such as our own Centro Cefalee): ties that were already established when we met in this city for the Symposium on Bradykinin in 1962. That time also, many guest speakers had gathered here from foreign lands, and the proceedings were published in English.

Speakers at this symposium come from many foreign countries and from Italy and have actively contributed to research work as well as to the discussion here. I take special pleasure in recalling the name of Erspamer and several young Italian investigators: Beretta-Anguissola, Migone, Piovella, Galletti, Panuccio, Procacci, Arcangeli, Ruol, Pepeu, and Mannaioni, beside, of course, Sicuteri and his associates, who have integrated pharmacodynamic concepts and methods relative to polypeptides with the functional point of view, and thus bridged the gap between theory and human physiopathology. To all I extend my sincere admiration and praise.

From the lectures and discussions that we have heard, one is at once impressed by the technical difficulties involved in the isolation of the peptides from the blood and tissues, notwithstanding the precise chemical definition of such substances; and one is intrigued by the elusive, unsteady concentration of the inactive precursor, kininogen, precariously balanced between its probable genesis in the liver and its transformation into active kinins.

So far there seems to be no definite physiological pattern for the concentrations of the various substances and their respective actions. The vasoactive effects seem to vary from one place to another, with rapidly varying concentrations here and there, and with effects that might be diphasic or even irregular, whose significance can be grasped not so much in normal conditions as in spontaneous or artificially induced emergencies, such as hemorrhage, thrombosis, trauma, burns, necrosis, or circumscribed necrosis (e.g. pancreas, adrenal glands), and shock in a broad sense.

This is, and will remain for the time being, a field reserved for experts, from Frey, Kraut, and Werle's kallikrein and Rocha e Silva's bradykinin onward; a field with many pitfalls and booby traps, indeed, even for such experts as we have heard in the discussions and comments after each lecture. And yet, effective methods have been devised and results have been obtained, and useful discoveries have been made in the field of pathology and clinical medicine; more particularly, in our understanding of vascular and tissue pain, of shock and stresses, of intracranial bleeding, of pancreatic and pulmonary hemorrhages, and of other localized lesions associated with general repercussions.

At the border between physiology and pathology, Back's tabulation (see text) depicts actions which afford a comprehensive view of the functions of hypotensive polypeptides of the type of bradykinin.

An improvised speech such as this is only valid inasmuch as it is short. I felt that it was my duty to comment briefly on the scientific points at issue, and I felt that it was my privilege to express my appreciation and gratitude to you for gathering here and giving us the benefit of your knowledge, discussion, and moral support.

So I conclude this International Symposium on Hypotensive Peptides with many thanks to you all, and my best wishes for a pleasant journey home after this interlude which we have spent together in this luminous Florentine autumn, in the Villa Lemmi and in the Lecture Hall of the Clinic of Traumatology.

REPORT OF THE COMMITTEE ON NOMENCLATURE FOR HYPOTENSIVE PEPTIDES

COMPILED BY M. E. WEBSTER

The nomenclature for those polypeptides which depress blood pressure and are formed by enzymatic action from α-globulins in plasma has become so involved that the same enzyme, the same substrate, and the same polypeptides are known by at least two and frequently more different names. To clarify this situation, a meeting was held on October 27, 1965, in Florence, Italy, to investigate the possibility of evolving a compromise nomenclature.

Three main systems, i.e. the kallikrein-kallidinogen-kallidin nomenclature, the bradykininogen-bradykinin nomenclature, and the kininogenase-kininogen-kinin nomenclature, are currently in use. The term kallikrein, derived from the Greek word for pancreas (kallikreas), was given (Kraut et al., 1930) for a hypotensive substance found in urine and in pancreas. Subsequent investigation made it clear that the kallikreins exerted their hypotensive action by enzymatic release of a polypeptide, called kallidin, from a substrate in plasma called kallidinogen (Werle and Berek, 1948). Simultaneously, other investigators (Rocha e Silva et al., 1949) reported that trypsin and some snake venoms released a polypeptide from plasma. This peptide was named bradykinin; the substrate in plasma was called bradykininogen. The term plasma kinin was coined (Laki, 1943) to describe a thromboplastin-like activity in blood which plays a role in the conversion of prothrombin to thrombin. However, this nomenclature was never widely adopted. Later (Schachter and Thain, 1954) used the term wasp kinin or simply kinin for a hypotensive polypeptide from wasp venom which could be distinguished from kallidin and bradykinin. In 1956 (Miller et al.) proposed it as a class name for substances implicated in the cell-division phase of fruit growth. Recently it has been suggested that phytokinin (Dedolph, 1965) or cytokinin (Skoog et al., 1965) replace the term kinin for these plant materials. In 1958 (Lewis), the use of the term plasma kinins was proposed to apply to all bradykinin-like polypeptides derived from plasma proteins. This nomenclature has since been amplified by a number of investigators so that the substrate in plasma is called kininogen, and the enzymes that release kinins are called kininogenases. In 1961-63 the expression kinin-hormones or simply kinins was proposed (Rocha e Silva, 1962, 1963) to describe a group of local hormones that would include the angiotensins, the bradykinins (kallidins), and substance P.

With due consideration of the above historical facts, the committee on nomenclature recommends the following compromise names.

RECOMMENDATIONS

Kinin-hormones

The committee was in agreement that a class name for vasoactive polypeptides, particularly for the angiotensins and the bradykinins (kallidins), would be desirable since they are both liberated from α-globulins in plasma by endogenous enzymes. However, the use of the term kinin would be confusing since the terms kininogenase, kininogen, and kinins could then be construed to mean renin or kallikrein, angiotensinogen or bradykininogen (kallidinogen), etc. The committee recommends that this nomenclature not be adopted.

Enzyme Nomenclature

Kininogenase. It is recommended that the use of the term kininogenase be restricted to a general term indicating any enzyme that liberates a kinin from an inactive protein substrate. Thus the kallikreins, trypsin, pepsin, snake venom, bacterial proteinases, etc., are all kininogenases and should be referred to by their trivial name. Adjectival terms such as kininogenic or kininolytic should be avoided.

Kallikrein (3.4.4.21). A kallikrein is defined as an endogenous enzyme which rapidly and specifically liberates a kinin from kininogen. The term is understood to cover all those enzymes previously described (Frey et al., 1950). The inactive enzyme precursor should be called prekallikrein in conformation with the suggestions of the International Union of Biochemistry on Nomenclature (Enzyme Nomenclature, 1965). In view of the many known chemical and physical differences between the kallikreins, each kallikrein should be identified by species and source, e.g. hog pancreatic kallikrein.

Kininase. The term kininase may be used as a general term to describe enzymes that inactivate kinins. As the various kininases are isolated and the bonds they split are identified, a descriptive trivial name should be assigned by the investigator in an attempt to describe the particular bond split by the enzyme, e.g. carboxypeptidase N for the main kininase activity found in plasma of a variety of animal species (Erdös et al., 1963, 1964).

Substrate and Polypeptides

Kininogen. As the committee did not wish to superimpose on the existing nomenclature system yet another compromise name for kallidinogen (Werle and Berek, 1948) and bradykininogen (Rocha e Silva, 1949), it is recommended that all biologically inactive proteins that release a kinin by the action of proteolytic enzymes be called kininogens.

Kinins. The term kinin is a general name indicating a hypotensive polypeptide which contracts most isolated smooth-muscle preparations but relaxes the rat duodenum. Kinins increase capillary permeability, produce pain when applied to a blister base on human skin, and cause bronchoconstriction in the guinea pig. Alternatively the name may be applied to a polypeptide resembling bradykinin in its structure and pharmacological activity. It is likely that certain polypeptides prior to their isolation and identification will be classified as kinins when in fact they may not be structurally related. Ornithokinin, however, since it is hypotensive in birds and is released from bird plasma by ornithokallikrein, should be classified as a kinin.

Bradykinin. The term bradykinin refers to the nonapeptide (H-Arg-Pro-Pro-Gly-Phe-Ser-Pro-Phe-Arg-OH) as isolated by Elliott et al. (1961) from the action of trypsin on a bovine plasma fraction.

Kallidin. The term kallidin refers to the decapeptide (H-Lys-Arg-Pro-Pro-Gly-Phe-Ser-Pro-Phe-Arg-OH) formed by the action of the glandular kallikreins on kininogen (Webster and Pierce, 1963; Werle and Trautschold, 1963; Habermann, 1966).

Other kinins. The committee was unable to agree on a uniform policy for naming derivatives of bradykinin or kallidin. It was recommended that this decision be reserved to the author who first identifies or synthesizes the structure. Thus methionylkallidin should be referred to as methionyllsysylbradykinin (Met-Lys-Arg)^{1}-bradykinin) (Elliott and Lewis, 1965), and the derivative obtained by the action of trypsin on wasp kinin should be referred to as glycine^{1}-kallidin (Gly^{1}-kallidin) (Schröder and Hemple, 1964; Pisano et al., 1965). Since this latter polypeptide is also a decapeptide, it is recommended that expressions such as kallidin-9, kallidin-10, kinin-9, kinin-10, etc., be discontinued.

MEMBERSHIP OF THE COMMITTEE

Members who attended the committee meeting in Florence were as follows:

D. Armstrong	England	M. Schachter	Canada
N. Back	U.S.A.	E. Schröder	Germany
C. R. Diniz	Brazil	F. Sicuteri	Italy
E. G. Erdös	U.S.A.	I. Trautschold	Germany
E. Habermann	Germany	W. Vogt	Germany
J. Margolis	Australia	M. E. Webster	U.S.A., Chairman
H. Moriya	Japan		
J. L. Prado	Brazil	E. Werle	Germany

Corresponding members:

D. F. Elliott	England
C. A. Keele	England

REFERENCES

Dedolph, R. R.: Kinins, cytokinins, phytokinins. Science 149, 658 (1965).

Elliott, D. F., E. W. Horton, and G. P. Lewis: The isolation of bradykinin, a plasma kinin from ox blood. Biochem. J. 78, 60 (1961).

Elliott, D. F., and G. P. Lewis: Methionyl-lysyl-bradykinin, a new kinin from ox blood. Biochem. J. 95, 437 (1965).

Enzyme Nomenclature, pp. 38, 43. New York: Elsevier (1965).

Erdös, E. G., A. G. Renfrew, E. M. Sloane, and J. R. Wohler: Enzymatic studies on bradykinin and similar peptides. Ann. N.Y. Acad. Sci. 104, 222 (1963).

Erdös, E. G., E. M. Sloane, and I. M. Wohler: Carboxypeptidase in blood and other fluids. I. Properties, distribution and partial purification of the enzyme. Biochem. Pharmacol. 13, 893 (1964).

Frey, E. K., H. Kraut, and E. Werle: Kallikrein. Stuttgart: Ferdinand Enke (1950).

Habermann, E.: Enzymatic kinin release from kininogen and from low molecular compounds. This volume (1966).

Kraut, H., E. K. Frey, and E. Werle: Der Nachweis eines Kreislaufhormons in der Pankreasdruse. IV. Mitteilung uber dieses Kreislaufhormon. Hoppe-Seyler's Z. Physiol. Chem. 189, 97 (1930).

Laki, K.: The autocatalytic formation of thrombin and clotting effect of hemophilic blood. Studies Inst. Med. Chem. Univ. Szeged, Hung. 3, 5 (1943). Chem. Abs. 41, 1299g (1947).

Lewis, G. P.: Formation of plasma kinins by plasmin. J. Physiol. (Lond.) 140, 285 (1958).

Miller, C. O., F. Skoog, F. S. Okumura, M. H. von Saltza, and F. M. Strong: Isolation, structure and synthesis of kinetin, a substance promoting cell division. J. Amer. chem. Soc. 78, 1375 (1956).

Pisano, J. J., Z.Tamura, E. Furano, and S. Udenfriend: Vasoactive peptides in wasp venom. Fed. Proc. 24, 488 (1965).

Rocha e Silva, M.: Definition of bradykinin and other kinins. Biochem. Pharmacol. 10, 3 (1962).

Rocha e Silva, M.: The physiological significance of bradykinin. Ann. N.Y. Acad. Sci. 104, 107, 190 (1963).

Rocha e Silva, M., W. T. Beraldo, and G. Rosenfeld: Bradykinin, a hypotensive and smooth muscle stimulating factor released from plasma globulin by snake venoms and by trypsin. Amer. J. Physiol. 156, 261 (1949).

Schachter, M., and E. M. Thain: Chemical and pharmacological properties of the potent, slow contracting substance (kinin) in wasp venom. Brit. J. Pharmacol. 9, 352 (1954).

Schröder, E., and R. Hempel: Bradykinin, kallidin, and their synthetic analogues. Experientia (Basel) 20, 529 (1964).

Skoog, F., F. M. Strong, and C. O. Miller: Cytokinins. Science 148, 532 (1965).

Webster, M. E., and J. V. Pierce: The nature of the kallidins released from human plasma by kallikreins and other enzymes. Ann. N. Y. Acad. Sci. 104, 91 (1963).

Werle, E., and U. Berek: Zur kenntnis des Kallikreins. Angew. Chem. 60A, 53 (1948).

Werle, E., and I. Trautschold: Kallikrein, kallidin, kallikrein inhibitors. Ann. N.Y. Acad. Sci. 104, 117 (1963).

Author Index

Subject Index

MIX
Papier aus verantwortungsvollen Quellen
Paper from responsible sources
FSC® C105338

If you have any concerns about our products,
you can contact us on
ProductSafety@springernature.com

In case Publisher is established outside the EU,
the EU authorized representative is:
Springer Nature Customer Service Center GmbH
Europaplatz 3, 69115 Heidelberg, Germany

Printed by Libri Plureos GmbH
in Hamburg, Germany